Towards a Theoretical Neuroscience: from Cell Chemistry to Cognition

Springer Series in Cognitive and Neural Systems

Volume 8

For further volumes:
http://www.springer.com/series/8572

L. Andrew Coward

Towards a Theoretical Neuroscience: from Cell Chemistry to Cognition

Springer

L. Andrew Coward
Research School of Computer Science
Australian National University
Canberra, ACT, Australia

ISBN 978-94-007-7106-2 ISBN 978-94-007-7107-9 (eBook)
DOI 10.1007/978-94-007-7107-9
Springer Dordrecht Heidelberg New York London

Library of Congress Control Number: 2013945853

Printed on acid-free paper

Springer is part of Springer Science+Business Media (www.springer.com)

Preface

This book describes how a theoretical neuroscience can be established that creates understanding of how higher cognition is generated by neurochemical, physiological and anatomical processes. The route to this theoretical neuroscience lies through the application of the techniques for understanding extremely complex electronic systems. Despite the ubiquity of such systems, these techniques are not generally appreciated. The most complex systems contain many billions of components like transistors, each performing its functions in a way that results in the features of the system. Design of transistors, design of integrated circuits, design of printed circuit assemblies, design of frames and system design must all be integrated together and the hardware design must be integrated with the design of software and the design of compilers to translate software into a form that can act on the hardware. Many people have written software or content for computer systems such as those supporting the internet, without necessarily understanding the immense problems that have been solved to make such activities possible.

From 1969 to 1999 I was employed by a company that was one of the first to design extremely complex electronic real-time control systems. When I joined the company, it was called Northern Electric, when I left it was called Nortel Networks, and for a number of years the research and development activities in which I worked were moved to a separate subsidiary called Bell Northern Research (BNR).

Over my career, the company (which I will call Nortel) was the pioneer in shifting telecommunications networks from an older technology based upon electromechanical switches to the new technologies of silicon and software. Being the pioneer meant that Nortel needed to become proficient in all the required technologies. These technologies included design and manufacture of transistors, integrated circuits, printed circuit assemblies and systems with thousands of such assemblies. The company developed the design tools, manufacturing capabilities and testing capabilities for all these technologies. In addition, Nortel developed the languages, compilers and the design environments and tools for the required software. System architecture was heavily affected by the requirement that one system could not be totally out of service for more than 2 hours in 40 years, for example resulting in the requirement that software upgrades be made “on the fly” on actively operating systems. The first systems were introduced in the 1970s, and by the early 1980s one system had around four billion transistors packaged into integrated circuits. The initial design of such a system took about 5,000 engineering man years, and after introduction a system required constant evolution to add and modify features.

At different times in my career, I had responsibilities in most of the different hardware and software design areas. As a result, I obtained a good perspective on the many issues encountered in designing, constructing, testing, and modifying extremely complex systems. In the early 1980s I began to wonder about whether the techniques used to understand such extremely complex systems could be applied to the brain. This was not to suggest that there was any direct resemblance between electronic systems and brains, but rather that the approaches needed to understand the qualitatively different types of systems might have some common elements. My book on how such techniques could be applied to the brain was published in 1990 [1].

In 1999 I left Nortel to be able to devote myself full time to developing these ideas. After performing a number of simulations and writing a number of papers, a second book was published in 2005 [2] that made the system ideas more rigorous. In 2007 I created and began to teach a course at the Australian National University on this system architecture approach to the brain, integrating much more anatomical, physiological and neurochemical information into the architectural framework. The current book is the result of thinking about how this integration can occur.

I am indebted to the many people who worked at Nortel, and provided the remarkable intellectual environment which made complex system design possible. This intellectual environment was the stimulus leading to the system ideas about the brain. I am also indebted to the Computer Science department at the Australian National University for the opportunity to develop and teach the course on brain architecture, and especially to my long term collaborator Professor Tom Gedeon.

The book follows the presentation order of the ANU course. The intent is to provide enough psychology and neuroscience to allow computer scientists to understand the application of the architecture to the brain, and enough systems explanation to allow psychologists and neuroscientists to appreciate what the approach can contribute to integrating knowledge in those different fields.

Research School of Computer Science L. Andrew Coward
Australian National University
Canberra, Australia
April 2013

Contents

Introduction

Approaches to Theoretical Neuroscience

A vast body of experimental knowledge in the brain sciences exists. Much is known about the anatomy of different brain structures and the ways in which they are interconnected. Much is known about the morphology and electrical properties of individual neurons. Much is known about the chemical processes that take place within neurons. A primary task of a theoretical neuroscience is to integrate this knowledge in such a way that higher cognition can be understood in terms of brain anatomy, physiology and neurochemistry.

A discipline known as computational neuroscience has in recent years laid claim to the title of theoretical neuroscience [3]. Computational neuroscience has a number of approaches. One is the detailed modelling of physiology, with roots in the integrate-and-fire neuron model, the cable theory of electrical signal propagation in dendrites, and models of action propagation. Very detailed computational models of neurons and assemblies of neurons have been developed from these roots. A second approach is development of various general information processing concepts. Such concepts include the Hebb rule that when two neurons fire at the same time they tend to become connected, the efficient coding concept that sensory information is represented in the brain by the minimum amount of activity, and the concepts of feedback and feedforward architectures. A third approach is artificial neural networks. Such artificial neural networks include networks based on the backpropagation algorithm, and on Hopfield nets. One serious problem encountered with these networks as models for the brain is that new learning interferes severely with prior learning. The adaptive resonance theory (ART) network uses algorithms developed to address this problem to some degree. A fourth approach is the use of the Bayesian inference method for combining probabilities as a model for brain information processes. A fifth approach, such as adaptive control of thought–rational (ACT-R) uses pure computational models that associate conditions with commands (as in an electronic computer) to model brain processes.

In all cases heavy emphasis is placed on electronic simulations, often of neuron assemblies. However, there are a number of serious problems with these approaches. Firstly, there is a considerable gap between simulations and observations of higher level cognitive processes. Secondly, the approaches are unable to identify the information processes performed by different neural structures, except close to sensory inputs. Thirdly, a general characteristic of these approaches is to separate rapid performance of tasks from slower learning of tasks. Since learning is often intermingled with performance in the brain, it is not clear that this is a valid separation.

This book develops the concept that there is a well founded scientific approach to understand higher cognition in terms of brain anatomy, physiology and neurochemistry. The appropriate paradigms for achieving such understanding are available and can be applied to the vast body of neuroscience knowledge. This approach makes clear the nature of the information processes being performed by different anatomical structures, and demonstrates that learning and performance are deeply intertwined in a way that makes separation of the two processes seriously misleading. This approach therefore addresses the primary tasks of a theoretical neuroscience.

Complex Systems

Physics is often thought of as the paradigm for a true science, and the mathematical precision of physical theories a target for other sciences. However, high mathematical precision is only possible for a very limited range of phenomena. Quantum mechanics achieves the highest quantitative accuracy, but its modelling is generally limited to a few elementary particles. Most physical theory uses more approximate higher level models, but with a clear understanding of how such models can in principle be mapped to more precise levels. Even these higher level models tend to be quantitative only for some simplified and tightly specified physical systems. Understanding very complex systems like the human brain, where every brain is unique to some degree, requires a somewhat different approach. Such a different approach can be found in the technologies for understanding very complex electronic systems. There are minimal direct resemblances between computer systems and brains, but the techniques for understanding computer systems can be modified and adapted for the theoretical neuroscience problem.

In the mid-twentieth century, electronic systems providing combinations of services to human beings began to be developed. By the 1970s, some systems needed to provide many different services with high reliability and with no significant delays, resulting in the design of the first extremely complex real-time systems. Examples include the systems that manage the networks providing telecommunications services.

Such systems include multiple billions of individual components like transistors. It is clearly not possible for one human being to understand the operation of a feature of such a system by simultaneously imagining the activity of all the participating transistors. Nevertheless, human beings have an understanding of such systems, as demonstrated by the abilities to design such systems, build them, test and repair them, and modify their features. This understanding requires organisation of the physical information processing resources of a system into some very constrained architectural forms.

Suppose for a moment that there were many computer systems available, and people had developed great skills in using those systems, but no-one understood how the systems worked at a more detailed level. In response to this situation, experimental sciences to study computer systems could develop. Computer-psychology might study the phenomenology of computers. Standard tasks could be defined, and the times taken by different computers to perform those tasks measured. The observation could be made that when a computer system was first turned on there was a period of time when there was clearly considerable activity, but no useful responses. This "boot-phenomenon" could be investigated and theories about its role proposed. Computer-anatomy could disassemble machines, identifying large and smaller structures like printed circuit assemblies and integrated circuits and the connectivity between them. Integrated circuits could be opened to reveal the detailed patterns of metal and silicon. Computer-physiology could investigate the patterns of impurity in the silicon, or insert probes into computers performing different tasks to measure voltage fluctuations on various timescales and levels of detail. These sciences could accumulate large bodies of experimental data, but the problem would be to synthesize an understanding.

With computers, we know the answer. The information processes of a computer can be understood on many levels of detail using the same information paradigms. At many different levels, information processes can be understood as either instructions or data read/write processes. This consistency makes it possible for human beings to understand the overall operation of a system in such a way that any element of that understanding can be mapped to a more detailed understanding, any element of that more detailed understanding mapped to an even more detailed understanding and so on, all within the limits of human information handling capacities.

At an intermediate level, computers perform instructions in a processor, and data reads and writes in a memory. At a more detailed level, an instruction can be viewed as a sequence of more detailed instructions to open and close gates on transistors. A data read or write can be viewed as a combination of more detailed information reads and writes at the level of transistors or magnetic domains. At higher levels, user interactions with computers can be viewed as commands that can be broken down into sequences of instructions, or as accessing/recording of information that can be broken down into combinations of data read/writes. The physical resources of a computer system are organized into a modular hierarchy of transistors, cells on integrated circuits, integrated circuits, printed circuit assemblies, and shelves. In this hierarchy, modules on one level are made up of more detailed modules, and the operations of a module on any level can be understood as instructions or data read/writes. This organisation makes a hierarchy of description possible, in which the operation of the system can be described in terms of high level modules, or more detailed modules or yet more detailed modules in ways that are consistent with each other.

The term "module" has been given many different definitions, and it will be important for much of the later discussion to identify what is meant by a module in this book. In electronic systems and in the brain, a module is a set of resources that have been optimized to perform a specific type of information process very efficiently. Thus in a computer system there is a processor module that executes instructions efficiently, and a memory module that performs data access and retrieval operations efficiently. The resources of a module will therefore be shared across all the features of the system that require information processes of the module type. Use of modules therefore reduces the overall system requirements for resources, but they have some disadvantages that result in some secondary characteristics. One disadvantage is that a change to a module to implement changes to some features may have undesirable side effects on other features. The way in which such disadvantages affect module form is discussed in detail in Chap. 7.

Of course, in electronic systems these modular hierarchies have been imposed by human intellectual control over design. The question is, does such a hierarchy exist for a human brain that has an architecture defined by natural selection? To address this question, consider another computer caricature. Suppose that rather than designing a computer system, the approach was to collect a very large number of transistors, and select and connect transistors at random to create a system. If the system did not have the desired behaviours, random changes would be made to transistors and connectivity. Eventually, a system emerged that had the desired behaviours. There are some serious problems with this scenario. Firstly, consider the specifications for building such a system. Each component and connection at the most detailed level is individually specified and the construction process will be very complex and error prone. Secondly, if a construction error is made, or if a component failed during use, how could the failed element be identified, other than by a component by component replacement until the system began to work? Thirdly, if there was a need to change a feature of the system, how could the component changes needed to implement such a change without interference with other system features be identified?

In human designed systems, these requirements to construct, repair and modify are major drivers of the architectural forms that make understanding possible. The point of this caricature is that natural selection will also tend to result in architectural forms that make possible a relatively simple construction process, an ability to recover from construction errors and damage, and an ability to modify a behaviour in desired directions without excessive interference with other behaviours. As described in Chap. 7, the need to address these practical considerations tends to result in the emergence of architectural forms that support understanding. However, for a learning system the architectural forms are qualitatively different from the forms for a system designed under external intellectual control. As discussed in Chap. 7, there are some common information models, but for a learning system these models are condition definition/detection and behavioural recommendation. Anatomical structures on many different levels of detail can be understood as implementing processes of these types, and the information model consistency makes it possible to map between levels in a manner analogous with the mapping in computer systems.

Computational neuroscience has advocated massive electronic simulation of large parts of the brain. Although there is undoubted value in some simulations, caution is necessary. Suppose that a computer simulation was constructed that modelled the operations and connectivity of all the neurons in a brain, with neurons modelled with some degree of physiological realism. Suppose also that when the model ran, higher cognitive behaviours were observed. The problem is that such a simulation is just another very complex system that generates cognitive behaviours in ways we do not understand. For human understanding there is no substitute for a hierarchy of descriptions which makes it possible to describe system behaviour at high level in a way that is both comprehensible and mappable to more detailed levels. Individual detailed elements must be comprehensible, and must fit together to form a higher level picture.

Approach in This Book

In the biological sciences distinctions are drawn between language models and mathematical models. Language models are essential because they can cover large domains of experiment, but are approximate. Mathematical models reduce a complex system to minimal parts, but simulation results can be quantitatively compared with experiment. Although both approaches are important, mathematical models and the simulations that implement them require a framework within which consistent modelling is possible. Without such a framework their results can be misleading, and such frameworks are established by language models. To give an example from the history of science, prior to the concept of a solar centred solar system, the model of an Earth centred system was modelled in great mathematical detail, and could predict the apparent motions of the sun and planets in the sky with high precision. When the solar centred model was first popularized by Copernicus, it was less precise that the Earth centred model. However, the solar centred word model was superior, and eventually led to superior mathematical models.

This book is primarily concerned with establishing a framework within which consistent modelling is possible. The way in which natural selection pressures create such a framework in the brain will be described. These pressures constrain the types of information processes performed by the brain and the way in which they are partitioned between different anatomical structures. Although the types of processes performed by different anatomical structure as described in this book have been computer simulated, the focus of the book is on the development of language models for the information processes performed by brain structures on different levels of detail, and for the combinations and sequences of these information processes that implement higher cognitive processes. In addition, the integration of a very wide range of experimental observations into a consistent picture is described.

Plan of the Book

In Chap. 1, the nature of scientific understanding will be explored, and the limitations to different approaches discussed. In Chap. 2, the major cognitive phenomena that we would like to understand in terms of anatomy, physiology and neurochemistry are described. In Chap. 3 there is an overview of brain anatomy and physiology, followed in the next three chapters by more detailed descriptions of current knowledge in some key areas: properties of neurons in Chap. 4; neurochemical processes within neurons in Chap. 5; and the structure and connectivity of the major anatomical structures in Chap. 6.

Chapter 7 establishes the theoretical framework for integrating this knowledge. In particular it describes how natural selection pressures tend to result in some specific architectural forms and information processes, and how these forms and processes support the creation of the hierarchies of description needed for human understanding. Chapter 8 then demonstrates that these forms and processes are present in the brain. Chapters 9 and 10 describe how hierarchies of description from psychology to neurochemistry can be constructed for some primary cognitive phenomena including attention, semantic and episodic memory, procedural memory, priming memory and working memory. Specifically it is demonstrated that key parts of high level descriptions of these phenomena can be mapped to anatomy, then to physiology, and then to neurochemistry. In Chap. 11 it is shown that various examples of more complex phenomena like prospective memory, speech, consciousness and self awareness can be understood as combinations of the information processes also used by the primary phenomena. Finally, Chap. 12 reviews how the approach described in earlier chapters constitutes a theoretical neuroscience, and how the approach needs to be embedded in experimental and computational work on the brain.

Chapter 1
The Nature of Scientific Understanding

This book addresses the issue of how higher cognition can be understood in terms of the neuroanatomy, neurophysiology, and neurochemistry of the human brain. Much is known about the chemical processes within and between neurons, and this knowledge has, for example, led to practical applications in the area of drugs to treat brain diseases and deficits. Much is also known about the interactions between neurons, leading to the modelling of interacting groups of neurons in anatomical structures. Much is known about anatomical structures in the brain, and the different deficits that can result from damage to different structures. Finally, a great deal is known about cognition, the vast range of perceptual, memory and behavioural phenomena that can be observed. Within each of these domains there is understanding of the processes of the domain, but between domains the understanding is much more restricted.

As a first step it is important to consider what we mean by "understand". Do you understand how television works? Even a small child understands that if you press one button on the remote the television turns on; pressing other buttons changes the channel; pressing yet other buttons changes the volume. This level of understanding works for many situations. So the child understands how television works in the sense that he knows the actions that will cause various desired results with the television. However, for some circumstances this level of understanding is not adequate. For example, if the television is not connected to electricity, it will not respond when buttons on the remote are pressed. It is necessary to know that electricity powers the television. If the picture is poor, it is necessary to know that a strong incoming signal is required, perhaps acquired by connecting to cable. If the television does not operate even if these factors are correct, it is necessary to know how the different components contribute to creating the picture and sound, in order to repair it. To design a television, a yet deeper knowledge is needed, even deeper if the electronic components of the television must be designed.

In this example there are therefore many different levels of understanding. The highest level of understanding means knowing that pressing a button on a remote will result in a program being played. At more detailed levels, understanding means

L.A. Coward, *Towards a Theoretical Neuroscience: from Cell Chemistry to Cognition*, Springer Series in Cognitive and Neural Systems 8, DOI 10.1007/978-94-007-7107-9_1,

knowing that the sound and visual images are coded as electrical signals that are broadcast or carried on a cable, and the electronic circuitry of the television extracts the visual and sound information and displays them. To design a television the extraction and display of information must be described in terms of the behaviours of transistors within integrated circuits and other components making up the television. To design the transistors, their behaviour must be understood in terms of atoms and electrons, using quantum mechanics.

There are a number of critical aspects of understanding which we can extract from this simple example. *Firstly*, understanding is hierarchical. In other words, we understand something at different levels of detail, with different levels being appropriate for different types of situation. *Secondly*, understanding is causal. In other words, when we understand something, we know the behaviour that will cause specific events to occur. We can predict what will happen when something is done. We know that pressing a button on the remote will change the channel. However, *thirdly*, understanding is more approximate at the higher levels, in other words it is incorrect in some situations. In the television example, the understanding that the television will turn on if a specific button on the remote is pressed is usually correct, but incorrect if the electricity is not connected (or if we were dealing with a television that does not use a remote). The dependence on electricity comes in at a more detailed level of description. Causal descriptions at high level use a relatively small number of very complex units (remote, button, television) while descriptions on detailed levels use very large numbers of relatively simple units (transistors, diodes, capacitors and resistors; or at an even deeper level electrons, nuclei and photons). It therefore takes a lot more information to describe the same event at more detailed levels, and the use of a more detailed level of description to determine what to do to achieve a result at a higher level will often be impractical. This impracticality reflects the limited information handling capacity of the human brain. Even for a television set we are not able to simultaneously imagine the activity and interactions of all the transistors and other components as they evolve over time. Hence the *fourth* aspect of understanding is that we must know when a description at a higher level is too approximate, and the *fifth* aspect is that we must know how to map between levels of description to achieve a higher level of accuracy when required. Thus, for example, when pressing the button on the remote does not result in the television coming on, we can shift to a deeper level and check that the television is plugged into the electricity or try replacing the batteries in the remote. However, the shift to a deeper level occurs for only a small segment of the higher level description at a time, because we would not be able to hold the complete detailed description active in our mind.

In summary, understanding is a complex concept which involves the existence of a hierarchy of causal descriptions at different levels of detail. The higher levels are more approximate but there are rules indicating when a shift to a more detailed level is required, and well defined mechanisms for making the shift. If a complete hierarchy of this type is in place we intuitively feel we “understand”.

1.1 Understanding in the Physical Sciences

It is important to recognize that this meaning of understanding is also present, although not always acknowledged, in the physical sciences. Quantum mechanics utilizing entities like electrons, protons, neutrons and photons can predict the outcome of measurements with great accuracy, and quantum mechanics is viewed as the foundation of modern physics. However, most physical phenomena are successfully modelled without direct reference to quantum mechanics. In practice, the number of entities like electrons that participate in almost any observable phenomenon is so large that a complete model of the phenomenon in terms of quantum mechanics that followed every individual electron would be completely incomprehensible to any human mind.

Higher level entities are therefore defined which can be viewed as groups of more detailed entities. Atomic nuclei are made up of protons and neutrons. Atoms are made up of nuclei and electrons. Molecules are made up of sets of different atoms. Chemicals are made up of large numbers of identical molecules. Materials are made up of mixtures of different chemicals. The full interaction between higher level entities is the sum of all the interactions between all the detailed entities of which they are composed, but this full interaction is approximated by a simpler interaction. Macro phenomena are described in terms of the appropriate higher level entities and their approximate interactions, with much less information content than a description in terms of all the detailed entities. Such higher level causal descriptions are less precise than quantum mechanics, but the nature of the approximations is well understood, and the ways to map a higher level description into a more accurate, more detailed description are also well understood.

The key requirement is creation of a consistent set of models on multiple levels. At each level, mental manipulation of the models is possible within a human mind, and there are clear rules for mapping from parts of a high level description to a more detailed description and from a detailed description into part of a higher level description. A critical issue is whether such a hierarchy of descriptions can actually be constructed. There is no obvious guarantee that such a hierarchy exists for all possible phenomena.

1.2 Complex Physical Systems and Chaotic Behaviour

Any physical system is made up of an extremely large number of electrons, protons, neutrons and photons, and perhaps a few other entities. Understanding requires some higher level entities which can be used to construct adequately accurate causal descriptions. For many physical systems there are problems with such construction.

The problem arises because very small differences in the initial conditions of just one higher level entity can make very large differences to the later overall state of the system as a whole. The prototypical example of this problem is the atmosphere, and the difficulty of weather prediction. Weather prediction depends upon conceptually breaking up the atmosphere into physical cells, setting initial conditions for each cell, and calculating the later state of the atmosphere. The problem is that the smaller the cell entities, the larger the required data collection and calculation, but as the timescale for prediction increases the initial state of even a very small cell can be critical.

The caricature of this situation is that the way a butterfly in Brazil flaps its wings at one point in time may make the difference several months later between a storm in Europe happening or not. This very sensitive dependence on initial conditions is known as chaotic behaviour, and is common in complex physical systems. Chaotic behaviour means that constructing a hierarchy of descriptions within human information handling capabilities may not be possible.

If the brain is an example of a complex chaotic system on timescales relevant to cognitive behaviours, then there might be no possibility of intuitively satisfying understanding.

1.3 Complex Control Systems

Human beings have designed some very complex electronic control systems. Such systems receive a high volume of information from their environment, and control a large number of behaviours that can act on that environment. These complex control systems are also very complex physical systems, but differ from chaotic physical systems in that their behaviour aims to meet certain objectives. These objectives include maintaining a degree of homeostasis. For example, a telecommunications network is an extremely complex system with the objectives of providing links between telephones and computers on demand with very high reliability. Unlike a chaotic physical system, a small difference in initial conditions will rarely make a large difference in later system state.

Furthermore, these complex electronic systems are understood in the sense defined earlier. For example, it is possible to modify the way the system is experienced by a human user (i.e. change a feature). It is possible to design such systems, manufacture and test them, and when components fail it is possible to identify the failed component and replace it. Clearly, human understanding of a user feature is not based upon the ability to simultaneously imagine the evolution of the activities of all the billions of transistors that are participating in the performance of that feature. Rather, there are hierarchies of description which make it possible to think about the system as a whole, or a small part of the system in somewhat more detail, or a small part of that part in yet more detail and so on.

The physical structure of the electronic system reflects this hierarchy of description. Within such a system there may be multiple printed circuit assemblies, each assembly having a different role within the system. Each printed circuit assembly is made up of many integrated circuits and other components, each such circuit or component having a different role. Each integrated circuit may be made up of a number of areas called cells, each cell with a different role. A cell is made up of many transistors, resistors, capacitors etc., and again each of these components has a different role. A system function could be thought about at the level of interactions between printed circuit assemblies. If that level of detail is inadequate for the purpose, a small part of the function can be thought about in terms of interactions between the integrated circuits on one assembly. At a yet more detailed level, an even smaller part of the function can be thought about in terms of the interactions between different cells on one integrated circuit and so on.

It is important to note, however, that the roles of components at any level do not map into different types of functions as perceived by a user of the system. In the relatively simple case of a personal computer, word processing and web access both use many of the same components at many different levels of detail. As a result there are two different types of architecture for complex electronic systems: the physical architecture and the functional architecture. The physical architecture describes how the physical resources of the system are separated into components, each component performing a different type of information process. This physical architecture, which can also be called the information architecture, describes the processes performed by different components and the interactions between the components. The functional architecture describes the features of the system as perceived by a user, and the interactions between them. A user manual is closely related to the functional architecture, but the relationship between the physical architecture and the functional architecture is very complex. However, it is the physical architecture that is critical for sustaining the hierarchy of descriptions.

As discussed in more detail Chap. 7, the physical architecture is shaped by a number of practical requirements. One requirement is to utilize as few physical resources as possible. A second requirement is to be able to modify any one feature without introducing undesirable side effects on other features. A related third requirement is to be able to identify a failed component in order to replace it. A fourth requirement is to be able to build copies of the system at reasonable cost. The requirements to be able to construct, modify and repair the system are essentially the same as the need to understand the system, and when combined with the need to make economical use of resources place some strong constraints on the form of the hierarchy of descriptions.

One such constraint is that the hierarchy makes use of two common information models for operations at every level. These models are instruction and condition (or data) read/write. These models reflect that the fundamental operations of any complex control system are detection of conditions in the information available to the system and association of conditions with behaviours. In a system designed under external intellectual control, behaviourally relevant conditions are precisely specified by design, and detection of a condition or combination of conditions can therefore be associated with an instruction or command to perform a behaviour.

At every level, a description is made up of combinations of operations of these two types. For example, at high level, a description of a process might be an instruction to connect with a particular website. At a very detailed level, the description would be a large combination of transistor operations which could each be viewed as either a command (e.g. open the gate of a logic transistor) or a condition read/write (access/change the state of a memory transistor). The high level description is an approximation, for example the display that appears on the computer screen could be derived from information located on many different physical sites. The common information models make it possible to map from the high level description to whatever level of detailed description is required to solve a particular problem.

1.4 The Brain and Hierarchies of Description

In the case of electronic systems, the understandable hierarchy of descriptions is of course created under human intellectual control. The brain is a complex control system that receives large amounts of information from the senses and uses that information and records of past sensory information to select an appropriate behaviour from within a large range of possibilities. Unlike a chaotic system, a brain is required to meet objectives including homeostasis, body survival and generation of offspring. Hence like an electronic system, conditions must be detected in the information available to it, and condition detections must be associated with behaviours. However, in a brain many of the conditions and their associations with behaviours must be defined heuristically.

Although there is minimal direct resemblance between a brain and an electronic system, natural selection imposes on brains many pressures analogous with those that shape the electronic system physical architecture. A species with a brain that can learn to perform a given set of behaviours with fewer neurons than required by the brain of another species will have a significant natural selection advantage. A species with a brain that can learn new behaviours with less interference to previously learned behaviours than another species will also have a natural selection advantage. A species in which the process for building a brain from DNA "blueprints" is less error-prone than in another species will have a natural selection advantage.

In Chap. 7 we will discuss in detail the ways in which these natural selection advantages tend to result in a brain with a physical architecture which supports the creation of a hierarchy of descriptions, analogous with but qualitatively different from the hierarchy in an electronic system. This brain architecture, which tends to result in any system which must learn a complex combination of behaviours with resources that are not unlimited, also makes use of two common information models. However, in the case of the brain, these two models are behavioural recommendation and condition definition/detection. These information models differ from those in a system designed under intellectual control because conditions are defined heuristically and can constantly change. Conditions can therefore only be associated with behavioural recommendations.

1.5 The Basis for Understanding the Brain

Understanding of higher cognition in terms of anatomy, physiology and chemistry requires a hierarchy of descriptions. At the highest level in such a hierarchy, a cognitive process would be described causally, in other words in terms of a cognitive circumstance leading to another circumstance and then another. At a more detailed level, any segment of the cognitive description could be described causally in terms of information operations within major anatomical structures. At a yet more detailed level, any segment of the major anatomical structure level description could be causally described in terms of information operations in anatomical substructures. At even more detailed levels segments of higher level descriptions could be described in terms of interactions between neurons, and segments of a neuron level description could be described in terms of chemistry within a neuron.

As in the physical sciences, the key property of such a hierarchy is the ability to know when the accuracy of a higher level description is inadequate, and to know how to map between levels. An adequate hierarchy of description is what a scientific theory delivers.

To understand any particular problem, the highest possible level of description must be used. A shifting to more detailed levels only occurs when greater precision is required than is available on the higher level. To remain comprehensible to a human mind, such a shift can only be made for a small part of the higher level description.

The concept we will develop through this book is that the behaviour recommendation and condition definition/detection information models imposed on the brain by natural selection make it possible to create hierarchies of description analogous with those for electronic systems. A cognitive process can be described at high level, a comprehensible description of any small element of the high level description can be described at more detailed level and a comprehensible description of any small element of the detailed level description can be described at a yet more detailed level.

The development of consistent hierarchies of description is the main focus of this book. However, to give a preliminary idea of how the process works, consider the attention process. *At high level* the process will in due course be described in terms of detecting conditions in the overall visual field, each condition recommending a group of different attention behaviours. Acceptance of the predominant recommendation results in detection of a much wider range of conditions within just the object of attention, each having recommendation strengths in favour of a range of behaviours appropriate in response to the presence of the object. *At an anatomical level*, part of the high level process will be described in terms of the cortex heuristically defining conditions, and detecting any of its conditions that are currently present. The basal ganglia interprets each currently detected cortical condition as a recommendation in favour of a range of behaviours, and selects the behaviour most strongly recommended across all currently detected conditions. The thalamus implements the behaviour selected by the basal ganglia. The hippocampus defines and detects conditions that recommend changes to conditions elsewhere in the cortex. *At a detailed anatomical level*, part of the anatomical level process will be described in terms of cortical columns detecting groups of conditions that we will call receptive fields, the structure of a column making it possible to manage receptive field changes to minimize undesirable side effects on the behaviours recommended by the current column receptive field definition. *At the neuron level*, part of the detailed anatomical level process will be described in terms of a pyramidal neuron defining and detecting a group of conditions called the neuron receptive field. Neuron receptive field detections indicate detections of the column receptive field and/or recommend changes to the column receptive field. *At the dendrite level*, part of the neuron level process will be described in terms of a terminal branch on a dendrite defining and detecting one condition within the neuron receptive field, and an output from the branch recommending detection of the neuron receptive field. The dendritic structure defines and implements the integration of branch recommendations. *At the synapse level*, part of the dendrite level process will be described in terms of a synapse defining and detecting one element in a terminal branch condition, the presence of an input recommending the detection of the condition. *At a neurochemical level*, part of the synapse level process will be described in terms of different kinase cascades recommending different types of changes to the weight of a synapse.

In this example, a higher level description simplifies and approximates the much greater information content of a more detailed description. The use of the information models means that translation between levels is possible, and the approximations introduced are well understood. These are exactly the requirements for a scientific theory as discussed earlier.

1.6 Computer Modelling and Its Limitations

The question might be raised, why not just employ massive computer simulations to support the descriptions on the more detailed levels? The issue with this approach is that it does not lead to actual human understanding. In the extreme, suppose that it was possible to create a computer model for every neuron in a brain, the connectivity between them, and the way in which each neuron responded to its inputs. Then suppose that this computer model actually exhibited cognitive phenomena. The problem would be that we simply have another system in which the link between component activity and cognitive phenomena is not understood.

Computer modelling to verify detailed models is certainly valuable. Design and implementation of a general artificial intelligence system will also require computer modelling. However, just as for regular electronic systems, development of such a system will only be possible through a hierarchy of descriptions that makes it possible for human beings to comprehend the design.

Chapter 2
Higher Cognition

Our brains play a role in almost everything we do, but in this book the focus is on activities under the general label of higher cognition, including memory and learning, speech, emotions, intellectual disciplines, planning, consciousness and self awareness.

Our brains are general purpose learning systems: they can learn to perform any of a very wide range of different types of behaviour. With very limited exceptions current computer systems do not learn. In contrast with such electronic systems brains are very good at defining and finding patterns and creating and recalling relevant associations between different objects and situations. Sometimes these pattern finding and association creating capabilities are misleading, such as when we see faces in clouds or in groups of rocks on the surface of Mars, or associate numbers like 13 with unpleasant situations.

Unlike our brains, electronic systems can have permanent, unchanging, reliable memories: the picture stored in the computer will look exactly the same in all details several years in the future. Brains on the other hand can extract meanings from memories, and select the more important information to be remembered.

Electronic systems are very good at precisely following tightly defined rules, such as the rules of arithmetic. Brains, on the other hand, find arithmetic much harder, making all sorts of errors. It has been tellingly pointed out that it takes from birth to about 5 years old for a human infant to learn a language, and at the end of that period speech is accurate and sophisticated. It takes another 5 years to learn arithmetic, and at the end of that period there are still lots of errors [4].

One oddity about brains is that they go off line for a significant proportion of each day into a mode called sleep that superficially looks passive. Unlike computers, which can actually be put in a passive mode (or "put to sleep"), sleep for the brain is actually a very active combination of many different processes. During sleep our brains go through several types of processing (stages), as indicated by attaching electrodes to the scalp and measuring electrical activity. This electrical activity occurs at predominant frequencies, with different frequencies in each of three or four different sleep stages, only one of which has frequencies

L.A. Coward, *Towards a Theoretical Neuroscience: from Cell Chemistry to Cognition*, Springer Series in Cognitive and Neural Systems 8, DOI 10.1007/978-94-007-7107-9_2,

resembling activity in an alert brain. Electronic systems perform perfectly well if rarely turned off or put to sleep, but brain sleep appears to be important for effective brain performance [5].

We do not always realize how much of our awareness is an artifact constructed by our brains. Our eyes have lenses that focus light on the retina, much like the lens in a camera focuses an image on the CCD detector. If we look at the image detected by a camera as we move the camera around we see change, the positions of all the objects move rapidly. If we move our eyes or tilt our head, the image projected on the retina is changing in the same way. But what we are aware of seeing is a stable environment, with all the stationary objects fixed in position and moving objects with the correct relative motion. This stable environment is not the image on the retina, it is constructed by our brain using inputs from the eyes and inputs from muscles in the eyes and body that indicate movement.

At a higher level, it is possible for eyes to be working perfectly, and the brain to be unable to understand what is being "seen". For example, certain types of brain damage can result in the patient being able to recognize visual features but being unable to put them together to recognize an object. A rose might be described as "about 6 in. in length; a convoluted red form with a linear green attachment" [6]. However, the patient does not perceive that he has any problem when describing things this way. If given a photograph and asked what he sees, he may describe nonexistent scenes in detail, believing that the picture is being described, a process known as confabulating. A river, a guest house and people dining on a terrace may be confabulated from a picture of sand dunes in the Sahara desert [6]. But again, he does not see any fault in his description.

At a yet higher level, there are patients who lose the ability to create new memories, and yet appear superficially quite normal. This type of amnesia, called anterograde amnesia, occurs in patients with damage to one or more of a number of brain structures. One example is Korsakoff's syndrome [6], where damage to a brain structure called the mammillary bodies (see next chapter) sometimes results from long term excessive alcohol consumption. You could visit someone with Korsakoff's syndrome, have a normal conversation for a few minutes, but if you leave and come back 5 min later they have no memory of ever having met you. Such a patient has no memory of how they arrived at the place where they are currently located. A patient in hospital meeting a doctor may speak initially as if he was in his grocery store before the amnesia occurred, and address the doctor as a customer. When this story breaks down the patient will immediately jump to seeing the doctor as a pre-amnesia friend, then a different pre-amnesia friend, then a doctor he has not met before, then back to a customer in his store again. The patient remembers nothing for more than a few seconds, each confabulation is an instinctive creation a current "story" in which to locate himself, and he has no awareness that his story is changing abruptly from moment to moment. This frantic jumping from story to story reveals the importance of such "stories" to our sense of identity, and the degree to which this sense of identity is an artifact constructed by our brains from our memories. In some ways these amnesic patients seem to an outside observer to have lost their identity or self.

Our personality is a large part of who we are, but patients with brain damage often show personality changes. Such changes include uncontrolled anger, inability to take responsibility, socially inappropriate behaviour, and impaired self awareness. An investigation of subjects with a history of recurrent attacks of uncontrolled rage with little or no provocation and no known reason (i.e. with psychotic, psychopathic, profit motivated, drug related, mental retarded cases excluded) found that in almost all cases there was evidence of brain damage [7].

The most famous example of personality change following brain damage is Phineas Gage, who in the middle of the nineteenth century sustained major damage from a spike passing through the front part of his brain. As the doctor who treated him over a long period of time reported the situation:

> Previous to his injury, although untrained in the schools, he possessed a well-balanced mind, and was looked upon by those who knew him as a shrewd, smart businessman, very energetic and persistent in executing all his plans of operation. In this regard his mind was radically changed, so decidedly that his friends and acquaintances said he was 'no longer Gage'. [8]

We will now consider in more detail some different aspects of the cognitive capabilities exhibited by our brains. We will start with two fundamental aspects of all cognitive activities. The first is attention, or the ability (and necessity) of focusing on a limited range of information selected from what is currently available. The second is memory, or the ability to recall information derived from past experiences and/or to use such information to influence current behaviour.

2.1 Attention

The environment in which we operate is very complex and responses to many different objects or combinations of objects in that environment could be appropriate. Generation of a response requires processing of the information derived from an object, and it is therefore necessary to select objects within the environment in order of priority for such processing. Selection of and processing of information from objects is known as attention.

In the case of visual attention, there are three stages to the attention process. For example, presented with a visual scene, the brain drives a guided scanning process in which the eye and head are moved to centre the sensitive central part of the retina (the fovea) on different objects. This scanning process is made up of rapid jumps from one object to another (called saccades), sometimes mini-saccades to correctly centre on a target object, and pauses of one or two hundred milliseconds at each object. Subjects cannot pay attention to one object and make a saccade to a different object [9]. Once the gaze has been focused on an object, simple behaviours can be generated with respect to that object, for example naming or avoiding the object. A further attention step is called conscious awareness [10]. This step includes a much richer subjective awareness, during which it becomes possible, for example, to talk about the object. Not all object attention steps are followed by conscious awareness.

2.2 Memory

In our daily lives, we remember some things for only a few seconds, other things for hours and days, yet other things for all our lives. Some of our memories are purely factual, like the name of the capital city of the country in which we are living. Other memories are for what happened at a particular time and place, and can include recall of the emotions felt at the time, such as an embarrassing, annoying, or highly successful personal event. Yet other memories are how to perform various skills like riding a bicycle, driving a car, or playing a piano. Another memory capability is holding a telephone number in our mind just long enough to dial it. Yet another capability is remembering to do something at a particular time and place. The ability to do this is sometimes called prospective memory, and failure to perform this task correctly is one of the most common reasons for complaining about our memory.

In order to study memory scientifically it is necessary to select relatively simple tasks that can be measured consistently across different subjects. An important part of such studies is imaging the brain during the tasks, and determining which physiological structures are active during different memory activities. Another source of information is observing how patients with damage to different brain regions perform on the various memory tasks. Yet another source is study of the brains of animals on tasks which have a degree of similarity to human memory tasks.

Such laboratory studies have led to the conclusion that there are five independent systems in the brain supporting different types of memory. These five systems are semantic memory, episodic memory, working memory, procedural memory and priming [11]. Each of these systems involves a set of memory tasks that are different from the tasks performed by the other systems. The reasons for believing the systems are separate is that if the circumstances in which memory tasks are performed is changed, some changes affect one type of memory task and not another type.

One particularly important set of examples of changed circumstances are observations that local damage to one part of the brain affect memory type A with no effect on memory type B, while damage to a different part of the brain has the opposite result. These type of combined observations are called double dissociations. We will be describing examples of double dissociations later in the chapter.

However, in real life an actual cognitive task will typically be a process involving many of the different memory systems. Learning to drive a car involves learning motor skills, but will involve a fair amount of memory for facts and events, especially in the early stages. Working memory will also be required. Remembering the sequence in which memory and other cognitive processes must be performed can itself be viewed as another type of memory.

In the following sections we will consider the different memory systems separately, but much later in the book we will discuss real life cognitive tasks and how they are achieved by sequences of different memory processes.

2.2.1 *Semantic Memory*

Semantic memory is the memory for facts, with no memory for the circumstances in which the facts were learned. We generally know that there are planets in the solar system called Mercury, Venus, Earth, Mars etc. but do not remember when or where we learned that there are such planets and what they are called. The meanings of words are important examples of semantic memory, and we generally have no memory of when we learned the words. It is possible to make mistakes, where we retrieve a fact or word meaning which is incorrect but which we believe to be true. Such mistakes appear to be very rare in normal people. It is not always possible to immediately recall a fact or word which we have learned, and sometimes we have a "tip of the tongue" experience in which we feel as though the memory is in some sense very close to being retrieved but not quite there.

In the laboratory, a frequently used test of semantic memory is category verification. A subject is given the name or picture of an object, and the name of a category, and asked to respond yes/no that the object is/is not a member of the category. An interesting observation is that the time to respond for an object that is clearly a member of the category (e.g. hawk – bird) is about the same as for a clear non-member (e.g. walnut – bird), but the time for an atypical member of a category (e.g. ostrich – bird) is somewhat longer. The difference is slight but significant. Over lots of trials, it takes the average subject about 1.3 s to respond for a typical category member or a non-member, and about 1.5 s for an atypical category member [12]. Other tests measure the speed with which semantic memories can be accessed. For example, subjects may be asked to think of five letter words beginning with the letter M, and the number that they can retrieve in 1 min is measured. This measure can vary considerably between different subjects, from under 8 to over 12 in one study [13], although almost all the retrieved words are known to all subjects. Yet other tests investigate memory for well known facts in the subjects culture, such as the names of famous people. Many of these tests are used to compare subjects with memory deficits with normal people, or older subjects with younger [14].

The "tip of the tongue" experience has also been investigated, and it is found that if we have such an experience when, for example, trying to think of the correct word corresponding with a definition, we can often identify the first letter, the number of syllables or a rhyming word even while the actual word is still inaccessible [15].

2.2.2 *Episodic Memory*

Episodic memory is the memory for events, often events in which we were personally involved, in which case it is called autobiographic memory. Episodic memory can reconstruct the sequence in which things happened and the emotions felt at the time. Recalling an episodic memory is almost like reliving the event, except that the raw sensory inputs are not there, in other words the recall is not an auditory and visual hallucination.

However, our memories of an event are selective. For example, if we are feeling strong emotion (such as fear) during an experience, our memory for the general situation during the experience (the gist of the event) is enhanced, although memory for more sensory aspects of the event is unaffected [16]. If an event is novel, we are more likely to be able to recall the event later, especially if there was strong emotion at the time. An extreme example of very enhanced memory is the phenomenon of flashbulb memory. When a person recalls a particularly dramatic event (such as the assassination of President Kennedy, the moon landing, or the 9/11 attack on the World Trade towers), it is often possible to also recall many details of where the person was and what they were doing and feeling when they first heard the news [17].

At the other extreme, if our attention is focused on one aspect of a situation, we may have no memory for another aspect. A striking example of this phenomenon is the "gorilla movie" experiment. In this experiment, subjects are shown a movie of a group of people passing a ball between them. Some people have white shirts, some black shirts, and the viewer is asked to concentrate on counting how many times the ball is passed between black shirted participants. When people see the movie for the first time under these conditions, about half fail to notice or remember the person in the gorilla suit who walks through the players, pausing at one point to face the camera and thump its chest [18].

Even if we can recall a memory, it is not always a reliable reconstruction of the original experience. For example, if someone is shown a picture of an often encountered scene (such as a beach or a classroom), and later asked to recall the picture and indicate if various objects were present, objects characteristic of such scenes but not actually present in the picture are often "remembered" [19]. Talking about an event can change our memory of the event. If someone witnesses a crime scene, and afterwards is asked to verbally describe the criminal, they are less likely to accurately select the criminal from a photo lineup than someone who was not asked for a verbal description [20].

Although remembering an actual past event and imagining some future event may seem to be very different activities, the imagining process uses many of the same parts of the brain as remembering an actual past event [21]. It is even possible to create false memories for events which never happened. Subjects were asked to read paragraphs describing various events from their past experience, some of which actually happened, some of which definitely did not happen. These descriptions were created in consultation with family members. A common false event was becoming lost in a shopping centre as a child. After reading each paragraph, the subjects were asked what they could remember of the events. Several times in the next few weeks they were shown the paragraphs and asked again what they could remember. At the end of this process, a number of subjects were convinced they could remember the non-existent event [22]. It is believed that false memories of childhood abuse may have sometimes been inadvertently introduced by a process of this type [23].

Another aspect of episodic memory is the ability to fit events into a timeframe. We can judge whether one event happened before or after another event. We have a subjective sense of how long an event lasted even if we do not have a memory of the actual start and finish times. However, this subjective sense is not precise: for example, most of us feel that time seems to have passed more slowly when we were younger. This subjective sense also tells us how long a current event has been under way, but again the feeling is not fully reliable relative to the clock: time seems to pass more slowly during a boring event than during a novel, stimulating event.

In the laboratory, both long term and short term episodic memory is measured. In one type of test of relatively short term recall, subjects read a story, and later are asked to recall what they can of that story. In another type of test a subject is given pairs of objects (which may be pairs of words or pairs of pictures), and later shown one object and asked to recall the other. Longer term testing includes being given a group of words, and asked what autobiographical memory is generated. As for semantic memory, these tests are often used to compare subjects with and without memory problems.

2.2.3 Declarative Memory and the Distinction Between Semantic and Episodic Types

The content of semantic and episodic memories can be described verbally, and both these memory types are therefore known as declarative memories. An interesting aspect of declarative memory is that it is much easier to recognize that we have seen something before than it is to recall a mental image. Our higher ability to recognize familiarity implies that we are constantly recording large volumes of information, much of which is not available for easy later recall. In one study, subjects were shown several thousand photographs, each for a few seconds. Several days later they were shown pairs of photographs, one new and one previously seen. They could pick out the familiar photograph more than 90 % of the time [24]. It is much easier to recognize that we have visited a place when we return than it is to imagine the place when we are not there. However, an indication of limits to recognition of familiarity is the déjà vu experience, in which a new experience feels as if it is familiar, along with an awareness that it is in fact novel.

In healthy people, there are declines in declarative memory capabilities with age, but these declines appear to be limited to the acquisition and retrieval of new information, not in the retention of older memories [25]. With a disease like Alzheimer's, all declarative memories can be degraded [14].

Brain damage provides some important evidence that semantic and episodic memories are supported by different brain systems. One example is the extensively studied patient Henry Molaison, known in the many papers written about him as HM until his death late in 2008. In 1953 at age 27, he had experimental surgery to treat severe epilepsy [26]. This surgery removed significant sections of his hippocampal system (see next chapter). Like the patients mentioned earlier in this chapter, he lost all ability to create new declarative memories, but retained normal intelligence and conversation skills. The separation between semantic and episodic memory is indicated by the observation that although he also lost his autobiographical memories for a period of 11 years prior to his surgery [27], he retained semantic memories for the same period, such as words which first came into use in those years [28]. HM also exhibited normal working [26], procedural [29] and priming memory [30]. In contrast with HM, other patients with damage to different brain areas can show a general deterioration in semantic memory but no effect on episodic memory [31].

2.2.4 *Working Memory*

Working memory refers to the number of different things we can be actively thinking about at the same time. When we look up an unknown telephone number in order to dial it, the task is more difficult if the number is longer. However, if part of the number is known from another context (such as the year we were born), the task is easier. If we are interrupted while dialling, the number may disappear from our mind and we have to look it up again. Imagine the visual appearance of cat. Now try to retain the cat image, and also imagine the visual appearance of a pineapple. Now try to retain the cat and pineapple image and also imagine a book. Then try to retain the three images and also imagine a chair. The task becomes more and more difficult. However, it is easier to imagine a scene in which a cat is sitting on a book located on a chair and scratching a pineapple than to imagine four unrelated objects.

Forming objects into a group in which they are associated in some way (four digits making up a familiar year; four visual objects relating to each other) is known as chunking, and in general chunking increases the number of objects available for immediate mental processing. The familiar process of reading depends on a hierarchy of chunking, we put words into phrases, phrases into sentences, sentences into paragraphs.

Laboratory tests of working memory attempt to quantify these observations. Subjects are given lists of unrelated words and a short while later are asked to recall them. Early work suggested that the number was the "magic number" seven [32]. However, this earlier work did not take chunking into account, and later experiments suggest that the actual number is in the range of three to four. For example, rather than measuring the average number of words that can be recalled, experiments can measure the largest list size for which the subject never makes errors [33]. If the subject is given a different verbal task during the delay between hearing the words and attempting to recall them, the number of words recalled is reduced.

There is evidence that we have several separate working memory stores, including a verbal store, a visual store, and a spatial store. The verbal store is tested by word lists as described in the previous paragraph. The visual store is tested by showing subjects a visual array of rectangles, some filled and some not (see Fig. 2.1a) and after a short delay asking the subject to indicate the filled blocks on an empty array. The spatial store is tested by the experimenter tapping on a set of blocks in some random order, and after a short delay asking the subject to reproduce the tapping order (see Fig. 2.1b). The typical subject can correctly fill in about nine rectangles, and reproduce five or six taps.

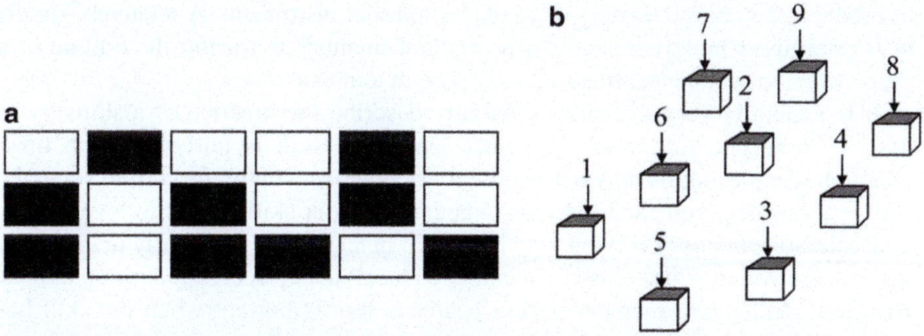

Fig. 2.1 Measuring visual and spatial working memories. (**a**) In visual working memory experiments, a subject views an array of rectangles for a few seconds. Arrays contain 4–30 rectangles in which about half are shaded. After a short delay, the subject is given a copy of the array with all rectangles unfilled, and asked to fill in the shading as on the previous array. Typical subjects can correctly recall up to about nine shaded rectangles. (**b**) In spatial working memory, a subject watches an investigator tap his finger on a sequence of two to eight wooden blocks in a random arrangement of nine such blocks. After a short delay, the subject is asked to reproduce the tapping sequence. Typical subjects can correctly reproduce sequences of up to five or six taps

The reason for believing there are separate stores is that if the delay between verbal presentation and recall is filled with a visual or spatial task, the verbal memory is less affected than if the distracting task is verbal. If the delay between visual presentation and recall is filled with a verbal or spatial task, the visual memory is less affected than if the distracting task is visual, and similarly for spatial memory [34].

Evidence that working memory is a different brain system from semantic and episodic memory comes from observations that one type of brain damage or deterioration strongly affects declarative memory with limited effect on working memory, and another type affects working memory with little effect on declarative memory. Important examples of declarative memory deficits with no effect on working memory are HM [26] and patients with Korsakoff's syndrome as discussed earlier. Patients with Huntington's disease show the opposite symptoms of relatively intact declarative memory combined with working memory deficits [35]. Patients with the frontal variant of frontotemporal dementia show working memory deficits but unaffected semantic memory, while patients with the temporal variant show severe semantic memory deficits but unaffected working memory [36].

2.2.5 *Procedural Memory*

Procedural memory refers to learning and performing skilled behaviours. Skills learned early in life include walking, running, descending stairs, and drinking from a cup. Some very complex skills are the generation of sequences of muscle movements to produce speech and writing or to play a musical instrument. A relatively simple skill often used in experiments on procedural memory is tracing the outline of a shape when the shape and trace are only seen in a mirror.

It is generally very difficult to verbally describe the practice of a skill: try to describe how you ride a bicycle. Once learned, a skill is fairly stable in time, although complex skills may deteriorate if not practiced. Only the current state of a skill is accessible, you cannot decide to go back to your skill level, say, 3 years ago.

Declarative memory is often very important in learning a skill. Early in the learning process you may use episodic memory to recall the correct sequence of actions. However, declarative memory access becomes less important when the skill has reached a high level. In fact, requiring an expert to verbally describe their skill can result in their performance deteriorating [37].

Furthermore, in some cases a skill can be learned without any apparent declarative memory, or even awareness that the skill is being learned. For example, subjects can be given a task that requires rapidly identifying the location of an object in a complex picture presented on a computer screen [38]. A sequence of pictures is presented with the object in different locations, but with a complex rule defining how the object shifts from picture to picture. Subjects learn to become faster in finding the object, but have no awareness of the rule or of having learned it.

Different deficits observed following different types of brain damage or deterioration supports the idea that declarative and procedural memory are supported by different brain systems. In the case of HM, declarative memory was strongly affected but past skills remained intact, and there was a retained capability to learn new simple skills such as mirror tracing that did not require initial assistance from declarative learning. In such simple tasks, skill would improve over a number of sessions, but with no memory of attempting the task earlier. Furthermore, patients with Parkinson's disease show degradation of procedural skills such as walking and loss of some abilities to learn new skills [39] but relatively little loss of semantic or episodic memories.

2.2.6 *Priming Memory*

If a message "drink Lipton Ice" is flashed on a screen during performance of another task, and the message is visible so briefly that the subjects have no awareness of having seen it, the subjects nevertheless have a later preference for the Lipton Ice drink, provided that they are already thirsty [40]. This phenomenon is much more limited than the exaggerated claims for effective subliminal advertising, but does demonstrate that there can be memory (i.e. an effect on later behaviour) when there is no conscious awareness.

In another laboratory test for priming, subjects are shown a series of line drawings of different objects and asked to identify the object [41]. Each drawing is visible for less than 50 ms, and is followed by a meaningless pattern with strong contrast to prevent retinal afterimages. The first time a drawing is seen under these conditions, the object is identified about 13 % of the time. However, if the same drawing is seen again under the same conditions a little later, the object is identified about 35 % of the time. In other words, somehow the exposure to the drawing, even though it could not be identified at the time, increases the chance of its being identified later.

Priming memory appears to be affected by damage to the visual areas of the cortex [42], but not by damage or deterioration that affects any of the other types of memory [43].

2.2.7 *Prospective Memory*

Prospective memory refers to our ability to remember to do something, when the intention to do it was formed earlier. Typical examples are remembering to go to an appointment with the dentist, remembering to pay a bill on time, remembering to get milk at the supermarket on the way home and so on. Prospective memory failures make up the bulk of everyday forgetting and are very common. Such failures become much more frequent with, for example, patients with Alzheimer's disease. Prospective memory failures are generally much more disruptive to daily life than failures to remember words or facts or past events [44]. Constant failures to remember appointments with friends or professional colleagues can be very disruptive to social and working life.

Prospective memory tasks are separated into time based or event based types. A time based task requires a behaviour at a specific time (e.g. put the meat in the oven at 5 p.m.) while an event based task requires a behaviour in response to a cue (e.g. mention something to a friend when you see them). Tasks are also classified as regular or irregular. A regular task is an habitual activity (e.g. take medicine every day with breakfast). At irregular task is occasional (e.g. water the garden tomorrow when you get home). A distinction can also be made between short term prospective memory (e.g. I forget what I was going to say) and longer term.

Prospective memory is typically regarded as a sequence of activities which can draw as required on other memory systems. The first step is a planning to develop an intention, the second is maintaining the intention while being distracted by other tasks, the third is returning to the intention at the appropriate point, the fourth is carrying out the original intention. Some parts of prospective memory could be working or episodic memory tasks. Study of brain activation during prospective memory tasks indicate overlaps with areas active during episodic and working memory [45].

2.2.8 *Information Models for Different Types of Memory*

The differences between memory types suggest different information models. Thus as illustrated in Table 2.1 semantic memory is learned by the repeated presence of two elements at the same time, and accessed by the occurrence of one element. This suggests

Table 2.1 Examples and information models for different types of memory

	Semantic	Episodic	Working	Priming	Procedural
Definition	Facts	Events	Number of objects available to processing at same time	Recent experience subconsciously affects behaviour	Skills
Example	Meanings of words	What did I do at specific time	Remembering telephone number while dialling	Brief visual exposure influences later behaviour	Writing
Method of learning	Repetition	Just happens, but stronger for novel situations	No apparent change to capability through learning	Recent activity, and no long term learning	Repetition and reward
Method of access	One item of information triggers another	A few items trigger recall	Put into working memory by senses or other type of memory access	Sensory input	Performing skill
Information model	Activation on the basis of frequent past simultaneous activity	Activation on the basis of past simultaneous recording	Limits on how many items available in same resources	Activation on the basis of recent simultaneous activity	Conditions linked to behaviours; past rewards determine strengths of links

an information model in which information is accessed on the basis of frequent past simultaneous activity. Episodic memory is sensitive to the degree of novelty in an experience, and novelty will tend to require more information recording. This suggests access on the basis of simultaneous past information recording. The information model for working memory is simply the number of objects which can be independently maintained in the same resources at the same time. The information model for priming memory is activation on the basis of recent simultaneous activity. The procedural memory model is reward modulated access to behaviours from sensory conditions.

As discussed in later chapters, activation on the basis of recent activity, frequent past simultaneous activity and simultaneous past information recording represent the three primary ways in which information recorded in the past about the environment can be accessed, and procedural memory represent the way in which current and past information can be linked to appropriate behaviours. Working memory represents a limit on the simultaneous information accessing processes that can occur in the same resources.

2.3 Speech

The ability to understand speech and to generate appropriate speech are fundamental to human cognition. Reading and writing are important extensions of these capabilities. Speech allows the thought processes of one person to be reproduced in the mind of another person, at least approximately. The second person can in turn modify the thought processes, and return them to the first person again. Speech and writing make it possible to experience situations that we have never directly experienced, including situations that do not exist in the real world. Frequently, even our own internal thought processes are experienced as sequences of words.

A newborn baby has the capability to learn any human language. The baby goes through a series of stages. The first stage, usually starting around 6 months, is babbling [46]. Use of individual words, usually nouns, follows by around 12 months. Meaningful use of word pairs is generally achieved by 2 years. Speech is completely intelligible by about 5 years. However, if a baby is not exposed to a particular language in the first 5 years, accent, grammar and style will not be fully mastered by later learning. There will be permanent major deficits in language use in general if no language is acquired by age 12 years [47]. Interestingly, many species of songbirds show a similar learning pattern, there are stages of babbling and subsong, with auditory feedback needed [48].

An infant hears a large amount of speech. However, in terms of the complexity of the language system which is learned, the total heard is in fact remarkably small. This suggests that the human brain must have some source of knowledge about the structure of language in general, which can be used to guide the learning process [49].

Understanding speech begins with a sequence of sounds. Within a language, there are sound units called phonemes. Each phoneme is a group of sounds, such that changing one sound to another within the group will make no difference to the

Table 2.2 Some ways in which different types of structure determine the meaning of a group of words

Word order
North of the lakes are the mountains
The mountains are north of the lakes
The lakes are north of the mountains
The north mountains are of the lakes
Lakes the mountains are the north of
The first two mean almost the same, except for a slightly different emotional tone: the second sentence is more factual. The third has the opposite meaning. The fourth has a somewhat obscure meaning that the mountains are associated with the lakes in some way. The meaning of the fifth is not clear
Longer pauses at certain points
there is no way I can do the task I am working on Thursday
there is no way I can do the task *pause* I am working on Thursday
The first states that I cannot do a task scheduled for Thursday, the second that I cannot do a task because I am working on Thursday

meaning. For example, in kit and skill as usually spoken, the k sound is slightly different: in kit the sound is aspirated, or ends with a tiny puff of air. In English, one sound could be exchanged for the other without changing the meaning, and the two sounds are therefore both in the same phoneme. In, for example, Icelandic, changing from one of these sounds to the other would change the meaning of the word containing them, and they are therefore parts of different phonemes. There are about 40 phonemes in English. A 6 months old baby can detect the difference between any two sounds that can be in different phonemes in any human language, but by 12 months can only discriminate between phonemes in the languages the baby has been hearing regularly [50].

Out of these sound units our brains extract words, phrases, sentences and meanings on larger scales. Individual words can communicate meaning. To generate meaning on a larger scale, multiple words must be presented within a structure known as syntax. In English this structure defines word order, longer pauses at certain points, relative stress on different words, the way the tone changes within a word, and even the detailed form of the word. If the structure is violated, the meaning is changed, difficult to grasp, or absent. For examples see Table 2.2.

The meaning of an individual word can change depending on the context in which it is used. One word may have multiple unrelated meanings, such as "light" in the three sentences: *I will light the fire*; *This backpack is fairly light*; and *The moon provides enough light*. A word may have different but related meanings, such as "window" in the two sentences: *I cleaned the window* (an object) and *The bat flew in through the window* (an opening). A word may have different related and unrelated meanings, such as "bank" in the three sentences: *The bank offered a no fee credit card* (financial institution); *The bank was destroyed in the fire* (building); and *There was a rabbit on the bank of the river*.

Understanding speech is so natural that it is hard for us to listen to the physical sounds of our own language. For example, spoken English contains a high proportion of "ss" sounds, but it is hard for a native speaker to observe that by listening, although it is obvious from inspection of written English.

Generating speech is equally natural, the required sequences of muscle movements are produced without any attention, and we are often not conscious of the words we are going to say until we say them, only (sometimes) of a general internal image that drives speech production. Nevertheless we are able to produce accurate, grammatical speech with very few errors.

The kind of speech deficits that result from damage to different brain areas can provide significant information about how the processing of speech is organized in the brain. However, it is important not to exaggerate the conclusions that can be drawn from such associations between damage and deficit. For example, if there is a type of local damage that affects the production of speech but not the understanding of speech, this does not necessarily mean that production and comprehension are supported by completely different brain regions. However, it does imply that there is a brain area specialized in an information process that is needed by speech production but not comprehension.

There are four major types of speech deficit (or aphasia) observed in patients with brain damage to certain brain regions. One is a deficit in the production of speech. A second is in the ability to understand speech and produce meaningful utterances. A third is if the ability to repeat sentences spoken by someone else. The fourth is difficult or unclear articulation of speech that is linguistically normal [51].

In a speech production deficit (one example is Broca's aphasia [52]), spontaneous speech is slow, laboured and choppy. On average, only two words are generated at a time, sometimes only one word, even though an intended meaning is present. Utterances include nouns, verbs and adjectives, but function words like articles, conjunctions and prepositions are not generated. An example might be an intent to take pride in a son who is attending university coming out as "son … university … smart … boy … good … studies … good".

In a speech comprehension deficit (one example is Wernicke's aphasia [52]), incomprehension leaves the patient unable to generate meaningful utterances, although the articulation and syntax remains normal. For example, a verbal discourse by a patient might be made up of sentences like "I told my cow with the radio but did not understand the fence". All spoken sentences are grammatically correct but devoid of meaning to the hearer.

A deficit in the ability to accurately repeat sentences is also present in both Broca's and Wernicke's aphasias, but in Transcortical Sensory aphasia [53] there are Wernicke-like production and comprehension deficits but no repetition deficit.

There is a specific brain region known as Broca's area, which often results in Broca's aphasia if it is damaged. Brain imaging results show that Broca's area is active during speech production [54]. However, a limited degree of speech comprehension deficit can sometimes be present with damage to this area, and brain imaging indicates that it is also active during language comprehension [55]. This region is also active during prospective memory [56]. Furthermore, a brain region very similar to Broca's area is also present in the brains of the great apes [57]. The implication is that although this brain area performs an information processing function that is very important for human speech generation, its function is also important for other cognitive capabilities including in the great apes with no (or extremely limited) language capabilities.

2.4 Arithmetic

As mentioned earlier, many people have difficulty with arithmetic, even though the need for some skill in this area is ubiquitous in society. Errors are often made in the mental performance of the basic manipulations (addition, subtraction, multiplication and division). These errors are common in children learning arithmetic, but still occur surprisingly frequently in adults. Given the effort put into the formal teaching of arithmetic, the high level of errors suggests that there is something about the information processes needed for arithmetic that do not match well with the capabilities of our brains. As Marr commented "I have no doubt that when we do mental arithmetic we are doing something well, but it is not arithmetic" [4].

Many different strategies may be used to perform the same manipulation including counting and memorisation of a fact portfolio. For example, to find the result of $4+3$, a child may start at 4 and count off the next three numbers in the standard number sequence. Alternatively, the answer may be known from memory. For more complex calculations we make use of various memorized rules.

Another aspect of number knowledge is the ability to look at a group of objects, and immediately know how many objects are in the group without counting. Human beings are able to do this for groups of up to five or six objects, but not for larger groups.

More fundamental is the capability called number sense, meaning an intuitive understanding of numbers, their relative sizes, and how they are affected by different arithmetic processes. Number sense is revealed in the ability to rapidly assess the relative sizes of 3/7 and ½, to immediately see that 100/0.97 is larger than 100, or if asked to multiply 7×99 to immediately recognize that the answer is $700-7$, a much easier problem. Number sense includes the ability to detect when the answer to a calculation does not "make sense", in other words when the result is wildly different from an intuitive estimate. Mistakes are more likely to be noticed and corrected if there is a good number sense. Interestingly, brain damage can sometimes make the performance of even simple arithmetic problems slower and less accurate, but leave number sense (such as judging which is the larger of two numbers) intact. Such a patient might be able to estimate answers but unable to calculate the exact answer [58].

2.5 Face Recognition

Most of us are able to remember and recognize the faces of a thousand or more different individuals, including family, friends, coworkers, and casual acquaintances. Furthermore, it is very rare to see a stranger among the many thousands we encounter and falsely identify their face as someone known to us. Compared with most of the other visual objects we need to be able to discriminate between, faces are very similar to each other. If we are brought up exposed only to black faces, all white faces look alike, and vice versa [59]. The detailed discrimination between faces is thus something we learn. In fact, it has been shown that if 6 months old babies are exposed to different monkey faces they learn to discriminate between monkey individuals [60].

There is a particular area of the brain called the fusiform face area (FFA) which is always active during face recognition tasks [61]. Damage to this area results in prosopagnosia, or the loss of ability to recognize faces, but the ability to recognize other types of object is unaffected [61]. However, the FFA seems to be also associated with different types of strong visual expertise. For the average person, the area is not active when recognizing birds, or dogs, or cars. Some people have a strong expertise in one of those areas, able to recognize many different species of birds, to distinguish between many different types of dog, or to identify the make and year of many different cars. In such people, the FFA is also active when performing tasks in their area of expertise [62].

Human beings also have a capability to recognize the emotion of a person from their facial expression. There is evidence that facial expressions indicating anger, disgust, happiness, sadness (or distress), fear, and surprise are recognized across widely different human cultures, from Japanese and American to preliterate Papua New Guinea [63]. Brain damage to specific structures can affect recognition of one or more types of emotion from facial expression [64]. An ability to recognize the emotional implications of facial expression independent of culture would imply that such expressions are to some degree programmed genetically, although it is also clear that cultural experience can modify any such genetic programming, in particular the interpretation of the degree to which an emotion is indicated by a given expression [65].

2.6 Emotions

Human beings can experience a range of emotions that are both reactions to circumstances and general influences on behaviour. An emotion can also label a personality trait ("He is an angry person"). There appear to be large numbers of emotions labelled by different English words, perhaps well over 100, and it has been argued that they can be organized to some degree on the basis of similarity into major categories, subcategories and sub-sub-categories [66]. If has also been argued that more complex emotions are combinations of simpler emotions [67]. For example, remorse could be a combination of sadness and disgust, and contempt a combination of anger and disgust.

Table 2.3 provides a list of the names of some widely accepted emotions. For the basic emotions, there is evidence that the emotion is defined under significant genetic control. Thus as discussed in the previous paragraph, the

Table 2.3 Some different emotions

Physical	Basic	Complex
Hunger	Anger	Anticipation
Thirst	Disgust	Disappointment
Sexual desire	Fear	Envy
Panic	Happiness	Hate
	Pride	Horror
	Sadness	Indignation
	Surprise	Longing
		Love
	Contempt	Pity
	Embarrassment	Remorse
	Shame	Trust

For basic emotions there is evidence (weaker in the case of the italicized group) that there are genetically defined facial expressions and body language associated with the presence of the emotion

physical human response to the presence of anger, disgust, happiness, sadness, fear, or surprise is partially independent of human culture. Pride also seems to be expressed in culture independent ways [68]. For other emotions like contempt, embarrassment and shame there is weaker evidence for cross cultural commonality [69].

2.7 Hobbies and Intellectual Disciplines

Human beings are able to develop sophisticated skills in a wide range of intellectual areas from astronomy through history and physics to zoology. Sometimes proficiency involves a high degree of specific motor skills, such as sports of different kinds. Sometimes the skill is in complex cognition within a defined rule framework, such as the games of chess or go. An intellectual discipline may require the ability to create associations within a very large body of factual knowledge, such as history or zoology. Alternatively, a discipline may require the ability to define a small set relatively simple starting points and develop an extensive range of logical consequences, such as mathematics and physics.

Anecdotally, individuals appear to be better at some types of skill than others. However, some types of skill often go together in the same individual, such as mathematics, physics, music and chess. For some skill types, the most creative contributions tend to be made earlier in adulthood, while for others creativity can sometimes steadily increase with age. Physical scientists do their best work at age 40 ± 10 [70], but historians generally do their most significant work when late in adulthood [71]. Poets and novelists often peak in their 30s with some differences depending on the type of writing [72].

2.8 Tool Making

Human beings can create tools to assist in achieving behavioural and cognitive objectives. A simple tool could be a stone that happens to be available locally, used as a hammer or as a weapon thrown at a target. At the other extreme, an airplane is a tool for travelling long distances, and a spacecraft is a tool for acquiring information about distant objects which cannot even be perceived by unaided human senses. Human tools can thus be extremely complex. The task aided by the tool may occur far away from where the tool is made, it is not necessary for the tool maker to be where the tool is needed. Tools can even be made for tasks in locations that have never been directly experienced such as the surface of another planet. Imagining the task and location is adequate to guide tool making. The ability to make a tool can be passed to another individual, and the tool improved by that individual.

It was once thought that toolmaking was unique to human beings, but it has been found that some forms of this capability are present in a number of animal species, from crows to chimpanzees.

Some crows are able to modify objects available on the spot to use as tools. One example is bending a straight piece of wire to extract food that is out of direct reach in a deep container [73]. Another example is the snipping of leaves followed by tearing to make tools for probing crevices containing insects. Study of the variation in these probing tools suggest that crows are able to pass the ability to make the tools to other crows and build up a technology across generations [74].

Chimpanzee tool making also seems to be limited to close to the time when the tool is used. However, observation of chimpanzees making stick tools to probe underground termite nests showed that although the tools were made (stripped of leaves and shortened) at the location where they were used, raw branches were collected at some distance from that location [75]. Chimpanzees also seem able to learn from other chimpanzees, even from videos of other chimpanzees making the tool [76].

2.9 Consciousness and Self Awareness

The topic of consciousness is controversial, and even the definition of consciousness is far from being universally agreed. At a very basic level, distinctions can be observed between normal wakefulness, being stunned (by a physical blow), being asleep, and being in a coma with the extreme of a persistent vegetative state. These different states of consciousness can also be observed in animals.

The more interesting question for higher cognition is whether there is something qualitatively different about the internal experience of human beings, also labelled "consciousness" or perhaps "human consciousness". In the nineteenth century, the psychologist William James described his concept of "stream of consciousness" as the unique and characteristic property of human consciousness [77]. This stream of consciousness concept refers to the experience of streams of mental images that have no apparent connection to current sensory inputs, and may be of situations never actually experienced or even impossible in reality. However, these mental images have some similarities to direct sensory experiences, although generally without the sensory detail that would make them hallucinations. The experience is marked by points at which the mental images are sharp, separated by periods of much more vagueness [77].

More recently, emphasis has been placed on the observations that some experiences can be reported verbally, while others can influence behaviour but are not accessible for verbal description. The ability to report verbally is called access consciousness [78]. One characteristic example of unconscious influence is the phenomenon of dichotic listening [79]. In dichotic listening experiments, a subject wearing headphones hears one meaningful text being read in one ear and a different meaningful text in the other ear, having been told to pay attention to the right ear. Such subjects can afterwards describe the material read to the right ear, but not the other ear. The material read to the left ear is ignored and is not remembered later. However, if part way through reading, the texts are switched between ears, the subject follows the continuation of the text that started in the left ear, without realizing that a switch has occurred. The content arriving at the unattended ear is thus available to influence behaviour despite being consciously unavailable. This laboratory observation is similar to the cocktail party phenomenon: one conversation can be followed in a room full of people, and generally a memory for the one conversation and no others is retained. Nevertheless, words indicating strong personal relevance in an unattended conversation can be detected and attention shifted at least briefly to that conversation.

Another possible aspect of human consciousness has been labelled phenomenal consciousness [78]. The concept of phenomenal consciousness has its origins in the observations that although our experience of something (such as the colour red) may be very vivid, it is hard to describe the full vividness of that subjective experience in words. It is therefore difficult to assess whether one person's experience of the colour red is similar to or completely different from the experience of someone else, or even if individual experiences of red are the same on different occasions. The term 'qualia' is used in philosophy to refer to these vivid, subjective experiences [80], and it has even been argued that scientific understanding of the subjective experience of qualia is not possible.

A final important aspect of human consciousness is self awareness. The stream of consciousness includes images of self, behaving and experiencing. These self images can be memories of past personal sensory experiences, actions and emotions. They can also be imaginings of future experiences by self, including plans for possible experiences and impossible fantasies. There are "I" and "me" type self images [81]. If you are asked to imagine the last time you swam in a lake, the image that comes to mind could be a view of yourself, standing at the edge of a lake, and perhaps diving in. This self viewed from the outside is a perspective never seen in real life, and has been labelled the "I" self image. Alternatively, you could imagine the water splashing on your face, where the perspective is self looking out. This self view is labelled the "me" perspective. Both perspectives are useful, for example in developing plans for the future.

Are these conscious capabilities unique to human beings, or are they shared by other animals to some degree? It can be argued that human consciousness and self awareness are dependent on speech capabilities, and on the basis of evidence from human literature developed only within historical times [81].

2.10 Individual Differences and Complex Society

There is relatively little difference between different individual flies. Differences between individual lizards are difficult to find. There are some detectable differences between individual birds, and even more differences between individual chimpanzees. The differences between individual human beings are very large compared with any other species, and a significant proportion of these differences are the result of different learning histories.

These differences are valuable to a complex society, and make it possible to undertake tasks requiring the participation of large numbers of individuals, each making a unique individual contribution. One major value of large cities is that they bring into close proximity the vast range of different human skills needed, for example, to create a new industry [82]. However, the combination of individual differences and the need to coordinate the activities of large numbers of individuals presents a problem. How can it be ensured that the efforts of each individual fits with the efforts of all the other individuals as required to achieve the common purpose?

For primates other than man, the maximum size of group is about 40 [83]. This limit is set by the need for enough interaction between members of the group (such as grooming) to maintain social coherence. If a group exceeds this size, there are consequences such as increased risk from predators. The group size for human Pleistocene hunters has been estimated at 25 [84]. However, around 12,000 years ago, archaeological evidence suggests settlements of about 200 people. As pointed out by Julian Jaynes [81], such a settlement size means that, for example, a leader cannot have a face to face encounter with each person every day. The coordination problem must be solved some other way. Groups of over 200 people need a hierarchical structure to enforce rules of cooperation. Still larger groups need rule enforcement involving the threat of violence.

In modern societies, the need to coordinate the activities of millions of unique individuals is addressed by moral standards, culturally defined behavioural norms, religious beliefs, legal and governmental systems, etc. All these systems are creations of higher cognition. Julian Jaynes has argued that the major driving force that led to human consciousness and self awareness is the need to make coordination of very large groups of human beings possible [81].

2.11 Art, Music and Literature

Animals can create remarkably complex structures, from spider webs to beaver dams. A particularly unusual example is the ornately decorated construction of the New Guinea bower bird. Male bowerbirds construct large structures of wood or straw, and decorate them with small stones, glass fragments, and bones. The purpose is to attract female bowerbirds. Bower quality and number of decorations have a strong effect on male success [85].

While human art may have "practical" purposes, such as attracting members of the opposite sex, propitiating natural forces supposed to be under the control of a god, or earning a living, the fundamental difference from animal examples is that a primary objective is to generate imaginative experiences for both the artist and the audience [86].

There is evidence for the existence of hominid art several hundred thousand years ago, such as the rough stone "Venus" carvings of Berekhat Ram and Tan-Tan, and the petroglyphs at Auditorium Cave, Bhimbetka. Examples of art become much more abundant from around 20,000 years ago with hominids that are anatomically Homo sapiens.

Similarly with music, animals like birds generate songs that appear to be assertions of territorial rights or attractors for females [87]. Human music may well include these uses, but another result is the generation of an imaginative, emotional experience.

There is nothing in animals that is analogous with fictional literature, and again a major phenomenon in human beings is the generation of an extremely complex imaginative experience.

2.12 Higher Cognition and the Brain

In this chapter we have briefly outlines some of the major components of what is viewed as higher cognition in human beings. The brain appears to be the anatomical organ that provides most of the biological support to higher cognition. The question to be addressed in later chapters is how activity in the observed complex assembly of interconnected neurons making up the brain corresponds with all the higher cognitive processes we have discussed.

Chapter 3
Brain Anatomy

In the previous chapter we looked at the phenomena of higher human cognition. The brain is the organ where much of the activity supporting higher cognition takes place, and in this chapter we will provide an overview of major brain anatomy structures, and a simplified overview of neurons and their physiology within those structures. In Chaps. 4 and 5 we will describe the physiology and chemistry of neurons in much more detail. Then in Chap. 6 we will describe the anatomy and physiology of the seven major structures and a number of smaller structures in more detail.

3.1 Major Anatomical Structures of the Brain

An image of the whole brain is shown in Fig. 3.1. Underneath the upper skull are the two cerebral hemispheres, left and right. Connecting between the middle levels of the two hemispheres is a thick band of nerve fibres called the corpus callosum. In between the two hemispheres in the region under the corpus callosum is a mass of tissue separated into a number of different structures such as the thalamus and basal ganglia. Under this mass of tissue, below the base level of the cerebral hemispheres, is a sequence of other structures, with the spinal cord emerging at the bottom of this sequence.

Many different structures are visible in the brain. Seven structures are prominent: the cortex, the hippocampus, the thalamus, the basal ganglia, the amygdala, the hypothalamus, and the cerebellum. These structures are distinguished because they are physically separate, have different types of neurons, and/or different connectivity with other structures. Each of these structures is divided up into substructures with smaller differences between them. For example, the cortex is a thin sheet of neurons organized into a number of layers. The cortex is divided up into different areas all with similar neurons but different patterns of layering. Each area contains many columns vertical to the sheet, with more connectivity within one column than between columns. The thalamus is divided up into many separate clumps of neurons called nuclei, with similar types of neurons but different connectivity with the cortex.

L.A. Coward, *Towards a Theoretical Neuroscience: from Cell Chemistry to Cognition*, Springer Series in Cognitive and Neural Systems 8, DOI 10.1007/978-94-007-7107-9_3,

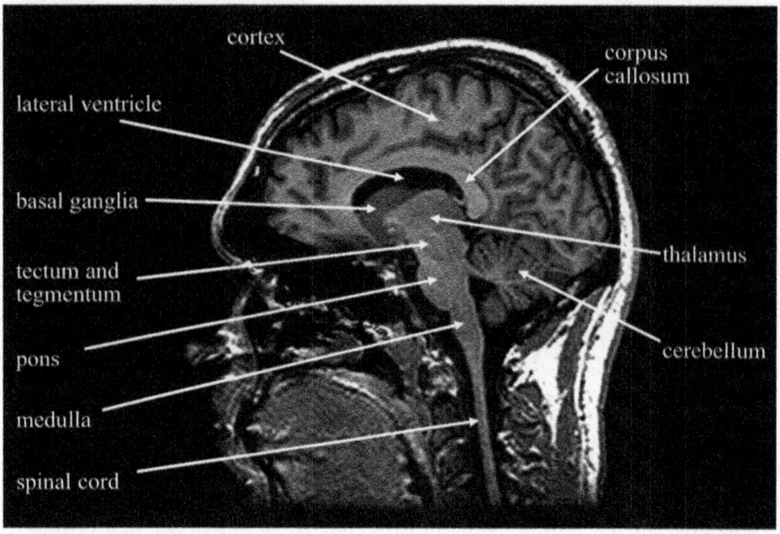

Fig. 3.1 An MRI image of the head with the complete brain. At the top of the brain are a pair of large structures called the cerebral hemispheres, almost completely covered by a layer of tissue called the cortex. The cortex is intricately folded to give it a much larger surface area than if it covered the hemispheres smoothly. In the centre of the brain is a vessel filled with liquid called the lateral ventricle which has structural but not information processing functions. The lateral ventricle is dark in the image. Extending over the top of the lateral ventricle is a bundle of nerve fibres called the corpus callosum. This nerve bundle is the primary information connection between the two cerebral hemispheres. Under the lateral ventricle and between the two hemispheres is a complex of structures including the thalamus, the basal ganglia, the amygdala and the hypothalamus. Under these structures are a sequence of anatomical components that lead down to the spinal cord. Off the back of one of these components (the pons) and extending under the back of the cerebral hemispheres is a large, convoluted structure called the cerebellum (Image from the Michigan State University Brain Biodiversity Bank. Funded by the National Science Foundation. Reproduced with permission)

The seven major structures make up the bulk of the tissue in the upper part of the brain. All of the structures are bilateral, in other words each is made up of two very similar structures on the left and the right sides of the brain. Lower down towards the spinal cord there are other structures, these will play a smaller role in the discussion of cognition. There are also many smaller structures in the upper parts of the brain that are not regarded as parts of the seven. Discussions of higher cognition will focus on the seven, but also touch on the roles of some of these smaller structures. The anatomy and connectivity of the major structures will be discussed in more detail in Chap. 6.

If the top of the skull is removed, a mass of red tissue is revealed. The red colour is the result of the rich supply of blood going to all parts of the brain. In a brain preserved after death, two types of tissue can be observed: grey matter and white matter. Grey matter is made up of cells called neurons which perform the basic brain information processes. White matter is made up of connections between neurons, called nerves or axons.

There are a number of ways to investigate the internal structure of the brain, three of the most important are physical sectioning, magnetic resonance imaging (MRI), and measurements of brain electrical activity such as electroencephalography (EEG) and magnetoencephalography (MEG). Physical sectioning means slicing a postmortem brain into flat, parallel layers, and staining the layers (called sections) to bring out different types of detailed structure. Some types of staining can make neurons more visible, others make axons more visible. MRI uses a combination of strong magnetic fields and radio signals to create images of section-like layers in a living brain. The resolution of MRI is relatively poor, it falls far short of being able to distinguish individual neurons. EEG records electrical activity outside the scalp. MEG records magnetic fields produced by electrical currents within the brain. Both EEG and MEG have better temporal resolution than MRI, but MEG is somewhat less sensitive to skull impedance etc. and therefore in principle has a spatial resolution comparable with MRI. However, relating EEG and MEG measurements to physical locations in the brain is a complex problem [88].

The brain is roughly the shape of a sphere that is extended towards the front and back of the head (i.e. technically an approximate ellipsoid). It is divided into two hemispheres, left and right. Wrapped around each of these hemispheres is a thin sheet of tissue, 2–4 mm thick, called the cortex. The cortex covers the outside of each hemisphere, and extends around to cover the areas of each hemisphere where they face each other, except where the corpus callosum connects the hemispheres. The cortex is not a smooth surface, it is crumpled up (or folded) with deep fissures (called sulci) so that its area is much larger than that of the surface it covers. Beneath the cortex wrapper, the two hemispheres are made up largely of nerve tissue, the nerves mostly connecting different parts of the cortex.

Beneath the middle of the two hemispheres, they are pushed apart to make room for a complex of subcortical structures, including the thalamus, the basal ganglia, the amygdala and the hypothalamus. Each of these major subcortical structures are made up of a number of nuclei, each nucleus being a clump of neurons.

Further underneath come a series of structures: the tectum and tegmentum, the pons, then the medulla, and finally tapering down into the spinal cord. All these structures are unitary: they are single structures rather than left and right pairs. Extending backwards out from the pons and under the cortex is a structure called the cerebellum which is bilateral.

Figures 3.2, 3.3 and 3.4 show the relative positions of these major structures in the physical slices stained to reveal neurons and in MRI images. Figure 3.5 summarizes this information in conceptual form.

3.2 Roles of Major Anatomical Structures

The roles of the cortex, hippocampus, thalamus, basal ganglia, amygdala, hypothalamus and cerebellum in higher cognition have been intensively investigated. It is clear that the cortex plays a role in almost every cognitive process, with

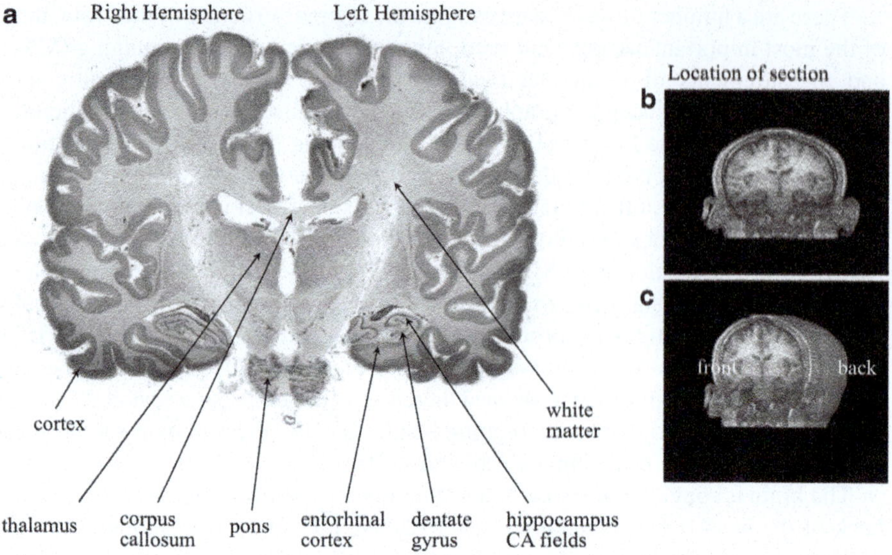

Fig. 3.2 (**a**) Coronal slice of the human brain with neurons stained dark. The location of the slice in the brain corresponds with a vertical plane that notionally cuts the brain into two roughly equal parts, with the eyes and front of the brain on one side, the back of the brain on the other side, passing symmetrically through the left and right ears. This slice is illustrated in MRI in (**b**), and its location in a 3-D image of the head is illustrated in (**c**). The two similar but not completely identical left and right cortical hemispheres can be seen in cross section. The major information link between the hemispheres is the bundle of nerves called the corpus callosum. The darker band that goes most of the way around the outside of each of the two hemispheres is the cortex. Much of the volume inside the cortex is white matter, or nerve fibres mainly connecting between different parts of the cortex. The cortex thickness is fairly uniform, but at the bottom of the brain it curls around in an area called the entorhinal cortex, then thins substantially to an area called the hippocampus. This thinner cortical sheet ends in an area called the dentate gyrus, which is V-shaped in this section. Below the corpus callosum the hemispheres are further apart, and in that space a number of structures are located. In this section the thalamus is visible. Below the thalamus is a series of structures (one of which is the pons, visible in the section) and at the bottom of the series is the spinal cord (Image from the Michigan State University Brain Biodiversity Bank. Funded by the National Science Foundation. Reproduced with permission)

some areas being more important than others for some types of process but with no one-to-one correspondences between area and cognitive function. The hippocampus plays an important role in memory, but again no one-to-one correspondences between hippocampal substructures and cognitive functions. The thalamus plays an important role in managing access of sensory inputs to the cortex, but also plays an important role in communications between cortical areas. The basal ganglia plays a role in behaviour selection and the cerebellum in managing frequently used behaviours. The amygdala and hypothalamus play roles in emotion, but the cortex is also important.

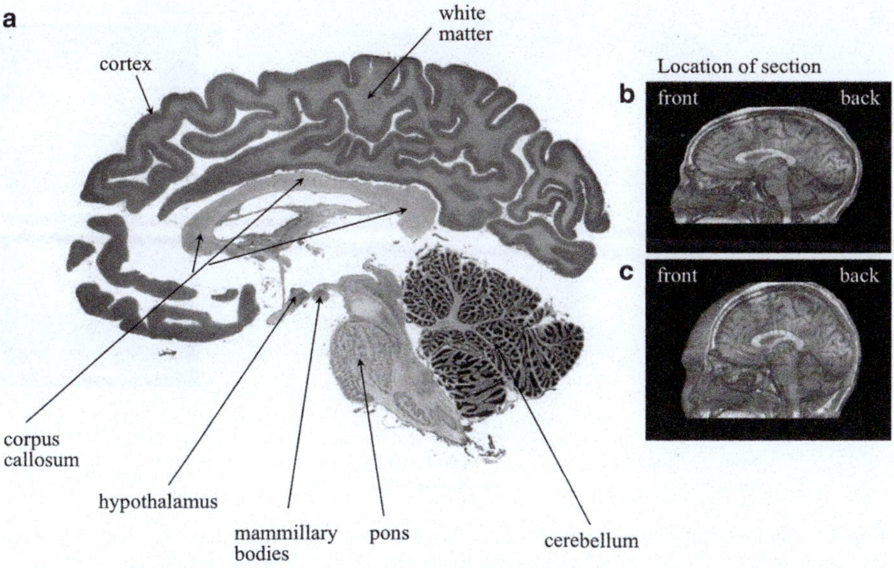

Fig. 3.3 Central sagittal slice of the human brain with neurons stained dark. The slice in image (**a**) is vertical from front to back along the centre of the brain as shown in the MRI images (**b**) and (**c**). The convoluted folding of the cortex can be seen. More detail of the large structure called the cerebellum is visible in this slice. Like the cortex, the cerebellum is made up of a relatively thin outer sheet of neurons, even more folded than the cortex. Inside this cerebellar cortex is white matter, and at the centre of the cerebellum is a set of different clumps of cells called the cerebellar nuclei (Image from the Michigan State University Brain Biodiversity Bank. Funded by the National Science Foundation. Reproduced with permission)

Later in the book the different information processes performed by these structures will be described. Understanding these information processes makes it possible to understand the different roles they play in higher cognition.

3.3 Neurons, Axons, Dendrites and Synapses

The brain, like all the rest of the body, is made up of large numbers of cells, the universal building blocks of life that are too small to be seen without the aid of a microscope. All cells are completely enclosed by a wall called the cell membrane that maintains some specific differences between the chemical environments inside and outside. A range of structures and chemical processes are protected inside the cell membrane. The membrane actively moves some chemicals in or out, allows a few to permeate relatively freely, allows a small number to pass in or out under specified circumstances, and blocks all others.

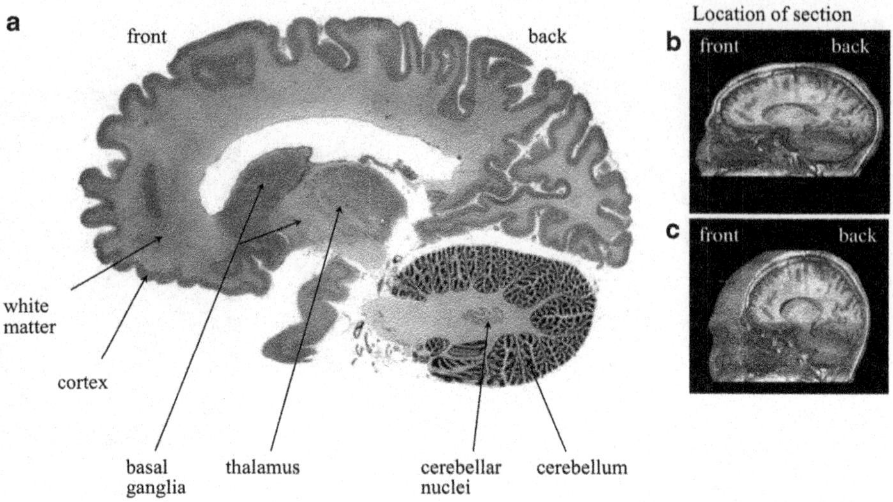

Fig. 3.4 Medial sagittal slice of the human brain with neurons stained dark. A slice of the brain from front to back, located in a plane parallel to the one illustrated in Fig. 3.3 but shifted to the left so that it passes between the nose and the left eye is illustrated in (**a**). The location is illustrated by the MRI image of the slice in (**b**) and a 3-D image in (**c**). The basal ganglia and thalamus are mainly located to the left and right of the central plane, and are therefore more visible than in the central slice shown in Fig. 3.3. The basal ganglia are located in front of and under the thalamus, and also wrapped around the side of the thalamus that is facing away from the centre. Some cerebellar nuclei are also visible in the centre of the cerebellar cortex in this section (Image from the Michigan State University Brain Biodiversity Bank. Funded by the National Science Foundation. Reproduced with permission)

A type of cell called the neuron performs the information processes supporting cognition. There are also other cells in the brain such as glia cells that play physiological support roles including supplying nutrients to neurons, recycling chemicals used by neurons, and isolating neurons from each other. The number of neurons in the human brain is difficult to determine. Estimates range from a few times 10^{10} neurons up to 10^{12} neurons. A high proportion of these totals is in the cortex and the cerebellum. The cerebellum is smaller than the cortex but contains very large numbers of small neurons called granule cells. Hence the cerebellum probably has more neurons than the cortex.

Neurons vary considerably in size, form, and properties. However, there are some general characteristics that can be identified, although the reader is warned that there are always a small proportion of neurons that are exceptions to almost any of the descriptions that follow.

Anatomical structures in the brain are concentrations of neurons. Such a concentration is separated from other concentrations by neuron-sparse regions, and has a different pattern of neuron connectivity from neighbouring structures. Major structures also differ in the types of neuron they contain.

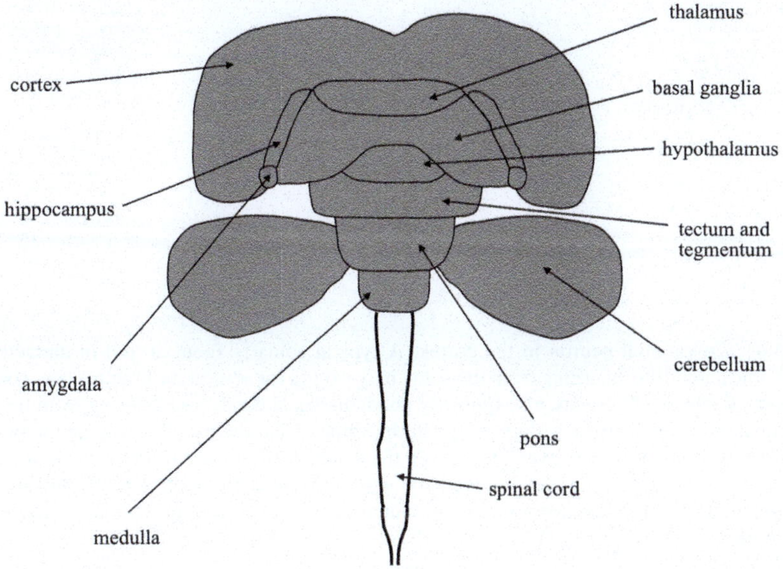

Fig. 3.5 Conceptual arrangement of major brain structures. At the top of the brain are the two cerebral hemispheres, which are a thin sheet of mostly neuron tissue (the cortex) wrapped around a body of nervous tissue. A region that resembles a thinned down area of the cortex is called the hippocampus. Underneath the cerebral hemispheres are the thalamus, basal ganglia, amygdala and hypothalamus, all of which are bilateral or duplicated in the left and right brain. Below these subcortical structures are a series of structures that lead to the spinal cord. These structures are the tectum and tegmentum, the pons and the medulla. Extending out from the pons is a further bilateral structure, the cerebellum. The tectum is a relay for auditory and visual information. The tegmentum includes nuclei (the substantia nigra and ventral tegmental area) that are functionally part of the basal ganglia. The cerebellum includes both a cortex like layer and inner nuclei. The pons, the tegmentum and the medulla all include nuclei that are important for communication with the cerebellum

Dimensions

The micron is a useful unit of length for describing neuron physiology. There are 1,000 μm in a millimetre, or one million microns in a metre.

The diameter of the period at the end of the previous sentence is about 200 μm, ten times bigger than the body of a typical neuron.

The nanometre is a useful unit for describing the components of neurons. There are 1,000 nm in a micron.

The thickness of the cell membrane surrounding and enclosing a neuron is in the range 3–10 nm or 0.003–0.01 μm

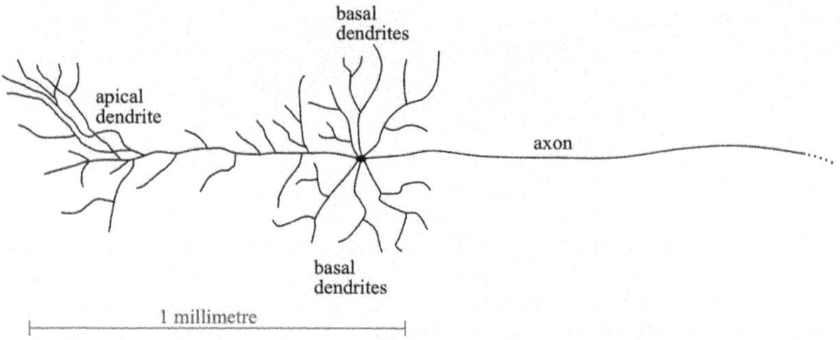

Fig. 3.6 A pyramidal neuron in the cortex. A typical soma is about 20 μm in diameter. The single apical dendrite emerging from the soma may extend for several millimetres (i.e. 100 times the soma diameter). There are also typically about six basal dendrites emerging from the soma. The one axon emerges from a slight bulge in the soma called the axon hillock, and may extend a number of centimetres (i.e. over 1,000 times the diameter of the soma) across the brain. On the same scale as the rest of the drawing the axon would extend a metre or more off the page, with multiple branches emerging towards its end. Some axons that extend through the spinal cord are even longer

The body of a neuron (called its soma) can range in diameter from about 4 to 50 μm or more. This diameter is not untypical of cells elsewhere in the human body. However, the distinctive physical characteristic that makes neurons different from any other type of cell is that they have extensions out of their somas called neurites. As illustrated in Fig. 3.6 there are two types of extension: axons and dendrites. Each neuron has one axon and several dendrites. An axon is a thin thread that emerges from a slight bulge in the soma called the axon hillock. The thread is about 1 μm in diameter and can extend several centimetres or more. In other words the axon can be over 1,000 times longer than the diameter of the soma. Towards its end an axon may branch into several thousand different threads, each thread at its end contacting another neuron.

Dendrites are generally several times thicker than axons, and begin to branch close to where they emerge from the soma, to give them a tree like structure. The dendritic tree can extend up to several millimetres from the soma in many different directions. The diameter of the volume in which dendrites can be found can therefore be 100 times the diameter of the soma from which the dendrites come.

The pyramidal neuron illustrated in Fig. 3.6 is the most common neuron type in the cortex. The drawing is roughly to scale. The dendritic tree can be illustrated, but if the axon were fully drawn on this scale it would extend a metre or more off the page with extensive branching towards its end. The axon and dendrites are all enclosed in the cell membrane, so the neuron somas make up less than 1 % of the volume of neuron tissue.

The axon from one neuron can branch and terminate on thousands of different other neurons. Most of the terminations arrive at different points on the dendrites of the target neurons, either close to the soma (called proximal connections) or further away from the soma (called distal connections). Some terminations can actually target the soma directly, or even the axon close to where it leaves the soma. Connections are not formed indiscriminately, each neuron connects with specific other neurons or groups of neurons, and the identity and strength of these connections are the primary way in which information is stored in the brain. At the end of the axon there is a slight swelling which almost touches a swelling out from the target neuron. These two swellings form a synapse, which is the point of contact between the two neurons.

Information is carried in the brain in the form of electrical impulses travelling along axons. These impulses indicate the activity of the neuron that is the source of the axon to all of the neurons targetted by that axon. Information processing occurs by neurons determining whether or not to become active and generate an electrical impulse on the basis of all the indications of activity they are receiving from other neurons.

A neuron acts upon its target by releasing a chemical called a neurotransmitter from the axon part of the synapse towards the dendrite part. For any one neuron, the same neurotransmitter (or sometimes a small group of neurotransmitters) is released from all its synapses on to other neurons. In many cases, the one neurotransmitter has similar effects on all the targetted neurons, either increasing or decreasing their tendency to produce their own output electrical impulse. Neurons that increase the tendency of their targets to become active are called excitatory, neurons that decrease that tendency are called inhibitory. Pyramidal neurons are the most common excitatory neuron in the cortex. Another excitatory cortical neuron is the spiny stellate cell which can be regarded as a simplified pyramidal neuron. There are many different types of inhibitory neurons in the cortex, generically called interneurons.

3.4 Ion Pumps

In all cells including neurons there is a difference in the concentrations of various types of ion between the inside and the outside of the cell membrane. These concentration differences are maintained by active passageways through the membrane called ion pumps.

One example is the sodium-potassium pump. Each of these pumps is actually a very complex molecule that interacts with sodium and potassium ions. These ions are atoms with one electron missing, and they therefore have one unit of positive electrical charge. Ions are the most common state of sodium and potassium atoms everywhere in the body. The sodium-potassium pump goes through a cycle, illustrated in Fig. 3.7, which moves three sodium ions (Na^+) from the inside to the outside of the neuron and two potassium ions (K^+) from the outside to the inside, resulting in a single net negative charge inside. One pump can go through over 100 cycles per second, and one neuron has over a million such molecular pumps. As a result of the operation of the sodium-potassium pumps, there is a lower concentration of Na^+ ions and a higher concentration of K^+ ions inside the neuron. Furthermore, a net excess of negative charge inside the cell in maintained. The net

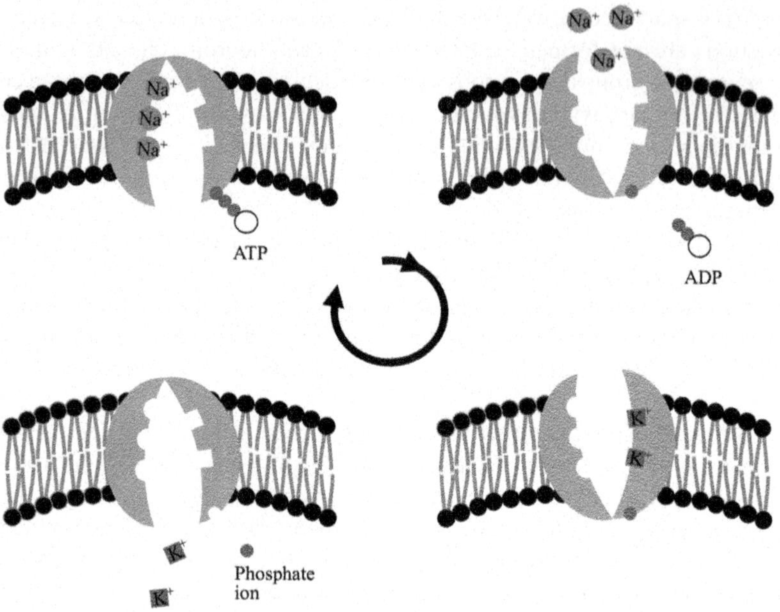

Fig. 3.7 The sodium-potassium pump that maintains the voltage inside a neuron below the level outside. The cell membrane is a double layer of oily molecules, less than 10 nm thick, that is impermeable to most chemicals present in the body. Sodium-potassium pumps are complex molecules spanning the cell membrane that use energy in the cycle illustrated to remove three sodium ions (Na^+) from the inside of the cell and inject two potassium ions (K^+) into the cell. Energy is provided by the universal source of energy within cells: the fracture of an adenosine triphosphate (*ATP*) molecule into an adenosine diphosphate (*ADP*) molecule and a phosphate. The pump has a number of receptors, or sites that have chemical affinities for specific chemicals that are generally available circulating in the environment. In the *top left*, the pump is open to the inside of the cell, and an ATP molecule attaches itself to a specific receptor on the pump that has an affinity for ATP. The attachment triggers the pump to activate three Na^+ ion receptors, resulting in adherence of three Na^+ ions. In the *top right*, the adherence of the Na^+ ions triggers a chemical reaction that opens the pump enclosure to the outside of the membrane. The energy needed for this reaction is provided by the breaking of ATP into ADP and phosphate. Another result of this reaction is that the three Na^+ ion receptor sites are deactivated, releasing the ions outside the cell, and two K^+ ion receptor sites are activated. In the *lower right*, K^+ ions attach to their receptors, triggering the opening of the pump to the inside of the cell. In the *lower left*, the switch in pump configuration triggers release of the K^+ ions and the phosphate into the cell interior, returning the pump to its original state ready for another cycle. The operation of the pump results in a deficit of Na^+ ions inside the cell and a rather smaller excess of K^+ ions. The net deficit of positive ions inside the cell results in a negative voltage potential of about 70 mV in the inside relative to the outside

excess results in a voltage difference between interior and exterior called the resting potential. The resting potential is generally in the range −65 to −85 mV, a common value is close to −70 mV.

Other types of ion pumps discussed in Chap. 4. The operation of these pumps results in the interior of the neuron having a lower concentration of chloride ions

(Cl^-), a higher concentration of HCO_3^- ions, and a very low concentration of calcium ions (Ca^{++}).

In specific circumstances the voltage difference can change or even collapse dramatically at points on the cell membrane of the dendrites, soma or axon. Collapses at different points can spread and reinforce each other. If a collapse reaches an area of the soma called the axon hillock where the axon emerges, it is launched along the axon. It then propagates all along the axon and arrives as a spike of voltage at the synapses on to all the targets of that neuron. This spike of voltage, called an action potential, influences the target neuron. Action potentials are the primary mechanism for information transfer within the brain, but to understand the generation of action potentials in a little more detail it is first necessary to describe molecules called ion channels.

3.5 Ion Channels

As described earlier, the sodium-potassium pump actively moves Na^+ and K^+ ions through the cell membrane, creating a deficit of Na^+ ions and an excess of K^+ ions inside the cell, relative to the outside. There are also pumps that maintain a deficit of Cl^- ions inside the cell membrane and calcium pumps that maintain a very low level of Ca^{++} ions inside the cell. If there is a higher density of ions in one region relative to another, there will be a strong tendency for ion flow to even out the density. However, the cell membrane largely blocks such flows.

Ion channels are molecules embedded in the cell membrane. An ion channel is conceptually illustrated in Fig. 3.8. Such an ion channel can be in one of two different states. In the open state a channel allows one or more types of ion to pass freely in or out of the cell. In neurons there are ion channels largely specific to sodium, potassium, calcium or chloride ions. In the closed state an ion channel does not allow ion passage. If a channel is in its open state, a flow occurs that will tend to even out any ion concentration differences between the two sides of the cell membrane. There can be very large numbers of ion channels present in certain areas of the cell membrane, and the more of these ion channels that are open and the longer they are open, the more rapidly the internal concentration close to the membrane will approach the external concentration. If they are all closed, the normal concentration difference will rapidly be restored by diffusion to or from deeper inside the neuron.

Ion channels are generally closed unless opened by a specific stimulus. Some ion channels are opened if a particular chemical appears in its environment outside the neuron and attaches itself to a specific site (called a receptor site) on the ion channel molecule. Such a chemical is called a neurotransmitter. Other ion channels open in response to a change in the electrical voltage difference across the membrane in which they are embedded. Opening of other channels depends on some combination of voltage and neurotransmitter.

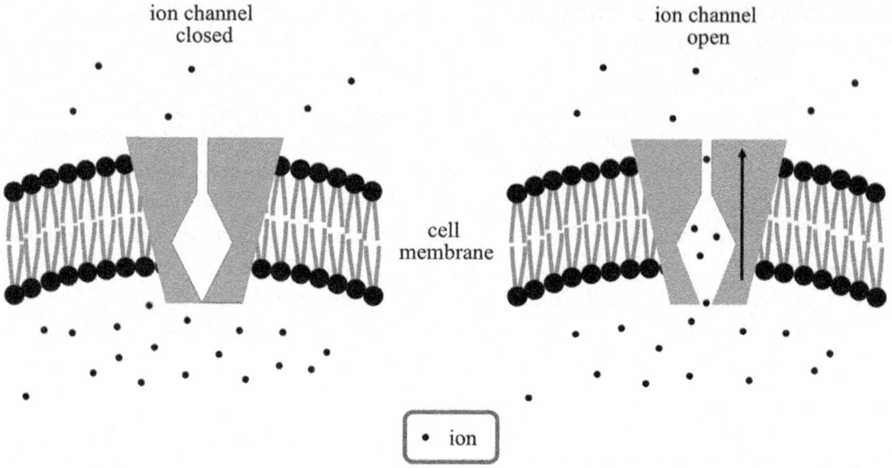

Fig. 3.8 An ion channel that when opened allows passage of just one type of ion. The cell membrane is a double layer of oily molecules that is impermeable to most chemicals present in the body. An ion channel is a molecule that spans the cell membrane. In its closed state, the ion channel is also impermeable. In its open state, the ion channel allows passage of one type of ion, which could be Na^+, K^+ Ca^{++} or Cl^-. In general there is a difference in concentration of these ions between the inside and the outside of a neuron. If the channel is open then ions of its type will pass from the region of high concentration to the region of low concentration. There are two ways in which the opening of an ion channel can be triggered. One is by a specific change in the voltage in the vicinity of the channel. The other is by attachment of one specific molecule (called a neurotransmitter) to a sensitive point on the ion channel called a receptor. Different ion channels, even for the same ion, may be sensitive to different neurotransmitters

Ions

An ion is an atom with one or more electrons missing or one or more extra electrons. Ions therefore have a net electrical charge of one or more positive or negative units. There are a number of atoms that are present in an ionized state in the brain and that are important to neuron functions. These include sodium and potassium ions with single positive charges, calcium atoms with two positive charges and chlorine atoms with one negative charge. These ions are represented in diagrams etc. by their atomic symbol followed by the appropriate number of $^+$ or $^-$ symbols to indicate their charge, i.e. a sodium ion as Na^+, a potassium ion as K^+, a calcium ion as Ca^{++} and a chlorine ion (usually called a chloride ion) as Cl^-. If the distribution of negative ions is uniform, an excess of positive ions in a region will result in a positive voltage in that region, while a deficit of such ions will result in a negative voltage. If unconstrained, ions will tend to drift towards locations where they will cancel out any electrical potential: positive ions towards where there is a negative voltage, negative ions towards where there is a positive voltage. Furthermore, even in the absence of any voltage difference, there is a strong tendency for ions to flow from regions where it is highly concentrated towards regions where the concentration is lower.

3.6 Synapses and Their Response to an Incoming Action Potential

The synapse is where the axon from one neuron contacts the dendrite of another neuron. Such a synapse is illustrated in Fig. 3.9. The axon of the source neuron (the presynaptic neuron) and the dendrite of the target neuron (the postsynaptic neuron) both swell slightly towards each other, but do not quite touch, they are separated by a narrow space called the synaptic cleft. This cleft is about 20 nm (0.02 μm) across.

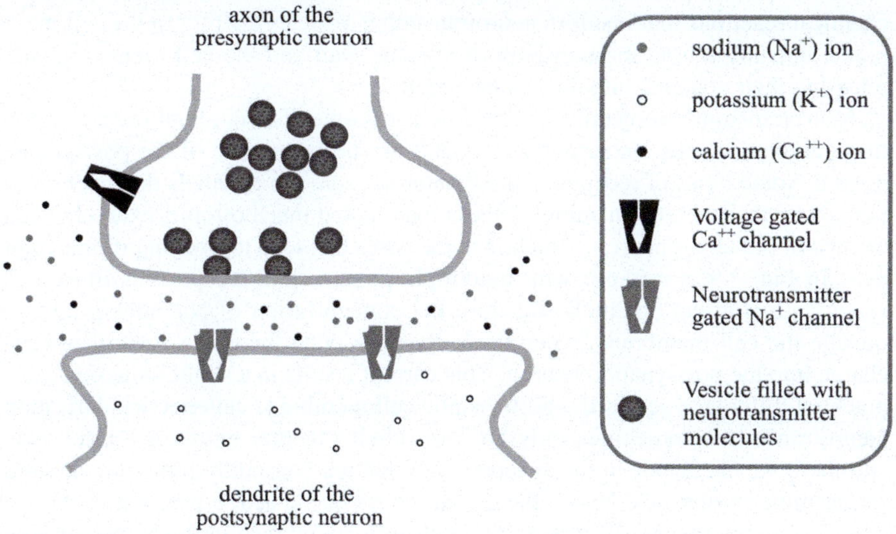

Fig. 3.9 Conceptual drawing of a synapse. The sizes of ions, ion channels and neurotransmitter molecules are greatly exaggerated relative to the size of the synapse, and all are much more numerous than illustrated. Part of the synapse is formed by swelling of the membrane of the axon of the presynaptic neuron and another part is formed by swelling of the membrane of the postsynaptic neuron. The two swellings are separated by a narrow gap about 20 nm wide called the synaptic cleft. In both parts of the synapse there is a deficit of Na^+ ions and a surplus of K^+ ions, resulting in a negative voltage potential relative to the outside. In addition there is a deficit of Ca^{++} ions relative to the outside. There are numerous calcium channels in the vicinity of the presynaptic axon swelling that if open would allow entry of these ions. The opening of these channels is triggered by collapse of the potential difference in the axon resulting from an incoming action potential. Also in the presynaptic volume there are a number of containers (called vesicles) filled with a neurotransmitter. Entry of Ca^{++} ions results in some of these vesicles moving to the cell membrane and releasing their contents into the synaptic cleft.

In the cell membrane of the dendrite of the postsynaptic neuron next to the synaptic cleft there are a number of Na^+ ion channels. These ion channels open if one of the neurotransmitter molecules released by a vesicle into the synaptic cleft attaches to its receptor site. A neurotransmitter only acts on a channel for a limited period of time, and there are chemicals in the environment outside the cells that are constantly removing any neurotransmitter, so the number of Na^+ ions that flow into the postsynaptic neuron as a result of the incoming action potential is limited, depending on the number of neurotransmitter gated Na^+ channels in the postsynaptic neuron

There are numerous ion channels embedded in the cell membrane in the synaptic swelling of the presynaptic neuron. These ion channels initiate the effect of an incoming action potential on the postsynaptic neuron. The presynaptic synapse contains a number of vessels (called vesicles) filled with a neurotransmitter chemical. The neurotransmitter is sealed within the wall of the vesicle, and the vesicle is inside the membrane of the synapse. The arrival of an action potential from the presynaptic neuron means the collapse of the voltage difference between the inside and the outside of the membrane on the presynaptic side. Calcium channels that exist in the membrane in this area are triggered to open by such a collapse, and because of the low concentration of Ca^{++} ions inside the membrane there is a rapid inward flow. The presence of Ca^{++} ions inside the membrane triggers a series of chemical reactions that result in neurotransmitter vesicles drifting to the cell membrane, opening a hole through both the vesicle wall and the cell membrane, and releasing their contents into the synaptic cleft.

Neurotransmitter molecules can diffuse across the cleft in about 0.2 ms, where they can bind to sites on receptor molecules in the membrane of the postsynaptic neuron. Some types of receptor, called ionotropic, are ion channels that open when bonded by their neurotransmitter. Other types, called metabotropic, exert chemical effects on the target neuron, which in some cases can lead to opening of ion channels. In Fig. 3.9, a synapse with ionotropic postsynaptic receptors is illustrated, specifically sodium channels. Because the concentration of Na^+ ions is greater outside the cell membrane, when these channels open there is a flow of positive charge into the postsynaptic neuron. This current results in a slight reduction in the potential difference across the postsynaptic cell membrane called a depolarisation. Neurotransmitter molecules only attach to their receptor sites for a brief time, and there are chemicals in the synaptic cleft that are constantly removing the neurotransmitter molecules. Hence the overall effect of the action potential is a small reduction in the potential across the postsynaptic membrane in the neighbourhood of the cleft. The concentration of Na^+ ions deeper in the cell is not affected, and flow of such ions from the interior restores the resting potential over a period of a few tens of milliseconds as illustrated in Fig. 3.10a.

If the ion channels opened in response to the neurotransmitter are chloride channels, there will be a small flow of Cl^- ions into the cell, which will result in a small increase in the magnitude of the negative potential across the postsynaptic membrane. The resting potential will again be restored by diffusion from the interior of the cell.

The small changes in voltage resulting from an incoming action potential to a synapse can be regarded as injections of postsynaptic potential. As explained in the next section, an injection of positive postsynaptic potential by a flow of Na^+ ions increases the tendency of the postsynaptic neuron to produce its own action potential. Such an injection is called an excitatory postsynaptic potential (EPSP) and synapses of this type are called excitatory. An injection of negative postsynaptic potential by a flow of Cl^- ions reduces the tendency to produce an action potential. Such an injection is called an inhibitory postsynaptic potential (IPSP) and synapses of this type are called inhibitory.

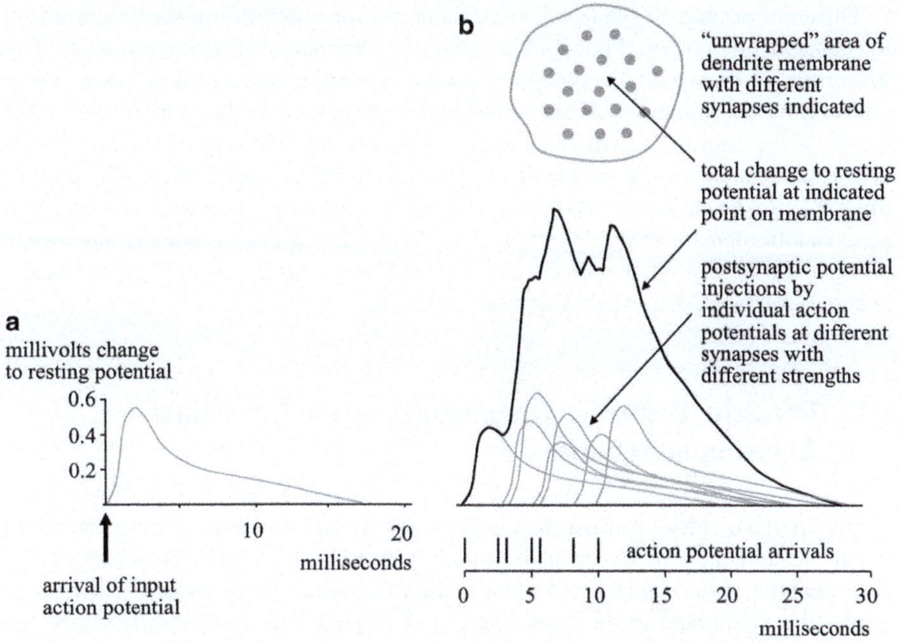

Fig. 3.10 Dynamics of postsynaptic potential injection at synapses. (**a**) Change in the membrane potential at a synapse in the postsynaptic neuron following the arrival of one action potential from the presynaptic neuron. Soon after the arrival of the action potential, there is a gradually increasing change in potential, which peaks after a couple of milliseconds at something of the order of millivolts, and then decays to 0 over the next 20 ms. The size of the peak depends on the synaptic strength, which is determined by the number of neurotransmitter gated ion channels in the postsynaptic neuron adjacent to the synaptic cleft. The decline is the result of diffusion of ions between the local membrane and neuron interior. If the ion channels opened are Na^+ channels, the potential change will be depolarising, reducing the magnitude of the –70 mV resting potential and therefore making the interior of the neuron less negative relative to the exterior. If the ion channels are Cl^-, the potential change will be hyperpolarising. The magnitude of the change and the time taken to restore the resting potential may differ for different synapses. (**b**) Integration of separate postsynaptic potentials over time. An area of the membrane wall of a dendrite is illustrated at the top. An action potential arriving at any one of the synapses injects a postsynaptic potential which rises initially to a peak value and then decays. Action potentials arrive at different synapses and at slightly different times, and the postsynaptic potential spreads from the arriving synapse into the local region. The postsynaptic potentials injected at different synapses due to arrival of different action potentials at different points in time are illustrated, and the total potential that could result from diffusion of the individual potentials to a point on a dendrite membrane central to the group of synapses is also illustrated. If at some point in time this total potential exceeds the threshold for opening voltage gated Ca^{++} channels, there will be a further depolarisation of the membrane potential which can spread. Such a further depolarisation will only occur if enough action potentials arrive within a short enough period of time for them to reinforce each other and reach the threshold for collapse. This period of time is called the integration window, and is closely related to the time within which an individual postsynaptic potential decays to a low level

Different excitatory synapses on the same neuron can differ in the magnitude of the postsynaptic potential injected in response to an incoming action potential. This magnitude is known as the synaptic strength or weight, and in many cases can be adjusted by combinations of incoming action potentials to the synapse and other inputs to the neuron elsewhere. Such adjustments are common in excitatory synapses, and such synapses generally have an asymmetrical appearance, with a much thicker structure on the postsynaptic side, known as the *postsynaptic density*. This postsynaptic density is the location of the chemical machinery managing weight changes. Inhibitory synapses have a more symmetrical appearance. Synaptic weight changes are the basis for learning from experience.

3.7 Dendrite Branches Integrating Action Potentials Arriving at Synapses

A branch of a dendrite is illustrated in Fig. 3.11. In addition to the neurotransmitter gated Na^+ channels there are also voltage gated Ca^{++} channels. A voltage gated Ca^{++} channel on a dendrite will open if the magnitude of the potential difference across the membrane in the area where it is located falls significantly below the resting potential of −70 mV. An incoming action potential to one synapse injects potential in the neighbourhood of the synapse which briefly reduces the magnitude of the resting potential by an amount of the order of millivolts by opening Na^+ channels. Such reductions can spread around the synapse injecting them. If a number of action potentials arrive at a group of synapses on the same branch within a time period significantly less than the 20 ms in which each postsynaptic potential decays (Fig. 3.10a), they can reinforce each other as shown in Fig. 3.10b. This time period can be viewed as an integration window within which action potentials must arrive if they are to be mutually reinforcing. If the depolarisation reaches the threshold at which the voltage gated Ca^{++} channels open, Ca^{++} ions will flow into the neuron in the vicinity of the channels, causing the potential across the membrane to drop further.

The larger drop in the membrane potential in one small area due to Ca^{++} influx will spread and reduce the magnitude of the membrane potential in neighbouring areas. If the reduction in a neighbouring area is large enough, Ca^{++} channels in that area will open also. Thus as illustrated in Fig. 3.12, reductions in the membrane potential can spread across a dendrite, depending on the distribution and timing of incoming action potentials. Such spreads are known as calcium action potentials. Unlike the sodium action potentials that propagate down an axon (see next section), the propagation of a calcium action potential may or may not occur depending on a number of factors. One factor is the local distribution of voltage gated Ca^+ channels: a high concentration of such channels will make propagation easier in that location. Another factor is whether calcium action potentials from different branches arrive at a common point at the same time and therefore reinforce each other. Other factors include the distribution of other types of neurotransmitter or voltage gated ion

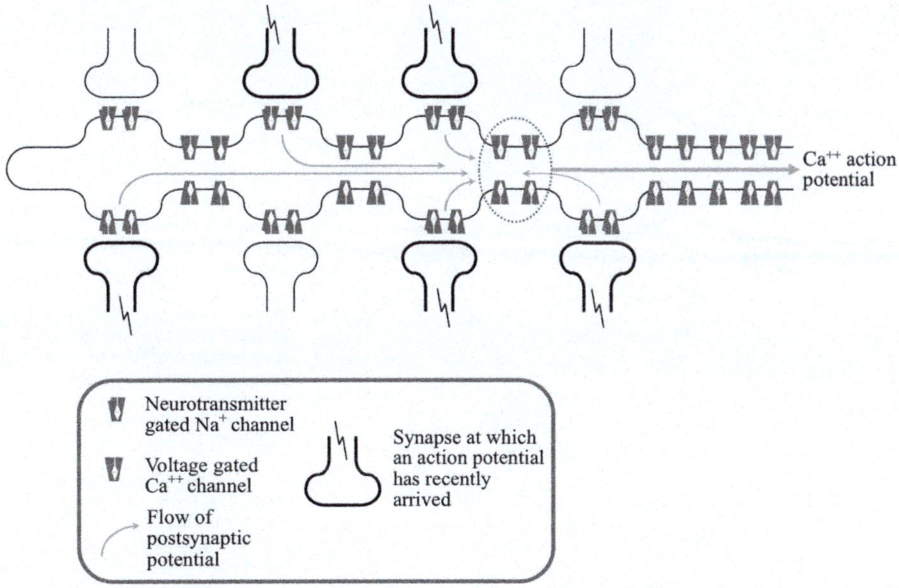

Fig. 3.11 A branch of a dendrite. There are a number of synapses on a branch, eight are illustrated. The synapses are from different source neurons. Each postsynaptic membrane contains a number of neurotransmitter gated Na^+ channels. Distributed over the branch and from the branch towards the neuron are numbers of voltage gated Ca^{++} channels. The voltage gated Ca^{++} channels open if the magnitude of the voltage difference across the membrane in their vicinity is sufficiently depolarised (i.e. becomes more positive, reducing the magnitude of the –70 mV resting potential). If an action potential arrives at one synapse, the total injected potential change due to the neurotransmitter gated Na^+ channels only results in a depolarisation of the order of millivolts in the neighbourhood of the synapse, and the resting potential is restored within 20 ms. If another action potential arrives at the same or a different synapse before the potential change from the first has decayed away completely, the two injected potentials will add to some degree. If enough action potentials arrive at the branch within a short enough period of time, the total depolarisation in an area of the branch membrane can reach the point at which the voltage gated Ca^{++} channels will open in that area. Injection of Ca^{++} further depolarises the local membrane. This further depolarisation can trigger opening of voltage gated Ca^{++} channels in neighbouring areas. The spreading depolarisation is known as a calcium action potential. Unlike a sodium action potential down an axon, a dendritic calcium action potential can decay as it propagates depending on various factors such as the local concentration of Ca^{++} voltage gated ion channels

channels. For example, an open Cl^- or K^+ channel will reduce the local membrane potential and therefore the chance of propagation of the calcium action potential. Such a channel could be opened by an inhibitory neurotransmitter derived from another neuron, or by a combination of voltage and presence of a neurotransmitter released by yet another neuron.

The overall effect is that calcium action potentials generally decay in magnitude as they propagate. To initiate a sodium action potential down the neuron, calcium action potentials propagating from one dendritic branch must combine with the effects of other branches, and the action potential resulting from the combination

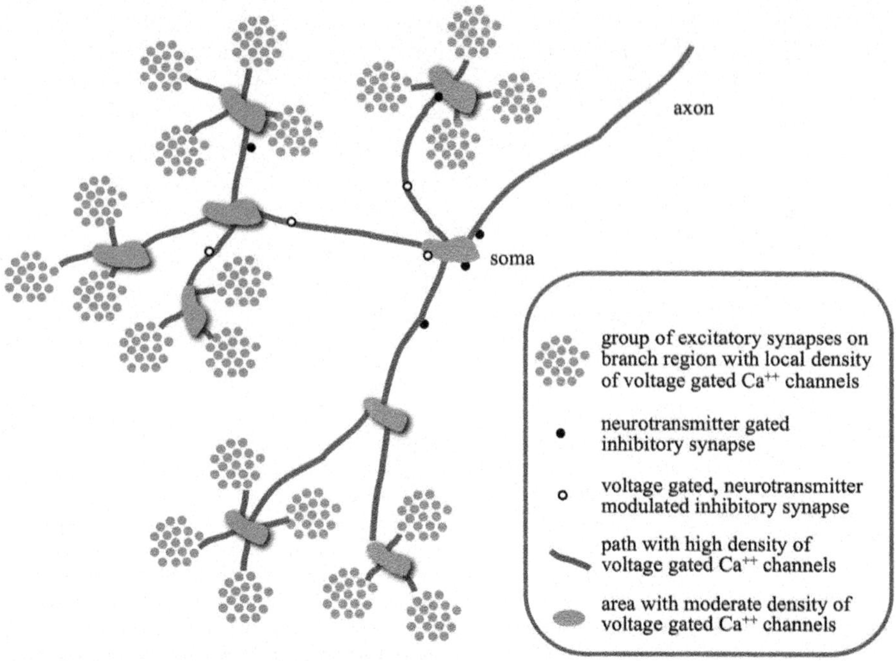

Fig. 3.12 Staged integration across a neuron. An information model for the generation of an output by a neuron is illustrated. Action potentials arrive at different synapses within a closely collocated group on a branch. If enough such inputs occur within a short period of time, a local calcium action potential is generated that proceeds towards a further integration point along a path defined by a high density of voltage gated Ca^{++} channels. Such a local calcium action potential indicates that the condition programmed on the branch by the group of synapses has been detected. This condition is defined as outputs within a short period of time any significant subset of the presynaptic neurons providing inputs to the group. Local calcium action potentials indicating the detection of different conditions arrive at a common integration point. If the total potential at the integration point exceeds a threshold defined by the moderate level of voltage gated Ca^{++} channels, a calcium action potential is generated that proceeds towards a higher level integration point, again along a path defined by a high density of voltage gated Ca^{++} channels. Local action potential propagation can be reduced or blocked by the presence of open Cl^- or K^+ ion channels. Such channels may be opened by neurotransmitters released by other neurons. They may also be opened membrane potential variations, which could for example limit the size of the calcium action potential propagated from a particular source. They could also be opened by a combination of voltage and external neurotransmitter, which could limit the size of the calcium action potential only if some other neuron was active. This second order integration indicates that a more complex condition defined by a group of simpler conditions has been detected. Further stages of integration result in the neuron producing an output that can be interpreted as detection of a very complex condition, defined as any significant subset of all the more detailed conditions programmed by different groups of synapses

propagate into the trunk of dendrite and then into the soma and deliver an adequate depolarisation to the axon hillock. Hence the distributions of many different types of ion channel over the surfaces of the dendrites and soma support very complex algorithms by which a neuron integrates its inputs to determine its output [89].

However, in information terms a dendritic branch detects a simple condition defined by a weighted combination of synapses. A dendrite detects a more complex condition defined by a weighted combination of branch conditions. A neuron detects a very complex condition (or receptive field) defined by a weighted combination of dendritic conditions. At all levels the weights can be adjusted dynamically, and the adjustment factors can be viewed as additional conditions that contribute to the receptive field definition.

3.8 Generation of Output Action Potentials

Injections of potential into the soma from dendrites and from synapses direct on the soma can result in spreading potential reductions and collapses on the membrane of the soma. If there are sufficient potential reductions to generate a collapse in the axon hillock region where the axon emerges from the neuron, a collapse results in the first part of the axon. Along the axon, there are sufficient voltage gated Na^+ channels that collapse in the first part reduces the magnitude of the potential in the next part by enough to induce a collapse. Collapse in that part results in collapse in the following part, and so on all along the axon. In other words, the propagation of the action potential along an axon resembles the collapse of a line of dominoes.

This domino-like collapse mechanism is superior to simple propagation of an electric current, because such a current would be subject to resistance and the signal would decay. The collapse mechanism results in the same signal reaching all the endpoints of an axon equally. However, it is much slower that simple current flow, travelling no faster than about 100 m/s.

3.9 Different Types of Neurotransmitter

Neurotransmitters are the chemicals released by synaptic swellings at axon terminations, that affect targetted neurons in various ways. Generally, a neuron generates the same neurotransmitter at all of its synapses onto target neurons. However, different neurons can generate different neurotransmitter chemicals, including glutamate, GABA, dopamine, acetylcholine, serotonin, noradrenaline and about 50 others. Neurons and axons are often labelled by the type of neurotransmitter they generate (e.g. glutamatergic, GABAergic, dopaminergic, cholinergic, serotonergic, noradrenergic neurons and axons). The neurotransmitter affects the target neuron by binding to sites on molecules called receptors that are embedded in the membrane of the target. This binding changes the properties of the receptor in various ways.

The neurotransmitter chemical released by synapses of pyramidal neurons to other neurons is glutamate. This neurotransmitter binds to a site on AMDA receptor molecules located in the membrane of the postsynaptic neuron adjacent to the synaptic cleft. These AMDA receptors are Na^+ ion channels. The ion channel open only when glutamate attaches to the site, and glutamate neurons are therefore excitatory. The neurotransmitter released by inhibitory interneurons in the cortex is called GABA. This neurotransmitter triggers the opening of Cl^- channels located in the postsynaptic membrane and is therefore inhibitory.

This picture is made more complex by the fact that neurons may have different receptors for these same neurotransmitters, and the different receptors have different effects on their targets. For example, there is also an NDMA receptor for glutamate that is a Ca^{++} channel that opens only in response to glutamate binding plus complete local membrane depolarisation. NDMA opening plays a role in triggering changes to synaptic strengths. In the basal ganglia, in some cases the same dopaminergic neuron can excite neurons with one type of receptor, and inhibit other neurons with a different type of receptor. Acetylcholine comes to cortex from subcortical structures. It is excitatory for pyramidals in cortical areas involved in control of body movement, but inhibitory elsewhere. Serotonin and noradrenaline also come to the cortex from subcortical structures, and often have a general modulatory effect on their target neurons. In some cases neurotransmitter releases from one synapse can diffuse to affect many different postsynaptic neurons in the vicinity. All these complexities are discussed in more detail in Chaps. 4 and 5.

3.10 Flow of Information Between Neurons

The predominant flow of information is thus primarily via an action potential initiated in the soma of a source neuron which travels along its axon to synapses on the dendrites (or sometimes the soma or axon) of target neurons. However, there are some exceptions to this flow direction. On a local scale, chemicals can travel back from the postsynaptic to the presynaptic neuron to make changes to the synapse. On a larger scale, chemicals can flow back along an axon from the target to the source neuron and influence the source [90].

3.11 Electrical Activity of the Brain

The various types of action potentials in dendrites and axons result in measurable electrical activity in the brain. EEG and MEG measurements can detect this activity. The cortex is closest to the skull, and pyramidal neurons are arranged parallel to each other. Hence pyramidal neuron activity contributes most strongly to such external measurements.

It is observed that the electrical activity varies in time, and analysis reveals a number of different frequency bands in which there is correlation between power in the band and cognitive state of the brain. For example, the delta band (up to 4 Hz) is often detected in slow wave sleep. The theta band (4–8 Hz) may be related to memory. The alpha band (8–13 Hz) is detected when the brain is drowsy or idling. The beta band (13–30 Hz) is present in an active, thinking brain. The gamma band (30–90 Hz) is associated with attention and working memory. Some bands (such as gamma) appear to reflect axonal activity, others (such as beta) may reflect activity in dendritic trees [91].

3.12 Descriptive Gaps

In this chapter we have briefly described the highest level anatomical structures in the brain, and outlined some detailed neuron physiology. The large descriptive gaps between the cognitive phenomena covered in Chap. 2, the anatomy covered in the first part of this chapter, and the neuron physiology covered in the second part illustrate the understanding problem. In the next two chapters we will describe neuron physiology and neuron chemistry in more detail, and begin to discuss what this physiology suggests about function. Then in Chap. 6 we will describe the major anatomical structures of the brain in detail, again beginning consideration of their functional roles.

Chapter 4
Neuron Physiology

In this chapter we will describe the morphology and physiology of neurons in much more detail, with particular emphasis on the ways in which molecules spanning the neuron membrane result in detection and integration of signals received from other neurons. At each point in time a neuron receives large numbers of external signals, most derived from other neurons. The internal chemistry of neurons results in the detection and integration of all the external signals and determines their current outputs and any changes to their integration algorithms. Hence understanding of how signals are received by neurons and integrated by chemical processes to influence current and future outputs is the detailed basis for understanding neural information processing.

4.1 Neuron Morphology and General Electrical Properties

In the brain, the most important cells from a functional point of view are neurons. Other cells, called glia cells, support neurons by the provision of chemical and structural resources and have some limited information processing roles but discussion of glia cells in this book will be limited. The barrier wall or membrane that partially isolates a neuron from its environment is illustrated in Fig. 4.1. The membrane is made up of two layers of phospholipid molecules. Phospholipid molecules have a water soluble (hydrophilic) head and a water repellent (hydrophobic) tail, and in the membrane the molecules are aligned in a double layer with hydrophilic heads facing out. The membrane is a barrier to most hydrophilic molecules, and allows both chemical and electrical differences to exist between the interior and exterior of the neuron.

L.A. Coward, *Towards a Theoretical Neuroscience: from Cell Chemistry to Cognition*, Springer Series in Cognitive and Neural Systems 8, DOI 10.1007/978-94-007-7107-9_4,

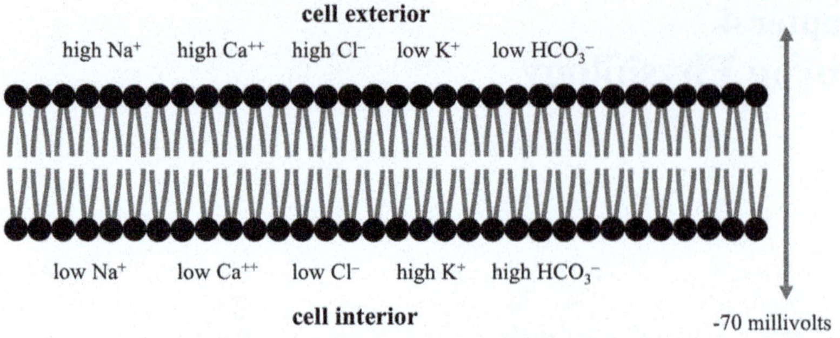

Fig. 4.1 The neuron membrane. This membrane separates the internal cytoplasm from the extracellular environment, and is a barrier to many chemicals, including ions like Na^+, Ca^{++}, K^+, Cl^- and HCO_3^-. The impermeability makes it possible to maintain concentration gradients in these ions between the inside and outside of the neuron, sustaining a steady state electrical potential difference across the membrane

4.1.1 Major Physical Components of Neurons

Inside the neuron membrane is a viscous fluid called the *cytoplasm* permeated by a large number of different chemicals. As illustrated in Fig. 4.2, a number of structures are suspended in this fluid. Different structures perform specialized functions and are often called organelles, by analogy with organs in the body. One organelle is the *nucleus*, which is the cellular genetic material (DNA) surrounded by a membrane which isolates the genetic material from the rest of the cell. There are large pores in this membrane which actively transport specific molecules in or out of the nucleus, and allow some smaller molecules to diffuse relatively freely. DNA is a long molecule in which specific segments (genes) code specific proteins needed by the cell. These proteins can be structural, functional, or catalysts for chemical processes (in which case they are called enzymes). Generation of a protein from DNA occurs in a series of stages. When a chemical signal indicating the need for the protein is received in the nucleus, a series of chemical steps lead to production of messenger RNA molecules (mRNA) which are transported out of the nucleus through pores to other organelles called *ribosomes*. Ribosomes are the structures where proteins are produced. Proteins required by the nucleus must be created at ribosomes and transported back into the nucleus through pores. mRNA can also be transported to ribosomes located in distant parts of the neuron, to support protein synthesis for specific local needs. For example, there are ribosomes close to individual synapses [92] where inputs arrive from other neurons.

Ribosomes can be free floating in the cytoplasm, in which case they generally produce proteins for internal use by the neuron. Alternatively, ribosomes can be attached to an organelle called the *endoplasmic reticulum*. The endoplasmic reticulum is an extensive set of folded membranes that are an extension of the membrane around the nucleus. Ribosomes attached to these membranes produce proteins for the use of the external cell membrane or for use outside the neuron. For example, in

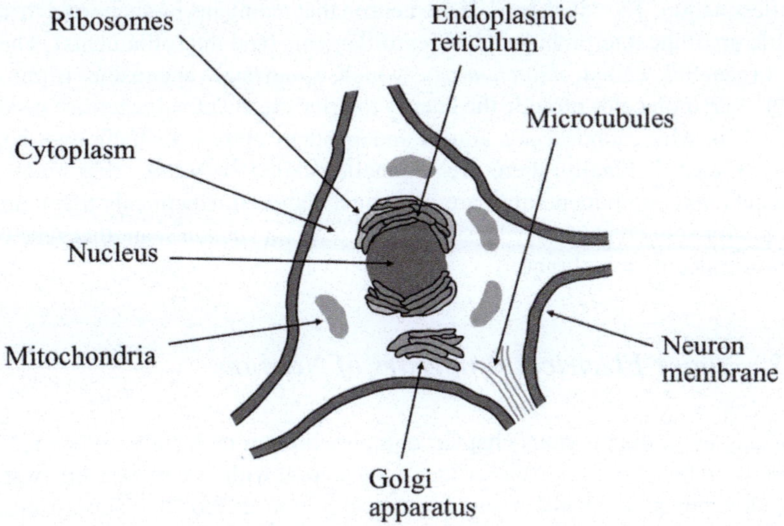

Fig. 4.2 The internal structure of a neuron. As in any cell in the body, neurons have a range of internal structures which are themselves surrounded by membranes. The nucleus contains the genetic information. The endoplasmic reticulum is the location at which proteins are synthesized by ribosomes, and forms a network which can extend throughout the neuron. Many neurotransmitters are synthesized and packaged in the Golgi apparatus. The mitochondria provide energy to the neuron in a form in which it can be utilized. The microtubules form a skeleton which supports the shape of the neuron and also provides a means for distributing chemicals to the locations in which they are required

a synapse there are vesicles, which are bubble like structures with a concentration of neurotransmitter(s) surrounded by membrane. Neurotransmitters can be relatively small molecules (up to about 50 atoms), or larger molecules (small proteins with 3–100 amino acid residues). Enzyme proteins that catalyse the synthesis of small molecule neurotransmitters are transported to the synapses where the synthesis occurs. Pump molecules located in vesicle membranes insert the neurotransmitter into the vesicles. Each pump molecule is specific to one neurotransmitter type, and depending on what pump molecules are present, multiple neurotransmitters can be inserted into the same vesicle. For large molecule neurotransmitters, precursor proteins and enzymes are synthesized from mRNA and packaged into vesicles in a special purpose part of the endoplasmic reticulum called the *Golgi apparatus* located in the soma. Vesicles are then transported to the synapses where they are required. In this case the synthesis of the neurotransmitter occurs during transport to the synapse. Additional small molecule neurotransmitters are often inserted into these vesicles at the synapse.

Transport of molecules and vesicles to the location where they are needed is performed by components called *microtubules*. These microtubules are long segmented threads, and molecules or vesicles are passed from segment to segment to their destination. The "skeleton" of the neuron that maintains its shape and flexibility is made up of the microtubules plus neurofilaments (and microfilaments). There are also organelles called *mitochondria* which synthesize adenosine triphosphate (ATP). ATP molecules provide the energy to drive the chemical reactions needed by the neuron. Mitochondria are almost independent cells with their own separate genetic material. Finally, there are organelles called *lysosomes* that break down chemicals and components that are no longer required. Chemicals that cannot be recycled for some reason may be stored long term on *vacuoles*, another type of vessel surrounded by membrane.

4.1.2 Major Electrical Properties of Neurons

As discussed in the previous chapter, complex protein molecules embedded in the external cell membrane link the inside of the neuron with its environment in specific ways. Ion pump molecules maintain concentration differences and an electrical potential difference across the membrane. In the equilibrium state this potential difference has a maintains the interior of the neuron at −70 mV relative to the exterior. This potential difference is called the resting potential. Ion channel molecules in their open state allow flows that can reduce concentration differences and change the potential difference. If the magnitude of the voltage difference is reduced (i.e. the interior potential becomes less negative) it is called a depolarisation, if it is increased (i.e. the interior potential becomes more negative), a hyperpolarisation.

4.1.3 Structure of Neurons

Neurons have neurites, or very large extensions from the soma, called axons and dendrites as illustrated in Fig. 3.6. A neuron has one axon, which carries the output of the neuron to other neurons. This axon generally comes out of the soma from an area called the axon hillock. An axon can branch to carry the neuron output to many other different neurons. The output of a neuron is a spike of voltage lasting a couple of milliseconds and called an action potential. The spike travels down the axon to influence other neurons. Under a constant input stimulus, different types of neuron generate different patterns of output action potentials [93]. In some neurons, called *regular spiking*, spikes lasting ~1.5 ms [94] are generated individually and relatively independently of each other with an average frequency that varies considerably over time. Such neurons often exhibit the phenomenon of adaptation, in which with a constant stimulus the average rate of action potential generation declines with time. Other neurons, called *intrinsically bursting*, produce clusters of action potentials, either singly or repetitively. Prolonged stimulus of a burst firing neuron results in a complex pattern of bursts and single spikes. Yet other neurons, called *fast spiking*, produce sequences of shorter (~0.5 ms) spikes [94], at a characteristic frequency which is different for different types of fast spiking neurons [95]. Under prolonged stimulation fast spiking neurons show no adaptation. The different output types are determined by the detailed structure of the neuron.

Some neurons have many dendrites that emerge separately from the soma. For example, the pyramidal neurons (Fig. 3.6) common in the cortex, hippocampus and amygdala have two types of dendrite: apical and basal. The one apical dendrite emerges from the soma on the opposite side to the axon. This apical dendrite typically has branches from its main shaft, and ends with a tuft made up of a number of branches. The multiple basal dendrites emerge from the soma in many directions perpendicular to the direction of the axon.

Each dendrite can branch extensively, and the degree of branching varies considerably. For example, in the CA1 area of the rat hippocampus, the typical pyramidal neuron [96] has an average of 17 oblique branches from its apical dendrite, each of which bifurcates once on average. The apical tuft has about 15 branches. An average of 5 basal dendrites each branch several times, forming about 40 terminal branches. So in this structure there can be on average 34 terminal branches on the apical shaft, 15 on the apical tuft, and 40 on the basal dendrites. However, even within this structure, the number of terminal branches can vary by a factor of two either way. Between structures the average degree of branching also varies, for example, pyramidal neurons in the prefrontal cortex have three times the branching complexity of pyramidal neurons in the primary visual area [97]. The branching pattern of a dendrite correlates with its electrical properties. For example, apical dendrites with many side branches have different electrical responses from those with few such branches [98].

4.1.4 Speed of Communication Between Neurons

Most inputs to a neuron from other neurons arrive at a synapse on a dendrite and are carried by a neurotransmitter across the synaptic cleft to receptors on the postsynaptic neuron. However, some synapses occur on the soma or even the axon, and a small proportion of receptors are located on the membrane distant from any synapse. Neurotransmitters are stored within vesicles in the presynaptic neuron. When required, the vesicle membrane fuses with the exterior synaptic membrane and the vesicle contents are released from the presynaptic synapse and diffuse across the synaptic cleft. Despite this intervening diffusion step, when the receptors are ionotropic (i.e. ion channels) the time delay between the start of an incoming action potential and the commencement of a postsynaptic response can be as little as 150 μs [99]. The minimum delay between an incoming action potential and an action potential output response by the postsynaptic neuron is of the order of 2 ms.

There are also electrical synapses with an even shorter delays, that appear to be important for achieving almost exact synchronisation between the activity of different neurons [100]. Such electrical synapses are most common in the olivary nucleus, the thalamic reticular nucleus, and between inhibitory interneurons in the cortex and hippocampus.

When the postsynaptic receptors are metabotropic (i.e. triggering complex chemical changes in the target neuron) the time delay between incoming action potential and detectable neuron response is longer, and in some cases long term structural changes to the target neuron can result, such as modifications to synaptic weights.

4.1.5 Physical Differences Between Neurons

Neurons differ in their morphology, i.e. in the sizes of their somas and in the sizes and general shapes of their dendrites and axons. Types of neuron are defined by similarities in these morphologies, and each brain structure contains a limited number of different types. Another major difference between neurons is that the axons of some neurons (projection neurons) travel outside the brain structure within which they are located, while the axons of other neurons (interneurons) only project locally. Often the projection neurons and interneurons within one brain structure have different morphologies.

Another major difference between neurons is that in some neurons many synapses are located on spines which protrude out from the shaft of the dendrite, while in other neurons the dendritic tree is more smooth. Each spine receives one excitatory input, inhibitory and other inputs are sometimes located on the shaft [101]. Spines have a narrow neck that emerges from the dendrite and ends with a larger head, and the synapse terminates on the head. It is believed that spines have the role of chemically isolating the synapse from the rest of the dendrite [102]. Although spines are formed, eliminated and change in size, many spines in an adult brain last for the lifetime of the animal [103]. Spine size varies widely, and correlates with synaptic size [102].

Yet another difference between neurons is in the neurotransmitters they release. Most neurons have a primary neurotransmitter, plus one or more additional neurotransmitters that often modulate the effects of the primary neurotransmitter in various ways [104]. Additional neurotransmitters may be packaged in the same vesicles as the primary, sometimes in varying proportions, or in separate vesicles.

4.1.6 Internal Chemistry of Neurons

As in all cells, there are complex, interacting chains of chemical reactions which implement the behaviour of the neuron. Some of these chemical reactions will be discussed in more detail in the next chapter, but some principles critical for general control will be described here.

For a specific chemical reaction to occur in a cell, at least three factors must be present. Firstly, energy must be available to drive the reaction. Secondly, appropriate catalysts must be present to enable the reaction to occur. Thirdly, the starting point chemicals and catalysts must be present at the same location at the same time.

Energy is generally provided by adenosine triphosphate (ATP) molecules. These molecules have three phosphate groups linked in a row to an adenosine molecule. The links between the phosphate groups require energy to create, and this energy is available to drive other chemical reactions by adding a phosphate group to another molecule.

Very few chemical reactions in a cell will proceed without a catalyst. A huge number of catalysts is therefore required. These catalysts are large proteins called enzymes, each enzyme targetting a different set of reactions. Some enzymes have an active state in which they can catalyse their target reaction, and one or more inactive states. The conversion of an enzyme into its active state is often implemented by addition of a phosphate group at a particular site on the enzyme, and the enzyme can be inactivated by removal of the phosphate group. These activation and deactivation processes themselves need to be catalysed by different activated enzymes. Enzymes called *kinases* activate other enzymes by phosphorylating them at specific sites, and enzymes called *phosphatases* inactivate other enzymes by dephosphorylating them [105]. A kinase may be needed to activate another kinase, resulting in long chains of chemical processes. Such chains may be initiated by messenger molecules that are released inside the neuron as a result of some external signal. An enzyme may require activation at more than one site, by multiple catalysts and/or messenger molecules, resulting in complex networks of chemical reactions as illustrated in Fig. 4.3. More details on these reactions will be described in Chap. 5.

Enzymes and smaller messenger molecules can catalyse or initiate more than one chemical process. Which process will actually occur at one point in time is often determined by anchor proteins [106]. Anchor proteins have binding sites for all the catalysts, messenger molecules and participating chemicals in a specific reaction. They therefore hold most of the reaction participants together in one place, and when the missing catalysts or messengers arrive, for example by diffusion from a release point elsewhere in the neuron, the reaction is initiated.

4.1.7 *Permanence of Neurons*

In most organs of the body, new cells are constantly generated to replace older cells. However, in the mammal brain, the vast majority of neurons exist at birth, with few subsequent additions. This presumably reflects the unique information captured by individual neurons through learning, which generally cannot be preserved if the neurons are replaced. The only exceptions to this general rule appear to be that new interneurons are generated in the olfactory bulb and new granule cells in the dentate gyrus of the hippocampus [107].

4.2 Molecules Embedded in the Neuron Membrane

A wide range of protein molecules are embedded in the neuron membrane and span that membrane to link the exterior and the interior in different ways. Four important categories of embedded molecule are ion pumps, ion channels, G-protein receptors

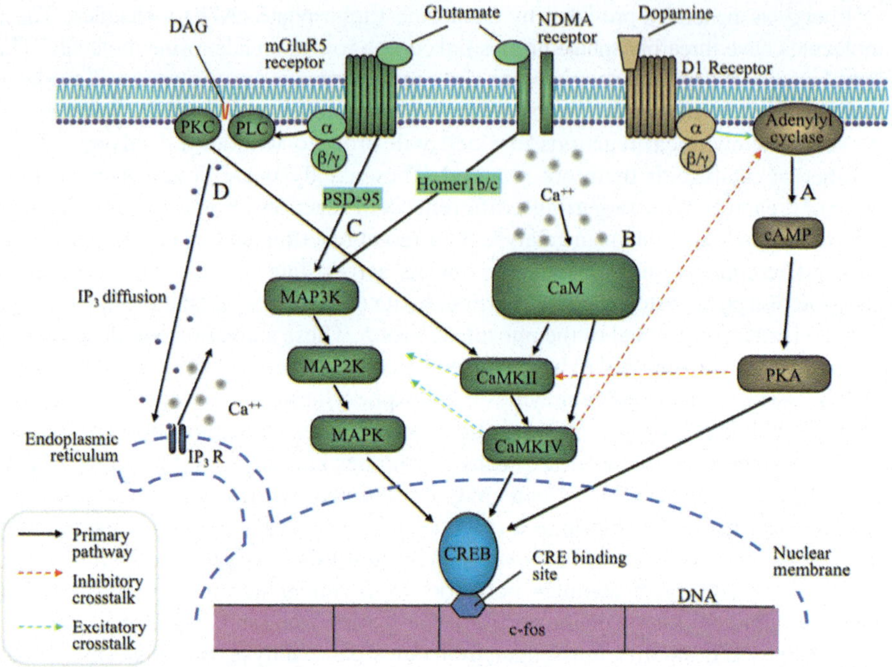

Fig. 4.3 A simplified view of some separate but interacting pathways that can result in gene transcription. *A* – The neurotransmitter dopamine targets a D1 G-protein receptor and triggers separation of the α-subunit from the β/γ subunits. The α-subunit then activates the membrane bound protein adenylyl cyclase (AC). AC catalyses conversion of ATP molecules into cAMP molecules. cAMP molecules catalyse the activation of the kinase PKA, which can activate the CREB transcription factor. Activated CREB triggers transcription of the c-fos gene. *B* – The neurotransmitter glutamate opens an NDMA channel, allowing diffusion of Ca^{++} ions into the cytoplasm. These ions bind to molecules of calmodulin, and the bound complex can activate the kinase CaMKII. This kinase can activate CaMKIV which in turn can activate CREB. *C* – Simultaneous activation of NDMA and mGluR5 receptors by glutamate results in an interaction between proteins bound to them in the postsynaptic density (Homer1b/c and PSD-95 respectively). The bound complex activates a MAP kinase, triggering a cascade of kinase phosphorylation leading to the activation of CREB. *D* – Activation of an mGluR5 receptor by glutamate results in the α-subunit activating PLC. PLC hydrolysed a membrane phospholipid into two components: DAG which remains in the membrane and IP_3 which is released to diffuse through the cytoplasm. DAG activates a membrane bound kinase called PKC. PKC can activate CaMKK, which leads to CaMKIV activation and again to CREB activation. IP_3 activates channels in the endoplasmic reticulum leading to release of Ca^{++} stores into the cytoplasm. Ca^{++} can trigger further reactions. Crosstalk between these pathways includes PKA in path 1 inhibiting CaMKK in paths 2 and 4; CaMKIV in paths 2 and 4 inhibiting AC in path 1; CaMKK and CaMKIV in paths 2 and 4 inhibiting the kinase activation chain in path 3

and tyrosine kinase receptors. Ion pumps are constantly active and force specific ions in one direction across the membrane, maintaining a concentration gradient of that ion between the inside and outside of the neuron. Ion channels, G-protein receptors and tyrosine kinase receptors have a resting or inactive state, and are switched to an active state by some stimulus. This stimulus is often the presence of some relatively small neurotransmitter molecule outside the neuron which activates the embedded molecule by binding to it. Small molecules that activate other molecules by binding to them are often labelled *ligands*. In the active state an ion channel is open to allow specific ion types to diffuse freely through the channel between the inside and outside of the membrane, while in the inactive state it is closed. In the case of ligand gated ion channels, opening is triggered by the presence of a neurotransmitter in the extracellular environment which binds to the ion channel. Voltage gated ion channels are opened if the membrane potential is in a specific range. Yet other ion channels require both a neurotransmitter and a specific membrane potential value. G-protein receptors respond to external neurotransmitters by triggering changes within the neuron which can vary from the opening of ion channels to transcription of genetic information in the nucleus. Tyrosine kinase receptors are activated by an external ligand, and when activated can phosphorylate proteins such as enzymes within the neuron, initiating chains of chemical reactions.

4.2.1 *Ion Pumps*

As discussed in Chap. 2, large numbers of protein molecules called sodium-potassium (Na^+/K^+) pumps are embedded in the neuron membrane. This Na^+/K^+ pump molecule constantly goes through a cycle powered by ATP molecules which forces three Na^+ ions from the inside of the membrane to the outside and two K^+ ions from the outside to the inside. The effect of the operations of large numbers of Na^+/K^+ pumps is to make the internal neuron environment lower in Na^+ ion concentration than the outside environment and higher in K^+ ion concentration. The concentration differences also result in a potential difference between the exterior and interior of the neuron in the range −65 to −85 mV.

Two other pump molecules move calcium (Ca^{++}) ions from the inside to the outside of the neuron [108]. One is the Plasma Membrane Ca^{++} ATPase (PMCA) molecule that uses the energy of one ATP molecule to move one Ca^{++} ion across the membrane. This molecule pumps at a relatively slow rate, but over time maintains a very low concentration of Ca^{++} ions within the neuron. A second molecule, the Na^+/Ca^{++} exchanger, uses the energy implicit in the low concentration of Na^+ ions within the neuron to shift one Ca^{++} ion to the outside in exchange for three Na^+ ions moving in. This pump operates only if there is a high concentration of Ca^{++} ions inside the neuron, but in this situation it operates much faster than the PMCA molecule.

There are also ion pump molecules that move chloride (Cl^-) ions across the cell membrane. These cation-chloride co-transporters (CCCs) are neutral in their electrical effects because any Cl^- ion movements are accompanied by an equivalent number of Na^+ or K^+ ion movements. Different CCCs can move Cl^- ions in either direction across the membrane, but in neurons there is generally a deficiency of Cl^- in the interior. Osmotic pressure will therefore move Cl^- ions into the neuron if a Cl^- channel is opened. Yet another ion pump exchanges Cl^- and bicarbonate (HCO_3^-)

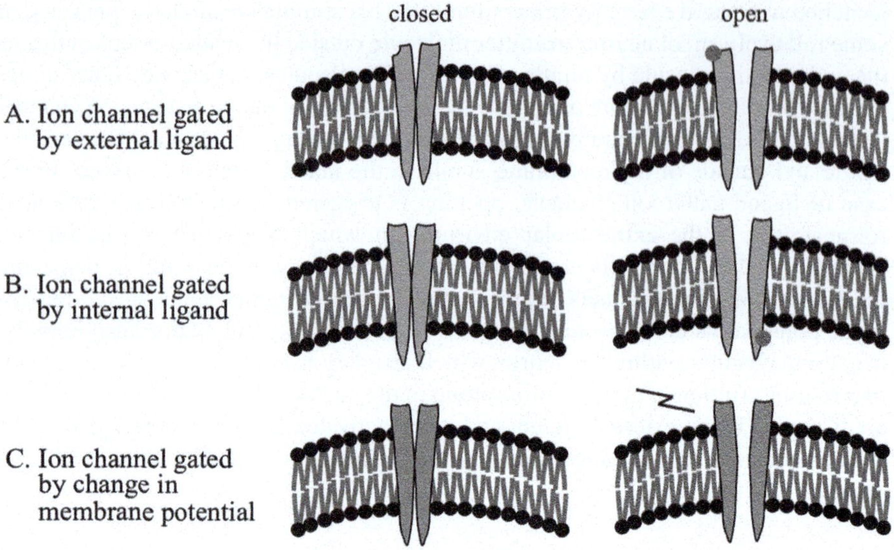

Fig. 4.4 The cell membrane is penetrated by various proteins. Some of these proteins serve as channels for ions to flow across the membrane. Ion channels can be fairly specific for one type of ion (generally Na^+, K^+, Ca^{++} or Cl^-), or can allow passage for multiple types (e.g. Na^+, K^+, and Ca^{++}, or Cl^- and HCO_3^-). Ion channels have an open state which allows passage of their ion type, and closed that blocks passage. The opening and closing of channels can be controlled by three types of mechanism. (**a**) A specific chemical called a ligand in the extracellular environment opens the channel by binding to a specific receptor site. (**b**) A chemical in the intracellular environment opens the channel by binding to a specific receptor site. (**c**) The channel is opened when the voltage across the membrane in the vicinity of the channel is within a specific range of values. Often but not always the range does not include the resting potential. Some ion channels are opened by combinations of these mechanisms. Ion channels of different types are widely but unevenly distributed over the neuron membrane, including soma, dendrites and axon

ions [109]. This Cl^-/HCO_3^- exchanger generally results in a higher concentration of HCO_3^- ions in the interior of the neuron.

The difference in concentration of ions means that there is an osmotic pressure in favour of a flow of ions to equalize the concentrations, but the phospholipid membrane largely blocks such flows.

4.2.2 *Ion Channels*

Different types of ion channel can pass Na^+, K^+, Ca^{++}, Cl^- or HCO_3^- ions. Some channels can pass more than one type of ion, for example cation channels that pass Na^+, K^+ and Ca^{++} ions to varying degrees.

As shown in Fig. 4.4, the opening of a channel molecule can be triggered in a number of ways. Some channels, called neurotransmitter- or ligand-gated, open if a specific chemical (the neurotransmitter) outside the neuron binds to a particular site on the molecule. There are many different neurotransmitters that can open different channels, and generally multiple different types of channel opened by the same neurotransmitter. Although in most cases a channel molecule bonds to molecules of just one neurotransmitter, in some cases an additional neurotransmitter will affect opening in some fashion. Ligand gated ion channels are generally concentrated in the post-synaptic density, the region of the postsynaptic neuron opposite the synaptic cleft.

In addition to open and closed states, some ion channels have additional states, for example a closed state in which they have a higher than normal resistance to being opened. Ion channels can also differ in the lengths of time they remain open.

The opening of an ion channel affects the membrane potential close to the channel in different ways. Because there is a much higher concentration of Na^+ and Ca^{++} ions outside the membrane, opening of Na^+ or Ca^{++} channels will result in an inflow of positive ions that will depolarise the membrane, exciting the neuron. Because there is a higher concentration of K^+ ions inside the membrane, opening a K^+ channel will result in an outflow of positive ions that will hyperpolarise the membrane and inhibit the neuron. In most neurons, there is a higher concentration of Cl^- ions inside the membrane. Although the concentration gradients across the membrane for Na^+, K^+ and Ca^{++} ions do not vary significantly between different types of neuron, inside some neurons the concentration of Cl^- ions is maintained lower [110], and opening a Cl^- channel will cause depolarisation [111]. In addition, Cl^- channels are often able to pass other negative ions like HCO_3^-, which may have opposite concentration gradients across the membrane [112], and channel opening can be inhibitory or excitatory depending on circumstances.

Voltage-gated ion channels are triggered to open by the membrane potential reaching some value (generally different from the resting potential) which is the threshold for the channel. The threshold is usually a depolarisation but can sometimes be a hyperpolarisation (e.g. the cation channel I_h) from the membrane potential. Depending on the ions passed by the channel, openings may reinforce or oppose the membrane potential change that caused them to open. When the threshold is reached, a channel takes a characteristic time to open, and remains open for another characteristic time, or in a some cases until the membrane depolarisation reaches some other voltage threshold. There are a large number of different types of voltage gated ion channels, that differ in the ions they pass, the time taken to open and close, and the ion current they pass when open at a given voltage (i.e. their conductance, the inverse of electrical resistance). Some channels pass ions more effectively one way than another and are called rectifying (e.g. the inward rectifying K^+ channel).

Some voltage gated channels are affected by factors in addition to the membrane potential, such as the presence of neurotransmitters. Many voltage gated channels are affected by the presence of chemicals called second messengers like cAMP [113], which are influenced by G-protein receptors. In one important K^+ channel type the primary gating is viewed as being by Ca^{++} ions but there is strong dependence also on membrane voltage.

Most ligand gated channels are located close to a synapse. Voltage gated ion channels can be located anywhere on a neuron, each type of neuron has a specialized selection of channels which are distributed in an uneven fashion over its soma, dendrites and axon [114]. For example, there can be large differences in concentration of one channel type and relatively uniform distribution of another channel type between soma and dendrites of the same neuron [115]. There is evidence for considerable variation in the numbers of channels between similar dendritic locations [116, 117]. Non-uniform distributions of K^+, Ca^{++} and I_h channels have been observed in the dendrites of many different types of neuron [114].

Later in the chapter we will discuss examples of the many different types of ion channel. To understand the significance of the many different types, consider a pyramidal neuron in the cortex. Such a neuron has large numbers of excitatory synapses from other pyramidal neurons, and these synapses contain Na^+ channel receptors. As discussed in Chap. 3, these synapses can be viewed as defining a number of different conditions instantiated on dendritic branches. The neuron receptive field is detected if a significant proportion of the conditions are present. A neuron therefore integrates detections of combinations of conditions. The communication of detections occurs largely by dendritic action potentials propagating along the membrane. The huge range of different ion channels make it possible to regulate these action potentials, and therefore the detection of a receptive field, in many different ways.

Voltage gated ion channels can define a reference integration algorithm, placing relative weights on different branches and dendrites through the strengths of the pathways carrying the dendritic action potentials. Voltage gated ion channels can also regulate the pattern of output axonic action potentials, determining for example if the neuron is regular, burst or fast spiking. Neurotransmitters targetting voltage gated ion channels and modulating their responses can regulate the reference integration algorithm in different ways. Inhibitory synapses allow their source neuron to reduce the probability of receptive field detection by the target neuron, regulate the relative influence of different dendrite or branch conditions, or even the relative influence of different synapses within one branch. Hence the detection of the primary pyramidal neuron receptive field within the activity of other pyramidal neurons can be regulated by the activity of other types of neurons in the cortex and in many other anatomical structures. The challenge is to describe this regulation without requiring huge and incomprehensible degrees of detail. Such a description will be approximate, so it must also be possible to know where approximations are being made and to map parts of the description to more detailed levels if required.

4.2.3 G-Protein Receptors

G-protein receptors are molecules that span the neuron membrane. Outside the membrane they are sensitive to specific neurotransmitters. Inside the membrane they can trigger the activity of G-proteins that are embedded in the membrane but only interact with the cell interior. G-protein activity triggers cascades of chemical

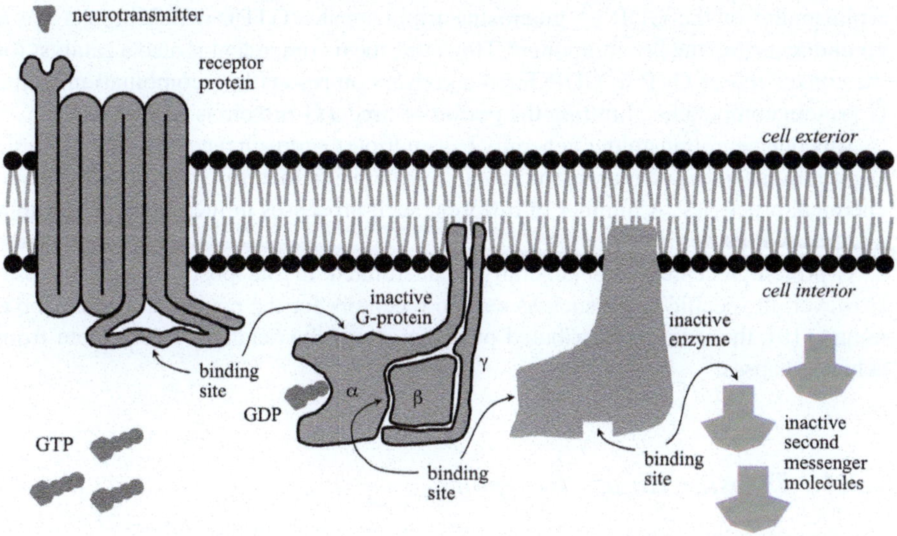

Fig. 4.5 G-protein receptor systems. G-protein receptor systems include a receptor with one or more extracellular binding sites, a G-protein, and various second messenger molecules. The receptor is a protein that spans the neuron membrane. The G-protein is made up of α, β and γ components that in their inactive state are bound together, with the α component also bound to a guanosine diphosphate (*GDP*) molecule. In a typical activation sequence, the receptor is bound at its extracellular site by a neurotransmitter; the bound receptor activates the G-protein to a state in which it dissociates from GDP and is bound by guanosine triphosphate (*GTP*). The G-protein then splits into an α subcomponent and a β/γ subcomponent, either of which can catalyse further chemical reactions. For example, the enzyme activated by the α subcomponent may catalyse conversion of molecules in an inactive state present in the intracellular environment into an active form which can diffuse within the neuron and catalyse yet other reactions. These active molecules are called second messengers

reactions that can have many different effects on the target neuron. Although there are many different G-protein receptors and G-proteins, they have a typical structure and mode of operation as illustrated in Fig. 4.5.

Outside the membrane, each G-protein receptor molecule has a binding site for one specific neurotransmitter. If the neurotransmitter binds to that external site, a site on the receptor inside the membrane is activated, which in turn activates a G-protein. In response to activation, the G-protein splits into two components, α and β/γ. The α component has a binding site that activates a membrane-bound enzyme. This enzyme in turn activates many molecules of an internal chemical messenger (called a second messenger). Second messengers can initiate sequences of chemical processes within the neuron, leading to results ranging from opening of ion channels to transcription of genetic information. The primary role of the β/γ component is to recombine and therefore inactivate the α component, but the β/γ component can sometimes have an additional role, such as directly opening ion channels [118].

In more detail, the G-protein is inactive when the α component is bound to a small molecule called guanosine diphosphate (GDP). The receptor protein catalyses replacement of the GDP by guanosine triphosphate (GTP), which breaks the α component from the β/γ component. However, the α component is also a catalyst for the conversion of GTP to GDP. Such a conversion results in recombination of the G-protein components, limiting the period of time a G-protein is activated.

Any one neurotransmitter generally has multiple G-protein receptors, but multiple G-protein receptors can target the same G-protein and therefore modulate the same chemical processes. A number of different such processes influenced by G-protein systems will be described below.

G-protein receptors are generally concentrated in the postsynaptic synapse. However, in specific cases such as certain receptors for the neurotransmitter dopamine [119], they can be distributed over regions of the dendritic tree remote from actual synapses.

4.2.4 Tyrosine Kinase Receptors

Tyrosine kinases, when activated, phosphorylate other proteins at tyrosine amino acid positions in the target protein. Some tyrosine kinases, called tyrosine kinase receptors, are bound to the cell membrane and are activated by an external ligand molecule. The most common ligands for tyrosine kinase receptors are a category of chemicals called neurotrophins or growth factors. Activation of tyrosine kinase receptors by neurotrophins plays a key role in managing cell division and differentiation of cells into different types. However, neurotrophins can also activate tyrosine kinase receptors in mature neurons to affect changes to synaptic strengths [120].

The four neurotrophins are nerve growth factor (NGF), brain derived neurotrophic factor (BDNF), neurotrophin-3 (NT-3) and neurotrophin-4 (NT-4). There are three types of tyrosine kinase receptors: TrkA; TrkB and TrkC. Each receptor can be activated by one or more of the neurotrophins. As discussed later, activations lead to triggering a number of signalling pathways which can lead to gene activation [121].

4.3 Electrochemical Signals Within and Between Neurons

Much of the communication within and between neurons is by electrical signals. However, these electrical signals are not simply currents. Simple currents would be fast, but would decay because of electrical resistance. The electrical signals between neurons are carried by chemical processes that regenerate the signal at a sequence of points. This regeneration allows the same signal to reach all the different synapses down the branches of the axon at considerable distances from the neuron. However, although the same signal reaches each synapse, its effect on that synapse can be different, allowing a source neuron to have different effects on its targets.

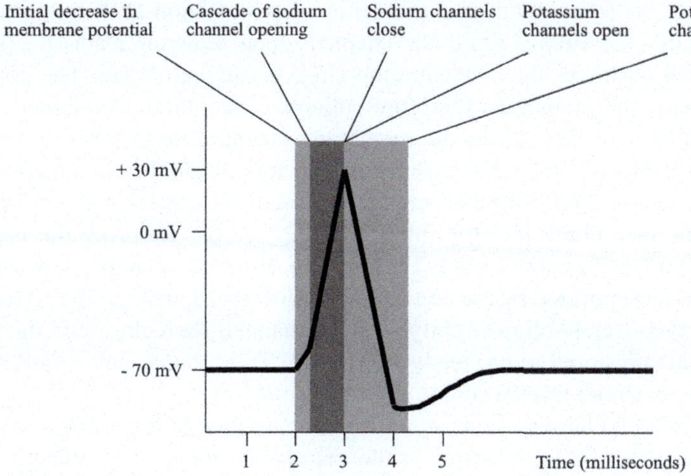

Fig. 4.6 Variation in membrane potential during an action potential. Slight decreases from the resting potential can occur at some area of the membrane, for example by opening of Na^+ channels in the vicinity. If such a decrease, called a depolarisation, reaches a threshold value, voltage gated ion channels that are able to pass Na^+ ions open, forcing a much larger depolarisation that collapses the membrane potential. Once collapsed, the Na^+ channels close and the membrane potential is restored, with a slight overshoot (hyperpolarisation) due to the opening of K^+ channels. The whole process takes of the order of 2 ms. The Na^+ ions can diffuse to the neighbouring region of the membrane, where they can trigger another collapse. Hence the collapse can propagate along the neuron membrane. Such a propagating collapse is called an action potential. Note that one effect of the hyperpolarisation is to prevent transmission of the action potential back towards the neuron

The effects of individual synapses can be separately managed based on presynaptic and postsynaptic neuron activity.

Within a neuron, electrical signals can be simple currents carried by diffusion of ions, but they can also be chemically regenerated electrical signals carrying longer distances across the dendritic tree.

4.3.1 Action Potentials Between Neurons

Neurons communicate with each other by means of action potentials. Such an action potential is a collapse of the voltage across a small area of axon membrane for a brief period of time, which propagates along the axon. The propagation of an action potential can be understood from Fig. 4.6. There is a high concentration of voltage gated Na^+ channels and of voltage gated K^+ channels all along the axon. The resting membrane potential is −70 mV, the Na^+ channels open if the membrane is depolarised to about −55 mV, and the K^+ channels open if the membrane is depolarised to +30 mV. If a calcium action potential initiated in the dendritic tree and/or external inputs direct to the soma result in a membrane depolarisation at the axon hillock reaching −55 mV, the voltage gated Na^+ channels open, allowing a massive flow of Na^+ ions which equalizes the concentration close to the membrane. The potential difference across the membrane therefore collapses. The unaffected excess of K^+ ions in the interior in fact causes the membrane potential to go positive to about +30 mV. This positive potential level is the trigger for closing the Na^+ channels and opening K^+ channels. With the Na^+ channels closed, the excess Na^+ ions diffuse away from the membrane into the interior of the neuron, restoring the negative potential within a millisecond or so. The opening of the K^+ channels removes the positive potential supported by the concentration difference, making the restoration faster and in fact overshooting slightly. The K^+ channels then close and the membrane potential is restored to its steady state value. The collapse and restoration of the membrane potential takes a couple of milliseconds.

If the potential collapses at one point in the axon membrane, diffusion of Na^+ ions means that there will be drops in the membrane potential in adjacent axon areas. This leads to a similar Na^+ channel driven collapse in the membrane potential in that adjacent area and the collapse propagates along the axon to all the targets of that axon [122]. The brief hyperpolarisation prevents reverse propagation of the action potential. The membrane collapse mechanism means that the action potential is regenerated at each point using the energy stored in the steady state membrane potential. Hence unlike simple conductance of an electrical current the magnitude of the potential does not decay as it propagates. Simple conductance of current does occur within neurons, but such currents decay with distance from point of origin.

The speed with which the collapse propagates along an axon has been measured in mammals to vary between 10 and 100 m/s [123]. One major factor determining this speed is the diameter of the axon, larger diameters resulting in higher speeds. The speed for a given diameter can be increased by about a factor of ten by a myelin coating [124]. Glia cells coat stretches of the axon with myelin, with the coated stretches separated by uncoated sections (called nodes of Ranvier). Myelin conducts electricity, and the distance between two nodes is short enough that an action potential is carried from one node to the next with enough strength to generate a collapse in that next node. The action potential is therefore carried much faster, jumping rapidly across myelinated sections and being regenerated fully at the nodes. The myelin coat has the additional advantage that it reduces the total flow of ions and therefore the energy to restore the axon to resting potential. Myelin is white, and volumes of brain tissue largely made up of myelinated axons therefore appear white, hence for example the white matter inside the cortex.

There is a general similarity between all action potentials along axons, but the frequency with which they can be generated and their detailed shape can differ between neurons. As described earlier, part of the process that terminates an action potential is the opening of voltage gated K^+ channels. The resultant membrane hyperpolarisation typically lasts for 1–10 ms, and it is much more difficult for a neuron to generate another action potential during this period. In some neurons this fast hyperpolarisation is reinforced by K^+ channels that are opened by the presence of Ca^{++} ions inside the neuron [125]. Such Ca^{++} gated K^+ channels are described below. In some cases there are also two longer term hyperpolarisations lasting from several hundred milliseconds to several seconds [125]. The medium component of this longer term hyperpolarisation begins within <10 ms of the action potential and lasts between 50 and several hundred milliseconds. The slow component is often only observed after several action potentials, has a slow rise and can last for several seconds. The longer term hyperpolarisations are produced by different types of Ca^{++} gated K^+ channels. The Ca^{++} ions involved in both fast, medium and slow hyperpolarisations are present inside the neuron as a result of the opening of voltage gated Ca^{++} channels during the action potential.

The detailed shapes of the fast, medium and slow hyperpolarisations depends on the concentration and type of the various ion channels, and influences the frequency and shape of the action potentials generated [126]. The adaptation phenomenon described earlier, in which the probability of spike production declines after a sequence of spikes, is the result of such hyperpolarisations.

4.3.2 Calcium Action Potentials Within Dendrites

Calcium action potentials tend to occur within dendritic trees, at least in pyramidal neurons in the cortex. Such action potentials reach their peak in 5–10 ms, slower than Na^+ action potentials [127]. They also differ from Na^+ action potentials along axons in that they are not fully self sustaining, often propagating only for relatively short distances across the dendritic membrane and not reaching the soma [128]. However, even if the action potential does not reach the soma, some current can reach the soma by passive conductance from the region in which the action potential occurred.

4.3.3 Backpropagating Action Potentials

When an action potential is initiated down the axon of a neuron, some potential will propagate back into the dendritic tree. This propagation can be passive, in which case it decays relatively rapidly. However, backpropagation into the dendritic tree can also occur actively by an action potential. A backpropagating action potential [129] depends upon voltage gated Na^+ channels but may also involve voltage gated Ca^{++} channels and can be blocked by K^+ channels. The degree to which it propagates depends upon the densities and activation states of these channels over the surface of the dendrite, which varies between different dendritic areas within the same neuron [130] and between different types of neuron [129]. For example, backpropagation is almost passive in Purkinje cells, which are a type of neuron located in the cerebellum. A backpropagating action potential penetrates actively to all parts of the dendritic tree of projection neurons in the substantia nigra (part of the basal ganglia). Backpropagation in cortical pyramidal neurons lies between these two extremes.

Such backpropagating action potentials communicate to the dendritic tree that an output action potential has been generated down the axon. This information is used to determine whether weight changes are made to recently active synapses that contributed to generation of the output.

4.4 Synapses and Synaptic Strengths

4.4.1 Synaptic Strengths

At an excitatory synapse, neurotransmitter released by the source neuron attaches to binding sites on neurotransmitter gated Na^+ channels in the membrane of the postsynaptic neuron resulting in flow of Na^+ ions into the target neuron. This flow of ions is an excitatory postsynaptic potential (EPSP), which depolarises the membrane of the target neuron in the vicinity of the synapse.

Chemical processes in the synaptic cleft constantly remove neurotransmitter, so it is only available for a brief period. The size of the EPSP is determined by the number of opened Na^+ channels. An action potential arriving at a synapse with more Na^+ channels and more neurotransmitter release will therefore have a larger effect. Such synapses are said to be stronger.

The pyramidal neurons common in the cortex release the neurotransmitter glutamate at all their synapses on to other neurons. Postsynaptic neurons have molecules of a protein called the AMPA receptor embedded in the membrane opposite the postsynaptic cleft. The AMPA receptor is a Na^+ channel opened by glutamate, which is therefore excitatory. The number of AMPA receptor molecules in a synapse is therefore a key parameter in determining the synaptic strength.

4.4.2 Synaptic Strength Changes on Different Timescales

If the magnitude of neurotransmitter release or number of ionotropic receptors changes, the strength of the synapse changes. If the strength increases, the change is called potentiation, if it decreases, depression. Changes to synaptic strengths can occur in response to specific types of neuron activity. Investigation of such changes typically involves electrical stimulation of an axon bundle carrying inputs to some structure, and measuring the postsynaptic potential generated by the target structure. One heavily investigated example is the Schaffer bundle of axons carrying inputs to CA1 from CA3 in the hippocampus. The electrical stimulation is in the form of a sequence of pulses with a general similarity to action potentials. Initially, a low level of stimulation is used to determine a baseline postsynaptic response. Different types of stronger inputs are then imposed, and changes to the baseline response measured. For example, one type of input is the theta-burst stimulation (TBS) protocol in which one unit of input is 10 bursts of 5 pulses delivered at 200 ms intervals, in a general imitation of the theta band of the EEG. Multiple TBS trains can be delivered at 30 s intervals. The change in postsynaptic response typically depends on the number of trains delivered [131]. In addition, relatively long term changes can depend on whether the postsynaptic neuron produces an output action potential, and the relative timing of the synaptic input and target neuron output [132].

Changes can be very short term, lasting for periods of the order of milliseconds to a minute or two until they relax back to the original strengths [133]. Very short term changes are believed to be supported by changes in the presynaptic neuron [134]. In some cases these very short term changes may be driven by receptors in the presynaptic neuron that are sensitive to the neurotransmitter released by the presynaptic neuron [135]. These receptors then influence the magnitude of neurotransmitter release in the immediate future. In other cases, residual Ca^{++} concentrations in the presynaptic neuron remaining from a previous action potential can influence the response to a later action potential [133]. These very short term changes will be labelled very short term potentiation (VSTP) for increases and very short term depression (VSTD) for decreases.

Longer term changes can persist more than a few minutes, much longer that the neuron activity that induced them. Changes lasting longer than a few minutes, but returning to original weight within 30–60 min are sometimes called short-term potentiation (STP) or depression (STD). Increases or decreases in synaptic weight change that are maintained for longer than 30–60 min are called long term potentiation (LTP) or long term depression (LTD) [136]. LTP and LTD are further subdivided by the length of time beyond 60 min that they persist. LTP and LTD have mainly been investigated in excitatory synapses on to both excitatory and inhibitory projection neurons, but have also been observed in inhibitory synapses made by interneurons. LTP and LTD have been observed in many different brain structures [137]. However, there is evidence that in live brains (*in vivo*), LTD is less common than LTP in many [138, 139] but not all anatomical structures. The cerebellar cortex is a specific example where LTD is commonly observed [140].

One very important type of LTP and LTD modulates AMPA receptor channels in the postsynaptic membrane adjacent to the synaptic cleft. Chemical modifications can change the size of the ion currents carried by AMPA receptor channels. Changes to the number of AMPA receptor channels, which can be accompanied by enlargement of the synapse [141], can occur in response to specific types of neuron activity [132]. Pre-existing AMPA receptor channels can diffuse from the nearby dendritic shaft, or can be released from a vacuole to the spine [142]. LTD occurs by removal of AMPA receptor channels from the postsynaptic membrane to the interior of the neuron [143].

Three different timescales for AMPA receptor channel based activity dependent LTP have been defined: short term LTP (STP or LTP1); early LTP (E-LTP or LTP2) and late LTP (L-LTP or LTP3) [144]. LTP2 develops later that LTP1, and LTP3 later than LTP2. Hence the initial increase in synaptic weights depends on LTP1, as LTP1 decays it is replaced by LTP2, and as LTP2 decays it is replaced by LTP3. If the later types are not induced, the synaptic weight increase decays again without being replaced [145].

In some cases LTP and LTD require the presence of other neurotransmitters such as dopamine in order to take place [146]. STP requires transient modifications to locally available enzymes, receptors and/or ion channels, but not synthesis of new proteins [147]. E-LTP requires some protein synthesis, but only utilizes mRNA that is already available locally, close to the synapse [148]. L-LTP requires structural modifications involving gene transcription in the cell nucleus and transport of new proteins or mRNA to the synapse [149], and generally require the presence of modulating neurotransmitters such as dopamine at an appropriate time [150].

Long term changes to synaptic weights are the physical substrate for memory [151, 152]. Such changes require sequences of chemical processes on a number of different timescales ranging from milliseconds to months [153].

4.4.3 Management of Long Term Average Neuron Firing Rate

Over time, pyramidal neurons in the cortex can grow and add synapses to their dendritic trees, and therefore receive more action potentials. The weights of individual synapses can change. However, the long term average frequency of a neuron firing is maintained within a fixed range [154]. When a pyramidal neuron in the cortex fires, it releases a neurohormone called brain derived neurotrophic factor (BDNF). BDNF in the extracellular environment reduces the strength of excitatory synapses on to pyramidal neurons, scaling the strengths of all synapses down by removing a proportion of AMPA receptors. BDNF also increases the strength of excitatory synapses from pyramidal neurons on to interneurons that inhibit local pyramidal neurons, also reducing average pyramidal firing rates. A reduction in extracellular BDNF increases the strengths of excitatory pyramidal synapses. These changes scale all the synapses on a neuron by the same multiplicative factor, and require hours or days of altered activity to have an effect [155]. Hence extracellular BDNF operates to stabilize long term pyramidal neuron average firing rates.

As discussed later, the functional significance of this regulation is that a neuron is detecting some circumstance defined by its receptive field. Its primary value is discriminating between circumstances in which different behaviours are appropriate. If a neuron fires too often, its discriminatory value is less.

4.5 Specific Molecules

A range of specific molecules have important roles in communication between neurons and in changes within neurons in response to such communications. In the rest of this chapter some of these molecules will be discussed under five categories. Firstly, neurotransmitters and their associated ion channel and G-protein receptors.

Secondly, various second messengers that carry amplified neurotransmitter signals within neurons. Thirdly, the different types of G-protein receptors that activate second messengers. Fourthly, voltage gated ion channels. Fifthly, kinases and phosphatases that activate and inactivate enzymes and other proteins will be mentioned, but more fully discussed in Chap. 5.

There are very large numbers of different molecules in most categories. In human beings there are hundreds of different G-proteins, which can be organized into five or six classes. There are over one hundred voltage gated ion channels, many hundreds of ligand gated ion channels and many hundreds of kinases. Not all of these G-proteins, ion channels and kinases are present in neurons. Five second messengers that have been found in neurons, and it is estimated that there are over 50 neurotransmitters.

Some examples from each category will be described, including molecules that will form part of the discussions of chemical processes within neurons in Chap. 5. One objective is indicate the extent of the range of chemical capabilities. In Chap. 6 some processes that support some key neuron information processes will be described, and a second objective is to provide information about the molecules that feature in these processes.

4.5.1 Neurotransmitters

A neurotransmitter is generally defined by a number of characteristics. Firstly, it is a chemical that is packaged in vesicles located within axon terminals. Secondly, Ca^{++} entry into the axon terminal triggers release of the neurotransmitter from the vesicles into the extracellular environment. Thirdly, the neurotransmitter binds to specific sites on receptor molecules embedded in the membrane of target neurons, producing changes to those neurons. Fourthly, there are mechanisms for removing the neurotransmitter from the external environment after release. There are more than 50 known chemicals that act as neurotransmitters in the human brain, each of which has a set of different receptor molecules to which it binds [156]. Neurotransmitters can range in size from individual ionized atoms, through relatively small molecules with up to perhaps 50 atoms to much larger neuropeptides. Zinc ions are an example of individual atoms. Small molecules are often either amino acids or molecules of a type called amines. These molecules contain under 50 atoms. Examples include the amino acids glutamate, aspartate, GABA, glycine and serotonin, and the amines dopamine and norepinephrine. A small molecule neurotransmitter that does not fit in either of the two categories is ATP, which is also the primary energy transfer mechanism in all biological processes. Neuropeptides are small proteins that can vary from 3 to 100 amino acid components. For example, oxytocin contains nine amino acids, a little under 150 atoms.

There are also some chemicals that are counted as neurotransmitters even though they do not have all the defining characteristics. In particular, gases like carbon monoxide (CO) and nitric oxide (NO) can be synthesized in one neuron and diffuse through the membranes to have effects in another neuron. NO will be discussed further below.

Most individual neurons produce one primary neurotransmitter, but often also produce additional neurotransmitters. Such additional neurotransmitters may be packaged in the same vesicles as the primary, in which case they will tend to be released in the same circumstances. In other cases, additional neurotransmitters are packaged in separate vesicles and released in somewhat different circumstances [157].

Receptor molecules on target neurons determine the effect of a neurotransmitter, and there can be many different receptors for any one neurotransmitter. The effects of receptor molecules on the target neuron can differ in both the type of change and the timescale over which the change initiates and persists.

Some very common neurotransmitters bind to ionotropic receptor molecules, i.e. ion channels. Binding causes them to open. Such binding occurs for a limited period of time and the neurotransmitter molecule is then released and the channel closes. Major parameters which differ between different ionotropic channels include the *types of ions* that are conducted; their *conductance* or the ion current that passes when they are open; their *activation time* or the time they take to open after binding; and their *deactivation time* or the time they take to close. In addition, some receptors no longer open after continuous exposure to their neurotransmitter beyond a certain time (*desensitisation time*), and once desensitized take time to recover (*recovery time*). All these time constants are collectively known as the kinetics of the channel.

Most neurotransmitters have more indirect and/or more complex effects, binding to molecules called metabotropic receptors and triggering a sequence of chemical processes within the neuron [158]. Such sequences generally involve a step using a G-protein which triggers release of a chemical messenger molecule into the interior of the neuron. This chemical messenger is called the second messenger because it is activated after the external neurotransmitter which is the first messenger, and such sequences are therefore called G-protein second messenger (GPSM) systems [159].

One major difference between GPSM systems and systems in which neurotransmitters act directly on ion channels is that GPSM systems are generally slower acting. Direct opening of ion channels can occur in a fraction of a millisecond, while GPSM systems take from 100 ms up to several minutes or more to complete their actions. Another difference is that GPSM systems amplify the effect of a single external neurotransmitter molecule. There are many different GPSM systems, and one neurotransmitter may have effects on more than one system, although for one type of neuron the effect on one system may be predominant. Furthermore, in addition to effects on multiple different GPSM systems, one neurotransmitter may have a direct action on an ion channel.

It is also possible for the presynaptic neuron to have receptors for its own neurotransmitter, which can regulate subsequent release of that neurotransmitter. There are also neurotransmitters which can act from the target neuron back on to a source neuron. For example, nitric oxide (NO) discussed below can diffuse through the neuron membranes from the postsynaptic to the presynaptic neuron.

4.5.1.1 Glutamate

Glutamate is a very common neurotransmitter, and its usual mode of action is to excite postsynaptic neurons by directly opening cation channels. There are three major types of glutamatergic ionotopic channels: AMPA receptor; NMDA receptor and kainate. These channels are permeable to Na^+, K^+ and Ca^{++} ions to different degrees. AMPA receptors mainly pass Na^+ ions, but some variants can pass significant Ca^{++} currents. The kainate channel is also a Na^+ channel, but again some variants can pass significant Ca^{++}. The kainate channel opens and closes somewhat more slowly than the AMPA receptor channel [160]. The NMDA receptor channel mainly passes Ca^{++} ions. The NMDA receptor channel is unusual in that it is generally blocked by the binding of a Mg^{++} ion. This ion is released if the membrane potential is depolarised in the vicinity of the channel. Hence an NMDA receptor channel requires both glutamate and depolarisation of the membrane potential to open.

In detail, the picture is rather more complex. All three receptor types can pass all three cations (Na^+, Ca^{++} and K^+) to varying degrees. Each receptor type is made up of subunits. In the case of AMPA receptor channels these subunits are called GluR1, GluR2, GluR3 and GluR4 and a channel is constructed from different combinations of four subunits. For NMDA receptor channels there are seven possible subunits (NR1, NR2A, NR2B, NR2C, NR2D, NR3A, NR3B) and a channel is again constructed from different combinations of four subunits. In the case of kainate channels these subunits are called GluR5, GluR6, GluR7, KA1 and KA2, and a channel is constructed from different combinations of five subunits. Each channel subtype varies in conductance for different types of cation, activation time, deactivation time, desensitisation time and recovery time [161]. Furthermore, AMPA receptor channels have four binding sites for glutamate, and conductance increases with the number of sites occupied [162].

Most synapses that show LTP and LTD are glutamatergic. The AMPA receptor channel based mechanism for LTP mentioned earlier can be understood at a deeper level as an interaction between AMPA receptor and NMDA receptor channels as illustrated in Fig. 4.7. NMDA receptors pass Ca^{++} ions when open. In the environment of the NMDA receptor channels, these Ca^{++} ions initiate the creation of new AMPA receptor glutamate-triggered Na^+ ion channels. The presence of such new channels means that more Na^+ ions will enter the postsynaptic neuron in response to an incoming action potential, resulting in a greater reduction in the negative membrane potential and a greater tendency to produce an action potential in the postsynaptic neuron. In other words, the strength of the synapse has been increased. NMDA receptor channels only open in response to a combination of two circumstances. One is that the glutamate neurotransmitter is bonded to the receptor site. The other is that the local membrane potential is depolarised by an action potential. These NMDA receptor channels are located in the membrane of the postsynaptic neuron, and the action potential must therefore be generated by the target neuron and back-propagated into its dendritic tree. However, the neurotransmitter must come from the source neuron. The overall effect is that synaptic strength will tend to increase if

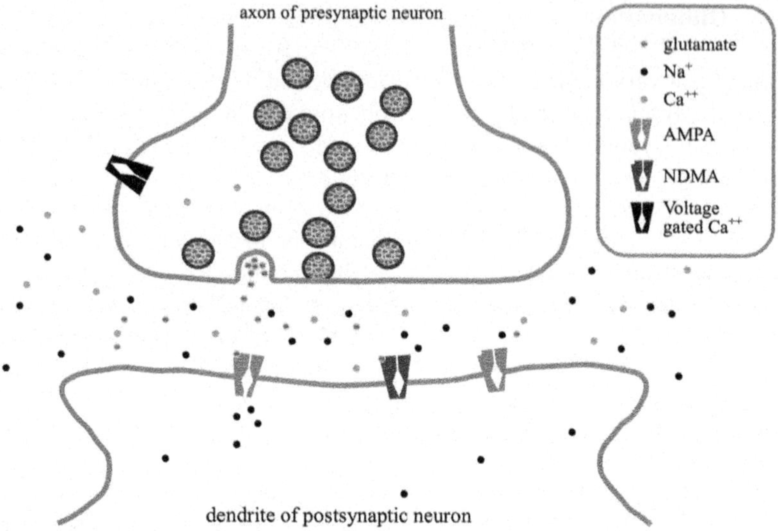

Fig. 4.7 Synaptic weight increases at a glutamate synapse. Receptors at glutamatergic synapses include AMPA receptors that are activated by glutamate alone and NMDA receptors that are only activated by a combination of glutamate and local membrane depolarisation. Open AMPA channels carry Na^+ and K^+ ions, open NMDA receptors carry Ca^{++} ions. Glutamate released by the presynaptic neuron binds to both AMPA and NMDA receptors, but only AMPA receptors open immediately. Hence synaptic weight is determined only by the number of AMPA receptors (plus the amount of glutamate release if this is a limiting factor). Glutamate binding lasts for a specific period of time. If the postsynaptic neuron fires, an action potential also backpropagates into the dendritic tree. If such a backpropagating action potential reaches the NMDA receptors while they are still bound to glutamate, they open to allow passage of Ca^{++} ions. These Ca^{++} ions trigger a cascade of chemical processes leading to a larger number of AMPA receptors, increasing the strength of the synapse. Silent synapses contain only NMDA receptors and have no synaptic weight until after an occasion in which an incoming action potential is immediately followed by a backpropagating action potential

an incoming action potential from the source neuron is followed shortly afterwards by an action potential generated by the target neuron, provided that action potential is backpropagated to reach the synapse [132].

A mechanism for LTD in AMPA receptor channel based synapses has been observed in Purkinje neurons in the cerebellum. In this mechanism [163], LTD results from removal of AMPA receptor channels. Such removal results when both AMPA receptors and mGluR1 GPSM receptors are activated by glutamate, and voltage gated Ca^{++} channels are also opened. The combination of these three events results in activation of a protein kinase (PKC) which phosphorylates the AMPA receptor, leading to its removal from the membrane.

AMPA receptors tend to be located in the centre of the postsynaptic density of the synapse, opposite glutamate release sites. Metabotropic glutamate receptors tend to be located on the periphery of the postsynaptic density [164]. Most glutamatergic synapses have a mixture of AMPA and NMDA receptors, and some synapses but not others also have kainate receptors and some synapses have only kainate receptors [160]. Kainate channels can be located in the presynaptic synapse, where they regulate the release of neurotransmitter. Some kainate receptors are located on the dendritic trunk, the soma and even on the axon [165]. Such receptors may be targetted by glutamate released by astrocytes, which would represent a functional role for glia cells.

There are also three groups of metabotropic glutamate receptors (I, II and III), and a number of variants in each group [166]. Group I receptors are generally located postsynaptically. They stimulate a range of cascades of chemical reactions including the classical pathway that begins with phospholipase C and can lead through Ca^{++} mobilisation and activation of protein kinase C to gene transcription required for L-LTD, as discussed later. Group II receptors can be either presynaptic or postsynaptic, and have a number of effects including inactivation of the enzyme adenylyl cyclase. Adenylyl cyclase catalyses conversion of ATP to a second messenger cAMP with a range of downstream effects including regulation of ion channels as discussed later. Group III receptors have largely similar effects to group II, but are predominantly located presynaptically. One effect of group II/III receptors is short term changes to synaptic strengths.

4.5.1.2 Aspartate

Aspartate is an excitatory neurotransmitter that is only found in neurons that also produce glutamate [167]. In some cases the aspartate appears to be packaged in the same vesicles as glutamate, but in other cases it appears to be in separate vesicles. However, even where aspartate and glutamate may be present in the same vesicles, circumstances in the neuron can vary the relative ratio in which the two neurotransmitters are released [168]. Aspartate weakly activates the NMDA receptors [169] targeted by glutamate, but appears to have no effect on AMPA or kainate receptors [168].

4.5.1.3 γ-Aminobutyric Acid (GABA)

GABA is also a very common neurotransmitter. There are two major receptor types for GABA, $GABA_A$ [170] and $GABA_B$ [171], plus a $GABA_C$ receptor mainly found in the retina. Although GABA is generally viewed as inhibitory, in fact its effects can be inhibitory or excitatory depending on the precise conditions in the neuron [172].

The $GABA_A$ receptor is an ion channel that passes both Cl^- and HCO_3^- ions [173]. The effect of a $GABA_A$ channel opening depends on the relative concentration of Cl^- and HCO_3^- ions inside and outside the membrane. In general, there is a higher concentration of HCO_3^- and a lower concentration of Cl^- inside the membrane. Hence opening of a $GABA_A$ channel results in a depolarising (excitatory) flow of HCO_3^- ions and a hyperpolarising (inhibitory) flow of Cl^- ions. The relative magnitude of these flows depends on the membrane potential. When the membrane is at resting potential, the net flow is depolarising, although not sufficient alone to generate an action potential. If the membrane has already been depolarised to some degree (for example by the opening of nearby Na^+ channels), flow of Cl^- ions predominates and the net effect is hyperpolarising.

Hence if $GABA_A$ channels are open at the same time and place as an existing depolarisation, the effect of the $GABA_A$ channels is inhibitory. However, if the $GABA_A$ channel opening is isolated temporally or spatially from another membrane depolarisation, the $GABA_A$ channels can reinforce that depolarisation even to the point of reaching the threshold for an action potential [172]. If a depolarisation resulting from another source occurs more than 5.5 ms after the opening of $GABA_A$ channels, there is a residual depolarising effect of the $GABA_A$ channels that adds to the following depolarisation. An example of spatial separation is that $GABA_A$ channel opening on the dendrites of a neuron can add to a depolarisation of the soma.

A further factor which can amplify the effect of the opening of a $GABA_A$ channel is its proximity to excitatory currents. Osmotic pressure will drive movement of Cl^- ions from the region of high concentration to the region of low concentration. If the concentration gradient is such that Cl^- ions must move towards a region with a more negative electrical potential, the flow will be reduced. A point can be reached at which osmotic pressure balances potential and there is no net flow. The potential at which this balance exists is called the reversal potential for the ion (E_{Cl}). If the resting membrane potential is close to E_{Cl}, then the effect of opening Cl^- channels is to lock the membrane potential at the resting level, overcoming the effect of excitatory currents in the vicinity. This locking, or shunting, of the membrane potential means that the effect of the opening of a $GABA_A$ receptor on nearby open excitatory currents is much larger that would be expected on the basis of simple addition of the positive and negative current flows [174].

Yet another complicating factor is that in some neurons, the concentration of Cl^- ions inside the neuron at resting potential is lower than outside. Opening $GABA_A$ channels will therefore depolarise the neuron.

Postsynaptically, the $GABA_B$ receptor is linked through a GPSM system to opening K^+ channels. Opening K^+ channels is inhibitive, but acts on a longer timescale than $GABA_A$. Presynaptic $GABA_B$ receptors, possibly a different subtype, down-regulate voltage activated Ca^{++} channels, reducing neurotransmitter release [171].

4.5.1.4 Glycine

Glycine is often inhibitory, because glycine receptors are Cl^- channels [175]. However, the presence of glycine is also essential for the activation of NMDA receptors, although glycine alone does not activate these receptors [176]. There is a glycine binding site on the NMDA receptor which is generally [177] but not always [176] occupied in vivo, suggesting that glycine plays a role in regulating this receptor. Absence of glycine in the extracellular environment blocks LTP [177]. There is a concentration of glycine transporters associated with glia cells in the neighbourhood of NMDA receptors [177, 178]. Glycine levels to support NMDA receptor activation could therefore be maintained by glia cells.

Glycinergic neurons are mainly located in the brainstem and cerebellum, and target neurons within those structures. However, they also target neurons in the hypothalamus, thalamus and basal forebrain, and more lightly target neurons in the cortex, hippocampus, amygdala and striatum [179].

4.5.1.5 Dopamine

Dopamine is another common neurotransmitter in the brain. However, rather than acting directly on ion channels, dopamine acts via GPSM mechanisms which can have multiple effects within cells. In addition, the mechanisms activated and the effects on the target neuron differ depending upon the type of dopamine receptor that is present in the target neuron membrane. As a result, the presence of dopamine can have many different effects on target neurons [180].

There are at least five different types of dopamine receptors (D1 through D5), divided into two families (D1-like and D2-like). The D1-like receptors are D1 and D5, while the D2 like receptors are D2, D3 and D4. All these receptors are part of GPSM systems. The effect of a neurotransmitter acting through a GPSM mechanism can be very complex. For example, activation of D1 receptors causes an increase in the activity of another protein embedded in the cell membrane called adenylyl cyclase. Activation of adenylyl cyclase triggers sequences of chemical reactions of which one effect is to block certain voltage gated K^+ channels. However, another effect is to reduce voltage gated Na^+ channel availability. The overall effect is to increase the responsiveness of the target neuron to sustained release of the excitatory neurotransmitter glutamate, but reduce the response to transient or uncoordinated glutamate release [181].

Because dopamine acts through different receptors into different GPSM systems, its effects on target neurons can be very different. For example, dopamine acts through D2 channels in similar neurons to reduce the activity of adenylyl cyclase, and therefore reducing their excitability in response to glutamate [181]. Thus depending on the receptor, dopamine can have opposite effects on otherwise similar neurons. In some brain structures many neurons have both D1 and D2 channels [182].

Dopamine plays a role in LTP in a wide variety of brain regions [183]. For example, activation of D3 channels in the hippocampus enhances LTP [184]. Dopamine can also determine whether LTP or LTD will occur. For example, in the striatum, activation of both D1 and D2 channels (on different neurons) encourages LTD, while activation of D1 channels and no activation of D2 channels is required for LTP [185]. In addition, dopamine acting through D1-type receptors is needed to convert activity driven changes to synaptic strengths (LTP and LTD) lasting an hour or two into much longer range changes (L-LTP and L-LTD) lasting days, weeks or years [149]. For example, activation of D5 channels in the hippocampus is required for conversion of LTP to L-LTP [186].

Dopamine receptors are not only found close to synapses. In the striatum, many receptors are distributed on dendritic shafts and spines, away from the locations of dopamine synapses [187]. Hence in this case the action of dopamine must occur through diffusion from dopaminergic axon terminations into the extracellular environment. In the striatum a background extracellular concentration of dopamine is created by the steady (tonic) firing of dopaminergic neurons in the SNc [188]. Burst firing of dopaminergic neurons creates higher concentrations of dopamine close to dopaminergic synapses, but is prevented from diffusing into the extracellular environment because the burst firing also opens channels around the borders of the synapse that recycle dopamine back into the axon of the source neuron [189]. Different dopamine receptors have different affinities for dopamine, affecting the degree to which they will be open at different dopamine concentrations [190].

4.5.1.6 Serotonin

Serotonin (also known as 5-HT) is another neurotransmitter with many different receptors, called 5-HT receptors. At least a dozen types have been identified in the brain [191]. The 5-HT_3 receptor is ionotropic, exciting the target neuron by direct opening of a cation channel that passes Na^+ and K^+ ions. The other 5-HT receptors are GPSMs, many acting by catalysing or blocking the production of the second messenger cAMP. The general effect on the target neuron is sometimes excitatory and sometimes inhibitory. For example, activation of serotonin receptors can prolong the opening times of voltage gated calcium channels, increasing the total calcium current at depolarised membrane potentials [192]. Neurons generating serotonin are almost entirely located in one brain structure called the raphe nucleus. Axons from this nucleus target neurons in almost all brain structures. However, different brain structures contain different types of 5-HT receptors [191].

4.5.1.7 Acetylcholine

There are two main families of acetylcholine receptors: nicotinic and muscarinic. Nicotinic receptors are ligand-gated cation channels which excite the target neuron [193]. Nicotinic channels pass Na^+ and K^+, but also significant Ca^{++} [194]. Muscarinic receptors are G-coupled protein receptors that can be found on both sides of the synaptic cleft (presynaptic and postsynaptic). Three types have been observed in the brain: M1, M4 and M5. M1 and M5 upregulate phospholipase C, while M4 decreases the intracellular level of the second messenger cAMP by inactivating adenylyl cyclase [195].

One interesting role of muscarinic receptors is the control of M-current [196]. M-current is a K^+ flow through voltage gated K^+ channels. These channels open when the membrane potential is depolarised near the threshold for action potential, reducing the probability of an action potential. Activation of M1/M5 receptors closes these K^+ channels.

4.5.1.8 Norepinephrine

The adrenoceptors [197] are a group of metabotrobic receptors that each bind to several neurotransmitters including norepinephrine (also called noradrenaline) and epinephrine (also called adrenaline). The degree of response generated by binding to the same receptor varies for the different neurotransmitters. Like serotonin, much of the norepinephrine in the brain comes from one structure, in this case the locus coeruleus, which targets almost all other brain structures [198].

Norepinephrine blocks the accommodation of spike discharge that normally occurs if similar inputs continuously excite a neuron, and therefore results in the neuron producing output action potentials for a longer period. Norepinephrine produces this effect by changing the opening probability of the ion channels that generate the slow hyperpolarisation following an action potential [199, 200].

4.5.1.9 Adenosine Triphosphate ATP

ATP, in addition to its ubiquitous role as energy source for chemical processes, also acts as a neurotransmitter. Often, vesicles at a synapse contain ATP co-stored with other neurotransmitters like glutamate, GABA or acetylcholine in the same vesicles. However, some synapses may have only vesicles containing ATP, and other synapses may have separate vesicles for ATP and, for example, glutamate [201]. Furthermore, when contained in the same synapse, the ratio of ATP to other neurotransmitters can vary [202].

A number of receptors for ATP are known, the ionotrophic P2X family with eight subtypes, and the metabotropic P2Y family with seven subtypes [203]. The P2X family excite their target neuron. They are all permeable to Ca^{++}, K^{+} and Na^{+}, but have different kinetics. All the metabotropic receptors are coupled to the same G-protein (Gq_{11}). Activation of these receptors activates phospholipase C, triggering a chain of subsequent chemical processes.

The release of ATP and GABA at the same synapses has been observed, with the two neurotransmitters probably stored in the same vesicle [204]. Although some synapses released a mix of ATP and GABA, others released only GABA. Synapses releasing only ATP were not observed in this case. It is interesting that the same synapse can therefore release both an excitatory and an inhibitory transmitter at the same time.

The ionotropic P2X receptors appear to play a role in LTP. Inhibition of these receptors facilitates LTP in area CA1 of the hippocampus [205].

4.5.1.10 Adenosine

Adenosine is different from many other neurotransmitters in that it is not delivered to synapses by "adenosinergic neurons" but becomes available in the extracellular environment as a by-product of both neurotransmitter release and of neuron firing

[206]. With increased neuron activity, more ATP is consumed by neurons and glia cells in maintaining homeostasis. Consumption of ATP leads to adenosine. Hence there is local fluctuation in adenosine concentration reflecting local neuron activity. In addition, ATP is itself a neurotransmitter that can be released from synapses (see previous section). Enzymes in the extracellular environment can catalyse conversion of this ATP to adenosine [207].

Two important adenosine receptors in the brain are A_1 and A_{2A} [208]. Both are GPSM systems. The A_1 receptor is found in a large number of brain areas including the cortex, hippocampus, cerebellum and striatum [208, 209]. It acts on presynaptic neurons to inhibit the release of other neurotransmitters. This inhibition has been observed for acetylcholine, glutamate, noradrenaline, serotonin and GABA. The A_{2A} receptor is found mainly in the basal ganglia and has a role in regulating the release of other neurotransmitters such as dopamine, glutamate and GABA [206].

Adenosine acting on non-synaptic A_1 receptors decreases general neuron activity and contributes to maintaining homeostasis. Adenosine generated from synaptically released ATP acts upon synaptically located A_1 and A_{2A} receptors to regulate release of other neurotransmitters. Small ATP releases result in a prevailing inhibition by A_1 receptors, large ATP release leads to A_{2A} activation which increases release of other neurotransmitters [207].

4.5.1.11 Neuropeptides

Neuropeptides [157] include the dynorphins such as dynorphin A and B; oxytocin; vasopressin; enkephalin; nociceptin; cholecystokinin; somatostatin; and many others. Neuropeptides often coexist in the same neuron with small molecule neurotransmitters. Small molecule neurotransmitters are stored in small, clear synaptic vesicles (SSVs), while neuropeptides are stored in large granular vesicles (LGVs). However, LGVs can also contain small molecule neurotransmitters. If there are multiple neuropeptides in one neuron, they are generally stored together in the same vesicles, and any differential release must be achieved by adjustment of relative proportions at synthesis. Release of small molecule neurotransmitters is triggered by a high Ca^{++} concentration close to the synaptic membrane where SSVs are docked, while release of neuropeptides is triggered by a smaller rise in Ca^{++} concentration in the interior of the synapse where LGVs are located. The effect of small molecule neurotransmitters on their target occurs rapidly after the triggering input (a few milliseconds) while release of neuropeptides takes longer (a few hundred milliseconds) and can continue after the triggering input has ended.

Neuropeptides can have many different effects. Often the small molecule neurotransmitter opens an ion channel while the neuropeptide changes the conductance or dynamics of the channel in response to further inputs. In some cases a neuropeptide like oxytocin can act to generally excite its targets. In other cases a neuropeptide can inhibit [210] or excite [211] LTP.

4.5.1.12 Endocannabinoids

Endocannabinoids such as anandamide (AEA) are lipids that bind to G-protein receptors. In the brain, the common receptor is CB1. Unlike a regular neurotransmitter, endocannabinoids are synthesized in a neuron, for example in response to Ca^{++} influx following depolarisation, or activation of a glutamate receptor [212]. The endocannabinoid then diffuses back across the neuron membrane [213], possibly assisted by a transporter [214], to bind to receptors on the presynaptic neuron. Hence the signalling is from postsynaptic to presynaptic neuron [212].

Endocannabinoids can have various effects, such as short term reduction in release of neurotransmitter by the presynaptic neuron allowing a postsynaptic neuron to regulate its inputs, and long term changes to synaptic weights [215]. The active ingredient in marijuana binds to CB1 receptors [212], and chocolate contains small quantities of AEA [216].

4.5.1.13 Zinc

Zinc (Zn^{++}) is sometimes present in glutamate vesicles [217, 218]. In some areas (such as from the dentate gyrus on to CA3) all synapses contain zinc. In other areas (such as from CA3 on to CA1) only a proportion of synapses contain zinc [219]. Within one synapse, zinc does not appear to be present in all vesicles [220]. Interestingly, synapses containing zinc have lower postsynaptic concentrations of AMPA receptors, while concentration of NMDA receptors is unaffected [219]. Zinc selectively affects ionotropic glutamate receptors: NMDA receptors are inhibited; kainate receptors are unaffected; while AMPA receptors are sometimes slightly excited [221]. The presence of zinc inhibits responses to GABA [222]. Zinc can also enter the postsynaptic neuron through AMPA receptor or kainate channels [223].

Zinc is sometimes required for induction of LTP, but this effect depends on entry of zinc into presynaptic or postsynaptic neurons rather than extracellular interaction with receptors [224].

4.5.1.14 Nitric Oxide (NO)

NO is a different type of neurotransmitter. One major difference is that it can freely diffuse through the neuron membrane, and it can therefore be released within one neuron and have effects on other neurons that are close to the source where it is released. For example, NO released postsynaptically can diffuse to produce modifications in the presynaptic terminal.

Some neurons contain a protein called NO-synthase which catalyses the release of NO. This protein is unevenly distributed [225], only occurring in some neurons in some structures, and even within the same pyramidal neuron may be present in the apical dendrite but not in the basal dendrites [226].

NO-synthase can be activated by a chain of events initiated by Ca^{++} ions entering a neuron following glutamate binding to an NMDA receptor. The released NO can diffuse out of the neuron to the surrounding environment, and enter neurons in the immediate vicinity (including other parts of the same neuron). Specifically, NO release in a postsynaptic neuron can diffuse into the presynaptic neuron that triggered its release. This feedback mechanism is believed to contribute to changes in synaptic strength [225], acting on the amount of neurotransmitter released in response to an action potential by the presynaptic neuron.

In addition, the diffusion of NO through a region can modulate the overall activity across a population of neurons by acting via Ca^{++} influx through NMDA receptors on voltage gated K^+ channels [227].

4.5.1.15 Neurotrophins

Neurotrophins are protein signalling molecules that are very important in the development and maintenance of the nervous system. The four neurotrophins are nerve growth factor (NGF), brain-derived neurotrophic factor (BDNF), neurotrophin-3 (NT-3) and neurotrophin-4 (NT-4). They are not fully typical neurotransmitters, being generally secreted by neurons, although at least in some cases BDNF is packaged into vesicles in presynaptic terminals like a regular neurotransmitter [228].

Neurotrophins exert their effects by acting upon receptors embedded in the neuron membrane. These receptors include the TrkA, B, and C tyrosine kinases [121]. They initiate a wide range of structural changes in neurons such as changes to axons, to dendritic trees or to synapses [229].

The overall picture is that there are a few neurotransmitters that act rapidly (~ milliseconds) to excite (e.g. glutamate) or inhibit (e.g. GABA) their target neurons by binding to ionotropic receptors. A much larger population of neurotransmitters, but including the fast action neurotransmitters, act on a longer timescale (~ tens to hundreds of milliseconds or more) to modulate their target neurons in different ways by binding on metabotropic receptors. Neurotransmitters with ionotropic receptors can have metabotropic receptors on the same neurons at the same synapses. Even within the same neuron, receptors for different neurotransmitters can target the same ultimate ion channel via different G-protein paths, and one neurotransmitter can target different receptors that act on different ion channels [200, 230, 231]. Many neurotransmitters are generated by large numbers of widely distributed neurons, each of which targets a widely dispersed population of target neurons. A few neurotransmitters are generated almost entirely by a limited number of neurons in one small structure in the brain, but these neurons target neurons in almost all other brain structures, with different effects on the different structures because of the presence of different receptors. A small number of neurotransmitters can diffuse into the extracellular environment and exert effects over a wide region.

Hence the firing of a neuron, or the change to the weight of a synapse, is influenced by a large number of different neurotransmitter signals derived from a very large number of different neurons.

4.5.2 *Voltage Gated Ion Channels*

As described earlier, voltage gated ion channels open when the membrane potential is at some threshold level, different for each channel. Voltage gated channels then remain open for some time period characteristic of the channel, or until the membrane potential reaches some different threshold value. Channels can differ in the ions allowed to pass when open, and in their conductance. Like ligand-gated ion channels, voltage gated channels can also enter desensitized states in which they are more resistant than usual to opening. In addition, the operation of voltage gated channels can be affected by other chemical factors within the neuron. Any one type of ion channel is present in only some types of neuron, and in those neurons will typically have a very non-uniform distribution over soma, dendrites and axon.

4.5.2.1 Voltage Gated Na^+ Channels

There are about nine different voltage gated Na^+ channels, all fairly similar in molecular structure, and all critical for the production of action potentials. Five channel types appear in various different types of neuron [232]. These channels have three states: closed, open and deactivated. Channels are closed when the membrane is at resting potential. Typically a channel opens in a fraction of a millisecond if the membrane potential is depolarised by a few tens of millivolts from resting potential. When the membrane potential is strongly depolarised, open channels move to the deactivated state. In this state the channel cannot open. After a few milliseconds, a deactivated channel moves to the closed state. One significant functional difference between the different types of channel is the exact membrane potentials at which they open and close [232].

4.5.2.2 Voltage Gated Ca^{++} Channels

There is more different types of voltage gated Ca^{++} channels than Na^+ channels [233]. Ca^{++} channels can differ in the voltages at which they open, the lengths of time they remain open, and their conductances (i.e. the magnitude of the ion currents they can pass) while open. One major role of the Ca^{++} channels is in supporting action potentials within dendritic trees. Another major role is to allow entry of Ca^{++} ions acting as second messengers to trigger chemical changes in the neuron.

One major type of channel is the low voltage activated (LVA) channel. This type opens with small membrane depolarisations, but only remains open a short while and passes a low current. A second type is the high voltage activated, moderate conductance (HVA_m) channel such as the L-channel. These channels require depolarisations reaching −10 mV before opening, generally open very briefly, but may reopen several times during the same depolarisation. A small proportion of HVA_m channels open for longer periods of the order of a millisecond. A third type of channel also requires high depolarisations (reaching 0 mV) to open, but have higher conductances and are therefore called HVA_l channels.

The activity of voltage gated Ca^{++} channels can be regulated by additional factors. For example [234], the L-type channel is affected by protein kinase C (see below).

4.5.2.3 Voltage Gated K^+ Channels (K_v Channels)

K^+ channels form the largest and most varied group of voltage gated channels. Twelve major types, $K_v1 - K_v12$ have been identified on the basis of the genes that code them, with many subtypes [235], although not all of these have been observed in the brain. These channels differ in the membrane depolarisation that causes them to open, the membrane potential at which they close, the speeds with which they open and close in response to membrane potential changes, and the speed with which they deactivate if a depolarisation is sustained. They also differ in the amount of current (I) which one channel will pass when open. Many voltage gated K^+ channels are also modulated by neurotransmitters.

Four functional types of voltage gated K^+ channels have been distinguished in pyramidal neurons. These are the A-type channels, M-current channels, D-current channels and delayed rectifier channels. These different types of channel have been identified by the types of current they support through the neuron membrane: I_A; I_M; I_D; and I_K for the delayed rectifier type. Each of these functional types can be associated with one or more genetic types or subtypes [235]. For example, Kv1.4 and Kv4.2 channels show A-type behaviour [236], many Kv7 subtypes show M-current behaviour [237], and Kv1.2 channels show D-type behaviour [238].

I_A type channels activate rapidly (in <10 ms) at slightly depolarised membrane potentials (around −60 mV). These channels deactivate completely in 50–100 ms with sustained depolarisation. Because these channels activate below the threshold for generation of an action potential, they will tend to delay the onset of an action potential in response to depolarising input until they deactivate. During an action potential they will activate strongly and contribute to spike repolarisation. Provided the appropriate receptors are present, I_A channels can be modulated or blocked by a wide range of neurotransmitters, including serotonin, glutamate and GABA [239] and by dopamine [240].

I_D type channels activate almost as rapidly (in 20 ms) at membrane potentials close to the resting potential. It takes several seconds for these channels to deactivate. The long deactivation means that these channels can cause long delays in the timing of the first action potential in response to a depolarising input, and can reduce the initial firing rate. Hence they inhibit burst firing of action potentials [238]. I_D channels can be inhibited by serotonin [241], by dopamine [242], and by glutamate via a G-protein pathway [243].

I_K type channels activate at membrane potentials around −40 mV, more slowly than I_D channels, and also take several seconds to deactivate under sustained depolarisation. These channels only activate during action potentials, and contribute to determining the duration of spikes. It is probable that there are neurotransmitters that affect I_K channels [236].

I_M type channels activate slowly (~100 ms) at membrane potentials only slightly depolarised from the resting potential, and do not deactivate while depolarisation is present. These channels are activated sufficiently during an initial action potential to inhibit subsequent action potentials. I_M channels are therefore an important factor in adaptation, in which the rate of action potential output of a neuron under constant depolarisation gradually declines. I_M channels are modulated by several neurotransmitters. For example, they are inhibited by acetylcholine [244] and serotonin [245], and also by glutamate acting by a G-protein path [246]. Such inhibition can change the response of a neuron from strongly adapting to tonically firing [237]. I_M channels are excited by a neurotransmitter called somatostatin [247].

The distribution of K^+ channels is very non-uniform. Channels that occur in one neuron type do not occur in another, and there are differences in distribution between axon, soma and dendrites of the same neuron [248]. For example, I_A, I_D, I_K and I_M have been observed in hippocampal pyramidals [236], I_A and I_K in large neocortical pyramidal neurons [249].

4.5.2.4 Inward Rectifying K^+ Channels

These channels allow K^+ ions to pass into the neuron much more easily than out, and therefore push the membrane potential back towards resting potential. These channels are important for adjusting responses if the neuron is short of energy. One type of channel is the K_{ATP} that is inhibited by the presence of ATP. If ATP levels are low (for example, as a result of low oxygen), these channels open, making the neuron less likely to be activated. This protects the neuron from damage in hypoxic circumstances.

4.5.2.5 Voltage Gated I_h Hyperpolarisation Channel

The I_h channel is a slow activating, non-deactivating channel which allows Na^+ and K^+ ions to pass into the neuron at membrane potentials hyperpolarised from resting potential [250]. The current increases with the degree of hyperpolarisation, up to −110 mV, and slightly depolarises the membrane. These channels influence how excitatory synaptic inputs are integrated in the dendrites [251], and limit the backpropagation of action potentials into the dendritic tree [252].

The presence of a neurotransmitter such as noradrenaline or serotonin considerably increases the current passed by these channels. The effect of such increases can be to reduce the ability of the neuron to generate rhythmic bursts of action potentials, changing the firing pattern to production of relatively independent spikes [230].

4.5.2.6 Voltage Gated Cl^- Channels

A number of voltage gated Cl^- channels are known [112]. One such a channel that is widely although inhomogeneously distributed in the brain is ClC-2 [253]. This channel is activated slowly by hyperpolarisation, does not inactivate, and closes when the membrane returns to resting potential [254]. It allows Cl^- ion flow out of but not into the neuron. One role of this channel could be to help restore the Cl^- gradient following repetitive activation of $GABA_A$ receptors. Another role could be amplification of the effects of opening GABA gated Cl^- channels [253].

The voltage gated ion channel types are different for neurons in different anatomical structures, and the channels are distributed in very uneven fashion across the soma, dendrites and axon of an individual neuron [255]. Hence the way in which overall neuron response to all its excitatory and inhibitory synaptic inputs through ligand-gated ion channels can be delicately adjusted in a way that differs for different neurons.

4.5.3 *Ca^{++} Activated K^+ Channels*

Another important type of ion channel is the K^+ channel activated by the presence of Ca^{++} ions inside the neuron [126]. The Ca^{++} ions which trigger activation are generally the result of the opening of voltage gated Ca^{++} channels. There are a number of general types of Ca^{++} activated K^+ channels found in neurons. Three important types are called the BK family, the SK (or I_{AHP}) family, and the sI_{AHP} family. The sI_{AHP} channel type may be a variant of the SK family [256].

BK channels have high conductances, and open in response to either Ca^{++} concentration or membrane depolarisation in the vicinity of the channel. The two effects reinforce each other, and the effect of depolarisation alone is relatively weak. The channels open rapidly in response to Ca^{++} concentration, and close rapidly when the Ca^{++} concentration disappears [257].

SK channels have low conductances and their activation is not affected by membrane potential. However, at hyperpolarised membrane potentials the Ca^{++} ion tends to remain attached to the channel, which therefore remains open for much longer than a BK channel. SK channels open ~100 ms after the development of the Ca^{++} concentration, and close in a period between 50 and several hundred milliseconds [257].

sI_{AHP} channels have similar conductances to SK channels. These channels open slowly, but can remain open for several seconds. This long open period could be due to release of internal stores of Ca^{++} within the neuron, triggered by the initial Ca^{++} influx. Such a release-from-store mechanism requires that the released ions, and therefore the channels, are somehow isolated chemically from other parts of the neuron also sensitive to Ca^{++} (such as the other Ca^{++} gated K^+ channels).

These three types of channel are important for the membrane hyperpolarisations that follow an action potential. These hyperpolarisations can occur on three different timescales: fast, medium and slow. The fast hyperpolarisation is part of the recovery from an action potential. The medium and slow hyperpolarisations occur after the action potential is complete, and are therefore referred to as after-hyperpolarisations (AHP). The fast timescale hyperpolarisation is affected by BK channels, which therefore influence the duration of individual action potentials. The medium hyperpolarisation is determined by SK channels, which therefore influence the frequency with which action potentials can be generated. Because the SK channels generate a Ca^{++} current (I) that determines an AHP, they are also known as I_{AHP} channels. The long timescale hyperpolarisation is the process by which the production of action potentials gradually declines when a neuron is exposed to the same stimulation for a period of time, the process know as adaptation. The slower sI_{AHP} channels therefore create this adaptation.

The activity of Ca^{++} gated K^+ channels is influenced by a number of neurotransmitters. These neurotransmitters therefore modulate the action potential generation of their target neurons in various ways. Voltage gated K^+ channels are the main factor driving the hyperpolarisation following an action potential, but in neurons in which BK channels contribute to this fast hyperpolarisation, inhibition of the operation of those channels by a neurotransmitter will result in broadening of individual action potentials. Neurotransmitters that inhibit the operation of I_{AHP} channels will increase the frequency with which action potentials are generated in response to the same stimulus. Neurotransmitters that inhibit the operation of sI_{AHP} channels will extend the period of time over which a neuron will produce action potentials in response to a constant stimulus.

For example, acetylcholine, noradrenaline and serotonin all block sI_{AHP} channels via a different G-protein coupled receptor [200], and will all therefore reduce the adaptation of their targets. Noradrenaline can also inhibit I_{AHP} channels, and therefore increase the target neuron firing rate in response to the same activation [258].

Ca^{++} gated K^+ channels have effects in addition to the management of action potential shape, frequency and adaptation. For example, SK channels play a role in changes to synaptic strengths [259].

4.5.4 *Second Messengers*

For most ligand- and voltage-gated ion channels, passage of ions affects the electrical potential across the membrane but has little influence on other chemical reactions within the neuron. Even the pH level (acidity/alkalinity) in neurons, which could be affected by passage of ions and have a major influence on chemical processes, is actively maintained close to neutral [260]. Hence passage of Na^+, K^+, Cl^- or HCO_3^- ions changes the local membrane potential but has little direct chemical effect.

Chemical effects are often the result of activation of metabotropic (G-protein) receptors, where activation of the receptor by an external neurotransmitter often leads to a cascade of chemical processes within the neuron. An early step in such cascades is production of many molecules of a chemical which diffuses within the neuron and triggers further processes. Such chemicals are referred to as second messengers, since they follow activation by the external neurotransmitter regarded as the "first messenger". The principal second messengers are calcium ions (Ca^{++}), the two cyclic nucleotides cAMP and cGMP [261], and the lipids DAG and IP_3. Because many molecules of second messenger are released into the cytoplasm in response to a single neurotransmitter binding externally, there is significant amplification of that initial signal.

4.5.4.1 Ca^{++} as a Second Messenger

One of the second messengers, Ca^{++}, is also important in communication of membrane potential changes, such as the calcium action potentials within dendrites described earlier. Such Ca^{++} enters the neuron by some ligand- and voltage-gated ion channels. Second messenger Ca^{++} is released as a result of some G-protein receptor activations, either by opening ionotropic channels that pass Ca^{++} or by triggering release of Ca^{++} from internal stores [262]. However, second messenger Ca^{++} can also enter by ligand- and voltage-gated ion channels.

Because of its role as a second messenger, ion pumps keep free Ca^{++} levels in the cytoplasm at an extremely low level. Furthermore, Ca^{++} ions are segregated into compartments within the neuron to isolate different effects from each other [263, 264]. Considerable calcium is stored within a neuron, bonded to specific proteins so that it cannot act as a messenger, but from which it can be released as required [265]. In some cases it can be released from such stores by another second messenger like IP_3 [265], providing a second stage of amplification.

In the cytoplasm are protein molecules of a family called calmodulin. A calmodulin molecule binds to a Ca^{++} ion, and in this bonded form can activate a kinase called CaMKII [266]. This kinase can in turn activate other proteins that do not themselves bond to Ca^{++}. The many different proteins that can be bonded can initiate many different chemical reaction chains [267]. Hence calmodulin can be regarded as a transducer which detects a Ca^{++} signal and initiates downstream chemical processes.

4.5.4.2 Cyclic Nucleotide Second Messengers

The two cyclic nucleotide molecules cAMP and cGMP also act as second messengers [261]. Neurotransmitters acting on G_s type G-protein receptors activate a membrane bound molecule called adenylyl cyclase, and activated adenylyl cyclase synthesizes cAMP from ATP. Neurotransmitters acting on G_i type G-proteins inactivate adenylyl cyclase. At least one type of adenylyl cyclase can also be directly activated by a kinase (CaMKIV) that is itself activated by Ca^{++} ions and calmodulin [268] providing cross-talk from Ca^{++} to cAMP second messengers. The neurotransmitter NO activates cytoplasmic guanylate cyclase, which when activated synthesizes cGMP.

cAMP activates kinases like PKA, and activation of PKA in turn activates other molecules by phosphorylation, leading to (for example) gene transcription by activation of transcription factors. cAMP can also regulate a family of hyperpolarisation-activated ion channels [269] to directly affect internal neuron signalling via membrane potentials. cGMP activates kinases, principally PKG, which has a cystolic type PKGI and a membrane bound type PKGII [270] that are generally not present in the same cells. cGMP acts on some of the same gene transcription factors as cAMP and also the same family of hyperpolarisation ion channels activated by cAMP.

Phosphodiesterases (PDEs) degrade both cAMP and cGMP [271], different types degrading cAMP, cGMP or both. PDEs in turn can be inhibited or activated by cGMP or cAMP. For example, PDE2A and PDE3A both degrade cAMP. cGMP inhibits PDE3A, increasing cAMP levels, but activates PDE2A, decreasing cAMP levels. PDE5A degrades cGMP, and is itself activated by cGMP [272]. Some G-proteins can inhibit phosphodiesterases. The PDE1 series of phosphodiesterases, that degrade both cGMP and cAMP, are activated by Ca^{++}/calmodulin [272], representing a further interconnection between Ca^{++} and cyclic nucleotide second messenger pathways.

There are thus multiple different mechanisms that regulate the synthesis and degradation of cyclic nucleotides, and these mechanisms operate on timeframes varying from milliseconds to hours [273].

4.5.4.3 Lipid Second Messengers

Lipids form a very large class of biological molecules. These molecules are fat soluble and either hydrophobic or amphiphilic. The amphiphilic lipids have part of the molecule that is fat soluble and another part that is water soluble. Amphiphilic lipids play an important role in cell and vesicle membranes. Important examples of lipid second messengers are the pair of messengers (IP_3 and DAG) generated by G-proteins.

G-protein receptors such as the glutamate receptor mGluR5 activate the membrane bound enzyme phospholipase C (PLC). This enzyme catalyses the hydrolysis of membrane phosphoinositide (PI) into the two second messengers DAG and IP_3. DAG remains bound to the membrane but IP_3 diffuses into the cell interior.

DAG acts directly on the membrane bound kinase PKC, but IP_3 opens ion channels that release Ca^{++} ions from internal stores in the endoplasmic reticulum.

The membrane bound enzyme phospholipase D (PLD) hydrolyses a different membrane lipid (phosphatidylcholine) to produce phosphatidic acid, which can itself be converted into DAG. This route does not result in IP_3. PLD is activated by a complex combination of factors including activation by Ca^{++} and kinases such as PKC [274].

4.5.5 G-Proteins

A wide range of neurotransmitters act upon receptors to activate G-proteins, resulting in activation of second messengers within the target neuron [275, 276]. These second messengers amplify the effect of the neurotransmitter within the neuron. Four main classes of G-proteins can be identified: G_s; G_q; G_i; and G_{12} [159]. All G-proteins have three subunits, α, β and γ. Binding of the appropriate neurotransmitter triggers dissociation of the G-protein into α and β/γ subcomponents. Both subcomponents can exert effects on second messengers, and also on a wide range of other molecules such as kinases [277].

Each class of G-protein activates or inhibits different second messengers [278]. These second messengers can initiate a wide range of chemical processes in the neuron [279]. One such process opens Na^+ or Cl^- channels, changing the membrane potential on a somewhat longer time scale than neurotransmitters acting directly on the ion channels. Another process opens Ca^{++} channels, which can trigger secondary effects including structural changes to the cell, because Ca^{++} ions are themselves significant second messengers. Yet other processes use sequences of kinase molecules, where a kinase is an enzyme that can add a phosphate group to another molecule (which could be another kinase). Kinase activation sequences can result in expression of genetic information in the nucleus.

Negative feedback can limit the activity of G-protein processes. For example, kinase molecules like PKA, PKC and GRKs can also act to desensitize the receptors themselves [280].

4.5.5.1 G_s Proteins

These proteins activate adenylyl cyclase, creating large numbers of molecules of cAMP in the cytoplasm. There are a number of different forms of adenylyl cyclase, at least six in the brain, with somewhat different properties, but all are excited by G_s proteins [281]. A cAMP molecule has at least three types of target [279]. Firstly it can activate a kinase protein called PKA. Secondly, it can activate a GTP-exchange protein called EPAC which can in turn activate kinases. Thirdly, it can open a type of ion channel called a cyclic-nucleotide-gated ion channel [282]. These ion channels are cation channels with different conductances for Na^+, K^+ and Ca^{++}. The conductances sometimes also depend on the local membrane potential.

The complexity of the network of chemical reactions which can be initiated or influenced is considerable [283]. Firstly, different neurotransmitters have G-protein receptors that target G_s proteins. Secondly, there are different forms of adenylyl cyclase. Thirdly, PKA can activate a wide range of different targets, but can be anchored to various locations by specific kinase anchoring proteins (AKAPs), which make its action more specific. Furthermore, there is considerable cross connection with other second messenger pathways such as those initiated by Ca^{++}. Many of the end points of cAMP pathways are gene transcriptions [284] which result in different structural changes to the neuron.

Examples of neurotransmitters that target G_s proteins are dopamine (D1 receptor) and serotonin (5HT4, 5HT6 and 5HT7 receptors).

4.5.5.2 G_i Proteins

These proteins inhibit adenylyl cyclase, and can also activate G-protein-coupled inwardly rectifying potassium (GIRK) channels. In addition, they can activate PI3 kinases [285], initiating the Ack (PKB) kinase cascade (see below). G_i proteins can also inhibit voltage gated calcium channels [286].

Examples of receptors that target G_i proteins are dopamine D2, $GABA_B$, glutamate mGluR2, 3, 4, 6, 7 and 8, and some adrenoceptors. Note that glutamate has both G_s and G_i receptors, and can therefore have opposite effects on a target neuron depending on the type of receptor that is present.

4.5.5.3 G_q Proteins

These proteins activate molecules of phospholipase Cβ (PLC) found in the neuron membrane. PLC hydrolyses a phospholipid component of the membrane (PIP2) into two molecules: diacylglycerol (DAG) and IP_3. DAG remains in the membrane where it activates a kinase called PKC; IP_3 is released into the cytoplasm where it releases Ca^{++} ions from stores in the endoplasmic reticulum. Such releases can increase the excitability of the target neuron. Some types of PKC require the simultaneous presence of Ca^{++} for activation. PLC can also close voltage gated potassium channels [287], resulting in increased neuron excitability. Activation of PKC can lead to gene expression [288] and is needed for neuronal plasticity and nerve growth [289]. Examples of receptors that target G_q proteins are glutamate mGluR1 and mGluR5, acetylcholine M1 and M3 receptors, the neuropeptide oxytocin and some adrenoceptors.

4.5.5.4 G_{12} Protein Family

There is a group of proteins called GTPases that are associated with cell growth and cell adhesion [290]. The activity of three members of this GTPase group (RhoA, Rac, and Cdc42) is influenced by the two known members (G_{12} and G_{13}) of this G-protein family. In the brain, the G_{12} serotonin receptor 5HT7 promoting increases in neurite number and length by action on the GTPase group [291], while the G_{13} serotonin receptor 5HT4 promotes decrease, acting on the same group [292]. Both receptors can also trigger gene transcription, which occurs by a path that is independent of the cAMP/PKA path activated by some other G-proteins. The serotonin 5HT7 receptor is also known to encourage REM sleep while having no effect on other sleep parameters [293].

4.5.6 *Kinases, Phosphatases and Phosphorylation*

Phosphorylation is the addition of a PO_4^{3-} group (a phosphate group) to a protein or other molecule. Most proteins can exist in unphosphorylated and phosphorylated forms, with one form being active, the other form inactivated. Depending on the protein, the phosphorylated version can be the active or the inactive form. A kinase is an enzyme which catalyses the addition of phosphate groups to specific other proteins. A phosphatase is a protein which catalyses the removal of phosphate groups from specific other proteins. Hence kinases and phosphatases can activate or deactivate their targets, and phosphorylation and dephosphorylation are very important ways in which the activity of proteins can be managed.

In the chain of amino acid residues that make up a protein, phosphorylation generally occurs at a serine, threonine or tyrosine residue. A protein may have multiple sites at which it can be phosphorylated and dephosphorylated, and there may be a mixture of serine/threonine and tyrosine sites. In some cases an enzyme may be partially activated by phosphorylation at one site and more activated if more sites are phosphorylated [294]. In other cases, phosphorylation at more than one site must occur before any activation [295]. An enzyme can even be activated by phosphorylation at one site but inactivated by phosphorylation at another site.

The serine protein kinase (SPK) class of kinases phosphorylate both serine and threonine residues. The tyrosine protein kinase (TPK) class phosphorylates tyrosine residues (tyrosine kinases), and kinases in the dual-specific class phosphorylate all three [296]. Similarly, the PTP class of phosphatases dephosphorylate tyrosine residues, and the PPM and PPP classes dephosphorylate serine and threonine residues, while a subclass of PTP, the dual specificity phosphatases, dephosphorylate all three [297].

The target protein of a kinase can itself be another kinase. There are numerous examples of chains of kinases (kinase cascades), with each member of the chain able to phosphorylate the next member. The end point of such a cascade can be a result like DNA transcription. Such cascades allow external influences at each step, and therefore permit management of the end point by complex combinations of influences. For example, there may be phosphatases and additional kinases which if activated can affect a cascade at various different points. Examples of such cascades will be described in the next chapter.

4.6 Chemicals and Information Processing

Neurons perform information processing on their inputs. This information processing is of two general types. One type is to determine whether to produce an output action potential. The other type is to change the algorithm by which future action potentials will be determined from inputs. The external inputs to a neuron are transient concentrations of neurotransmitters at different points on its membrane. These external inputs are detected by a range of different receptors. Receptors trigger combinations and sequences of chemical processes that implement the neuron information processes.

The very wide range of molecules make possible for chemical processes to implement very sophisticated information processing. In the next chapter, we will describe how the molecules interact to perform these chemical processes, and how these chemical processes implement the required information processes.

Chapter 5
Intracellular Message Chains

A neuron receives large numbers of input stimuli from other neurons and generates an output action potential in response to those inputs. In order to preserve past learning the integration algorithm that determines outputs from inputs must be relatively stable, but to implement short term or longer term learning it must be changeable on various timescales. Hence the algorithm and changes to the algorithm must be tightly controlled. There are two general types of change that can be made. One is change to individual synaptic weights, the other is change to the algorithm that determines the way in which the individual weights of currently active synapses are integrated.

The integration algorithm and changes to that algorithm must be implemented by chemical processes within the neuron. The available chemical processes are specified by genetic information in the form of DNA located in the nucleus of the neuron. This DNA codes for proteins, which are long chains of amino acid residues. The DNA code for all the proteins in the genome is present in all cells in the body, and a neuron therefore has potential access to this complete coding. For any one neuron, some of this coding is probably made permanently inaccessible by chemical processes occurring when cells differentiate into different organs and substructures, but a significant proportion is available.

Some proteins have functional or structural roles. For example, ion channels, G-protein receptors and some neurotransmitters are proteins, and the skeleton of the cell that defines its shape is made up of proteins. Other proteins are catalysts for specific chemical processes. Most chemical processes in a neuron will not occur unless catalysed by a protein, called an enzyme. Processes controlling the synthesis, distribution and degradation of functional and structural proteins, of small molecule neurotransmitters and other neurochemicals require enzymes. Furthermore, processes controlling the synthesis, distribution and degradation of enzymes also require enzymes.

A change to the behaviour of a neuron requires changes to some proteins or synthesis of new proteins. There must therefore be processes managing protein changes and access to genetic information. The nature of a change may be influenced

L.A. Coward, *Towards a Theoretical Neuroscience: from Cell Chemistry to Cognition*, Springer Series in Cognitive and Neural Systems 8, DOI 10.1007/978-94-007-7107-9_5,

by inputs to the neuron from many sources, and all these sources must be integrated to determine the precise changes to be implemented. Interacting cascades of chemical processes therefore connect external stimuli with protein changes and DNA expression.

In this chapter we will first discuss some of the processes by which genetic information is expressed. Then some of the multiple routes that link external stimuli to protein changes and control of gene expression will be described. Finally we will discuss how neuron integration algorithms are implemented, and the chemical pathways by which they are changed.

5.1 Gene Expression

Genetic information is recorded in DNA molecules in the cell nucleus. DNA molecules have a long backbone made up of deoxyribose sugar molecules linked together in a chain. Attached to each sugar molecule is a nucleotide molecule. There are four different nucleotide molecules that occur in DNA: guanine (G), cytosine (C), adenine (A) and thymine (T). The DNA is organized into sections, called genes. The sequence in which the four nucleotides occur in a gene includes the coding for the molecular organisation of a protein used by the organism, and can be used to direct the synthesis of that protein at the appropriate time. These proteins can be structural, or enzymes catalysing chemical reactions important to the organism. Other parts of a gene include sequences of nucleotides used to trigger access to the gene information. DNA is made up of two strands coiled around each other. There is a C nucleotide in one strand opposite every G nucleotide in the other strand, and an A nucleotide opposite every T nucleotide. This complementary structure makes duplication of DNA for cell division possible: the two strands separate and a complementary strand synthesized for each, with no loss of information.

A gene is a length of DNA in which the sequence of nucleotides ultimately codes for a biologically relevant protein. The names of the gene and of the corresponding protein are generally the same, and to avoid confusion, the gene is italicized, for example the gene *cFos* and the coded protein cFos.

There are a number of stages to the expression of a gene, and there can be regulation of the expression at any stage. First, RNA molecules are transcribed from the DNA template [298]. This RNA is a single strand molecule with a backbone of ribose sugar molecules and a nucleotide attached to each sugar. In RNA the same DNA sequence information is maintained, but a uracil (U) nucleotide replaces all the T nucleotides in the DNA original. Hence the RNA carries equivalent information to that coded in the DNA gene. RNA molecules as transcribed are then spliced together in various ways to form the messenger RNA (mRNA) which is the basis for protein synthesis [299]. Protein synthesis occurs at locations remote from the DNA, and mRNA is transported to those locations [300]. The mRNA is translated to protein [301] and the protein spliced [302]. Finally, mRNA is at some point broken down [303]. The genetic expression process can be regulated at each stage. However,

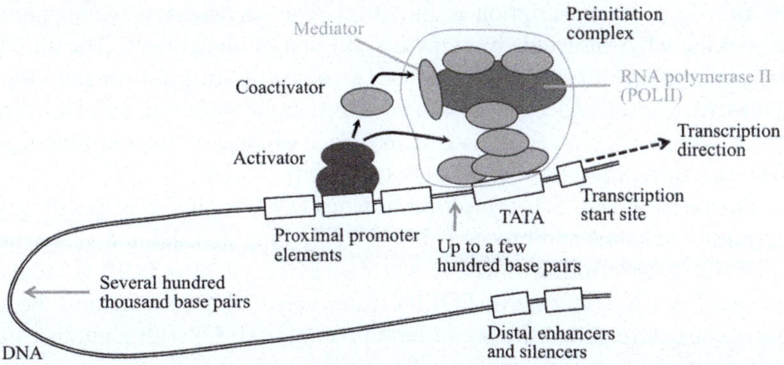

Fig. 5.1 Regulation of transcription. The transcription of mRNA from a region of DNA that codes a protein is extensively regulated. Transcription occurs in one direction, and the stretch of DNA upstream from the protein coding region is illustrated. In a common regulation scheme, there is a transcription start site at the beginning of the protein coding segment. Just upstream there is a DNA segment called the TATA box. Binding of a complete and activated preinitiation complex (*PIC*) to the TATA box initiates transcription. This PIC includes the enzyme RNS polymerase II (POLII) which catalyses transcription. The assembly and activation of the PIC is influenced by the binding of appropriate proteins, (called activators, enhancers and silencers) to sections of DNA further upstream. Coactivators also regulate the PIC under the control of activators. Many kinase cascades can influence activators, coactivators, enhancers and silencers

the regulation that occurs around the initiation of transcription from DNA is the best understood and most of the discussion here will be limited to this stage. The framework within which the regulation of transcription occurs is illustrated in Fig. 5.1.

5.1.1 Transcription of mRNA

DNA is generally encased in proteins called histones, and the complex tightly wrapped to form a structure called chromatin. This packaging prevents access of the chemical transcription machinery that creates mRNA. In different types of cell, different groups of genes are made more accessible by loosening of the chromatin packaging. Genes involved in the transcription process are accessible in a wide range of cell types, but other genes are only accessible in some types of cell [304]. For example, AMPA receptor mRNA, like many neurotransmitter receptors, is found only in neurons [305]. Thus neurons will have access to a limited group of genes appropriate to their functions. Within this limited group, an appropriate subset must be transcribed at each point in time.

Neuron activity can initiate the transcription of hundreds of different genes [306]. These genes can have a wide range of effects on brain architecture, neuron structure, and synapse behaviour [307]. For example, these genes can cause the creation or destruction of synapses, changes to the size of synapses, and changes to the numbers of receptors in the postsynaptic density.

The first step in transcription is initiation. The second step is elongation: the creation of the RNA molecule by stepwise addition of nucleotides. The third step is correct termination to release the new RNA molecule. In a fourth step, the RNA molecules as transcribed can be spliced together in different ways to form the mRNA on which protein synthesis is based. Each of these steps is closely regulated to reflect the current requirements of the cell [308].

As illustrated in Fig. 5.1, transcription from DNA begins at a specific site (the transcription start site) and proceeds in a specific direction defined as downstream from that site. Transcription of RNA for protein synthesis from DNA is catalysed by the enzyme RNA polymerase II (POLII). To be active, this enzyme must be assembled into a structure called the preinitiation complex (PIC) with a number of other molecular components, and positioned correctly relative to the initiation site. This positioning is achieved by binding of the PIC to the core promoter region, which is a nucleotide sequence in the DNA upstream from the initiation site, often including the TATA box sequence. The degree of transcription activity is also determined by the binding of activators to various nucleotide sequences, called proximal promoter elements, located a short way upstream from the core promoter region. Activators are proteins which bind to specific promoter element sequences. In addition there can be DNA sequences much further upstream which can also promote or block transcription if bonded by an appropriate activator or repressor. These distal enhancers and silencers are probably brought physically close to the core promoter region by DNA molecule folding.

Individual activators can increase the transcription rate, and multiple activators can increase the rate by more than the sum of the individual increases [309]. Even if an activator molecule is already bound to the enhancer site, the presence of additional activator molecules can increase transcription rate [310]. Activator proteins are themselves coded in genes. An activator protein as synthesized directly from mRNA may require modification to make it active, and in some cases inactive activator molecules are always present in the vicinity of or bonded to the DNA, just requiring modification to trigger activation of a target gene. Genes of this type are called immediate early genes (IEGs) [311]. IEGs can code enzymes, but often code for other transcription activators, and there can then be a second phase of genetic expression of delayed response genes.

Coactivators (corepressors) are proteins that increase (decrease) transcription without directly binding to DNA. Coactivators enhance transcription by binding to activators, enhancing the accessibility of the DNA and/or the activity of POLII [312]. The same activators, coactivators or corepressors can influence expression of many different genes. As illustrated in Fig. 5.2, one gene can have multiple promoter elements, and the rate of transcription is influenced by binding of each transcription factor to its element, binding of various activators to a transcription factor, and the activation state of each transcription factor and activator, plus the activity of repressors. The selection of the gene to be transcribed is made specific by the need for combinations of transcription factors to achieve strong transcription [313].

Neuronal activity can influence the activation states of transcription factors, coactivators and corepressors by many mechanisms [314]. Such influences can

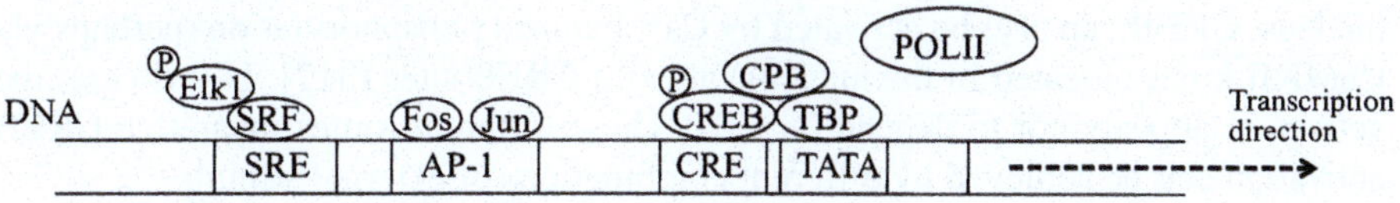

Fig. 5.2 Multiple transcription factors influence transcription activation. In this simplified diagram, the gene contains the SRE, AP-1, CRE and TATA sequences. The transcription factor SRF binds to the SRE sequence if activated by the phosphorylated protein product of the IEG *Elk1*. Elk-1 can also bind directly to SRE. The protein products of the IEGs *Fos* and *Jun* bind to the AP-1 sequence. CREB is generally bound to the CRE sequence, but must be activated by phosphorylation and then by binding to the coactivator CPB. When TBP is activated by CREB and CPB it binds to the TATA, and then recruits the rest of the preinitiation complex including POLII. A range of different chemical routes link external stimuli to activation of different transcription factors. Sometimes multiple factors must all be activated for gene transcription, and sometimes additional factors increase transcription. *CRE* cAMP response element, *CREB* CRE binding protein, *IEG* immediate early gene, *P* phosphate, *POLII* RNA polymerase II, *SRE* serum response element, *SRF* serum response factor, *TATA* is named after the thymine-adenine-thymine-adenine sequence at its beginning, *TBP* TATA binding protein

include changes to the activation states of coactivators and corepressors of prebound transcription factors, moving transcription factors into the nucleus where they can bind to their target genes, and acting on the histone packaging to change the accessibility of promoter element binding sites. These mechanisms can be triggered by neurotransmitter receptors and voltage gated ion channels at synapses and elsewhere in the neuron membrane, activating a range of chemical pathways that regulate the initiation of transcription [314]. Transcription of the same gene can be initiated by multiple such pathways [308].

As shown in Fig. 5.2, some promoter nucleotide sequences are cAMP response element (CRE), serum response element (SRE), AP-1 and the TATA box sequence mentioned earlier. There are indirect interactions between the corresponding transcription factors. For example, protein products of the IEGs *cFos* and *Jun* bind to the AP-1 sequence, but CRE is present as a promoter sequence in these two genes. CRE is also present in genes coding for some neurotransmitters.

The family of CRE binding (CREB) proteins are activators that can bind to the CRE sequence if themselves activated by phosphorylation [313]. Such phosphorylation can be the end point of kinase cascades, including cascades initiated by activation of the second messenger cAMP. The activator CREB is generally bound to CRE in accessible genes with that sequence, but in an inactive form. There are vast numbers of known routes by which CREB can be activated [315]. When phosphorylated, CREB binds to a coactivator CREB binding protein (CBP), and CBP further reduces the histone packing density and also contributes to activation of POLII. Another coactivator, CREB regulated transcriptional coactivator (CRTC1), also binds to CREB and promotes transcription, but does not appear to require CREB phosphorylation. In some neurons CRTC1 can be confined to synapses, but in response to calcineurin activation translocates to the nucleus where it can bind to CREB [316]. In this case, cAMP regulates the length of time CRTC1 persists in the nucleus. CREB can also be activated by Ca^{++} induced phosphorylation (perhaps via CaMKII kinase located in the nucleus) at extra CREB sites [317]. In this case, the activation appears not to depend on CBP. These different routes mean that CREB activation can be achieved by different combinations of external stimuli.

Another promoter sequence important in neurons is serum response element (SRE) which is bonded by two transcription factor proteins: SRF and ELK-1. Like CRE, SRE also occurs in the promoter region of IEGs like *Jun* and *cFos*. SRF can be activated by Ca^{++}-calmodulin and MAP cascades [318]. The transcription factor ELK-1 can be activated by MAP cascades, with crosstalk with the PI_3 cascade [319]. A number of other transcription factors can substitute for ELK-1 [320]. For maximum transcriptional response of *cFos*, both CRE and SRE must be bonded, but bonding of just SRE produces some response [321].

Other promoter sequences important in neurons are the calcium-response elements CaRE1 and CaRE2 [322]. CaRE1 is bonded by the calcium-response factor (CaRF) and other activators [323], and has been found in many genes active in neurons including the gene for BDNF [324]. CaRE1 and CaRE2 mediated transcriptions are activated by Ca^{++} by currently unknown routes that are independent of cAMP/PKA.

Although transcription initiation has been the main focus of this discussion, the elongation phase can also be regulated, in some cases by the same activators as initiation, in other cases by different activators. Absence of the appropriate elongation regulators results in a high production rate of unfunctional, truncated mRNA [325].

Following transcription of mRNA, the molecule must be transported to the location where protein synthesis can occur. This transportation is another closely regulated phase. For example, mRNA can be transported to ribosomes located in the dendrites of neurons, where translation of protein is under local control [326]. There can also be mRNA transport along the axon for presynaptic translation of protein [327]. This local control means that the concentration of some enzymes is independently managed at different synapses. mRNA present close to synapses includes mRNA coding for CaMKII [328], AMPA receptors [329], the cytoskeletal protein ARC [330]. This local management includes control of the transport of specific mRNA to specific synapses [330].

5.1.2 Translation of Protein

Proteins are sequences of amino acids, and the sequence for a specific protein is coded by the sequence of nucleic acids in the corresponding mRNA. Each possible sequence of three nucleic acids (a triplet) corresponds either with an amino acid or with a "punctuation" mark such as a start or end indicator. These triplets in the mRNA are called codons. Translation from mRNA to protein uses transport RNA (tRNA) units made up of an amino acid linked to a triplet of nucleic acids. The triplet in each type of tRNA will bind only to a codon corresponding with the amino acid in the mRNA. The process of protein synthesis moves along the mRNA. At each codon the appropriate tRNA links to the mRNA, its amino acid is added to the growing protein, and the tRNA remnant is discarded. Protein synthesis can be divided into three stages and can be regulated during each stage: initiation [331], elongation [332], and termination [333].

The synthesis process is performed by ribosomes. A ribosome has two subunits, a small subunit that binds to mRNA and a large subunit that generates protein. Initiation of translation occurs in three steps. First, a tRNA unit corresponding with the start codon on mRNA binds to the small ribosome subunit. Second, the small subunit binds to the start codon of the mRNA. Thirdly, the large ribosomal subunit binds to the small subunit and translation begins.

In human beings there are more than a dozen initiation factors (called eukaryotic initiation factors or eIFs). These eIFs bind to the ribosome, and their ability to initiate translation can be regulated. The assembly of the initiation factor eIF4F (made up of eIF4A, eIF4E and eIF4G) on the appropriate mRNA location is a key step in translation, and this assembly is regulated by both the kinase mTOR and the kinase MNK, which phosphorylate different targets [334]. In some cases, two kinases have additive effects on protein synthesis [335].

In the elongation stage there are at least three elongation factors (eEFs) [332]. These eEFs are regulated by a number of kinase cascades including PKC, Ca^{++}-calmodulin, PKA and MAPK cascades [328].

The termination phase that occurs when a stop-codon is reached in the mRNA involves two release factors (eRFs) [333], but whether these release factors are regulated by kinase cascades is not known.

5.2 Kinase Cascades

As described in Chap. 4, phosphorylation adds a phosphate (PO_4) group from an ATP molecule to the target molecule, and is catalysed by enzymes called kinases. Dephosphorylation (or phosphorolysis) is the removal of a phosphate group by hydrolysis, and is catalysed by enzymes called phosphatases. Kinase and phosphatase enzymes can themselves be the target molecules for other kinases and phosphatases. In many cases the phosphorylated version is the activated form in which the enzyme is able to perform its catalysis function, and dephosphorylation inactivates the protein, or less often vice versa. Phosphorylation is therefore a critical mechanism for controlling which chemical processes will take place.

In some cases, one kinase can phosphorylate another kinase and so on until an ultimate target such as the CREB transcription factor is phosphorylated. Long sequences of this type are called kinase cascades. Such cascades can be influenced at each point by other chemical processes including the action of phosphatases, allowing very detailed management of access to the end point.

There are many kinase cascades that can link an external stimulus such as arrival of a neurotransmitter to initiation of a change in the target neuron by gene transcription. We will first separately describe some of the important cascades that have been identified, then discuss some of the interactions (or crosstalk) between these cascades.

5.2.1 *Mitogen Activated Protein Kinase Cascades (MAPK Cascades)*

A mitogen is an extracellular chemical that encourages cell division in a target cell, and mitogen activated protein kinases (MAPKs) were first identified in the context of the regulation of cell proliferation, differentiation and triggering cell death. However, some MAPKs are present in mature neurons and have been demonstrated to play a role in synaptic strength changes. A MAPK cascade, when activated, catalyses some result such as activation of a genetic transcription factor by phosphorylation. The typical MAPK cascade [336] is made up of three kinases. The MAPK catalyses the target of the cascade, the MAP2K catalyses the activation by phosphorylation of the MAPK, and the MAP3K catalyses the activation by phosphorylation of the MAP2K. The MAP3K is activated by a (typically complex) combination of signals derived from extracellular stimuli. A number of MAPK cascades of this type exist [337].

In neurons, one MAPK cascade in particular, the ERK cascade, plays an important role in synaptic weight changes [338]. Examples of key aspects of this cascade are illustrated in Fig. 5.3. The reference way in which this cascade is initiated is by binding of a BDNF neurotrophin to a TrkB receptor. This neurotrophin binding triggers the bonding of phosphate groups to the receptor. The activated TrkB receptor in turn activates the Ras GTP-binding-switch-protein via Grb2 and SOS docking proteins. In this Ras activation process, the activation of SOS displaces a GDP molecule bonded to Ras, replacing it with a GTP molecule. The activated Ras then activates the Raf-1 kinase, which is the first kinase (the MAP3K) in this cascade. The Raf-1 kinase phosphorylates (and thereby activates) the MEK kinase, and the MEK kinase phosphorylates the ERK-1 and/or ERK-2 kinases. The ERK kinases can have a number of types of downstream effects, generally via a number of further intermediate stages. One effect can be changes to the effectiveness of existing ion channels, such as reduced conductance of AMPA receptor channels [339]. Another can be activation of protein synthesis by local mRNA [300]. A third type of effect can be activation of transcription factors like CREB and Elk-1, leading to gene transcription [340]. The Ras-MEK-ERK pathway can also inhibit gene transcription by activation of transcription repressors [341].

A number of factors can modulate this ERK cascade. One factor is the existence of anchor proteins. These proteins bring together the participants in a reaction, making the reaction much more likely. For example, in Fig. 5.3, the MP1 scaffolding protein bonds to both MEK and ERK-1 [342]. Activated Raf-1 activates MEK, then if the MEK molecule is held in proximity with ERK-1, ERK-1 will be activated without the need for MEK diffusion to the ERK-1 location. A second factor is the

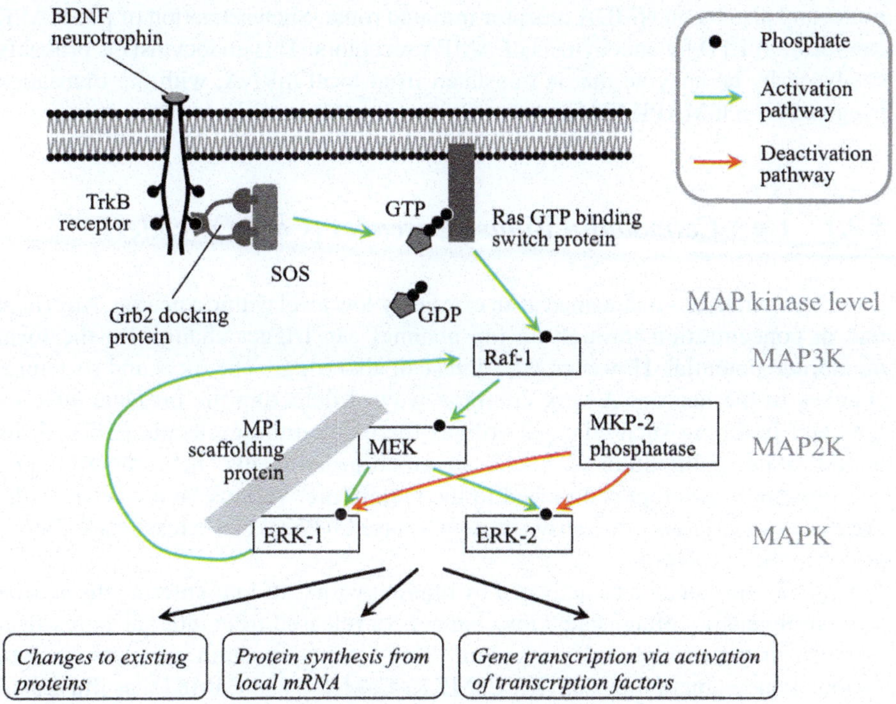

Fig. 5.3 MAP kinase cascades. In this type of cascade, a MAP kinase (MAPK) is activated by a MAP kinase kinase (MAP2K), and a MAP2K is activated by a MAP kinase kinase kinase (MAP3K). A MAP cascade can be initiated in a number of ways, such as the illustrated binding of extracellular BDNF to a TrkB receptor. MAPKs can act, often via further stages, to modify existing proteins, to trigger translation of new protein from locally available mRNA, or to trigger transcription of new mRNA. ERK-1 and ERK-2 are also known as $p44^{mapk}$ and $p42^{mapk}$

existence of phosphatases that remove the phosphate group from activated kinases, thus deactivating them. Specific phosphatases block the activity of specific kinases [343]. For example, as illustrated in Fig. 5.3, the MKP-2 phosphatase at certain concentrations inactivates both ERK-1 and ERK-2 [344]. A third factor is feedback within the cascade. For example, as also illustrated in Fig. 5.3, Raf-1 is activated by Ras, but if ERK-1 molecules are already active they can lock Raf-1 into its active state for a longer period [345].

In addition to these modulation factors, the ERK cascade can be initiated by other extracellular neurotransmitters, and can interact within the neuron with other kinase cascades that are themselves initiated by other neurotransmitters. These crosstalk considerations will be described after the discussion of other types of kinase cascades.

There are other MAPK cascades that also play a role in neurons. For example, the p38MAPK kinase is best understood as managing responses to cellular stress, but is also activated by a glutamate → G-protein → Rap-1 → MKK-3/6 → p38MAPK route, and also by an NMDA receptor initiated route. Such activation of p38MAPK can result in LTD by endocytosis of AMPA receptors. This endocytosis is probably catalysed by an enzyme that is translated from local mRNA, with the translation triggered by p38MAPK [346].

5.2.2 *Ca^{++}-Calmodulin Kinase Cascades (CaM Cascades)*

Ca^{++} concentration is maintained at an extremely low level within a neuron. Any local rise in concentration through an ion channel can trigger changes to the local membrane potential. However, Ca^{++} ions can also trigger chemical and structural changes to the neuron. A very common way of detecting the presence of Ca^{++} ions and initiating such changes utilizes the calmodulin molecule [347]. In its resting state, calmodulin is bound to other proteins like neuromodulin and neurogranin in a complex attached to the membrane. Various factors such as the presence of Ca^{++} ions or of active protein kinase C (PKC) can release the calmodulin into the cytoplasm.

Calmodulin can then be activated by binding with Ca^{++} ions entering the neuron through ligand or voltage gated ion channels or released from internal stores such as those in the endoplasmic reticulum. This Ca^{++}-calmodulin can then activate various kinases including CaMKI, CaMKK, CaMKII, and CaMKIV as illustrated in Fig. 5.4.

Once CaMKII is activated by Ca^{++}-calmodulin, it can autoactivate itself to prolong its own activation, although at a somewhat lower level than when directly bonded to Ca^{++}-calmodulin [348]. Activated CaMKII targets a large number of molecular substrates including AMPA and GABA receptors [349]. The kinase can also target transcription factors leading to structural changes in the neuron [350]. In addition, binding of activated CaMKII to a number of other molecules results in maintenance of its activity independent of Ca^{++}-calmodulin binding. One example is the NMDA receptor, where binding between the receptor and CaMKII influences synaptic plasticity [351].

CaMKK, CaMKI and CaMKIV form an interdependent group. All are activated by Ca^{++}-calmodulin, but although CaMKI and CaMKIV are potently activated by Ca^{++}-calmodulin, they are further activated by activated CaMKK [352]. The activation of these kinases can both modify existing proteins and activate transcription factors leading to new protein synthesis. CaMKK and CaMKIV are present in a neuron both in the nucleus and in the cytoplasm, while CaMKI is only present in the cytoplasm. Hence CaMKI does not appear to target transcription factors.

As examples of activation of existing proteins, CaMKIV is one of many factors that affect the activity of Oncoprotein 18 (Op18) [353]. Op18 influences neuron shape by regulating the equilibrium between microtubules and the free tubilin fibres from which they are constructed. Activation of two different variants of CaMKI (CaMKIα and CaMKIγ) can drive growth (respectively) of axons and dendrites [354].

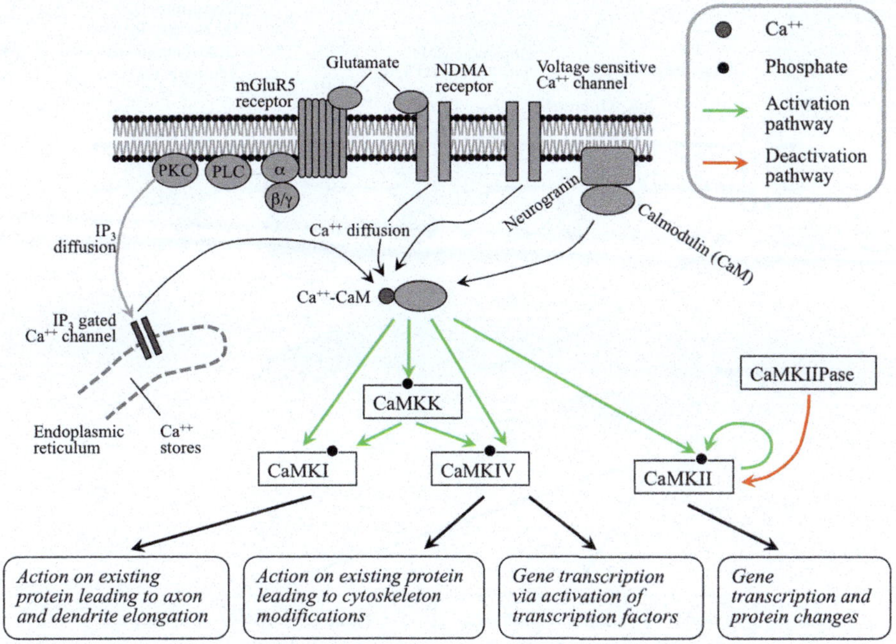

Fig. 5.4 Calcium-calmodulin cascades. Calmodulin is a molecule that is activated by binding with Ca^{++} ions which can enter the neuron through ligand gated or voltage gated ion channels or be released from internal stores in the endoplasmic reticulum by action of a second messenger like IP_3. The resulting Ca-Calmodulin (Ca-CaM) complex can activate a number of different kinases, which in turn can lead to downstream effects

CaMKIV can also target transcription factors such as CREB [352]. CaMKIV alone can activate CREB, but if both CaMKK and CaMKIV are present in the nucleus, transcription is enhanced [355] by a factor of about fourteen.

An important issue is how calcium signals can control so many different cellular functions which need to be invoked at different times. There are a number of factors that contribute to this capability for separate control [356]. *Firstly*, there are many different receptors in the external membrane, and many different isoforms of intracellular proteins. Hence even the same neurotransmitter can produce different Ca^{++} responses depending on the exact receptors and internal isoforms that are present. For example, glutamate targetting the mGluR1 receptor produces a single Ca^{++} release burst from internal stores, while GluR5 produces an oscillatory release from the same stores. *Secondly*, there are different pumps that remove Ca^{++} at different speeds, so duration will depend upon which pumps are present in the vicinity of the release. *Thirdly*, Ca^{++} release may be confined physically to a particular region, for example as in the case of dendritic spines. *Fourthly*, the presence of appropriate anchor proteins bringing together specific reaction participants in a location with access to the Ca^{++} release, for example the anchor proteins that tie

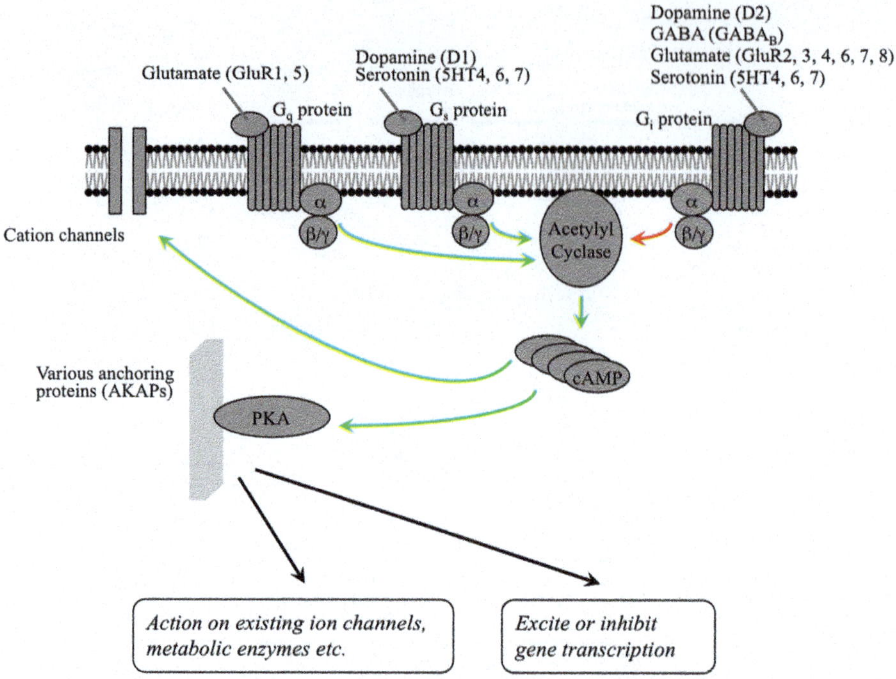

Fig. 5.5 cAMP/PKA cascades. A number of different neurotransmitters have G-protein receptors that lead to generation of cAMP molecules via activation of acetyl cyclase. cAMP can activate ion channels directly, or can activate the kinase PKA which can lead to a number of downstream effects

vesicles to the synaptic membrane in close proximity with voltage gated ion Ca^{++} channels and hence ensure neurotransmitter release in response to an incoming action potential.

Specific phosphatases can inactivate specific kinases or other participants such as Ca^{++} channels. For example, CaMKIIPase is a phosphatase that targets CaMKII but no other proteins [357]. Phosphatases can be present in the same protein complex as kinase and substrate to regulate the process.

5.2.3 cAMP-Protein Kinase A Cascades

cAMP is the second messenger activated by the activation of the membrane bound protein adenylyl cyclase [358]. As illustrated in Fig. 5.5, there are G-protein receptors for dopamine, serotonin and glutamate that excite adenylyl cyclase, and G-protein receptors for dopamine, GABA, glutamate and serotonin that inhibit adenylyl cyclase. Many cAMP molecules are synthesized from ATP by one activated adenylyl cyclase molecule. cAMP molecules can directly affect a number of different types of ion channels. For example, cAMP makes the threshold for activation of the hyperpolarisation cation channel I_h more positive [359].

However, the most diverse effects of cAMP [279] derive from its activation of protein kinase A (PKA). PKA exerts an extensive range of effects, including on existing ion channels [360], on release of neurotransmitters [361] and by phosphorylation of transcription factors [313].

5.2.4 cGMP-Protein Kinase G Cascades

cGMP is a second messenger with number of similarities to cAMP. It is synthesized from GTP, which like ATP is a molecule involved in energy exchanges within cells. Synthesis is regulated by guanylate cyclase, which is typically activated by the neurotransmitter nitric oxide (NO) [362]. cGMP in turn activates protein kinase G (PKG), which has numerous downstream effects including regulation of ion channels, and gene transcription [363].

5.2.5 G_q Activated Protein Kinase C Cascades

G-proteins of type G_q are receptors for neurotransmitters such as glutamate and acetylcholine. G_q type receptors activate an enzyme called PLC which cleaves the phospholipid PIP2 present in the membrane to create two second messengers: diacylglycerol (DAG), which is membrane bound; and the second messenger IP_3 which is released into the cytoplasm. DAG activates protein kinase C (PKC).

While inactive, PKC is located in the cytoplasm, but when activated it is anchored to specific locations (such as a region of the membrane) by anchor proteins called receptors for activated PKCs (RACKs) [364]. For some PKC isoforms, Ca^{++} is also required for activation [289]. In some cases, activation is further enhanced by the presence of other chemicals like lysophosphatidylcholine (LPC) and various fatty acids.

A major role played by PKC is modulation of channels and receptors in the postsynaptic membrane [289].

5.2.6 G_i Activated Protein Kinase B Cascades

G-proteins of type G_i are receptors for neurotransmitters such as glutamate, GABA and dopamine. G_i type receptors activate the kinase PI3, which when activated phosphorylates the lipid PIP2 to PIP3. Protein kinase B (PKB, also known as Akt) binds to PIP3, and as part of the bound complex can be activated by phosphorylation by the constitutively active kinase PDK1 [365]. Activated PKB can in turn activate or deactivate many downstream substrates, including mTOR [366]. mTOR is itself a kinase that regulates many cellular functions including gene transcription [367] and protein translation [368].

PKB can be phosphorylated at two different sites, and only one site is phosphorylated by PIP3. This phosphorylation leads to strong PKB activity, but for maximum activity a second site must also be phosphorylated [369]. This second site is phosphorylated by mTORc2. Protein translation occurs at ribosomes, and mTORc2 is activated by binding with ribosomes [370]. The G protein-PIP3-PKB-mTOR cascade is required for some synaptic strength changes, including changes requiring protein translation [371].

5.2.7 *Calcium-Calcineurin Cascade*

Cascades can also involve phosphatases as their primary route. For example, Ca^{++} ions entering by L-type channels activate calcineurin, which in turn activates the transcription factor NF-ATc by dephosphorylation [372]. One downstream gene of NF-ATc codes for a Ca^{++} channel. Although this is a relatively short cascade, there are many factors that influence it, such as a wide range of calcineurin inhibitors. Calcineurin can also act directly on a wide range of ion channels [373].

5.2.8 *Crosstalk Between Chemical Cascades*

A selection of cascades including MAPK, Ca^{++}-calmodulin, cAMP-PKA, cGMP-PKG, G_q-PKC and G_i-PKB cascades have been described in their simplified reference form. In many cases different cascades can target the same apparent endpoint. For example, the DNA transcription factor CREB is activated by phosphorylation, which can result from a number of kinase cascades including Ca^{++}-calmodulin and MAPK cascades [374]. Translation of protein from mRNA can be initiated by the kinases MNK and/or mTOR. Activation of MNK is controlled by a number of MAPK cascades [375], while the primary route for activation of mTOR is the PI_3-PKB cascade.

One way in which this relatively simple picture is made qualitatively more complex is by the existence of extensive crosstalk between the cascades, in which an activated component at some point in one cascade also affects a component in a different cascade [376]. For example, in the PI_3-PKB route to mTOR activation, PKB can also be activated indirectly by PKA [377]. A more complex example is illustrated in Fig. 5.6. The route from Trk receptors through ras, raf-1, and MEK to the activation of ERK was described earlier. As illustrated in Fig. 5.6, ERK can also be activated by CaMKI from a Ca^{++} initiated calmodulin cascade [378]. MEK can be activated by PKA from a G_s or G_q protein-cAMP cascade, via Rap-1 and B-raf [379], but PKA also inhibits raf-1 in the primary ERK cascade [380]. In addition,

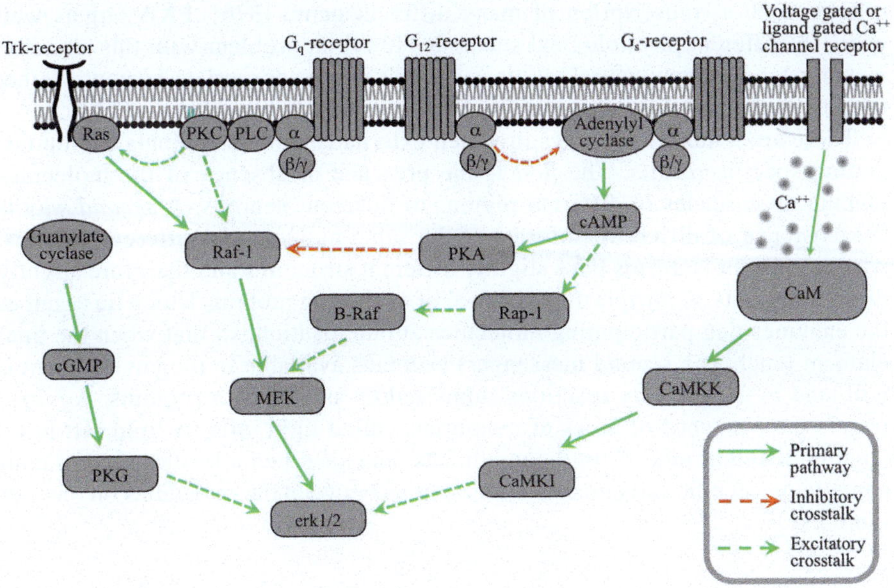

Fig. 5.6 Crosstalk between MAPK cascade and PKG, PKC, PKA and calmodulin cascades. In this example a number of different cascades can all influence the activation of the ERK kinase by an MAPK cascade via inhibitory and excitatory crosstalk

ERK activation is also promoted by the NO-guanylate cyclase-cGMP-PKG pathway [381]. There are thus multiple paths by which ERK can be activated. These paths can be initiated by voltage gated ion channels or by neurotransmitters like glutamate, noradrenaline, dopamine, acetylcholine, nitric oxide and BDNF acting on multiple receptor types. Not indicated in Fig. 5.6 are the many, often less well understood, ways in which phosphatases can be activated and act on the cascades.

The complexity created by the multiplicity of cascades and the interactions between them is only hinted at in the above description. There are over 500 kinases in the human genome [382], and there can be crosstalk between many levels in different primary cascades [383]. One problem is how a specific result is achieved given this extensive interaction network.

5.2.9 *Specificity of Cascade Results*

The same signal can initiate or affect many different cascades. For example, intracellular presence of Ca^{++} has a wide range of different influences [356]. Furthermore, the activation of one cascade can result in very many different neural changes. For example, activation of ERK can lead to modifications to many existing proteins, synthesis of many different proteins from existing mRNA, and/or transcription of many different genes [376]. PKA targets well over 100 different physiological substrates [384]. A problem with this situation is understanding how neural inputs can result in the appropriate response rather than all of the possible responses.

There are a number of ways in which externally initiated signals end up targetting specific results. The *first* is the presence or absence of the molecules making up cascades in different regions of different neurons. A *second* way is the existence of different isoforms of the same molecule in different neurons, where different isoforms have slightly different structures and therefore slightly different effects. A *third* is the existence of anchoring proteins which tie together the enzymes and participating molecules at one location, so that when the final element (such as a second messenger) becomes available in the environment it will tend to trigger only reactions supported by such anchor proteins. A *fourth* way is the existence of areas of membrane called lipid rafts. A lipid raft has a slightly different mix of lipid components, and specific molecules tend to concentrate on specific rafts, making interactions between the molecules on one raft more likely.

5.2.9.1 Distribution of Molecules

An example of the first way is that specific types of ligand receptors occur in specific areas of specific types of neurons. In cortical pyramidal neurons, AMPA receptors tend to be located towards the centre of the postsynaptic density of glutamatergic synapses, while NMDA receptors are located around the edges [164]. In the striatum, one population of projection neurons has D1 dopamine receptors, the other has D2 receptors.

5.2.9.2 Existence of Isoforms

Isoforms are very similar versions of a protein, coded on different genes or assembled in different ways after transcription. Many functionally significant molecules have different isoforms, sometimes with significant differences in chemical properties. One example is the different isoforms of the $GABA_A$ receptor. This receptor is made up of a combination of five similar subunits, most often two type α, two type β and one type γ. There are 19 forms of these subunits coded on different genes, six type α isoforms, three type β, and three type γ [385]. Different subunit combinations are found in different brain regions, and even in different synapses on the same neuron. The presence of a different isoform can have major effects, for example different anaesthetic effects of the drug etomidate depending on whether the β3 or β2 isosomes are present [386]. Another example of different isoforms is the various forms of PKA, which exist in different regions of the brain [387]. These isosomes have different chemical properties, and different regulation of genetic expression [388].

5.2.9.3 Anchoring Proteins

Anchoring proteins bind to a number of different molecules, ensuring that when, for example, a messenger molecule is released into the intracellular environment, there is a high probability that the reaction involving the bound molecules and the messenger will proceed. Anchor proteins associated with PKA have been extensively investigated [389]. These A-kinase-anchoring proteins (AKAPs) bind to PKA and to a subcellular environment such as a specific area of the membrane. In addition, some AKAPs also bind to kinases, to phosphatases, to phosphodiesterases that deactivate cAMP, or to ion channels [390].

AKAPs bring together the kinase PKA with other molecules that form a reaction in which PKA participates. Other proteins can perform an analogous role, such as scaffold proteins for MAP kinase cascades [391], accessory proteins in the case of G-protein activity [392], and perhaps HOMER proteins linking various types of receptors [393]. The anchoring of protein kinases in different cellular locations is a general mechanism for achieving specific chemical endpoints [364].

5.2.9.4 Lipid Rafts

The membranes forming the external cell wall and surrounding internal structures like the endoplasmic reticulum are made up of phospholipids. However, there are different phospholipids, and the mix present at different locations can differ. Microdomains known as lipid rafts exist which are enriched in particular lipids, and some proteins preferentially associate with rafts [394]. Proteins can be modified to anchor to a raft, and protein-protein or protein-lipid interactions could act to stabilize lipid rafts for long periods of time [395]. For example, CaMKII kinase binds to a lipid raft, and the resultant raft traps post synaptic density proteins that do not themselves interact directly with the kinase, stabilizing the post synaptic density [396]. Proteins associated with one chemical cascade can be concentrated on one raft, for example by concentrating the receptors, scaffolding proteins, kinases etc. and protecting the group from phosphatases that would otherwise interfere with the process [397]. For example, separate pools of cAMP can be maintained by associations of adenylyl cyclase, PKA and other proteins [398].

The difference between isosomes can affect binding to a raft. For example, two Ras isosomes (K-ras and H-ras) have very slight molecular differences, but H-ras binds to rafts and K-ras does not. The implication is that the two isosomes support different chemical processes [397].

5.2.9.5 Variation in Target Effects

The same molecule may be activated in different ways. For example, the CREB molecule which induces expression of genes has a three different phosphorylation sites (labelled Ser133, Ser142 and Ser143). Phosphorylation at Ser133 results from many different cascade paths, but phosphorylation at all three results from a more limited range [317]. Different patterns of gene expression may therefore by triggered by different cascade routes.

The overall picture of the cascades of chemical reactions within neurons is therefore that one neurotransmitter can initiate many signalling cascades and the same cascade can be initiated by many neurotransmitters. There is considerable crosstalk between cascades. The same cascade component can initiate many different end results, and there are complex mechanisms to target different such results. The endpoints of a cascade can be activation or inactivation of a structural or functional protein, synthesis of a structural or functional protein from already available mRNA, or synthesis of new mRNA from genes. The same endpoint can be targetted by different cascades.

5.3 Neuron Processes

In conclusion for this chapter, we will discuss three types of overall neuron processes that are critical for neuron information processing in the brain. One is the process by which synaptic inputs are integrated to determine neuron firing, the second is the way in which changes to synaptic weights are determined, the third is the way in which the algorithm for integrating synaptic inputs across the dendritic tree is changed. Neurons differ in the ways in which these processes are performed, and much of the focus will be on the most investigated example: pyramidal neurons in the cortex and hippocampus.

5.3.1 Integration of Inputs by a Neuron

An open ion channel allows a current to pass through the neuron membrane, changing the local membrane potential. Such changes can propagate and add together in a complex, time dependent fashion, and if as a result the membrane potential in the axon initial segment reached a threshold depolarisation, an action potential propagates down the axon, and may also propagate back into the dendritic tree.

One pyramidal neuron receives large numbers of glutamatergic inputs from other pyramidal neurons, many GABAergic inputs from multiple local interneurons, and inputs from neurons in various brain structures mediated by other neurotransmitters. In addition, a pyramidal neuron has many voltage gated ion channels that can be modulated to various degrees by external inputs. Interneurons also receive inputs from pyramidal neurons, from other interneurons, and from neurons in other brain structures, and have voltage gated ion channels.

In the central nervous system the primary inputs encouraging a neuron to produce an output arrive at glutamatergic synapses. A cortical pyramidal neuron can have ~10,000 such synapses, and as conceptually illustrated in Fig. 5.7, most of

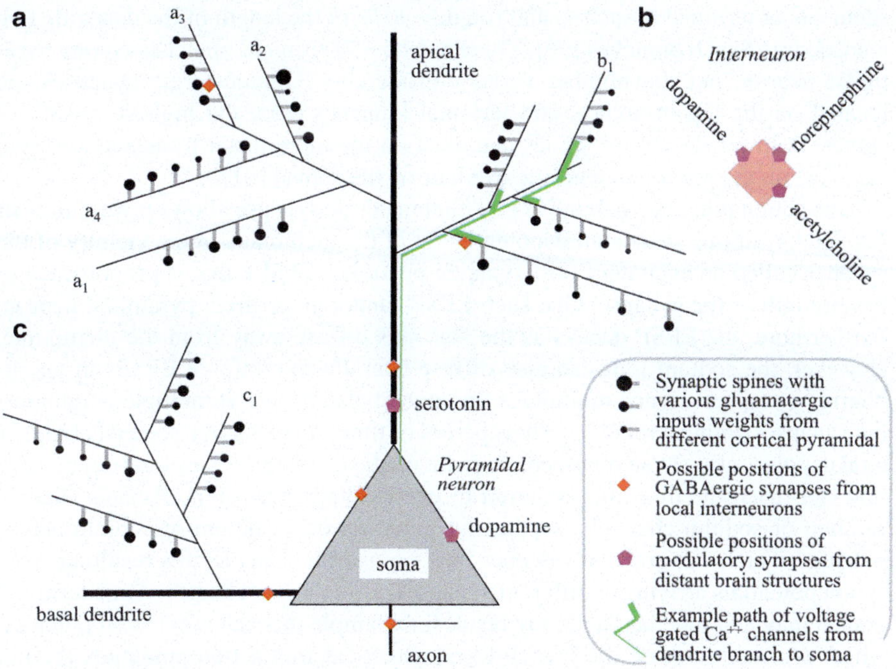

Fig. 5.7 Integration of pyramidal neuron inputs. Most of the inputs to a pyramidal neuron are excitatory glutamatergic inputs that arrive at synapses of varying weights on the dendritic tree. These inputs are organized on terminal branches, and if enough action potentials arrive at a group of sufficiently strong synapses on a terminal branch within a short period of time, a calcium action potential is initiated that propagates deeper into the dendritic tree. These calcium action potentials travel on paths defined by concentrations of voltage gated Ca^{++} channels, and a single such action potential tends to decay before reaching the soma. Ca^{++} action potentials generated by multiple terminal branches can reinforce each other and approach the soma, where sufficient potential results in an output action potential. This integration of glutamatergic inputs is regulated in a number of ways. GABAergic inputs from interneurons can inhibit the influence of individual synapses, terminal branches, the propagation of Ca^{++} action potentials, individual dendrites as a whole or the neuron as a whole. In addition, GABAergic inputs can regulate the timing with which the neuron produces outputs. Cholinergic inputs acting on interneurons also influence the timing of the outputs. Serotonergic inputs on proximal dendrites provide excitation. Norepinephrinergic inputs to interneurons reduce pyramidal activity by exciting interneurons. Somatic dopaminergic inputs can increase activity, but dopaminergic inputs to interneurons can indirectly decrease activity. Neurotransmitter inputs, perhaps to different locations, can have other effects such as regulating gene transcription and protein synthesis. A wide range of voltage gated K^+ channels (not illustrated) with variable distribution on the membrane inhibit the integration process at various points, and their influence is modulated by external neurotransmitters

these inputs are located on the dendritic tree, and are organized in a hierarchy. At the top level of this hierarchy there are different dendrites, including one apical dendrite and multiple basal dendrites, all with trunks that emerge from the soma. A dendritic trunk can fork into separate arms, and each arm can fork into different branches and so on. The final dendritic lengths that do not themselves fork are called terminal segments or terminal branches, and roughly 90 % of the length of the dendritic tree is made up of such branches [96]. There can be ~50 terminal branches on one pyramidal neuron, but the number varies considerably. Glutamatergic synapses are located on the top of spines, and terminal branches carry the highest density of spines. A lower density of spines can exist on the more distal trunk sections, but very few on the proximal dendritic trunk or on the soma [101].

One action potential arriving at one glutamatergic synapse allows inward flow of Na^+ ions, resulting in a small depotentiation of the membrane in the vicinity of the synapse called an excitatory post synaptic potential (EPSP), that is proportional to the strength of the synapse. One such EPSP cannot alone fire a pyramidal neuron. Furthermore, the EPSP decays as the Na^+ ions diffuse away from the membrane. However, the sodium ions can also diffuse laterally to reinforce EPSPs in neighbouring synapses. Hence multiple action potentials arriving at multiple neighbouring synapses within a relatively short period of time can result in a somewhat higher total depolarisation at an intermediate point on the membrane. This total depolarisation can reach the threshold required to open voltage gated ion channels such as calcium or sodium channels, supporting initiation of a calcium or sodium action potential which can propagate deeper into the neuron. Thus EPSPs resulting from action potentials arriving at different synapses within a small length of one terminal branch may add together, and may result for example in a calcium action potential which propagates along the branch [399]. Such calcium action potentials are not fully self-propagating (unlike sodium action potentials along an axon) and decay with distance travelled [114]. However, if reinforced by other active synapses along the branch, a calcium action potential may exit the branch and propagate along the dendritic trunk. If several such terminal branch calcium action potentials are initiated at a similar time, they add to each other and reach the soma [400]. Sufficient dendritic input to the soma could result in initiation of an output action potential. The pattern of propagation depends on the distribution of voltage gated calcium channels, which can vary significantly across the dendritic tree [233]. The distribution of different types of voltage gated ion channels therefore defines paths along which spikes will tend to pass, or in other words the integration pattern of the neuron [114]. EPSP initiated dendritic spikes can also be carried by voltage gated Na^+ channels, and voltage gated K^+ channels can impede such dendritic spikes [114].

5.3.1.1 Other Effects of Voltage Gated Ion Channels

There are other voltage gated ion channel mechanisms that adjust the relative influence of distal dendritic, proximal dendritic and somatic inputs. One such mechanism utilizes the I_h voltage gated ion channel. I_h channels are open if the membrane is hyperpolarised or close to resting potential, but deactivate in response to significant depolarisation. This behaviour results in an effective resting membrane potential slightly depolarised from the value it would have in their absence, but the effect is turned off when the depolarisation exceeds their threshold. A high density of such channels therefore amplifies the initial effect of any local depolarisations.

The density of these channels is very low in the soma but increases in dendrites with distance from the soma [401]. This arrangement gives greater weight to more distal inputs, and compensates for the greater distance they must travel to reach the soma [402]. The relative influence of proximal and distal locations can be adjusted by a number of such mechanisms, and varies in different types of neuron [251].

A-type voltage gated K^+ channels regulate the backpropagation of action potentials into the dendritic tree. Small clusters of A-type channels occur at different points on the soma, proximal dendrites and distal dendrites of neurons, with one neuron containing dozens of such clusters [403]. As described earlier, A-type channels are opened by depolarisation, but deactivate completely in response to sustained depolarisation. Single backpropagating action potentials are severely attenuated, but sequences of action potentials above a certain frequency penetrate much more successfully [404]. The existence of clusters suggests that the penetration of backpropagating action potentials to different regions of the dendritic tree is separately managed.

There are many different types of voltage gated ion channels [405], and the distribution of different types of ion channels differs between different types of neuron and different brain regions [255]. On the same neuron, gradually increasing or gradually decreasing concentrations of different types of ion channels have been observed going from proximal to distal regions of a dendritic tree. For other types of channel, local concentrations have been observed at different places on the dendrite [406]. The different types and distributions reflect the different types of integration being performed [407].

5.3.1.2 Interneuron Effects

Some additional factors regulate the integration of inputs in various ways. One critical factor is the inhibition provided by different GABAergic interneurons. As an example [408], one pyramidal neuron in layers II/III of the cortex has ~1,000 inhibitory synapses, ~15 % of the total synapse count. These inputs come from ~70 different interneurons. Approximately 70 % of these interneurons are in the same layer and column as the pyramidal neuron, ~15 % in the same column but a different layer, the rest in a different column.

Interneurons have been classified into different types by the physical appearance of their axonic and dendritic trees. Different types on this basis include basket, chandelier, bipolar, double bouquet, and Martinotti cells. A more functional based classification derives from the observation that different interneurons target different regions of pyramidal neurons [409]. Different regions are the axon initial segment, the soma, the proximal dendrites, the distal dendrites and the necks of individual spines. Some interneurons synapse only on the axon initial segment of pyramidal neurons. Other interneurons synapse mainly on one region, with some but fewer synapses on neighbouring regions. For example, some neurons synapse mostly on the soma, with a small proportion of synapses on the proximal dendrites and a very small proportion on the distal dendrites. Other interneurons synapse mostly on distal dendrites, with small proportions in other areas. Yet other interneurons synapse only on spines and distal dendrites in different proportions. The chandelier morphological type is the axon targetting neuron [410], basket cells favour somata and proximal dendrites, and double bouquet cells favour spines and distal dendrites [411]. Martinotti cells target distal apical dendrites, but are unusual in also targetting proximal dendrites and somata [412].

Most pyramidal neurons are regular spiking type, a small proportion are burst spiking, with burst spiking pyramidals occurring more frequently in specific cortical layers. Many cortical interneurons are fast spiking, but others are regular spiking and some are burst spiking [408]. Martinotti cells tend to be regular or burst spiking [412]. The burst spiking interneurons tend to occur in the same layer as burst spiking pyramidals. A few interneurons have firing patterns that do not fit these classifications, such as late spiking and irregular spiking types [408].

Interneurons tend to target a limited range of pyramidal neurons. As discussed later, cortical pyramidal neurons are arranged in layers, and in columns perpendicular to the layers. Some interneurons only target pyramidal neurons in the same column and layer, others only target pyramidals in the same column, yet others target pyramidals in neighbouring and distant columns within the same layer.

There are two types of GABAergic receptors: $GABA_A$ and $GABA_B$. $GABA_A$ is a fast acting, generally inhibitory Cl^-/HCO_3^- channel. $GABA_B$ is a slower acting G-protein receptor that inhibits by activation of K^+ channels [413]. $GABA_B$ can also inhibit voltage dependent Ca^{++} channels, with a presynaptic effect of reducing neurotransmitter release [414] and a postsynaptic effect of reducing dendritic excitability [415].

Hence there are four dimensions to the ways in which interneurons act upon the integration of inputs by pyramidal neurons. Firstly, selective targetting of different groups of pyramidal neurons means that interneurons can manage the relative activity levels between groups. Secondly, selective targetting of different regions of pyramidal neurons means that they can manage the output of an individual pyramidal neuron, the relative contribution of different dendrites, dendritic arms, or dendritic terminal segments, or the relative contributions of individual synapses. Thirdly, the different spiking types mean that different types of electrical effects can be imposed on their targets. Interneurons can precisely coordinate their firing by means of electrical synapses that make fast, direct electrical connections between them without an intervening neurotransmitter. These electrical synapses are between interneurons of the same spiking type [416]. Fourthly, different types of GABA receptors act on different timescales, and differ in their locations on the target neuron.

5.3.1.3 Effects of Other Neurotransmitters

Serotonergic, cholinergic, noradrenergic and dopaminergic neurons in other brain structures target cortical neurons, and have various effects including modulation of pyramidal neuron activity levels.

Serotonergic inputs target the proximal apical dendrites of pyramidal neurons, mainly by the receptor 5HT2A. Activation of this receptor inhibits the calcium gated potassium channels responsible for the slowly decaying component of the afterhyperpolarisation that follows an action potential [417] and therefore reduces the probability of another action potential immediately afterwards. Hence one effect on the target neuron is an increase in the firing rate [418] in response to glutamatergic inputs to the apical dendrite.

Norepinephrine depresses the activity of pyramidal neurons, by increasing the activity of interneurons [419]. Norepinephrine neurons do not appear to target pyramidals directly.

The effects of acetylcholine upon pyramidal neurons are complex and not well understood. Firstly, acetylcholine can act presynaptically or postsynaptically, directly via receptors on pyramidal neurons or indirectly via receptors on interneurons [420]. Secondly, receptors may be concentrated at synapses made by cholinergic neurons on to their targets or distributed over the surface of the target neuron to detect acetylcholine released by non-synaptic vascularisations on cholinergic axons [421]. Thirdly, there are both nicotinic (ion channel) and muscarinic (G-protein) receptors. Nicotinic receptors occur on interneurons where they are excitatory [422], but evidence of their occurrence on pyramidal neurons is contradictory [423]. Muscarinic (G-protein) receptors are present on pyramidal neurons and can have tonic excitatory but phasic inhibitory effects [424]. Fourthly, receptors are observed in both symmetric and asymmetric synapses [425]. There seems to be a tendency in pyramidal neurons for asymmetric synapses to occur on the heads of small spines and symmetric synapses on the somas, with both types occurring on dendritic shafts [426]. It appears that muscarinic receptors may be present in glutamatergic synapses [427]. In interneurons, synapses may generally be symmetric [426]. There is evidence that a class of interneurons that in some brain regions tends to target only other interneurons [428] and receives less cholinergic input than other interneurons [429]. Acetylcholine appears to play a major role in controlling the oscillations in neuronal outputs that are reflected in the appearance of gamma (30–80 Hz), theta (4–12 Hz) and other frequencies in the EEG [430].

Another neurotransmitter that can affect pyramidal neuron excitability levels is dopamine. In pyramidal neurons D1-type receptors are located on the somas, on dendritic shafts, and most heavily in spines close to glutamatergic synapses [431]. Dopaminergic receptors in different locations may have different functions. D1-type receptors close to spines have a role in regulating local protein translation [432], but D1-type receptors on pyramidal somas may stimulate action potential activity [433]. D2-type receptors on interneurons may act indirectly to reduce pyramidal activity [434].

5.3.1.4 Timing of Pyramidal Neuron Outputs

The activity of populations of pyramidal neurons exhibits significant temporal ordering. This temporal ordering is revealed in the frequencies found in the EEG. Two such frequencies that are important for cognitive processing are gamma-band

(30–100 Hz) and theta-band (4–10 Hz) [435]. Pyramidal neuron activity tends to occur at peaks in these frequencies when certain types of cognitive behaviours are under way, and in some situations both frequencies are present, with activity maximal when the two frequency peaks coincide [436].

Interneurons inhibit the overall activity of pyramidal neurons, but in addition impose these modulation frequencies on pyramidal activity. Different patterns of pyramidal neuron firing appear in different behavioural situations, and each pattern is imposed by the combination of different types of interneuron that tend to fire at different time points in the current modulation frequencies [437]. For example, axo-axonic and basket cells fire at gamma frequencies, but axo-axonic cells tend to fire preferentially just after the peak of theta frequency, while basket cells fire preferentially in the descending theta phase [438]. The primary inputs driving GABAergic interneurons are glutamatergic, arriving at AMPA receptor synapses. Spike timing is precisely controlled in such interneurons with the times during which glutamate is release and the times for removal of glutamate from the synaptic cleft being around ten times shorter than for AMPA receptor synapses on pyramidal neurons [439]. Hence pyramidal neurons integrate information over relatively wide temporal intervals, while interneurons regulate the timing of outputs carrying the integration results.

Interneuron firing imposes timing constraints by inhibiting pyramidal outputs at the arrival times of the GABAergic spike, but can also have more complex effects to encourage appropriate timing of pyramidal outputs. For example, as described in Chap. 4, $GABA_A$ receptors are inhibitory if the membrane is depolarised, but slightly excitatory at resting potential. If an action potential arrives at a $GABA_A$ synapse 5–10 ms before an action potential arrives at a glutamatergic synapse nearby, the GABA input will reinforce the depolarisation resulting from the glutamate input [172]. Hence not only will the pyramidal neuron be inhibited in phase with the GABAergic inputs, but will be excited to some degree when out of phase with those inputs. Acetylcholine inputs also play a role in influencing the timing of pyramidal inputs, and it is interesting that the muscarinic M1 receptor is excitatory in response to a steady background level of acetylcholine, but inhibitory in response to a spike of concentration [424].

5.3.1.5 The Factors Influencing Generation of Action Potentials

The production of an action potential by a pyramidal neuron is the result of the integration of a very large number of inputs. These inputs are generally action potentials generated by other neurons. These action potentials result in effects on different timescales, which are all integrated by the target neuron. There are many different combinations of inputs that can lead to an output action potential, and generally no one input is completely decisive. In addition, information that an action potential has been generated is fed back into the dendritic tree by a backpropagating action potential.

5.3.2 Management of Long Term Synaptic Weight Changes

Changes to synaptic strengths are the primary means by which the responses of a neuron to its inputs are adjusted. These changes must be appropriate to the circumstances, and must therefore be determined by external inputs modulated by internal states. The specific conditions under which synaptic weights change, and the mechanisms by which such weight changes are implemented, differ between different types of neuron and different anatomical locations. The most studied phenomena are the increases and decreases in synaptic weight (potentiation and depotentiation) of glutamatergic synapses following synapse and neuron activity in pyramidal neurons, especially in the hippocampus. These changes are implemented by conductance changes to AMPA receptors, addition and removal of AMPA receptors, and structural changes to spines. Conductance changes result from the phosphorylation or dephosphorylation of AMPA receptors, catalysed by locally available proteins [440]. Trafficking involves addition and removal of AMPA receptors from the postsynaptic density [305]. Structural changes to spines involve modification of the actin cytoskeleton, for example changing small stubby spines to large, mushroom shaped spines with larger postsynaptic densities and higher numbers of AMPA receptors [141]. The following discussion will centre primarily on synaptic weight changes by AMPA receptor related changes in pyramidal neurons.

Proteins in human cells are constantly being degraded and replaced by newly synthesized proteins. The half life of an AMPA receptor molecule in such recycling is ~48 h [441]. Furthermore, there is internal recycling of AMPA receptors, with receptors being removed from the membrane and later returned to the membrane. The half life for residence at the cell surface is ~30 h [441]. Hence the weight of a synapse can be managed by regulation of this trafficking process in different ways. For example, the half life of AMPA receptors in the postsynaptic density can be reduced to about 40 min by stimulation of endocytosis, substantially reducing synaptic weight [442].

Within the cytoplasm, AMPA receptor molecules (like all membrane proteins) are contained within vesicles. Insertion of AMPA receptor molecules into and removal from the membrane is carried out by vesicle exocytosis and endocytosis [443] in which a vesicle is fused with the membrane or budded off from the membrane. AMPA receptor mRNA exists in the dendritic tree [444], in some cases localized to spines [445]. AMPA receptors newly synthesized from this local mRNA are inserted into vesicles at the local endoplasmic reticulum and transported to the membrane of the dendritic shaft by the motor protein myosin Va [446]. Insertion is by membrane fusion, in which SM proteins catalyse the linking of a v-SNARE protein attached to the vesicle membrane with a t-SNARE protein attached to the target membrane [447] The complexin protein is essential for exocytosis associated with changes to synaptic strength but not exocytosis in support of regular recycling [448]. Following insertion into the dendritic shaft membrane, the AMPA receptor molecules diffuse into the spine [449]. Alternatively, there may be vesicle transport of AMPA receptor into the spine [450] with direct insertion into the membrane adjacent to the postsynaptic density [451], or perhaps even into the postsynaptic density [452]. Spine topology limits diffusion of AMPA receptors into the spine, but such diffusion is encouraged by a GTPase called dynamin [453]. In the postsynaptic density, AMPA receptors are held in place by various scaffold proteins such as NSF and GRIP [454]. For endocytosis of AMPA receptors, they must be dissociated from the scaffold proteins and diffuse into the membrane peripheral to the postsynaptic density [443]. NSF in particular inhibits such endocytosis of AMPA receptors [454]. Endocytosis involves pinching off an area of the membrane containing the AMPA receptors to form a vesicle, which can then either be used to reinsert AMPA receptors later or can be broken down by a lysosome. Endocytosis is catalysed by a complex of a membrane bound protein endophilin, the GTPase dynamin, and the protein ARC [455].

Spine size correlates with synaptic strength [456]. Spine size increases can be temporary, returning to original size within minutes, or longer term, lasting at least hours. Spine size increases are caused by polymerisation of G-actin to form long F-actin filaments, the addition of G-actin to F-actin filaments at the tip of the spine driving its expansion [456]. G-actin and F-actin are in a state of constant exchange in spines. There are several pools of F-actin in a spine: a dynamic pool in which the turnover time is ~40 s; an enlargement pool that appears when spine growth is triggered by Ca-calmodulin with a turnover time ~2–15 min; and a stable pool with a turnover time ~17 min that is proportional to the square of the spine volume [457]. Hence any modifications to the rates of G-actin/F-actin exchange can rapidly change spine size [458]. Important factors in such modifications are the actin binding proteins profilin and cofilin, which respectively encourage and discourage polymerisation [459]. The GTPases Rac, Rho and Cdc42 have a strong influence on the actin system [460], are activated by various receptors [461], including ionotropic glutamate receptors, serotonergic receptors, and neurotrophins, and influence spine morphology by multiple routes, including paths that target profilin [462] and cofilin [463].

Conductance, trafficking and spine size changes are catalysed by enzymes, and the activity of the enzymes is regulated in three different ways. Firstly, the enzymes can be constantly available in the vicinity of the AMPA receptors, with weight changes regulated by mechanisms to activate those enzymes. Secondly, the enzymes can be synthesized from locally available mRNA, with weight changes regulated by control of the synthesis process and also by regulation of any posttranslation modifications. Thirdly, mRNA transcription in the nucleus can occurs first, and the enzymes then synthesized from the new mRNA. This synthesis may be in the nucleus and followed by transport to the synapse, or by initial transport of the mRNA to the synapse followed by local mRNA translation. Weight changes in this third case are regulated by control of transcription, transport and degradation of mRNA, plus regulation of protein synthesis and postsynthesis modification.

Rather than measuring one synapse operating naturally in a learning brain, LTP is generally investigated in artificial ways that electrically stimulate the neuron in a manner that mimics natural conditions to some degree, such as the TBS protocol [464]. In such artificial experiments, it is observed that changes to synaptic weights

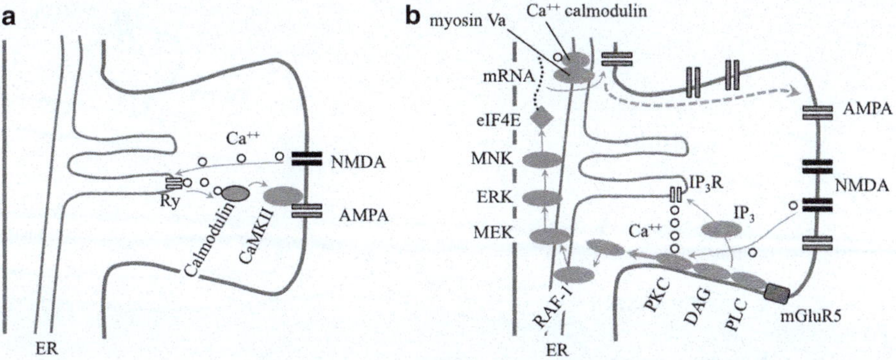

Fig. 5.8 Mechanisms supporting LTP1 and LTP2. (**a**) In LTP1, Ca^{++} ions that enter through NMDA receptors activate Ry receptors on the endoplasmic reticulum (*ER*) to release internal stores of further Ca^{++} ions. These Ca^{++} ions bind with calmodulin, and the Ca-calmodulin complex activates the kinase CaMKII, which in turn catalyses the phosphorylation of AMPA receptors to temporarily increase their conductance. (**b**) In LTP2, activation of glutamate metabotropic mGluR5 and NMDA ionotropic receptors initiates a cascade of reactions leading through PKC and ERK to the translation activator eIF4E. Activation of eIF4E triggers translation of AMPA proteins from mRNA available in the endoplasmic reticulum (ER) near the synapse. These AMPA molecules are transported to the dendrite shaft membrane, inserted into the membrane, and diffuse to the postsynaptic density where they increase the strength of the synapse for a period of hours

following activity can persist for different periods of time. As discussed earlier, for increases to synaptic weights lasting much longer than the stimulus that generates them, three different decay times can be discerned. For LTP1, the decay constant is ~ minutes to hours, for LTP2 ~ hours to days, and for LTP3 ~ weeks or more [137, 465]. In a typical TBS protocol, a sequence of one to eight trains of electrical spikes are provided as inputs to a neuron. A train is made up of ten bursts of spikes, with the bursts occurring at the EEG theta frequency (5 Hz) which is associated with memory. Each burst contains four or five spikes at 100 Hz. Between one and ten sequences of trains are applied, with trains separated by ~10 s or more. Immediately following a TBS protocol, synaptic strength typically more than doubles. However, the subsequent decay in synaptic strength depends on the number of trains, with one train typically resulting in LTP1, four trains in LTP2 and eight trains in LTP3 [144]. As illustrated in Figs. 5.8 and 5.9, it is believed that LTP1 is primarily implemented by modifications to already available proteins, LTP2 by translation of proteins from mRNA available near the synapse, and LTP3 by transcription of new mRNA in the nucleus followed by protein translation [144].

LTD also occurs in AMPA receptor synapses, induced by different stimulation protocols. There is evidence for a transient LTD involving modification of existing proteins and a persistent LTD based on mRNA transcription and/or protein synthesis [466]. There appears to be a form of LTD which develops within minutes of the triggering stimulus and depends only on local protein synthesis and not gene transcription [467]. However, LTD induction appears to activate gene transcription, which may be required for long term maintenance [468].

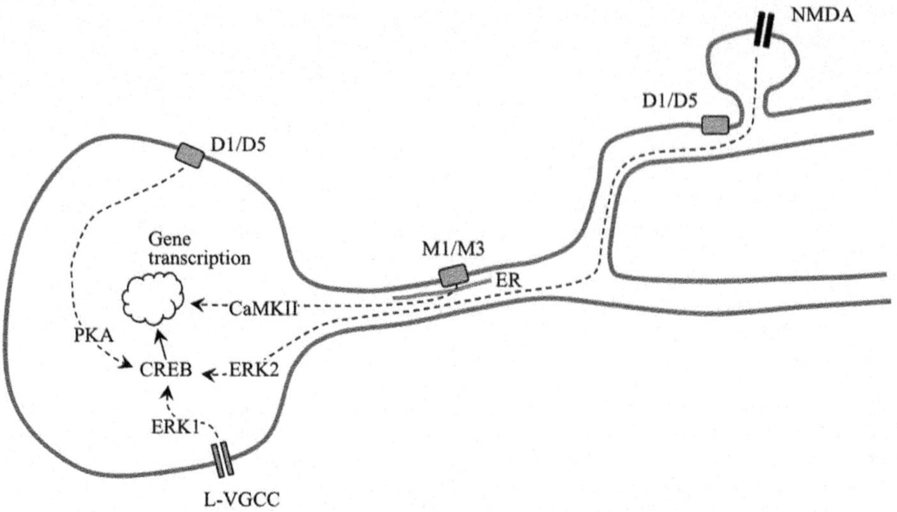

Fig. 5.9 Mechanisms supporting LTP3. Ca^{++} entry through NMDA receptors at the synapse initiates a kinase cascade leading to ERK2 activation; Ca^{++} entry through voltage gated Ca^{++} channels on the soma initiate a kinase cascade leading to ERK1 activation; dopamine D1/D5 receptors activate a PKA cascade. All these routes must be activated for full activation of the transcription factor CREB. In addition, cholinergic m1/m3 receptors activate a PKC cascade leading to CaMKII activation which also activates transcription factors. One effect can be transcription of mRNA to support synthesis of the cytoskeleton molecule actin. This mRNA may be transported to the synapse, where synthesis of new actin leads to a semi-permanent expansion of the synapse and increase in synaptic weight

We will now discuss how the three different options for changes to synaptic weights are regulated, using LTP1, LTP2 and LTP3 as examples. Many of the changes to synaptic weights are initiated by entry of Ca^{++} through NMDA receptor or voltage gated calcium channels. Different routes of Ca^{++} entry initiate different but overlapping chemical cascades, with different timescales for operation but extensive crosstalk allowing cross modulation [347].

5.3.2.1 LTP1 and Regulation of Synaptic Weight Changes Involving Locally Available Proteins

As shown in Fig. 5.8a, in LTP1, Ca^{++} entry through NMDA receptors activates Ry receptors in the endoplasmic reticulum (ER). Activation of these receptors releases Ca^{++} ions from stores in the ER. These ions activate calmodulin, and the Ca^{++}-calmodulin activates CaMKII kinase molecules attached to the postsynaptic density in proximity with AMPA receptors. The CaMKII phosphorylates the AMPA receptors, increasing their conductance [440], and therefore increasing the synaptic strength.

In LTP1, dephosphorylation of the AMPA receptors occurs with some decay time, returning the synapse to its original strength. Such dephosphorylation could be carried out by various phosphatases also associated with the CaMKII/AMPA receptor complex, some of which are constitutively active and others activated by Ca^{++}-calmodulin [469].

5.3.2.2 LTP2 and Regulation of Synaptic Weight Changes Involving Protein Synthesis from Already Available Local mRNA

LTP2 is induced by a somewhat longer sequence of electrical stimuli. AMPA and NMDA receptors are distributed within the postsynaptic density, with NMDA receptors clustered towards the centre [470]. mGluR5 receptors are concentrated in a ring around the periphery of the postsynaptic density [164], but also frequently occur in the neuron membrane well away from synapses [471]. Hence a longer electrical stimulation will mean that glutamate can also reach the mGluR5 receptors. As illustrated in Fig. 5.8b, activation of mGluR5 receptors means that activated PLC will hydrolyze membrane lipids to DAG and IP3. DAG plus Ca^{++} activates membrane bound PKC and releases it into the cytoplasm [472]. The Ca^{++} can come from glutamate activated NMDA receptor Ca^{++} channels or be released from stores in the endoplasmic reticulum through IP_3 receptors. PKC then activates a kinase chain RAF-1 → MEK → ERK → MNK. MNK activates the translation initiation factor eIF4E, triggering protein synthesis from existing mRNA. Newly synthesized AMPA receptors are transported to the membrane of the dendritic shaft by the motor protein myosin Va [446]. This motor protein also requires Ca^{++} for its activation by binding with Ca^{++}-calmodulin [473]. Insertion into the shaft membrane requires activation of PI3 kinase, in this case by Ca^{++}-calmodulin with the Ca^{++} derived from NMDA receptors [474]. Diffusion into the spine is regulated by the GTPase dynamin.

AMPA receptors can also be removed, and there is constant two way trafficking of receptors at the membrane [475]. Hence local synthesis of AMPA receptors can lead to a temporary increase in concentration at the postsynaptic density, but if for example the postsynaptic density remains the same size, the extra AMPA receptors will tend to diffuse away once local synthesis has stopped, returning the synaptic weight to its original value.

Spine size increases can be temporary, returning to original size within minutes, or longer term, lasting at least hours. Early spine increases do not require protein synthesis, but such synthesis is required for stable spine expansion [476]. It is observed that polyribosomes shift from the dendritic shaft to spines after the induction of LTP [477], suggesting that protein synthesis is targetted to individual spines.

Spine enlargement requires NMDA receptor Ca^{++} current leading to actin polymerisation [141]. Long term spine enlargement also requires CaMKI [478] and CaMKII activation [141]. It has been observed that NMDA receptor activation results in movement of profilin from the dendritic shaft into spines [479], which could be the mechanism for early spine enlargement without the need for protein synthesis.

5.3.2.3 LTP3 and Regulation of Synaptic Weight Changes Requiring Gene Transcription

LTP3 requires even more TBS trains than LTP2, and appears to be dependent on gene transcription [144]. For example, mRNA from the *Arc* gene associated with the cytoskeleton is delivered to dendrites and translated soon after stimulation, and is required for long term changes but not shorter term LTP [137]. Some of the processes by which such gene activation is managed are illustrated in Fig. 5.9. As shown in the figure, multiple external influences including glutamate, dopamine and acetylcholine plus the opening of voltage gated channels determine whether transcription occurs. Various signalling pathways in neurons lead from synaptic activity to activation of transcription factors like SRF and CREB [480]. SRF is activated by Ca^{++} initiated and neurotrophin initiated cascades [318], CREB by cAMP and Ca^{++} initiated cascades [481], and in at least some neurons by neurotrophins [482]. Activation of transcription factors can trigger transcription of IEGs. IEGs that can be transcripted as a result of neuron activity include both the genes for transcription factors like *Cfos* and *Zif/268*, and effector genes coding proteins which can change synaptic strengths such as *Arc* [455] and *Homer-1*. In some cases the newly transcribed mRNA is transported to recently active synapses [330]. There is a second wave of transcription of late response genes [483], at least partly dependent on transcription factors coded in IEGs. For example, the late response genes for neurotrophins like BDNF are transcribed [484]. Again, the newly transcribed mRNA may be transported to specific synapses on the basis of recent activity [484].

Activity dependent gene transcription is dependent on calcium influx through NMDA receptors [485], calcium influx through type L voltage gated calcium channels (L-VGCCs) [486], and activation of D1/D5 type dopamine receptors [183]. There is also a dependence on muscarinic acetylcholine receptors [487]. To promote gene transcription, calcium entry via NMDA receptors acts upon a MAP/ERK cascade. D1/D5 receptors act via PKA. L-VDCCs act via a MAP/ERK cascade, but may use ERK1, while the NMDA receptor path may use ERK2 [144]. Muscarinic acetylcholine receptors can act via PKC to activate CaMKII and also a MAP cascade. Expression of the growth related early immediate gene *Cyr61* is under the control of muscarinic receptors [488]. NMDA receptors are located at synapses. In pyramidal neurons D1/D5 receptors are located on the somas, on dendritic shafts, and most heavily in spines close to glutamatergic synapses [431]. Dopaminergic receptors in different locations may have different functions, such as modulating activity [433], protein translation [432], or gene transcription. L-VDCCs are distributed over the soma, dendrites and spines, with higher densities in smaller dendrites [489]. However, the L-VDCC derived Ca^{++} concentration relevant to LTP3 appears to be in the soma [465]. Muscarinic receptors are located on the main dendritic shaft, and the Ca^{++} released from internal stores diffuses to the nucleus [487]. Different sources of Ca^{++} are isolated from each other and can act differently. NMDA receptors access a pool of Ca^{++} which activates ERK2, both close to the receptors [490]. L-VDCCs activate a separate pool of Ca^{++} in the nucleus that leads to activation of ERK1. Muscarinic receptors activate yet another pool from stores in the endoplasmic reticulum which then diffuses to the nucleus.

CREB is one transcription factor know to be activated in LTP, and CREB can be activated by multiple Ca^{++} dependent and other routes [315]. There is evidence that PKA can augment transcription triggered by CREB activation by another route [491], and inactivation of muscarinic receptors reduces but does not block LTP [492]. Hence activation by a subset of the possible paths may be adequate for at least partial strength LTP.

SRF is a transcription factor for many of the genes related to the actin cytoskeleton, and SRF activity is regulated by the rate of actin turnover [493]. Hence there is a mechanism by which an adjustment to a relatively short term spine size change could be established long term by gene transcription.

Once transcribed in the nucleus, the mRNA must be targetted to the appropriate synapse. Inactive mRNA is transported in granules called ribonucleoprotein particles, the transport regulated by synaptic activity [300]. Alternatively, proteins may be synthesized in the nucleus from the new mRNA and transported to the appropriate synapse in vesicles. Active synapses are chemically tagged in some way to allow capture of gene products [494].

However, the maintenance of LTP3 may also depend on local protein synthesis from existing mRNA not requiring new transcription, for example the synthesis of AMPA receptors [329]. It is possible in some cases that LTP3 can be sustained by this protein synthesis using local mRNA [495].

Note that the picture is certainly much more complex that the simple view illustrated in Fig. 5.9. For example, the L-VGCC channel can be regulated by the kinase PKC. PKC is activated by G_q-proteins, and there are G_q receptors for glutamate, acetylcholine, ATP and oxytocin among others, allowing the possibility for cross-talk or additional regulation.

5.3.2.4 Effects of Timing on Synaptic Weight Changes

Across populations of neurons, high frequency synaptic firing drives LTP, while low frequency drives LTD [496]. However, at a more detailed level there are a number of factors that influence weight changes. Long term changes to a synaptic weight tend only to occur when an input to the synapse due to the firing of the presynaptic neuron occurs close to a time when the postsynaptic neuron itself fires. The relative timing of presynaptic input and postsynaptic neuron firing influences both the type and magnitude of the change [132]. Other factors affecting the change include the number of presynaptic and postsynaptic activity pairings (typically 60–100 are required for plasticity [497]), the presence of synaptic inputs from multiple presynaptic neurons [498], and local depolarisation in the neighbourhood of the synapse [499].

LTP and LTD tend only to occur at a synapse if there is a presynaptic input within 20–100 ms of the postsynaptic neuron firing. In pyramidal neurons, AMPA/NMDA receptor based LTP occurs when a presynaptic action potential occurs up to 20 ms before postsynaptic neuron firing, but AMPA/NMDA receptor based LTD occurs if neuron firing precedes postsynaptic input by up to 20–100 ms, with a sharp transition of a few milliseconds between LTP and LTD timing [132]. The functional implication is that an input which arrives prior to postsynaptic firing is assumed to have contributed to the firing and has its weight increased, while an input that arrived after the firing is assumed to be irrelevant and has its weight decreased. However, in other types of neuron, LTD occurs if the presynaptic input precedes neuron firing, provided an additional signal is received. One example is parallel fibre AMPA/NMDA receptor synaptic inputs to a Purkinje neuron in the cerebellum. If a synaptic input is active just before the neuron fires, its weight is slightly increased (LTP) but if an input from a mossy fibre to the Purkinje neuron is also received the input weights are reduced (LTD) [500]. The implication is that the additional signal is communicating that the recently active synapses were in some way inappropriate.

The firing of the postsynaptic neuron is communicated to the synapse by an action potential that backpropagates from the soma into the dendritic tree in parallel with the regular action potential that propagates down the axon [129]. This backpropagating action potential attenuates as it propagates through the dendritic tree, because of the activity of various types of voltage gated K^+-channels [501]. Because of the characteristics of these channels, the propagation is more effective if there is some dendritic depolarisation as a result of synaptic activity. The backpropagating signal will therefore tend to reach dendritic regions where there has been significant recent synaptic activity [502] as a result of inputs from multiple presynaptic neurons that target those regions and generated local depolarisation.

The dependence on relative timing is a result of the detailed properties of NMDA receptors. An NMDA receptor allows passage of Ca^{++} only if it is glutamate bound and if the local membrane is depolarised sufficiently to release a Mg^{++} ion that otherwise blocks the channel. Hence the Ca^{++} current required to trigger synaptic weight changes only occurs if the presynaptic glutamate release occurs close in time to the arrival of a backpropagating action potential adequate to release the Mg^{++} block. The difference between LTP and LTD may involve different NMDA receptor subtypes, difference in the Ca^{++} entry routes, and resultant differences in the direction of change [136]. LTP mechanisms were discussed earlier. In the case of LTD, one mechanism could be reduction in AMPA receptor conductance by dephosphorylation [503].

5.3.2.5 The Factors Influencing Synaptic Weight Changes

Overall, there are many signalling factors affecting whether there will be a synaptic weight change at a particular synapse, and the direction, magnitude and duration of any change. For one synapse, many of these factors but not necessarily all will play a role. Important factors are the relative timing and degree of presynaptic glutamatergic input and postsynaptic activity. Other factors are presynaptic activity resulting in release of other neurotransmitters, including dopamine and acetylcholine as discussed earlier, and also for example serotonin [504] and norepinephrine [505]. GABA also appears to play a role in LTP [506].

These signalling factors act via various kinase cascades to implement weight changes by various mechanisms, including action on existing proteins, new protein translation from existing mRNA which may be located close to the synapse, and transcription of new mRNA followed by protein translation. There is evidence in favour of a tendency for these different mechanisms to result in changes lasting for different periods of time, but this is not always the case.

Note also that LTP and to some degree LTD in pyramidal neuron AMPA receptor synapses has been discussed most fully, but there can be differences between such synapses in different brain areas. LTP can also occur in AMPA receptor synapses on to interneurons [507] in areas where pyramidal neurons are the principal neurons, and in AMPA receptor synapses on to other types of principal neurons. There can also be changes to the weights of GABAergic synapses, perhaps in support of LTP at glutamatergic synapses [508].

5.3.2.6 Other Types of Synaptic Weight Changes

The discussion has mainly been limited to synaptic weight changes to glutamatergic synapses on to pyramidal neurons. However, weight changes have been observed to GABAergic synapses on to pyramidal neurons [509]. Various types of projection neurons in the cerebellum, basal ganglia etc. show LTP and LTD on both glutamatergic [136] and GABAergic synapses [510]. LTP and LTD have also been observed in glutamatergic synapses on interneurons. [511] Frequently, the direction and magnitude of synaptic weight changes depend on the relative timing of synaptic input and target neuron firing [512].

Another factor in management of synaptic weight changes can be signalling from the postsynaptic neuron back to the presynaptic neuron. Such signalling is generally implemented by endocannaboloid synthesis in the postsynaptic neuron, with diffusion back to the presynaptic neuron [215]. For example, in some pyramidal neurons, LTD triggered by presynaptic spiking following postsynaptic spiking is implemented by changes to the presynaptic neuron [513]. The appropriate timing relationship is detected by endocannaboloid diffusion to the presynapse combined with activation of presynaptic NMDA receptors.

5.3.3 Changes to Neuron Integration Algorithm

Integration of synaptic inputs to determine neuron output is heavily dependent on the local dendritic excitability in response to combinations of postsynaptic potentials. This excitability is determined by the distribution and properties of different voltage gated ion channels over the surface of dendrites. A wide range of different sodium, calcium, and potassium voltage gated channels are present in the same dendrite [514]. Changes to the integration algorithm will involve changes to the properties and/or distribution of voltage gated channels of different types. Such changes could be implemented on different timescales, with shorter timescales implemented by changes to channel properties and longer timescales perhaps requiring gene transcription and protein synthesis or protein synthesis from pre-existing mRNA. In general the management of changes to the properties and distribution of voltage gated channels over the dendritic surface has not been investigated to the same extent as at the synaptic level. However, it is becoming clear that such changes can occur in response to neuron activity [515], and that changes to both excitability and synaptic weights can occur independently or coordinated in the same local region [516]. However, there are only a few experiments demonstrating long term activity-dependent changes to dendritic excitability, and much more work is required to determine how common such changes are [517].

A wide variety neurotransmitters including glutamate, acetylcholine, serotonin, noradrenaline, dopamine and NO can act via receptors to affect voltage gated ion channels [515, 518]. Opening of voltage gated Ca^{++} channels can also lead to changes to other ion channels. These changes are implemented via kinase and phosphatase cascades including CaMKII, MAPK, PKA and PKG. In many cases phosphorylation of the target ion channel alters the kinetics of the channel.

Regulation of activity dependent changes to dendritic excitability by local protein synthesis or mRNA transcription is believed to occur, but relatively few examples are yet known [517]. One example is that the gene for M-type K^+ channels can be activated by Ca^{++} entry through L-type voltage gated channels. Such Ca^{++} entry leads to movement of the NFAT transcription factor from the cytoplasm to the nucleus [519]. Other routes for accessing voltage gated ion channel protein synthesis or mRNA transcription are believed to include MAPK, PKA and PKC cascades [517].

Hence as for the management of synaptic weight changes, neuron activity plus multiple neurotransmitters acting via multiple but interacting chemical pathways can trigger changes to dendritic excitability.

Another aspect of dendritic excitability is the degree to which backpropagating action potentials are transmitted through the dendritic tree. The A-type K^+ channel is key in limiting backpropagation. Downregulation of the channel by PKA, PKC and MAPK kinase cascades increases backpropagation [520].

5.4 Multiple Paths Contribute to Any Neuron Behaviour

The determination of neuron behaviour from moment to moment, from production of an action potential to change to the weight of one synapse, can potentially be influenced by many inputs to the neuron acting via many different chemical paths with considerable crosstalk between the paths.

The behaviour selected at each point in time is determined the specific combination of current and recent inputs, and will be the result of contributions from multiple paths, generally with no one input being decisive.

Chapter 6
Major Anatomical Structures

6.1 Sources of Information

There are many different types of neuron organized into many different anatomical structures in the mammal brain. Many of these structures are connected together, sometimes by multiple routes, and often indirectly via other structures. Neuron physiology involves large numbers of different chemicals interacting by many complex pathways.

A vast amount of experimentation has been performed on the anatomy of the mammal brain. Direct investigation of human brain anatomy and physiology is limited because physically intrusive experiments are only possible post mortem, with the exception of very limited experiments with patients suffering from various brain illnesses. However, there are strong similarities between the human brain and other mammal brains. All the major structures discussed in this chapter have homologues in almost all mammal species. Primate brains are the most similar at a more detailed level, but strong similarities are also present in the cat and rat brains. Furthermore, modern imaging techniques make it possible to visualize major human anatomical structures directly, and follow their activity in time. These techniques do not allow resolution of individual neurons, and even many anatomical substructures are too small to be resolved. However, they make it possible, for example, to identify the structures that are active during different cognitive processes, and the structures that are damaged in patients with cognitive deficits. At the neuron physiology level, there are many similarities even in brains much more distantly related to human than other mammals.

Much of the information presented in this chapter is based upon experiments with monkeys, cats and rats, and to a lesser degree with other mammal species, confirmed as far as possible by imaging and post mortem anatomical work on the human brain

A further complicating factor is that although there are, for example, correspondences between structures in the rat brain and in the human brain, for historical reasons corresponding structures do not always have the same names.

L.A. Coward, *Towards a Theoretical Neuroscience: from Cell Chemistry to Cognition*, Springer Series in Cognitive and Neural Systems 8, DOI 10.1007/978-94-007-7107-9_6,

In the chapter, knowledge derived from investigations of non-human brains will not be explicitly identified, unless there is some experimental evidence that the human brain is significantly different. In other words, the human brain will be assumed to be similar to other mammal brains unless there is evidence to the contrary. Some theoretical reasons why this is a reasonable position will be described in Chap. 7.

6.2 High Level Structure of the Human Brain

The brain is subdivided into a number of anatomical structures. These subdivisions are identified in several different ways. One way is that different structures may contain neurons that differ in morphology, size, density, type of neurotransmitter(s) generated and/or types of neurotransmitter receptors. Another way is that different structures may be separated by regions that contain axons but few somas. A third way is that the inputs and outputs of one structure are different from those of another structure.

Major structures have substantial differences of all three types. However, each major structure has substructures with somewhat smaller differences, and the substructures may have yet finer "sub-sub-structures" with yet smaller differences and so on.

A list of some large and smaller structures, and their substructures where they exist, is given in Table 6.1. All of the structures in the table will be discussed in this chapter.

6.3 The Cortex

The cortex is a sheet of tissue with a thickness in the range 2–3 mm. The sheet is wrapped around the outside of the cerebral hemispheres but folded so that the total area is about 1,600 cm^2 [521], much larger than a smooth surface would be. The cortex contains ~$1–2 \times 10^{10}$ neurons [522]. In the human brain section in Fig. 6.1, it is the cortex that is visible in the upper half. The cortex is conventionally divided into a number of lobes that are usually separated by particularly deep folds. The occipital lobe is at the back, the frontal lobe is at the front, the temporal lobe is along the sides, and the parietal lobe is over the top. The inner part of the surface area where the two hemispheres face each other is sometimes called the limbic lobe. The part of the frontal lobe that is under the brain is called the orbitofrontal or orbital cortex.

The folding of the cortical surface results in regions that bulge out and regions where the surface folds in. A gyrus (plural gyri) is a region that bulges out, a sulcus (plural sulci) is a region that is folded in. Cortical folding differs between individuals, but some large gyri and sulci can be seen in many different brains. These major

Table 6.1 Brain anatomical structures

Major structure	Substructures and more detailed structures	Examples
Cortex	Areas	Primary visual
	Columns	
	Layers	I; II; III; IV; V; VI
Hippocampus	Regions	CA1; CA2; CA3; Dentate gyrus
Thalamus	Nuclei	Ventrolateral; reticular
Basal ganglia	Divisions	Dorsal; ventral
	Nuclei	Striatum; globus pallidus
	Regions	Striatal patch and matrix
Basal forebrain	Nuclei	Septal; nucleus basalis
Amygdala	Divisions	Cortex like; striatal like
	Nuclei	Cortical, central
Hypothalamus	Nuclei	Mammillary; paraventricular
Cerebellar cortex	Lobes	Neocerebellum; paleocerebellum
	Layers	Molecular; Purkinje; Granular layers
	Microzones	
Cerebellar nuclei	Divisions	Dentate; interpositus
Neurotransmitter distribution		Locus coeruleus; raphe nucleus
Other nuclei	Pontine nuclei; olivary nuclei; habenula	Columns in the inferior olive

The table lists the larger anatomical structures found in the mammal brain, their organisation into substructures, and some example substructures. Other less prominent structures will also be discussed in this chapter. All structures are made up of neurons, the number, type and connectivity of neurons can differ between different structures and substructures

gyri and sulci and often named by their position in a lobe (e.g. the superior frontal gyrus is the protruding area at the top of the frontal lobe). The names of some prominent structures are given in Fig.6.2.

A number of terms are important in relative positions in the brain (see Box 6.1). For example, the part of a gyrus or sulcus closest to the front of the brain is called anterior (e.g. the anterior part of the superior temporal gyrus). The part closest to the rear of the brain is called posterior. The part towards the top is called dorsal and towards the bottom or underneath is called ventral. The part towards the centre is called medial while the part away from the centre towards the side is called lateral.

There are areas of the cortex that differ from each other in physical structure and functional role. Some of these areas, often called Brodmann areas, are also shown in Fig. 6.2. For different individuals, these areas can be in different positions relative to the major gyri and sulci so the mapping in Fig. 6.2 is approximate. Observations of cortical activity during cognitive tasks are often described by giving the positions of active regions relative to these topological structures. These descriptions are useful, but to varying degrees are therefore only approximate in terms of more functionally significant cortical areas.

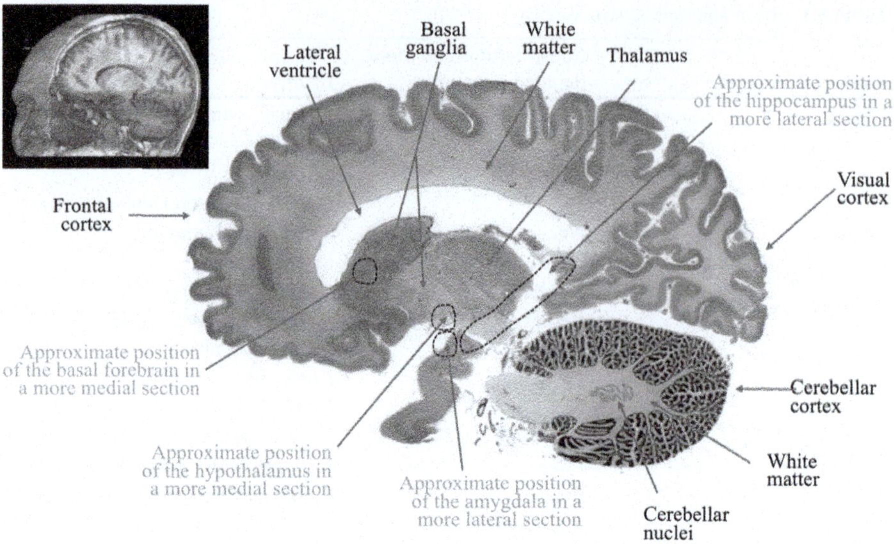

Fig. 6.1 Medial sagittal section of a human brain indicating positions of the seven major structures. The intricately folded sheet of neurons forming the cortex is visible around the outside. The paler white matter made up mainly of axons is located underneath the cell sheet. The lateral ventricle is filled with cerebrospinal fluid. Beneath the lateral ventricle are the subcortical nuclei including the thalamus and basal ganglia. The basal ganglia are located in front of and lateral to the thalamus. The hippocampus, amygdala, hypothalamus and basal forebrain are not visible in this section, but their approximate positions in sagittal sections medial to (closer to the midline of the brain) or lateral to this section are indicated. The cerebellum is located below the back of the cortex, and has a similar structure of intricately folded outer cell layer (the cerebellar cortex), white matter and subcortical nuclei (Image from Michigan State University Brain Biodiversity Bank. Supported by National Science Foundation)

Fig. 6.2 (continued) numbering by more recent work are shown, but there are further subdivisions that are not shown. (**b1**) Inside view of left hemisphere with right hemisphere cut away. The paracentral gyrus is partly in the frontal lobe and partly in the parietal lobe. The superior frontal gyrus continues over from the side view in a1, ending at the cingulate sulcus which separates it from the cingulate gyrus in the limbic lobe. The precuneus is regarded as being in the parietal lobe, separated by the parietal-occipital fissure from the cuneus in the occipital lobe. The inferior temporal gyrus continues over from the side view, ending at the inferior temporal sulcus. On the other side of the inferior temporal sulcus is the fusiform gyrus, regarded as part of the temporal lobe, and separated from the parahippocampal gyrus in the limbic lobe by the collateral fissure. The cortical area of the cingulate gyrus that faces the corpus callosum is called the retrosplenial cortex. (**b2**) The division of the inside view into Brodmann areas. Areas 29 and 30 face the corpus callosum and make up the retrosplenial cortex. Again, the correspondence of Brodmann areas with major landmarks varies between individuals, and there are more recently determined subdivisions within the original Brodmann areas. (**c1**) View of the left hemisphere orbitofrontal cortex from underneath the brain. The gyri and sulci in the orbitofrontal cortex are very variable between individuals, one of at least three different forms is illustrated. (**c2**) The division of the orbitofrontal cortex into areas. The orbitofrontal cortex received much less attention from Brodmann, and a number of the illustrated areas are not in his human cortex map. In particular, he did not identify the insular cortex (area Ia in the diagram) in the human brain he investigated. Many of the illustrated areas are in fact subdivided, including four subdivisions of the insular cortex

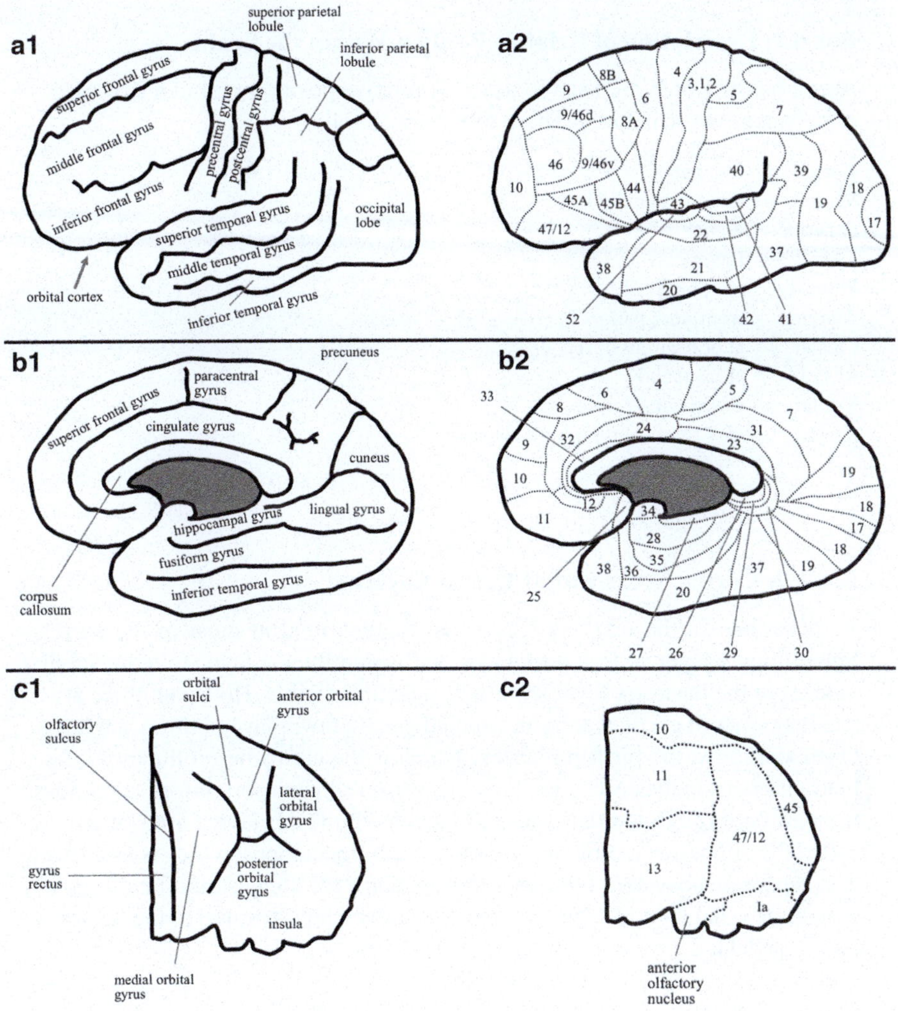

Fig. 6.2 The gross physical appearance of the cortex, approximately related to its organisation into different Brodmann areas. (**a1**) Side view of left hemisphere. The frontal lobe in this view contains four major sulci (the precentral sulcus and the superior and inferior frontal sulci), separating the frontal lobe visible in this view into four major gyri: the precentral; and the superior, middle and inferior frontal. The parietal lobe is made up of the postcentral gyrus, separated from the superior and inferior parietal lobules by the postcentral sulcus. The superior and inferior parietal lobules are separated by the intraparietal sulcus. The supramarginal gyrus wrapped around the end of the lateral sulcus (which separates the frontal and temporal lobes) is also regarded as part of the parietal lobe. The arcus gyrus is on the boundary between the parietal and occipital lobes. The occipital lobe in the side view is not subdivided. The temporal lobe is separated from the frontal lobe by the lateral fissure. The part of the temporal lobe visible in this view includes two major sulci (the superior and middle temporal sulci) separate three gyri (the superior, middle and inferior temporal). The angular gyrus wrapped around the end of the superior temporal sulcus is regarded as part of the temporal lobe. (**a2**) The division of the side view into Brodmann areas. The locations of the areas is not in exactly the same place for all individuals relative to the major landmarks in a1, but some typical locations are indicated. Some of the modifications to Brodmann's original

Box 6.1: Description of Relative Position Within the Brain

A number of words are used to describe relative position within a structure or between two structures. These terms are as follows:

Dorsal	top, upper
Ventral	bottom, underside
Superior	higher
Inferior	lower
Posterior	rear, hind end
Anterior	front, fore part
Caudal	at or near tail or rear end
Rostral	at front end
Medial	towards the centre
Lateral	way from centre towards the side

Box 6.2: Cortex Areas with Different Layering

As described in the text, six layers can be identified in most of the cortex. Sometimes a layer is less prominent, and sometimes a layer may be subdivided, but the basic six layer pattern is generally visible. However, there are a small number of cortical areas in which there are fewer layers, often just three. One example is the piriform cortex, which is the area where olfactory information arrives in the cortex. Unlike other senses, this sensory information does not go first through the thalamus. The areas with fewer layers are sometimes called the allocortex. Another example is the hippocampus, which is discussed as a separate subsystem later. In addition, there are some subcortical regions with cortex like layering. One example discussed later is some of the nuclei of the amygdala.

There are a number of different types of neuron in the cortex. The most common neuron is the pyramidal, which is an excitatory neuron with an axon that can extend long range across the brain. Another excitatory neuron is the spiny stellate, which is generally similar to a pyramidal but has axon outputs limited to its local region. There are a number of inhibitory interneurons, also with mainly local outputs.

The neurons in the cortex are arranged in an easily visible structure. With a few exceptions (see Box 6.2) the sheet is separated into about six layers, conventionally labelled by Roman numerals from layer I closest to the skull to the innermost layer VI. Different layers can be distinguished by the different numbers and sizes of pyramidal neurons they contain. The layer arrangement is not exactly the same across the entire cortex, in fact the well over 50 different cortical areas can be distinguished by a detailed layer structure that is constant within an area and transitions rapidly at a boundary to different structures in neighbouring areas.

Table 6.2 Frequencies observed in EEG measurements of cortical voltages during different cognitive states

Cognitive state	Name	Frequency band (Hz)
Memory	Theta	4–7
Relaxed	Alpha	8–12
Thinking	Beta	12–30
Attention	Gamma	30–100
Sleep stage I	Theta	4–7
Sleep stage II	Sigma	12–14
Sleep stage III/IV	Delta	Up to 3
REM (Dream) sleep	Similar to awake	

In addition, there are column structures that extend vertically through the cortex sheet, spanning all the layers. There is strong vertical connectivity between neurons in different layers of a column, and all the pyramidal neurons in the same column have some similarities in the circumstances in which they generate outputs. There are estimated to be 3–5 million such columns in the human brain. Below we will consider cortical neurons, layers, areas and columns in more detail.

The activity of the brain is reflected in varying electrical voltages which can be detected externally. Because the cortex is closest to the skull, electrodes attached to the scalp during electroencephalogram (EEG) measurements can detect cortical voltages at different points. It is found that cortical voltages vary in time in a number of characteristic frequency bands, and different frequencies appear during different types of activity as shown in Table 6.2. Voltage variations in the beta band (12–30 Hz) appear during thinking, variations in the gamma band (30–100 Hz) appear when attention is paid to something, and other frequencies appear when relaxed, drowsy or asleep. It is of interest to note that on the basis of these electrical signals, sleep is not a period of rest for the brain, but a period of complex activity very different from waking activity.

6.3.1 Cortical Neurons

Typical cortex neurons are illustrated in Fig. 6.3. The most common neuron in the cortex is the pyramidal neuron, which generates the neurotransmitter glutamate at all its synapses onto other neurons and therefore excites those neurons. As illustrated in Fig. 6.3, a pyramidal neuron has an apical dendrite emerging from its soma in the opposite direction to its axon, and a number of basal dendrites that emerge all around the soma in the directions perpendicular to the axon. The axon of a pyramidal neuron can project long distances within and sometimes outside the cortex, but branches of a pyramidal axon also terminate locally on neighbouring neurons.

In areas of the cortex that receive inputs fairly directly from the senses, there are many small glutamatergic (i.e. excitatory) neurons called spiny stellate cells. Spiny stellate cells can have dense dendrite projections in all directions, although

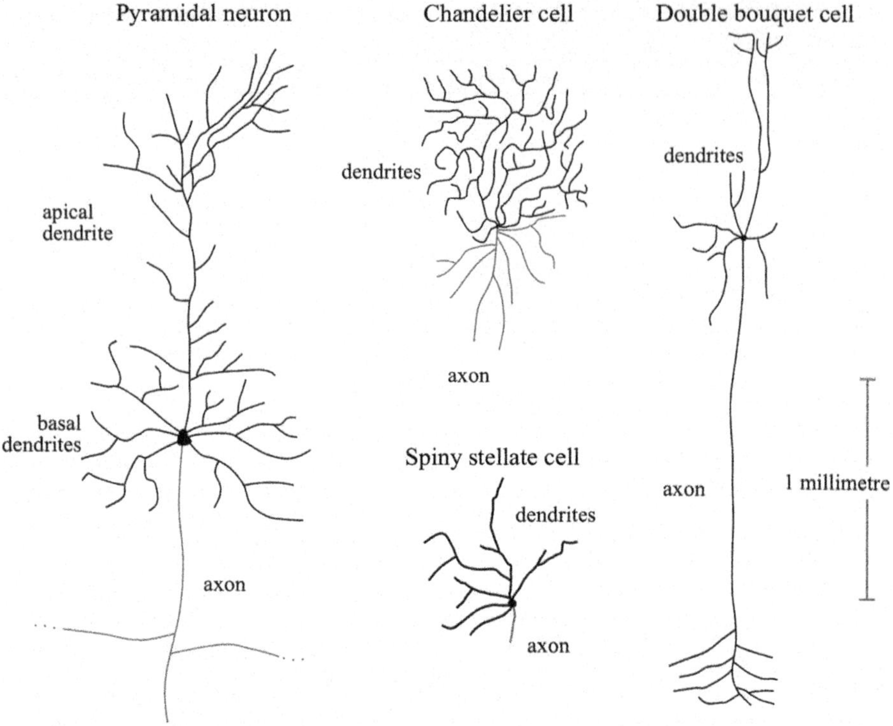

Fig. 6.3 Some different types of neuron in the cortex. The pyramidal is an excitatory neuron with a relatively large soma, an apical dendrite that extends vertically through the cortical sheet, and basal dendrites that extend more laterally. Spiny stellates have a star like dendritic tree, but also excite their targets. Pyramidal axons can extend for many centimetres, off the illustration, but a spiny stellate has an axon that branches extensively in the same volume as its dendrites. Both pyramidal and spiny stellate neurons generate the neurotransmitter glutamate at their synapses on to other neurons. Chandelier and double bouquet cells are two types of inhibitory interneuron. Interneuron axons generally only project locally, within about 500 μm of their source soma, and generate the neurotransmitter GABA at synapses onto other neurons

they sometimes exhibit an asymmetry with most dendrites emerging from one side. Spiny stellates lack the prominent apical dendrite seen in pyramidals, but can have a partial apical-like dendrite that extends only into the cortical layer immediately above the one where their soma is located. Spiny stellates only project locally, and are therefore called interneurons, although unlike most cortical interneurons they are excitatory. For a spiny stellate cell there is a strong overlap between the volume occupied by its dendritic tree and the volume occupied by its axon branches.

Small excitatory neurons, whether spiny stellates or very small pyramidals, are sometimes called granule cells, because they give the layers of the cortex in which they are predominant a rather granular appearance. Such layers (often layers II and IV) are sometimes called the granular layers.

Fig. 6.4 Pyramidal neuron receptive fields within neighbouring cortical columns in visual processing area TE of the visual cortex in the macaque monkey [526]. Probes detected the activity of individual pyramidal neurons in different cortical layers in area TE. The monkey given a wide range of visual stimuli, and the activity of each neuron in response each stimulus was measured. The receptive field of each neuron was determined by calculating the common denominator in all the different visual stimuli that produced activity. The neurons within a cortical column with diameter about 400 μm all had similar receptive fields. The receptive fields of neurons in adjacent columns were different, with no obvious similarities with their neighbours

The concept of receptive field has been very important in understanding the information role of pyramidal neurons. The receptive field of a neuron is classically defined as the area of the sensory input space within which a stimulus could elicit a response from the neuron. For example, the area of the retina influencing the output of a neuron in the visual cortex is referred to as the receptive field of that neuron. However, in higher visual areas, the receptive field defined in this way can be a significant proportion of the retina [523], and the neuron tends to respond to any size of a particular shape anywhere within that retinal area [524]. In the higher visual areas of the cortex that receive inputs mainly from the eyes, it is possible to assign a visual shape meaning for neuron receptive fields as illustrated in Fig. 6.4. In yet higher cortical areas in which neurons receive inputs derived from multiple senses, a neuron will respond to activity by different groups of the neurons providing its inputs, but no useful definition of receptive field in terms of a sensory domain is available. Such a pyramidal neuron responds in circumstances defined by combinations of the circumstances in which its input neurons respond.

The definition of the term "receptive field" in visual processing is often expanded to include not just the retinal area but also the structure within that area which will generate a response [525]. The definition can be expanded further to include the circumstances in which a higher level neuron will respond, in this case the receptive field is defined by the different subsets of its inputs which if active at the same time will cause the neuron to respond. This definition of the term "receptive field" will be used throughout the book.

There are also a number of small neurons which generate the neurotransmitter GABA and therefore generally inhibit all their target neurons [408]. Most of the outputs of these neurons are targetted locally within about 500 μm of their soma, and they are therefore called interneurons. However, their inputs may come from a larger region.

On the basis of physical shape it is possible to distinguish over a dozen different types of interneuron including Martinotti, double bouquet, bipolar, bitufted, basket, chandelier, etc. of which two are illustrated in Fig. 6.3. Somas of most of the different forms occur in all layers except I, where soma form is very variable and almost all axons and dendrites are confined to layer I [527]. The significance of form is that it reveals differences in the cortical layers from which interneurons collect their inputs and to which they deliver outputs.

For example, some interneurons collect most of their inputs within one layer, and deliver most of their outputs to the same layer. Other interneurons collect most of their inputs from one layer but deliver their outputs to two or three other layers. Still other interneurons collect inputs from two or three different layers but deliver their outputs to just one layer. There are some interneurons that collect inputs from the upper layers and unlike most interneurons deliver outputs over a wider area outside the usual 500 μm local range.

Another important difference between interneurons is the location on pyramidal neurons to which they synapse. Some interneurons target distal dendrites, others target proximal dendrites, others target somas, and chandelier (also called axo-axonic) neurons even target the early stages of the pyramidal axon. Such differences in position determine the type and strength of effect an inhibitory input will have. Although there is evidence that axo-axonic interneurons have an inhibitory effect on their target pyramidal neurons [528], there is also evidence for an excitatory effect [529]. An excitatory effect could be due to local differences in the concentration on a Na^+-K^+-$2Cl^-$ ion pump changing local Cl^-ion concentrations [530].

Among other functions, interneurons influence the frequency with which pyramidals produce outputs. The predominant frequencies observed in EEG measurements are believed to reflect the operation of interneurons. For example, interneurons like basket cells interconnected together and targetting close to the pyramidal somas support gamma frequency activity [531].

6.3.2 *Cortical Synaptic Weight Changes*

As described in Chap. 5, synaptic weight change mechanisms have been extensively studied in glutamatergic synapses on to pyramidal neurons in the cortex and hippocampus. Primary conditions triggering weight changes are arrival of action potential at a synapse shortly before the neuron fires leading to an increase in weight, and arrival of action potential at a synapse shortly after the neuron fires leading to a decrease in weight. Changes are fairly specific to the recently active synapse. This basic mechanism is modulated by various neurotransmitters including dopamine, acetylcholine, serotonin and norepinephrine. The presence of these neurotransmitters results from activity in other brain structures such as the basal ganglia, basal forebrain, etc. Modulation often includes an effect on the timescale over which the weight change persists. The detection of relative timing between synaptic input and target neuron firing generally performed by NMDA receptors. An important early step in the chemical processes leading to weight changes is often entry of Ca^{++} ions through NMDA channels, other ligand or voltage gated ions channels, or release of Ca^{++} ions from internal stores. Weight changes are generally implemented by changes to synaptic size and the number of AMPA receptors in the postsynaptic density of the synapse.

Weight changes can also occur at GABAergic synapses on to pyramidal neurons [509]. In this case, if the neuron begins to fire, and an input to a GABAergic synapse arrives within <300 ms, the weight of the synapse decreases. If the input arrives >400 ms after the start of firing, the synaptic weight increases. The weight change is dependent upon Ca^{++} entry triggered by backpropagating action potentials. Some sensitivity to neurotransmitters like serotonin has been observed [532].

6.3.3 Cortical Connectivity

Most of the outputs from neurons in the cortex target other neurons in the cortex. Interneurons target other cortical neurons that are relatively close. Pyramidal neurons can project locally and/or longer distances, but about 98.5 % of pyramidal projections are to other parts of the cortex within the same cortical hemisphere, about 1 % are between the hemispheres, and less than 0.5 % go to other brain areas. An area of the cortex will project strongly to some areas and much less to others [533]. Sometimes projections between two areas are reciprocal, sometimes predominantly one way. In general, the vast majority of connectivity within the cortex is excitatory, and almost all inhibitory connectivity is relatively short range.

The hippocampus is in some ways an extension of the cortex, and has a very high volume of reciprocal glutamatergic connectivity with the cortex. Many subcortical structures provide inputs to the cortex, and in general such subcortical structures receive reciprocal inputs back from the cortex. The largest source of subcortical inputs to the cortex is the thalamus, including sensory inputs that (with the exception of smell and taste) all reach the cortex via the thalamus. Thalamic inputs are glutamatergic. Almost all outputs from the basal ganglia and cerebellum only reach the cortex via the thalamus. However, there are a range of subcortical structures that send relatively light inputs to the cortex. The hypothalamus provides inputs to the entire cortex, probably the largest subcortical source of input after the thalamus and utilizing a wide range of different neurotransmitters [534]. The amygdala sends glutamatergic inputs to the cortex [535] that target spines on the dendritic trees of pyramidal neurons, but not dendritic shafts or somas [536]. Raphe nucleus neurons produce serotonin and extensively target both pyramidal and to a smaller extent interneurons [537], with differences between different regions [538]. For example, in the prefrontal cortex, serotonergic axons are present in all layers, with a slight reduction in density in layer III and differences in receptor type between layers [539]. The locus coeruleus also extensively targets the cortical neurons with the neurotransmitter norepinephrine [198]. Acetylcholine is produced by some basal nuclei neurons that target the cortex, both pyramidal neurons and interneurons, with again with some differences in targetting between layers [540]. Dopaminergic neurons in the ventral tegmental area of the basal ganglia also target the cortex. Receptors are found on both pyramidal and interneurons, with higher densities in deeper layers [539]

As discussed in Chap. 4 there are different receptors in cortical neurons for the same neurotransmitter, with different effects. One example is the two receptors for glutamate, one immediately excitatory and the other resulting in long term changes to synaptic strength. Another is the two types of receptor in the cortex for GABA, both inhibitory but acting at different speeds. Multiple receptor types are also known

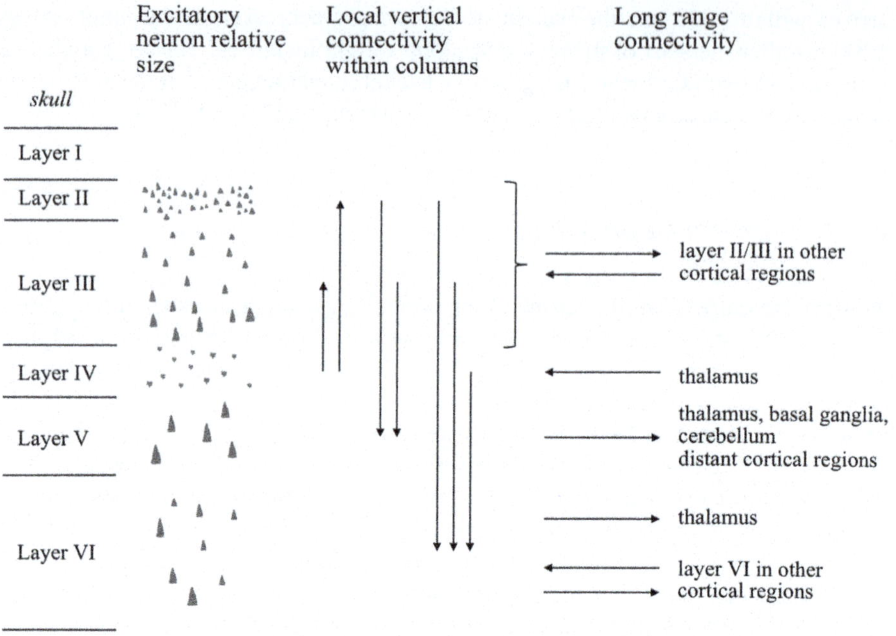

Fig. 6.5 Standard cortical layering. The illustrated pattern of principal neurons (i.e. pyramidals and spiny stellates) and of connectivity is visible with some variations in all areas of the cortex. Layer I has very few neuron bodies, it is largely made up of axons and dendrites. Layer II has a high density of relatively small pyramidal neurons, with the size of the neuron bodies increasing towards layer III, and increasing further across layer III. Sometimes layers II and III are regarded as a single II/III layer. Layer IV contains a high density of very small neurons, which may be pyramidals or spiny stellates. Layer V contains the largest pyramidal neurons, and layer VI contains a wide range of different sized pyramidals. There is a pronounced pattern of vertical connectivity between principal neurons, from IV to II/III, from II/III to V, and from II/III and IV to VI. There is also a pronounced pattern of connectivity laterally, from II/III to and from layer II/III in other cortical regions, from the thalamus into IV, to the basal ganglia, cerebellum and distant cortical regions from V, and from layer VI to and from the thalamus

for acetylcholine, norepinephrine, epinephrine, and serotonin. Because different cortical areas have different densities of receptor molecules of different types, sub-cortical structures are able to influence different areas in different ways.

6.3.4 Cortical Layers

Viewed in cross section, the cortex has a distinct layer structure, primarily made visible by differences in the size and numbers of pyramidal neurons. As shown in Fig. 6.5 there are typically six prominent layers that are conventionally numbered I–VI.

Layer I has very few neuron somas, it is made up mainly of axons and dendrites. Many of the axons in layer I originate from both pyramidal and interneurons in the cortex, some from relatively local pyramidal neurons and interneurons, others from pyramidal neurons in more distant cortical areas. Some of the axons in layer I come from outside the cortex. The dendrites in layer I are parts of the apical dendrites of pyramidal neurons in layers II–V. The limited number of somas in layer I belong to interneurons with no particular consistency of form. Their dendritic trees are limited to layer I, and their axons are also largely confined to layer I but in some cases can extend vertically through layers II/III and even into layer IV [527].

Layer II is made up of large numbers of small pyramidal neurons. Layer III contains medium sized pyramidal neurons. Sometimes layers II and III are regarded as one layer with gradually increasing pyramidal size from top to bottom. Layer IV contains very small excitatory cells which may be spiny stellates or pyramidals. Layer V contains the largest pyramidals, sometimes with an increase in size going from the boundary with layer IV to the boundary with layer VI. Layer VI contains a wide variety of both large and small pyramidals. The small size of neurons in layers II and IV give those layers a granular appearance, and they are sometimes called granular layers and their excitatory neurons called granule cells.

Pyramidal neurons in layers II and III can receive inputs from and send outputs to neighbouring cortical areas. The granule cells in layer IV receive inputs from the thalamus. Large neurons in layer V can send outputs to distant cortical areas and to the basal ganglia, cerebellum (via an intermediate structure called the pontine nucleus) and spinal cord. Large neurons in layer VI can send outputs to distant cortical areas and to the thalamus. There is also a pronounced local vertical connectivity from pyramidal neurons and spiny stellate cells in layer IV to pyramidal neurons in layers II/III, from pyramidals in layers II/III to those in layer V, and from those in layers II/III and IV to those in layer VI [541]. This vertical connectivity is an important part of the column structure discussed below.

Interneurons tend to have dendritic trees in specific layers, and axonal branching in the same or other specific layers [542] as illustrated in Fig. 6.6. Many interneurons have dendrites that only extend within their local column and are an important part of column definition. They therefore play a role in communicating the degree of activity between or within layers of one column. For example, some interneurons in layer II/III are observed to receive inputs from layers V/VI pyramidals [543]. Interneurons in layer I and some interneurons in layers II/III have axon branching in layer I that extends across many columns, and therefore play a role in regulating intercolumn activity.

All of these descriptions of cortical layers are to some degree approximate. As mentioned above, layers II and III are sometimes regarded as forming one layer. Layers IV and V are often separated into sublayers. As we discuss in a later section, the major way in which different areas of the cortex are identified is by such detailed differences in the layer structure.

6.3.5 *Cortical Columns*

There is a pronounced columnar organisation perpendicular to the cortical layers. A column is about 300–500 μm in diameter and contains several tens of thousands of neurons [544]. All the pyramidal neurons in one column tend to respond to

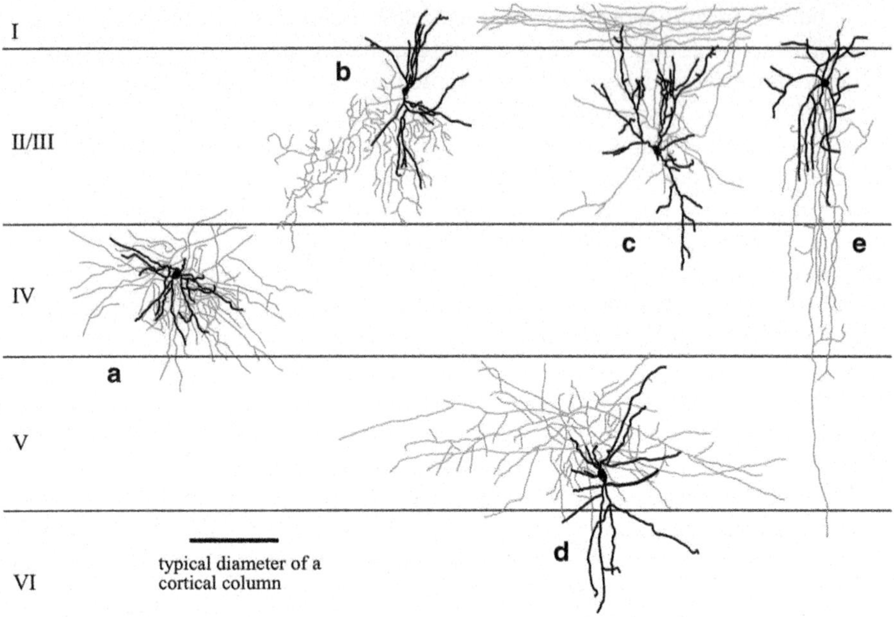

Fig. 6.6 Some different types of cortical interneurons. (**a**) Basket cells with somas in layer IV that have most of their dendrites and axons also confined to layer IV. (**b**) Chandelier cells in layers II/III with dendrites in layers II/III and I, but axon branches confined to layers II/III. (**c**) Martinotti cells in layers II/III have most of their dendritic tree limited to the same layers, but most axon branching in layer I and extending some distance in that layer. (**d**) Basket cells in layer V have dendritic trees extending over layers V and VI but axon branches only in layer V. (**e**) Double bouquet cells in layers II/III have most of their dendrites in the same layers but axon branching down through all deeper layers

similar circumstances, or in other words have similar receptive fields. For example, all the pyramidal neurons in a column with diameter about 400 μm within the cortical area TE tend to respond to a visual input containing the same shape, as illustrated in Fig. 6.4. These column structures are ubiquitous in mammals, even appearing in the mammal relative most distant from human beings, the duck billed platypus in the monotreme subclass [545].

Sensory inputs to a column from outside the cortex come via the thalamus. Many of these inputs arrive in layer IV, but there is no flow of connectivity from this layer back to the thalamus. However, there is a strong connectivity from layers V and VI to the thalamus [546].

Within a column, there is strong vertical connectivity between excitatory neurons as illustrated in Fig. 6.5. This connectivity goes from spiny stellate and pyramidal neurons in layer IV to pyramidals in layers II/III, from pyramidals in layers II/III to those in layer V, and from pyramidals in II/III and excitatory neurons in IV to pyramidals in layer VI.

Layers II/III send and receive connectivity from layers II/III in other columns in the cortex, including distant areas. Layer V sends outputs outside the cortex to the basal ganglia, cerebellum (via intermediate structures) and the spinal cord. Layer VI has reciprocal connectivity with layer VI in other cortical regions and with the thalamus.

There is also a high degree of local connectivity between neurons in a column. Most interneurons within a column target pyramidals and spiny stellates within the same column. There is also strong connectivity within a column between layer IV excitatory neurons, reflected in the asymmetry of the spiny stellate axonal trees.

In summary, within a column there is a connectivity flow thalamus → IV → II/III → V, and an additional flow IV and II/III → VI supplemented by V/VI → thalamus. There is also a high degree of inhibitory connectivity within the column, with more limited inhibitory connectivity between neighbouring columns. Outside a column but within the cortex there is reciprocal flow between layers II/III and II/III in other columns and reciprocal flow between layer VI and layer VI in other columns. There are inputs from the thalamus to layer IV, and between layer V/VI and the thalamus. Layer V generates outputs to other subcortical regions such as the basal ganglia and cerebellum.

6.3.6 Cortical Hemispheres

The two cortical hemispheres are similar in general appearance, but there are some significant differences. At an anatomical level, there may be some differences in the gyri and sulci into which the hemispheres are folded. At a physiological level, the corresponding areas in the two hemispheres may be different sizes and have slightly different detailed layer characteristics [547].

At a functional level, the left hemisphere receives sensory input from and sends motor commands to the right half of the body and vice versa. In addition, the two hemispheres play significantly different roles in some major types of cognitive processing. For example, in most of the human population, the left hemisphere predominates in speech processing [548], and the right hemisphere plays a larger role in music appreciation [549]. However, although one hemisphere may predominate in a particular task, activation is generally also observed in the other hemisphere, which is therefore playing some role.

6.3.7 Cortical Areas

Microscopic investigation of the cortical layer structure reveals that although the six layers characteristic of the standard layer structure described earlier are usually present, there are some significant differences between the layers in different areas.

In areas of the cortex with strong influence on motor behaviours (the motor cortex), the pyramidal neurons in all layers tend to be larger than average, and layer IV contains only moderate sized pyramidal neurons and no spiny stellates. Layer II and layer IV therefore do not appear granular, and areas of the cortex of this type are called agranular.

Areas of the cortex which receive inputs fairly directly from the senses (the primary sensory areas) have a very prominent layer IV (sometimes divided into sublayers) with large numbers of spiny stellate cells. Pyramidal neurons in all layers are smaller than average and layers II and IV appear very granular. Such areas are called granular cortex. These primary sensory cortices are also about 20 % thinner than typical.

Other areas of the cortex are closer to the standard layer descriptions, but there are still differences visible. These differences include the average sizes of neurons in different layers and the relative thickness of different layers. On the basis of such differences, Brodmann in 1909 identified 44 different cortical areas in the human brain and his numbering scheme is still widely used, with each area identified as BA followed by the number. A number of workers have more recently identified additions to and subdivisions of Brodmann's areas on the basis of the same neuron size and density criteria. Brodmann's areas with some of the more recent refinements are illustrated in Fig. 6.2. However, although every area appears in every human brain, there are considerable differences between individuals in area sizes and in their location relative to the major gyri and sulci. The cortex in the left and right hemispheres contain the same areas, although there are sometimes differences in the size and exact location between the hemispheres.

Areas can also be identified on the basis of the pattern of axon connectivity, the distribution of a neurochemical called neurofilament protein that influences axon and dendrite diameters, and the measurement of the density of receptors for different neurotransmitters [550]. The different types of measurement give results consistent with the areas defined on the basis of neuron size and density.

Cortical areas can be associated with different information processing functions by a number of techniques. One technique is observation of the types of cognitive deficits that result from damage to one or more areas. Another is imaging of the brain during different types of cognitive tasks and determining which areas are active during each type of task. A third technique is observation of the effects of stimulation of different areas during surgery. In general it is not possible with any of these techniques to make the physiological measurements that define areas. Hence active regions can only be identified relative to gyri and sulci landmarks. Because the precise mapping between areas and such landmarks varies between individuals, there can be some uncertainty in the precise areas that are active. However, enough is known to make it clear that the different areas of the cortex have different information processing roles.

In the following sections we will describe some examples of information about area roles that has been collected by a combination of cortex imaging during cognitive tasks, probing of living brains during surgery, determination of the cortical damage associated with different types of cognitive deficits, and physical examination of the cortex postmortem. There has been a very large amount of work of this type, leading to the conclusion that although different areas have different functions, most cortical areas support multiple different types of cognitive processes.

6.3.7.1 Areas Involved in Visual Processing

Information from the eyes goes first to the thalamus, and only from there to the cortex. It enters the cortex into area BA17 at the back of the brain. Information from the left eye enters BA17 in the right cerebral hemisphere, while information from the right eye enters the corresponding BA17 in the left hemisphere.

There are two pathways in which this visual information is processed across the cortex. One pathway climbs up the back of the cortex towards the top and is called the dorsal stream. This pathway takes a route BA17→BA18→BA19→BA40→BA7. The other pathway goes lower down along the side of the brain and is called the ventral stream. This pathway takes a route BA17→BA18→BA19→BA21 and some BA20. Although some of the same Brodmann areas appear in both routes, in BA17 and BA18 most of the information for the dorsal stream flows through different neurons from the neurons used by the ventral stream. The two streams use physically separate parts of BA19. However, although the paths are clearly separate, there is some information exchange between them.

These two paths have different cognitive roles. Brain activity is highest in the ventral path when a subject is identifying visual objects, while activity in the dorsal path is highest when locating visual objects in space [551]. Damage to the ventral path results in deficits in which a patient cannot name objects but can manipulate them correctly, while patients with damage to the dorsal path can name objects but cannot manipulate them. For example, the patient DF with damage to the parts of BA18 and BA19 supporting the ventral stream [552] made errors in naming everyday objects (identifying a cup as an ashtray, or a fork as a knife) but had no difficulty with visual tasks like opening doors or eating meals. In laboratory tests she was unable to name objects accurately but when asked to insert her hand into a rectangular slot could correctly adjust her hand as it approached the slot to any slot angle. However, her verbal judgements of slot angle in the absence of reaching were poor. On the other hand, a patient such as VK with damage to the ventral path [553] had no difficulty in verbally identifying an object, but when asked to pick it up had difficulty with reaching in the right direction.

The receptive fields illustrated in Fig. 6.4 are detected in a cortical area within the ventral stream, and are important for discrimination between different shapes in the visual field. Receptive fields in the dorsal stream discriminate between different types of motion [554].

6.3.7.2 Areas Involved Directly in Information from the Body and in Body Movements

Severe epilepsy and other neurological problems can sometimes be treated by surgical removal of defective parts of the cortex. Before any removal, the cortex is stimulated electrically at different points to determine the response and as far as possible to avoid incapacitating side effects. Such stimulation tests have led to the realisation that some brain areas map the surface of the body.

Information derived from touch, and information derived from muscles that indicates body position (called propioceptic information) enters the cortex in areas BA3, 1 and 2. These areas together are called the primary somatosensory cortex. Different regions within this cortex receive information from different parts of the body. Stimulation of different parts of this cortex resulted in sensations being felt in the corresponding part of the body [555]. In the same way as for vision, sensory information from the left side of the body enters the right cerebral hemisphere and vice versa.

It was also found that stimulation of BA4 (also called the primary motor cortex) or of BA6 (the premotor association area) resulted in body movements. Within both areas, stimulation of one subarea causes movement in a corresponding part of the body. As for sensory information, the motor areas in the left hemisphere influence the right side of the body and vice versa. Damage to the motor cortex results in the inability to initiate voluntary movement in the part of the body corresponding with the damaged area. However, some recovery of the use of the affected part of the body is possible in human beings over a period of months [556]. Hence there must be some ability for other parts of the cortex to take over the role of damaged parts.

BA4 and BA6 are also active when a movement is generated. Interestingly, there are similarities between the activity in BA4 when performing an action and the activity observed when someone else is observed performing the same action [557].

Thus the primary somatosensory cortex, the primary motor cortex and the premotor association area all contain a "mapping" of the body. In all of these "maps" the area corresponding with hands and mouth is much larger than the area corresponding with the rest of the body, reflecting the greater sensitivity and more detailed motor control of those body parts.

In a manner analogous with the processing of visual inputs, there may be two separate paths across the cortex processing somatosensory inputs [558]. One of these paths identifies objects on the basis of feel, the other locates objects relative to the body.

6.3.7.3 Areas Integrating Information from Different Senses

Area BA7 plays a role in integrating visual and somatosensory inputs to locate objects in space. It lies at the end of the ventral path for visual processing, Damage to BA7 results in deficits similar to those resulting from damage to earlier points on the ventral path: difficulties in reaching for objects.

In general, areas that receive almost all their information from just one sense are called monomodal, while areas receiving information derived from more than one sense are called polymodal.

6.3.7.4 Areas Influencing Speech

In the nineteenth century it was observed that brain damage to the cortex could result in speech deficits (called aphasias). One type of aphasia is Wernicke's, in which the ability to understand speech is lost. Verbalisations are still produced, with the rhythm and syntax of speech, but devoid of meaning. Another type of aphasia is Broca's, in which the ability to generate speech is lost. In Broca's aphasia, most comprehension is unaffected, although there can be difficulties with very complex sentences [559].

These aphasias were believed to be caused by damage to specific cortical areas in the left hemisphere (part of BA22 closest to the back of the brain for Wernicke's aphasia, BA44 and BA45 for Broca's aphasia). However, this classical picture is not confirmed by more recent work. An extensive study of the location of brain damage in patients with speech deficits found that a number of cortical areas in the left hemisphere influence speech. In the temporal cortex these areas include the middle temporal gyrus, the neighbouring superior temporal sulcus all the way back to include the angular gyrus, and across the superior temporal sulcus to the anterior part of the superior temporal gyrus. In the frontal cortex the areas include parts of the middle frontal and of the inferior frontal gyri. The classical Wernicke's and Broca's areas have limited effects, but are close to more significant areas [560].

Imaging of the brain during speech tasks has demonstrated activity in the areas indicated by the deficit studies, and in some additional areas [561]. A number of normal (i.e. undamaged brain) subjects performed a speech task while their brains were imaged. In this task, they heard a word which was the name of an animal, and they were asked to identify if the animal was *both* native to North America *and* useful to humans. They pressed a switch to indicate yes or no. They also performed a task in which they heard a sequence of tones and were asked to identify if one specific tone occurred twice in the sequence. In the tones task, most activity was in the right hemisphere, in the speech task, most activity was in the left. In the speech task, there were four separate large regions of the left hemisphere cortex that were active. These regions are illustrated in Fig. 6.7. *The first region* was in the temporal lobe with some spillover into the lower limbic lobe. *The second region* was in the frontal gyrus with some spillover into the upper limbic lobe. *The third region* was a separate location in the temporal lobe. *The fourth region* was a separate part of the cingulate gyrus plus neighbouring parts of the precuneus in the parietal lobe. These regions include a large number of different cortical areas. Both Wernicke's and Broca's areas were within active regions. Activity during the speech task was also observed in the left hemisphere thalamus and the basal ganglia, and in the right hemisphere cerebellum. These regions external to the cortex will be discussed later.

The picture that emerges is that for speech processing tasks, a considerable number of different areas play an information processing role. Some areas are so critical that their removal has a major effect on the ability to perform the task, others play a role but perhaps in cases of damage that role can be handled by other areas.

6.3.7.5 Areas Involved in Semantic Memory

Deterioration of specific areas of the cortex (BA20 and BA38 in the anterior temporal lobe) results in semantic dementia [31]. In this syndrome, there is a deficit

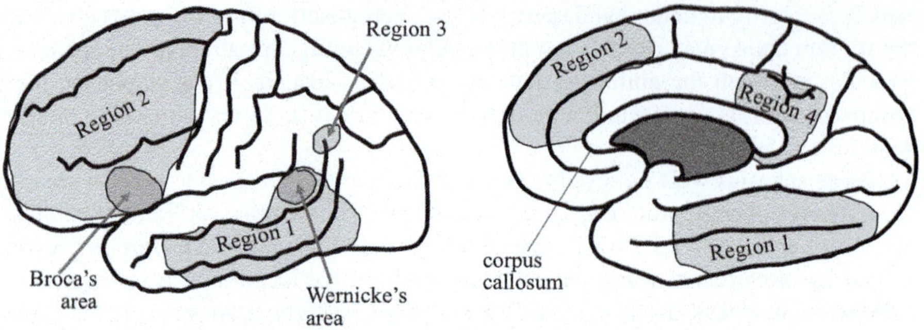

Fig. 6.7 Cortical areas activated during a speech task. Four separate areas were activated during a task involving hearing the name of an animal and pressing a button if the animal was both native to North America and useful to humans. *The first area* was almost all the middle temporal gyrus, spreading up across the superior temporal sulcus and down across parts of the inferior temporal gyrus, the fusiform gyrus and into the parahippocampal gyrus. *The second area* was the entire inferior frontal gyrus, plus the front and rear middle frontal gyrus (i.e. excluding the central BA9/46 area), most of the superior frontal gyrus including BA10, BA9, BA8B, and some spread of activation into the front of the cingulate gyrus. *The third area* was the angular gyrus in the temporal lobe. *The fourth area* included the rear part of the cingulate gyrus including the isthmus and the retrosplenial cortex, plus neighbouring parts of the precuneus

in all types of semantic memory tasks: picture naming; generating examples of a category; matching words and pictures; defining and drawing objects from their names; selecting an appropriate colour for an object (e.g. red for a tomato); and classifying objects by some criterion (e.g. living or non-living); etc. All types of category are affected, animals, plants, inanimate objects etc. A point to which we will return later is that episodic memory is unaffected. For example, patients can recognize that a picture has been seen before.

The picture obtained by brain imaging during semantic memory tasks is quite different [562]. The cortical activations observed during semantic processing are in the posterior temporal gyrus and in the left inferior frontal gyrus (Brodmann's areas 44, 45 and 47), with relatively little activation in the anterior temporal lobe areas that are the sites associated with semantic dementia. Furthermore, there is evidence that some cortical regions tend to be active during semantic memory tasks on some general categories and not others (e.g. animals or tools). Although the location of these sites varies between individuals, they are in the same place for the same individual observed at different times. However, judgements of the typicality of a category member occur in the same area (in the fusiform gyrus) independent of the type of category [563].

It is also interesting that semantic tasks involving words and tasks involving pictures tend to activate many of the same cortical areas, distributed throughout the inferior temporal and frontal cortex. Only a small proportion of activated regions were activated only by words (left anterior middle temporal gyrus and left inferior frontal sulcus) or only by pictures (left posterior inferior temporal sulcus) [564].

6.3.7.6 Areas Involved in Priming

In experiments when subjects attempted to identify brief glimpses of pictures that had either been seen recently (priming) or not seen recently (baseline), brain activity was observed in a wide range of cortical areas [565]. A few extra areas were active for baseline tasks.

6.3.7.7 Areas Involved in Planning

It is believed that the frontal lobe plays a major role in defining goals, planning and control [566], and it has been argued [567] that a sequence of areas across the frontal cortex participate in planning at different levels of complexity, with BA6 being involved in selection of simple responses, BA8 in selection of sequences of responses, BA9/46 in selection of complex sequences of responses and switching between tasks, and BA10 in setting more general goals. In other words, the sequence of areas played strong roles in increasingly abstract planning.

Imaging studies of the frontal cortex show that there is a group of areas that are active in a wide range of cognitive tasks including not only planning type tasks like spatial problem solving and suppression of inappropriate responses but also perception, working memory and episodic memory tasks [568].

6.3.7.8 Areas Involved in Face Processing

An area in the posterior of the fusiform gyrus in the right cerebral hemisphere called the fusiform face area (FFA) is active when a subject is performing a task involving human faces, and much less active when the task involves animals, inanimate objects, or even whole human bodies [569]. The higher activity is present for full face views, profiles, and even cartoon faces. Damage in this region can result in prosopagnosia, or the inability to recognize individuals from their faces [570]. It was therefore initially concluded that this area specialized in face recognition.

However, it is also found that in people who have developed a strong visual expertise, use of that expertise also results in strong activity in the FFA. For example, an expertize in different species of birds or an expertise in different models of cars results in FFA activity when the expert is using their expertise, although similar tasks performed by a non-expert do not generate FFA activity [62]. Recognition of visual material other than faces in patients with prosopagnosia can be intact, but in some cases highly discriminative visual skills such as species of birds or types of fruit are impaired [571]. Prosopagnia patients tend to have difficulties with discriminating complex visual configurations, especially those that cannot easily be verbalised [572].

Furthermore, has also been found that FFA activity is also present in tasks involving perception of social situations. Subjects were shown short movies in which geometrical shapes moved around, and they were asked to interpret the movement of shapes as social interactions (e.g. a game of hide-and-seek, a fight, or a love triangle)

[573]. Among other cortical areas, the FFA was strongly active during such tasks, but inactive when interpreting such movies as fairground "bumping car" interactions. Interestingly, it has been found that autistic individuals show depressed levels of FFA activity in response to faces [574].

6.3.7.9 Areas Involved in Episodic Memory and Imagination

Imaging of the brain during recall of actual memories and during imagining possible future events reveals some dramatic overlap in the cortical areas activated by the two types of task [21]. Subjects were given a cue, and asked to construct either a real memory or an imaginary future event by free association from the cue. The task was divided into a construction phase in which the subject came up with a concept for the event, and an elaboration phase in which the subject tried to recall or generate as much detail as possible about the event.

It was found that for both past and imagined events, 5 Brodmann areas were active during construction and 18 Brodmann areas were involved in elaboration. An additional ten Brodmann areas were active only during construction of an imaginary event, one Brodmann area was more active during elaboration of an imaginary event. No areas were more active for a real event than for an imaginary event.

In other words, any area participating in recall of a real event also participates in imaginary events. Coming up with the concept for an imaginary event requires significantly more cortical areas than the concept for a real event, but there is remarkable similarity between the cortical activity to elaborate on a real event and the activity to elaborate on an imaginary event.

A considerable number of cortical areas participate in recall and imagination. Other brain structures discussed later that participate significantly include the hippocampus and the cerebellum.

6.3.7.10 Functional Role of Cortical Areas

There are a number of implications of the preceding examples of cortical area involvement in different cognitive tasks. The first implication is that any one task requires the participation of many different areas. For example, a speech processing task requiring development of a response to individual nouns involved activation of many areas in widely separated cortical regions. The second implication is that any one area participates in different types of tasks. For example, remembering a real event and imagining a possible future event involves the participation of many of the same cortical areas. A review of the different Brodmann areas active during a wide range of cognitive tasks demonstrates the general truth of these two implications [575].

The third implication is that areas may be specializing in a type of information processing rather than a type of cognitive task. For example, the fusiform face area is active for a number of tasks, all of which require discrimination between a large number of very similar visual objects with different cognitive implications. Hence the role may include such discriminations. The fourth implication that one area can sometimes take over the functions of another area (for example an area damaged by a stroke), although the process may take many months.

6.4 The Thalamus

The thalamus is a mass of neurons located somewhat below the centre of the two cortical hemispheres. As with most brain structures, it is divided into a left and right thalamus, physically very similar to each other. The left and right thalamus are subdivided into a number of nuclei as illustrated in Fig. 6.8. There is a band of axon fibres (white matter) that appears Y-shaped from above and divides the anterior, medial and lateral nuclei from each other.

Nuclei are classified into four types: relay, association, reticular and non-specific. Relay and association thalamic nuclei (together called principal thalamic nuclei) tend to send large numbers of outputs to one or a small number of primary cortical areas, and to receive a high volume of inputs from these areas and a limited number of secondary areas. Primary cortical areas are indicated in Fig. 6.8. Relay nuclei in addition receive inputs from one sense – visual, auditory, or somatosensory – and with some minor exceptions this sensory information only reaches the cortex via that relay nucleus. The bundle of axons coming from the eyes (the optic tract) terminates in the lateral geniculate nucleus (LGN) of the thalamus. Information from the ears passes through a number of intermediate structures to the medial geniculate nucleus (MGN), and touch and propioceptic information goes to the ventral posterior (VP) nucleus. Each of these thalamic nuclei has strong reciprocal connectivity with the cortical area that performs the initial processing on the sensory information. Olfactory (taste and smell) inputs are the exception and provide inputs direct to the cortex. Association nuclei have connectivity with cortical areas but not sensory inputs.

The thalamic reticular nucleus (TRN) is a thin sheet of neurons around the outside of the principal nuclei. All axons between the cortex and principal nuclei pass through a region of the TRN specific to one nucleus or a small group of nuclei. The non-specific nuclei are located in two regions. The intralaminar nuclei are in the Y-shaped axon band, and the midline nuclei are located over the top of the thalamus.

Stimulation of thalamic reticular neurons results in gamma band oscillations in the corresponding cortical areas [576, 577]. There are neurons in the thalamus that fire in the theta rhythm [578], and the theta frequency in the cortex appears to be driven by the basal nuclei [579], which also provides inputs to the thalamus [580]. The thalamus appears to be closely involved in the generation and integration of the various EEG frequencies observed in the cortex [581].

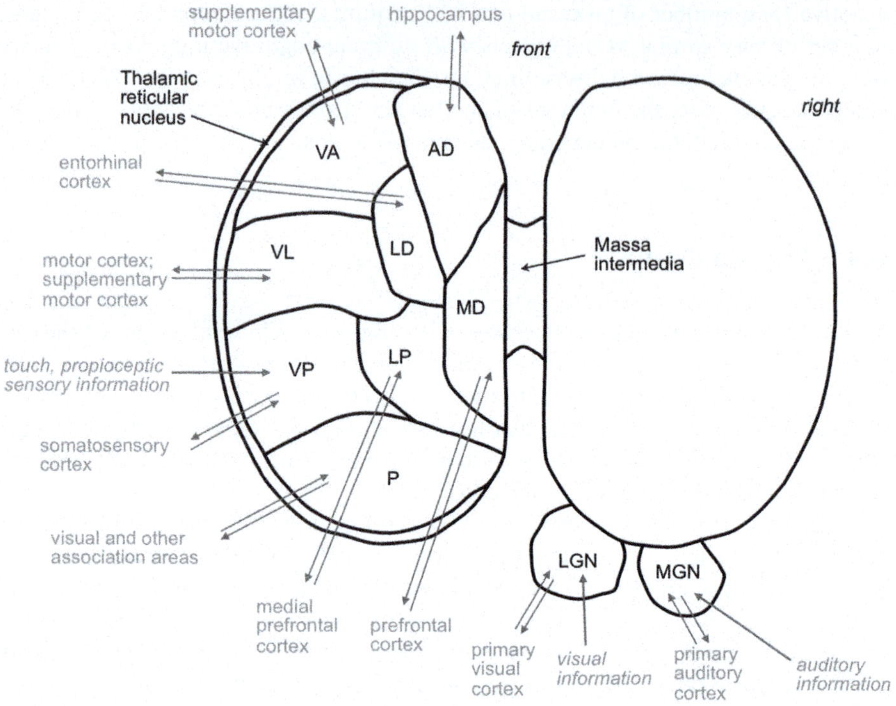

Fig. 6.8 The major nuclei and connectivity of the thalamus. The nuclei are the same in the left and right thalamus, but only the nuclei in the left thalamus are illustrated, except that the optical and auditory processing nuclei on the left are omitted, and illustrated on the right. Each nucleus has strong reciprocal connectivity with a limited number of cortical areas, examples are shown. Each nucleus also receives inputs from other areas but does not project back. The LGN and MGN extend off the back of the thalamus, receive inputs from the eyes and ears respectively, and have reciprocal connectivity with the corresponding primary sensory processing areas in the cortex. A nucleus called the thalamic reticular nucleus (TRN) is a thin sheet wrapped around the outside of most of the rest of the thalamus. All connectivity to and from the cortex passes through the TRN. Abbreviations: *AD* anterior, *VA* ventral anterior, *VL* ventral lateral, *LD* lateral dorsal, *VP* ventral posterior, *LP* lateral posterior, *MD* medial dorsal, *P* pulvinar, *LGN* lateral geniculate nucleus, *MGN* medial geniculate nucleus

6.4.1 Thalamic Neurons

Within the relay and association nuclei about two thirds of the neurons are glutamatergic thalamocortical projection neurons that mainly target the cortex. Most of the rest are GABAergic microneurons that target thalamocortical projection neurons within the same thalamic nucleus [582]. Within the TRN there is just one common type, a GABAergic neuron called the reticular cell [546]. Neurons in the non-specific nuclei are glutamatergic [583].

6.4.2 *Synaptic Weight Changes in the Thalamus*

Synaptic weight increases have been observed in glutamatergic synapses of cortical axons on to thalamic projection neurons [584]. LTP results from prolonged input, but does not require firing of the postsynaptic neuron. The mechanism appears to be presynaptic change rather than postsynaptic. It depends upon Ca^{++} entry, but is not NDMA related.

6.4.3 *Connectivity of the Thalamus*

The connectivity of a principal thalamic nucleus is illustrated in Fig. 6.9 [546]. Thalamocortical projection neurons often have many axon terminations in cortical layer IV, especially in the primary sensory areas. In some cortical areas most thalamocortical synapses are in layer II/III or in layer I [585]. In general, thalamocortical neurons target pyramidals in both layers II/III and IV [586] and possibly in some cases other layers [587]. There are return projections to the thalamus generally originating from layer VI [588], but in some cases layer V [587]. Regardless of detailed targetting, it is clear that any inputs to a cortical area, whether from a sensory organ or another cortical area, are regulated by the thalamus [589]. There appear to be two types of thalamocortical synapse, class 1 and class 2. Class 1 is ionotropic and occurs in layer IV and deeper, class 2 is metabotropic and occurs in layers II/III [587].

A primary interaction route with the cortex is thus that thalamocortical projection neurons excite pyramidal neurons in cortical layer IV, and are excited by pyramidal neurons in cortical layer VI, and a subset of projection neurons also receive strong inputs from pyramidal neurons in cortical layer V. Axons between the thalamic nucleus and the cortex pass through the TRN and side branches target reticular neurons, which in turn target thalamocortical projection neurons. One sector of the TRN generally contains axons from a number of cortical areas and from more than one principal nucleus.

Microneurons receive inputs only from pyramidal neurons in the cortex, and can target either the dendrites or the axon terminal of thalamocortical projection neurons in the same nucleus [590]. Another source of inhibition for thalamocortical projection neurons is GABAergic projection neurons in the substantia nigra (reticulata) and globus pallidus (internal) nuclei of the basal ganglia, which target projection neurons in all principal thalamic nuclei. Glutamatergic neurons in the cerebellar nuclei excite thalamocortical projection neurons in all principal nuclei [591]. GABAergic neurons of the basal nuclei target reticular neurons [592], while cholinergic neurons of the basal nuclei target thalamocortical projection neurons and microneurons [590]. Basal nuclei inputs are associated with generation of the theta frequency in the EEG. Histaminergic inputs from the

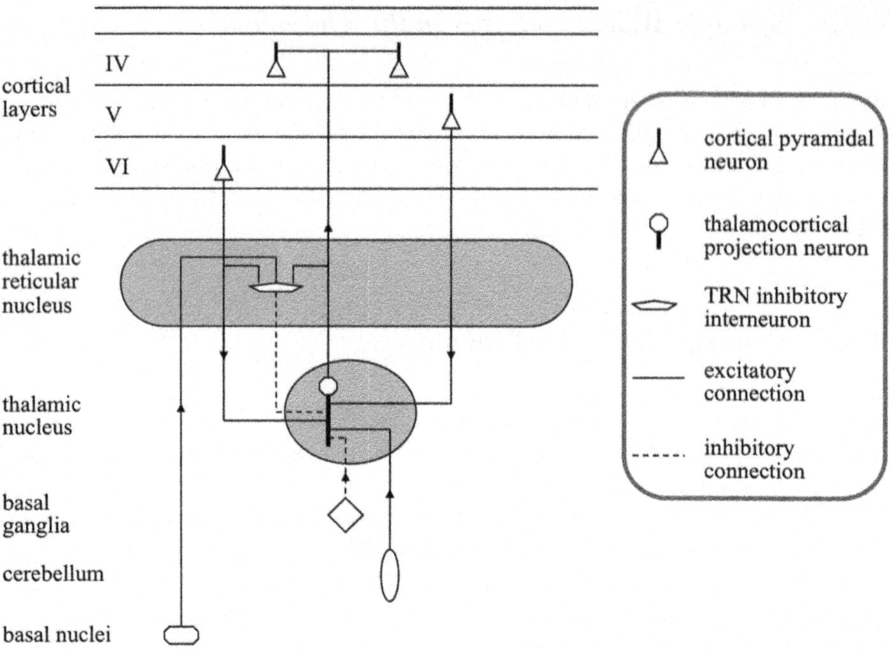

Fig. 6.9 Major connectivity of a typical higher order thalamic nucleus (i.e. one that does not directly receive sensory inputs) is illustrated. Thalamic projection neurons target layer IV of their associated cortical area. The thalamic projection neurons receive excitatory inputs from pyramidal neurons in layers V and VI of the associated cortical area plus a limited number of other areas. The inputs from the cortex also target thalamic microneurons. These microneurons are inhibitory, and only target the thalamic projection neurons in their own nucleus. Thalamocortical projection neurons also receive inhibitory inputs from the basal ganglia and excitatory inputs from the cerebellar nuclei. The axons of both pyramidal neurons and thalamocortical projection neurons pass through the TRN, and side branches from these axons target the reticular cells. These cells are inhibitory and target the thalamic projection neurons. The TRN inhibitory interneurons also receive inputs from the basal forebrain. In the case of thalamic nuclei that receive inputs directly from one of the senses (visual, auditory or somatosensory), thalamic projection neurons receive those inputs in addition to inputs from layer VI of a limited range of cortical areas including the associated primary sensory area. Outputs of the projection neurons go to layer IV of the primary sensory area, targetting spiny stellate cells that are a pyramidal cell variant

hypothalamus, noradrenergic inputs from the locus coeruleus and serotonergic inputs from the raphe nucleus are also present in the thalamus [590]. There is also a significant dopaminergic input from the substantia nigra (compacta) and ventral tegmental area nuclei of the basal ganglia and from various nuclei in the hypothalamus [593]. This dopaminergic input targets both the principal nuclei and the midline nuclei.

Neurons in the non-specific nuclei most heavily target the striatum in the basal ganglia, but also a wide range of cortical, hippocampal and other targets [594]. There is a tendency for most outputs from one nucleus to target a limited range of extrathalamic regions. Within the thalamus, non-specific nuclei generally target the TRN and sometimes other non-specific nuclei. Some non-specific nuclei target the anterior thalamic nucleus and a few other principal nuclei. The non-specific nuclei can disperse the activity of one principal nucleus across other principal nuclei and many cortical areas.

6.4.4 *Role of the Thalamus*

The role of the thalamus is closely involved in the role of the cortical area which is the target for its outputs. Damage to a thalamic nucleus results in similar cognitive deficits to those resulting from damage to the corresponding cortex. An example of this is the neglect syndromes [595]. The right cortical hemisphere processes sensory information coming from the left side of the body and vice versa. If there is damage to the visual association areas in the right cortical hemisphere, the patient will often fail to eat food on the left side of a plate. If asked to draw a clock, the drawing by such a patient will only show the numbers 12 and 1–6, with the left side of the clock shrunken and blank. Damage to the right pulvinar nucleus which interconnects with the right hemisphere visual association areas results in a similar neglect syndrome [596]. Similarly, as discussed later, damage to the hippocampus results in major deficits in declarative memory. Damage to the anterior thalamic nucleus that interconnects with the hippocampus results in similar deficits [597].

Visual attention is the selection of the visual information to be processed by the cortex, and the thalamus plays a central role in this selection [598]. Activity in the LGN correlates with whether attention is directed towards or away from a visual object [599]. The TRN is also strongly involved in the visual attention process [600], interacting with the LGN [601]. The gamma band EEG frequency is associated with visual attention [602], indicating that one role of the TRN is to impose a gamma band modulation on the activity of principal thalamic nuclei and cortical areas. Gamma band activity appears in a wide range of different cognitive activities [602], suggesting that this frequency modulation may be a general implementation mechanism for selection of sensory and cortical information by the thalamus to be further processed by the cortex.

The non-specific nuclei play a role in arousal and awareness of stimuli derived from different senses [594]. They can spread activity by influencing modulation frequencies across cortical areas and thalamic nuclei [603]. Synchronized gamma activity in the cortex depends upon interaction between principal, TRN and non-specific thalamic nuclei [604]. As discussed later, a major source of inputs to the thalamic principal nuclei is from the basal ganglia. The non-specific nuclei provide connectivity back to the basal ganglia which could influence the relative activity of different thalamic principal nuclei.

6.5 The Basal Ganglia

The basal ganglia is a group of nuclei which are separated physically and are much less similar to each other than the thalamic nuclei. However, there is a pattern of connectivity between the nuclei, illustrated for the dorsal basal ganglia in

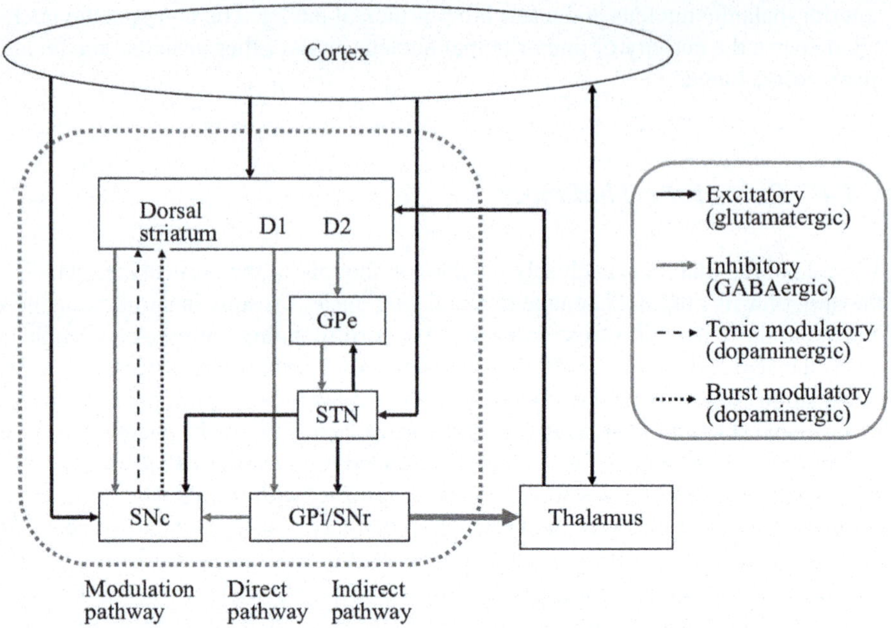

Fig. 6.10 The major connectivity paths of the dorsal basal ganglia, The main input to the basal ganglia comes from the cortex and targets the striatum. This input is excitatory. The main output from the basal ganglia comes from the SNr and GPi and targets the thalamus, which in turn has strong reciprocal connectivity with the cortex. The outputs to the thalamus are inhibitory and tonal, which means they are maintained at a steady level which applies a constant base inhibition to the thalamus. This constant base inhibition can be changed by inputs from other parts of the basal ganglia. There are two routes from the striatum input to the GPi/SNr output, called the direct and indirect pathways. The direct pathway is from a population of neurons labelled D1, and inhibits the GPi/SNr, hence reduces the inhibition of the thalamus. The indirect pathway is from a separate population of striatal neurons labelled D2. These D2 neurons are also inhibitory, but their outputs proceed via two intermediate nuclei and end up exciting the GPi/SNr and therefore increasing the inhibition of the thalamus. Hence ultimately the direct pathway excites the cortex and the indirect pathway inhibits the cortex. There is also a modulatory route from the SNc back to the striatum. This route uses the neurotransmitter dopamine, which excites striatal D1 neurons and inhibits D2 neurons, adjusting the balance between the direct and indirect pathways. Abbreviations: *GPi* globus pallidus internal segment, *GPe* globus pallidus external segment, *SNr* substantia nigra pars reticulata, *SNc* substantia nigra pars compacta, *STN* subthalamic nucleus

Fig. 6.10 and the ventral basal ganglia in Fig. 6.11. This connectivity pattern suggests that they can be regarded as a functional unit. The nuclei are arranged under and around the thalamus, and in some cases lower down in areas like the tegmentum. The major basal ganglia nuclei are: the dorsal striatum, separated into the caudate nucleus and the putamen, and often referred to as just the striatum; the globus pallidus (GP) separated into the internal segment (GPi) and external segment (GPe); the substantia nigra (SN) separated into the pars reticulata (SNr) and pars compacta (SNc); and the subthalamic nucleus (STN). These form the dorsal

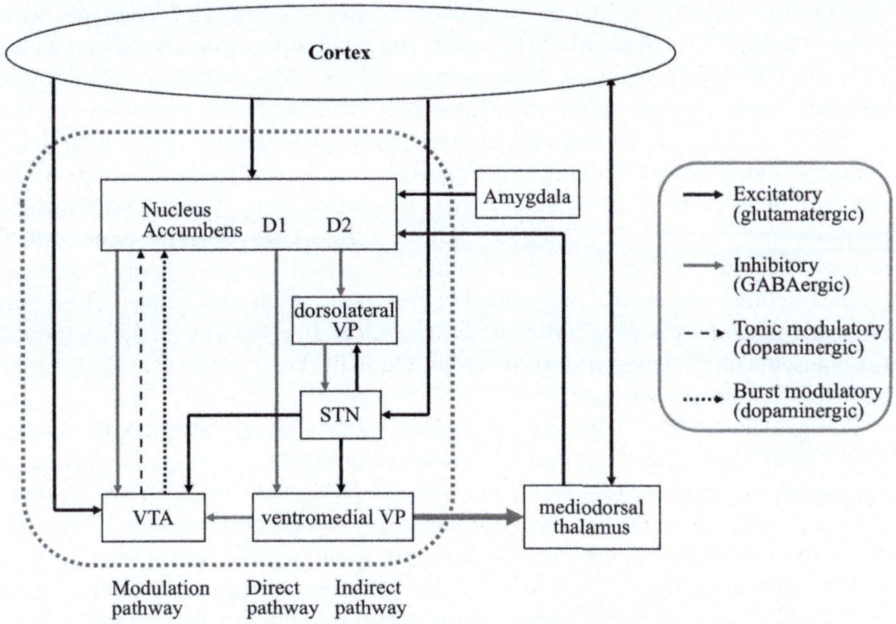

Fig. 6.11 The major connectivity paths of the ventral basal ganglia. The ventral striatum is often called the nucleus accumbens. The pattern of connectivity is very similar to that of the dorsal basal ganglia shown in Fig. 6.10. However, the nucleus accumbens has additional inputs from structures like the amygdala

basal ganglia. The GPi and SNr are very similar in appearance and they both receive inputs mainly from the same sources and project to the same targets, although there are some detailed differences discussed below [605]. To some degree they can be regarded as the same nucleus, accidentally separated by an axon bundle (the internal capsule). In dolphins they actually form one structure [606]. There is also a ventral part of the basal ganglia, with the ventral striatum (often called the nucleus accumbens) corresponding with and similar to the dorsal striatum, the ventral palladium (VP) corresponding with the GPi/SNr, and the ventral tegmental area (VTA) corresponding with the SNc.

The nuclei differ in the types of neuron they contain. However, in general each nucleus contains projection neurons that send outputs to specific other nuclei and/or outside the basal ganglia, and one or more types of interneurons which in most cases only send outputs within their own nucleus. The nuclei differ in the sources of their inputs and in the targets of their outputs. Some of the basal ganglia nuclei also have some internal subdivisions, visible as differences in neurochemical sensitivities and connectivity.

As illustrated in Figs. 6.10 and 6.11, the major source of inputs to the basal ganglia is the cortex, which projects mainly to the dorsal and ventral striatum, and to the subthalamic nucleus. Almost all cortical areas project to the striatum, major exceptions include the early sensory processing areas V1 and V2 (Brodmann areas 17 and 18) [607]. The thalamus is the major target of outputs from the basal ganglia, with the GPi/SNr generating those outputs for the dorsal striatum, the ventral pallidum generating the outputs for the nucleus accumbens (ventral striatum). The pattern of connectivity between the different basal ganglia nuclei links input information to output information. Specific cortical areas project to specific parts of the striatum, those parts of the striatum project to specific parts of the GPi/SNr, which project to specific thalamic nuclei, which in turn project heavily to the same cortical area that generated the inputs to the striatum. Some nuclei receive inputs from other brain structures such as the amygdala, hypothalamus and hippocampus. These patterns of connectivity will be further described below. In contrast with the cortex and thalamus, the highest proportion of neurons within the basal ganglia are GABAergic and therefore inhibitory.

A range of cognitive deficits associated with defects in the basal ganglia demonstrate that the basal ganglia play a role in motor control. Degeneration of neurons in the substantia nigra results in Parkinson's disease [608], with the symptoms of slowness of movement and difficulty in initiating movement. Degeneration of neurons in the striatum results in unwanted movements being inserted into normal activity, a syndrome called Huntington's disease [609]. Lesions to the subthalamic nucleus result in Hemiballism [610], with wild, uncontrollable flinging movements of arms or legs. Tourette's syndrome, in which the symptoms include involuntary repetitive muscle movements like head shaking, throwing, or speaking phrases, seems to be associated with a striatal pathology [611], and excess production of the neurotransmitter dopamine in the basal ganglia [612].

However, it is also clear that the basal ganglia play a role in a wide range of behaviours in addition to motor movements [613, 614]. Most Parkinson's patients exhibit some cognitive problems [615]. Such problems can include difficulties in allocating attention, in setting priorities, and understanding social situations. There can be loss of memory, especially how to carry out previously learned skills (i.e. procedural memory). In some cases, the syndrome progresses to include speech problems beyond just the motor movements for production of sounds, such as category naming [616]. It appears that within the dorsal striatum, the putamen is mainly concerned with motor control, and the caudate nucleus with cognitive processes. It has further been demonstrated that the ventral basal ganglia (the ventral striatum or nucleus accumbens, ventral palladium and ventral tegmental area) play an important role in reward and motivation [617].

The basal ganglia nuclei participate in the oscillatory electrical activity with the cortex and thalamus [618], implying that the various EEG frequencies also play information processing roles in these nuclei. Changes in the pattern of temporal activity are observed in Parkinson's disease [619].

Because of the differences between the different nuclei, the internal structure, neuron types and connectivity of each nucleus will be described separately, and then the predominant patterns of connectivity between the nuclei and with the rest of the brain will be discussed.

6.5.1 Dorsal Striatum

The striatum was named because the axon fibre bundles passing through it give it a striated appearance. The dorsal striatum is made up of two large nuclei, the caudate nucleus and putamen, with similar internal structure but different sources of cortical input. The putamen is associated with motor behaviours, and the caudate nucleus with cognitive behaviours.

The most common neuron in the dorsal striatum is the medium sized spiny neuron (MSN). This type of neuron projects outside the striatum, but unlike cortical and thalamic projection neurons is GABAergic, and therefore inhibits its targets. There are also a number of types of GABAergic interneurons that mainly target MSNs. Two electrical types have been identified [620]. One is a fast spiking (FS) interneuron which tends to lock to cortical oscillations. The other is the low threshold spiking (LTS) interneuron which produces bursts of a few spikes with interspike interval <10 ms. The FS tends to target the somas of MSNs, while the LTS tends to target dendrites [621]. Different types of LTS interneurons can be distinguished neurochemically, for example one type releases dopamine in addition to GABA [622]. There is also a large cholinergic interneuron that fires tonically. These cholinergic interneurons target some GABAergic interneurons by excitatory (nicotinic) synapses and thus regulate MSNs [623]. However, these interneurons also target MSNs directly by muscarinic G-protein coupled receptors [624].

The main glutamatergic inputs to the dorsal striatum come from layer V of the cortex and from the midline nuclei of the thalamus in roughly equal numbers [625]. There is also a high volume of dopaminergic inputs from the SNc. When targetting MSNs, cortical inputs synapse on spines, thalamic inputs mainly on dendritic shafts, and dopaminergic inputs sometimes on somas, but mainly on shafts and spines [626]. On spines, a dopaminergic input is invariably accompanied by an excitatory (glutamatergic) input, although a majority of spines have only a glutamatergic input [627]. Cortical inputs come from a very large number of pyramidal neurons that target different regions of the dorsal striatum in a heterogeneous fashion [625], while dopaminergic inputs come from a relatively small number of neurons and form an evenly spaced axon lattice distributed throughout the dorsal striatum [628]. As illustrated in Fig. 6.10, GABAergic MSN outputs from the dorsal striatum have three targets: the GPi/SNr, the GPe, and also the SNc. One is called the direct path because it directly targets the GPi/SNr. The second is called the indirect path, because it targets the GPi/SNr via the GPe and subthalamic nucleus. The third path targets the SNc.

Separate compartments can be distinguished in the dorsal striatum on the basis of the presence of different neurotransmitters and neurochemicals [605] and also the input and output connectivity of neurons. Within the dorsal striatum are zones, also called patches or striosomes, 200–500 μm in diameter. These striosomes occupy 10–20 % of the volume of the dorsal striatum, and were originally identified by their low level of acetylcholine compared with the striatal matrix in which they are embedded [629]. There are also compartments, or matrisomes, in the matrix, about the same size as striosomes.

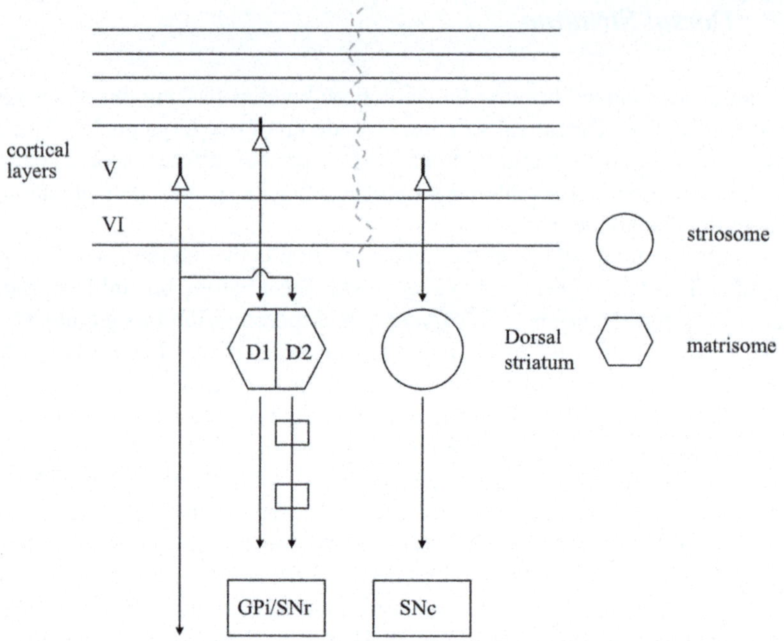

Fig. 6.12 Tendencies in connectivity of striosomes and matrisomes in the striatum. The striatum is divided into patches (called striosomes) within a matrix (called matrisomes). MSNs in the striosomes receive inputs from deep layer V in the cortex, and project to the SNc. MSNs in the matrisomes are of two types (D1 and D2) that (respectively) project into the direct and indirect paths. Inputs to D1 type MSNs come from superficial cortical layer V. Inputs to D2 type MSNs come from deep layer V, and these cortical neurons also project to the brainstem. The cortical neurons in one cortical region tend to target striosomes that are dorsal to the matrisomes targetted

As illustrated in Fig. 6.12, the three major projection paths by MSNs out of the dorsal striatum map into the matrisomes/striosome structure. Matrisomes project mainly into the direct and indirect paths, while striosomes project mainly to the SNc [605].

D1 type dopamine receptors are present in striosomes and to a lesser degree in matrisomes, while D2 receptors are more evenly distributed [630]. Dopamine receptors on MSNs in the direct path are mainly type D1, which results in the neurons being excited by dopamine. The MSNs for the indirect path contain mainly the receptor type D2, which results in neurons being inhibited by dopamine [181]. Receptors for dopamine occur in dopaminergic synapses, often located close to spines corresponding with glutamatergic synapses. However, dopaminergic receptors are also distributed on somas away from dopaminergic synapses, where they are only exposed to background dopamine in the extracellular environment [187]. The association of dopaminergic synapses with dendrites (especially spines) is found for both D1 and D2 populations [630]. The dopamine inputs to the matrisomes tend to come from somewhat different neurons from those to the striosomes [631].

There are different zones in the dorsal striatum which when activated result in different very specific body movements [632, 633]. It seems likely that these different zones correspond with different matrisomes [605]. Damage to striosomes on the other hand appear to be associated with mood disorders including irritability, anxiety, compulsive behaviour, and greatly increased activity level [634].

There are two types of projection from the cortex to the dorsal striatum. One type called intratelencephalically projecting-type (IT) originates in upper layer V or even layer III, and the axons end in the striatum. The other type is called pyramidal tract type (PT) and originates in deeper layer V or even layer VI, the axons ultimately target the brainstem but a branch targets the dorsal striatum [635]. There appears to be a tendency for IT neurons to target striatal neurons of the direct path, and PT neurons to target striatal neurons of the indirect path [636]. There is also evidence that neurons in layers II/III and superficial layer V project more to the matrisomes, while neurons in deeper layer V and layer VI project more to the striosomes [637]. Furthermore, any one cortical area tends to target striosomes in a more dorsal region of the striatum and matrisomes in a more ventral region [638].

One cortical area projects to many regions of the striatum, with a high volume of projections to a smaller but still extensive domain [639]. The striatal domains of high projection for different cortical areas can partially overlap. The projections from one cortical area are broken up into separate striatal zones, less than a millimetre in diameter. If two cortical areas have significant direct connectivity the zones in the domain to which they both project in high volume generally overlap, but do not coincide: for example the zone for one area may fall inside the zone for the other area. Two cortical areas that do not have significant direct connectivity tend to have less overlap in their heavily targetted striatal domains, even if they are adjacent. There is a strong tendency for zones heavily targetted by the cortex to be located in the matrix, with a much lower proportion of cortical inputs targetting the striosomes.

The IT neurons appear to provide the main excitatory input [640]. A pyramidal neuron projecting to the striatum can make synapses in both matrisomes and striosomes. One such neuron can project to a large region ~15 % of the striatum, making thousands of synaptic contacts, while another may occupy a smaller region and make fewer total synapses [641]. One MSN has ~10,000 spines, half from cortical neurons, but one cortical neuron makes relatively few (certainly <40, perhaps not more than 1) [625] synapses on any one MSN, and inputs from a large number of cortical neurons at the same time are needed for an MSN to fire [642].

The hippocampus also projects to the dorsal striatum. These projections arise from the subiculum, with different parts of the subiculum targetting different regions of the striatum [643]. The basolateral nucleus of the amygdala also projects to the dorsal striatum, although more weakly than to the ventral striatum. These amygdala projections have uneven distribution with respect to matrisomes and striosomes [644, 645]. Finally, there are some inputs to spiny projection neurons from a structure called the raphe nucleus which are serotonergic. These inputs increase the effect of dopamine [646]. Inputs from the dorsal raphe also excite the cholinergic (excitatory) interneurons, which target the spiny projection neurons as discussed below. Raphe nucleus inputs thus have a complex, modulatory effect on the spiny projection neurons.

Within the dorsal striatum, the GABAergic interneurons receive inputs from the cortex and thalamus and target the MSNs. The cholinergic interneurons are found only in the matrix, receive inputs from the cortex, thalamus and also spiny projection neurons, and target spiny projection neurons in the matrix. Cholinergic interneurons also have D1 and D2 receptors for dopamine inputs [647]. In addition, they have excitatory serotonergic inputs from the raphe nucleus [648], and are sensitive to norepinephrine inputs [649], perhaps from the locus coeruleus [650]. The cholinergic interneurons produce tonic (steady) outputs which are interrupted by sensory circumstances correlating with past rewards [651].

The axons of spiny projection neurons target other basal ganglia nuclei, but have extensive local branches that target (and therefore inhibit) other MSNs. There are two types of local axon branching: in the most common type, the axon branching is limited to not much more than the diameter of the dendritic tree of the source neuron, and there are some synapses on the neuron itself; in the other type, the branching begins much further from the neuron and does not overlap its dendritic tree [652]. It has been argued that this cross-inhibitory network is not strong enough to support a winner-takes-all type competition between spiny projection neurons [653]. However, total inputs to an MSN from other MSNs appear to be comparable in strength with total inputs from FS interneurons [654].

6.5.2 Subthalamic Nucleus (STN)

Unlike the other basal ganglia nuclei, the projection neurons in the STN [655] produce a steady tonal excitatory output, using the glutamate neurotransmitter. These projection neurons receive many excitatory inputs from the cortex and the midline nuclei of the thalamus, and substantial inhibitory inputs from the dorsal striatum and also the GPe. They mainly project [656] to the basal ganglia output nuclei (the GPi/SNr), back to the GPe, and to a small extent back to the striatum [657]. There are also smaller inhibitory GABAergic interneurons in the STN [658]. The sources of inputs to and targets of outputs from these interneurons are not known, although it is probable that they target STN projection neurons, and perhaps, by analogy with the striatum, receive inputs from both local projection neurons and from the cortex and thalamus.

An STN output to the GPi/SNr tends to uniformly excite a large number of neurons, in contrast with a striatal output that tends to inhibit selected subsets of GPi/SNr neurons [659].

6.5.3 Globus Pallidus External Segment (GPe)

The projection neurons [660] in the GPe are inhibitory (GABAergic) and receive a high volume of inputs from both the striatum and the STN. They also receive inputs from other GPe projection neurons. The projection neurons primarily target the STN, but have extensive local axon collaterals. There is also a small population of interneurons [661], but their inputs and targets have not been determined, although it is likely that they target GPe projection neurons.

6.5.4 Globus Pallidus Internal Segment and Substantia Nigra Pars Reticulata (GPi/SNr)

As discussed earlier, there is evidence that these two structures should be regarded as to some degree the same nucleus. The projection neurons in these nuclei are large GABAergic (inhibitory) neurons with a high resting discharge rate. Inputs to the projection neurons come from the striatum, GPe and STN. Their outputs target various nuclei in the thalamus, with no local axon collaterals. These outputs provide a steady tonal inhibition of their target thalamic nuclei, thus affecting the cortical areas with which these thalamic nuclei interact. One major difference between the GPi and SNr is that the SNc dendrites extend into the SNr, and release dopamine into the SNr [662]. A functional difference is that the GPi interacts via the thalamus with motor cortices while the SNr interacts via the thalamus with the prefrontal cortex [606]. Interneurons have not been observed in the GPi/SNr.

A small group of neurons in the striatum receives inputs from different cortical sources and projects to a small group of neurons in the GPi/SNr [652]. Neighbouring groups in the striatum do not project to the same groups in the GPi/SNr, but rather project to neighbouring groups [652]. Hence information is conveyed along multiple, segregated channels. These small groups of neurons in both structures correspond with specific behaviours [663], and neurons in both groups tend to be active only during those behaviours, for example specific body movements [664].

6.5.5 Substantia Nigra Pars Compacta (SNc)

The projection neurons of the SNc are dopaminergic. Excitatory (glutamate) inputs to these neurons come from the STN, inhibitory inputs (GABA) from the striatum and from the SNr [665]. The projection neurons heavily target the striatum with smaller projection to the GPe and STN [666]. It has been demonstrated that dopaminergic neurons projecting to the striatum can also release GABA at the same synapses [667]. One projection neuron may target several hundred thousand striatal neurons. In many cases a dopaminergic synapse on a striatal neuron is closely associated with a glutamatergic synapse derived from the cortex or thalamus.

There is also a small population of GABAergic interneurons [668, 669]. These interneurons have excitatory dopaminergic inputs from within the SNc [669], and perhaps GABAergic and glutamatergic inputs.

6.5.6 Ventral Striatum (or Nucleus Accumbens)

The nucleus accumbens is subdivided into core and shell components. The core is morphologically very similar to the dorsal striatum. It exhibits a patch and matrix structure similar to that observed in the dorsal striatum [670, 671], and contains a large population of GABAergic spiny projection neurons. It also contains smaller populations of cholinergic interneurons and GABAergic interneurons [672]. The shell does not exhibit the exact patch and matrix structure. However, the shell is affected by a wider range of neurochemicals and has a more diverse range of input sources and output targets than the core, and exhibits a substructure based on different neurochemical and connectivity differences.

Major inputs to the nucleus accumbens come from the prefrontal cortex, the basal complex of the amygdala, the cortical areas associated with the hippocampal system, various thalamic nuclei, and from the ventral tegmental area (VTA). Major outputs target the ventral palladium, the VTA and both SNr and SNc, with reciprocal inputs from all of these nuclei. However, there are differences in connectivity between core and shell. Both core and shell receive inputs from certain cortical areas and the basal nuclei of the amygdala, but core and shell also receive inputs from different, additional cortical areas. Core and shell receive inputs from different (although partially overlapping) groups of thalamic nuclei. In general, there is much more subcortical input to the shell than to the core. The shell receives strong inputs from the VTA, VP and hypothalamus. Outputs from the shell strongly target the VP. Outputs from the VP target thalamic nuclei that in turn target the cortical areas that provide input to the core but not the shell. Outputs from the core target the ventral palladium, the subthalamic nucleus, the VTA, and the SNr and SNr. Via the SNr and then the thalamus the cortical areas providing input to the core are targetted. There are differences in connectivity between patch and matrix. In the core, the patches project to the SNr, while the matrix projects to the SNc and the VTA [670].

Inputs and outputs to core and shell are distributed inhomogeneously [670]. For example, different nuclei of the basal amygdaloid complex terminate in different parts of the nucleus accumbens in a highly complex arrangement [673]. Both shell and core are made up of a mosaic of different groups of neurons, distinguished by higher cellular density, different neurotransmitter sensitivities, different sources of input and different output targets. This has led to the hypothesis that these different groups of neurons have different behavioural roles [674].

6.5.7 Ventral Pallidum

The ventral pallidum contains many GABAergic projection neurons, some cholinergic projection neurons, and a small number of glutamatergic projection neurons. The inhibitory GABAergic neurons project to mainly to the thalamus, but also to the ventral tegmental area, substantia nigra and the STN [675]. Interestingly, the excitatory glutamatergic neurons also project to the ventral tegmental area [676]. The excitatory cholinergic neurons project to the amygdala and cortex [675]. There are also projections to the substantia nigra. There are strong dopaminergic inputs from the ventral tegmental area, GABAergic inputs from the nucleus accumbens, and glutamatergic inputs from the STN, amygdala, thalamus and prefrontal cortex [677].

Like the nucleus accumbens and ventral tegmental area with which it has strong connectivity, it has been demonstrated that the ventral pallidum plays a role in rewards [678].

6.5.8 Ventral Tegmental Area (VTA)

The majority of neurons in this area are dopaminergic projection neurons, a small proportion are GABAergic neurons [679]. Many of the GABAergic neurons are also projection neurons, but there are some GABAergic interneurons. Furthermore, some dopaminergic projection neurons can also release GABA at target synapses [667]. Most excitatory inputs to the VTA come from the prefrontal cortex and the hypothalamus [680], most inhibitory inputs are GABAergic and come from the nucleus accumbens and the ventral pallidum [679]. However, the ventral pallidum also projects a smaller number of excitatory (glutaminergic) inputs [680]. The VTA also receives serotonergic inputs from the raphe nucleus and noradrenalergic inputs from the locus coeruleus [679]. GABAergic interneurons receive excitatory input from a small nucleus called the lateral habenula, and inhibit dopaminergic neurons. Activity of the lateral habenula correlates with negative rewards [681]. Activity of the GABAergic neurons also correlates with negative rewards, while activity of dopaminergic neurons correlates with positive rewards [682].

Both dopaminergic and GABAergic outputs strongly target the nucleus accumbens [683] and the prefrontal cortex [684]. Projections to the nucleus accumbens are largely to the shell but with some targetting of the core and even the dorsal striatum. Dopaminergic neurons that receive inputs from the prefrontal cortex project back to the prefrontal cortex, but GABAergic neurons receiving prefrontal cortex inputs project to the nucleus accumbens [684]. Dopaminergic neurons that project to the nucleus accumbens do not receive inputs from the prefrontal cortex [684].

6.5.9 Basal Ganglia Synaptic Weight Changes

The most investigated are changes at glutamatergic synapses from cortical axons on to projection neurons in the striatum. Synaptic weight changes have been observed at such synapses. In the striatum, appropriate relative timing of inputs and postsynaptic depolarisation can result in LTP or LTD [146]. The direction, magnitude and persistence of such weight changes depends on the presence or absence of various modulatory neurotransmitters including dopamine, acetylcholine and adenosine [685].

Another nucleus that is functionally part of the basal ganglia is the pedunculopontine nucleus (PPN). This nucleus receives most of its inputs from and mainly projects to various nuclei in the basal ganglia, although it also receives inputs from the habenula and cerebellar nuclei, and also projects to the thalamus [686]. Synaptic weight changes have been observed in glutamatergic synapses from the PPN on to SNc and VTA dopaminergic neurons. These changes have been observed following a combination of electrical stimulation and exposure to various addictive drugs [687].

6.5.10 Basal Ganglia Connectivity

The connectivity of the basal ganglia can be represented approximately by Figs. 6.10 and 6.11 [688, 689]. External input to the basal ganglia is excitatory and comes mainly from the cortex, heavily targeting the striatum and the STN. External output from the basal ganglia is inhibitory and is generated by the GPi/SNr, mainly targeting the thalamus. This output is tonal, in other words the GPi/SNr projection neurons produce a steady inhibitory output which is selectively increased or reduced by inputs from other basal ganglia nuclei. There is strong reciprocal excitatory connectivity between the thalamus and the cortex. Hence a reduction in the inhibition of a thalamic neuron will result in an increase in the activity in the cortex.

As illustrated in Figs. 6.10 and 6.11, there are two major routes by which information passes from the striatum to the GPi/SNr, called the direct and indirect pathways. Within the striatum are two populations of inhibitory projection neurons, labelled D1 and D2. The D1 population projects by the direct pathway to the GPi/SNr, inhibiting the projection neurons and therefore reducing their tonic inhibitory output to the thalamus. Activity of the D1 population therefore reduces the activity of the GPi/SNr, and hence reduces the tonic inhibition of the thalamus and increases cortical activity. In the indirect pathway the STN receives independent input from the cortex, and excites the GPi/SNr, increasing their inhibition of the thalamus. The activity of the STN is modulated by reciprocal connectivity with the GPe, which itself receives input from the D2 population in the striatum. The overall effect is that activity of the D2 population increases the activity of the GPi/SNr and therefore increases the inhibition of the thalamus.

There is a feedback loop from the indirect path back to the striatum, via the SNc. Neurons in both the GPe and STN project to neurons in the SNc. These SNc neurons are dopaminergic and project to the striatum. The dopaminergic neurons fire tonically at about 4 Hz [188]. This tonic firing maintains a background extracellular level of dopamine in the target striatal regions, which can be adjusted by indirect path activity modulating the proportion of dopaminergic neurons firing in the tonal mode. Dopamine excites D1 population neurons, and inhibits D2 population neurons [181]. Because the D2 channel has a higher affinity for dopamine than the D1 channel, the background concentration mainly affects the D2 channels [190]. If activity in the indirect path rises, the proportion of dopaminergic neurons in the SNc firing in tonal mode will be increased, resulting in a reduction in D2 neuron activity that shifts the balance of activity between the D1 and D2 populations in the striatum [690]. This background concentration can also be moderately increased by hippocampal activity acting via the striatum [691], also increasing the number of dopaminergic neurons firing tonically.

There is another way in which the dopaminergic neurons of the SNc influence the striatum. Dopaminergic neurons can also fire in bursts of 3–20 action potentials at an intraburst frequency of about 20 Hz [692]. These bursts depend upon frontal cortex activity, probably via the STN [693]. There is evidence that bursts also depend on activity of striatal projection neurons which inhibit GABAergic inputs to the dopaminergic neurons [694]. Burst activity will result in much higher concentrations of dopamine near synapses. Background dopamine concentration will also rise, but such rises are limited because increased opening of D2 channels also results in opening of dopamine reuptake channels which limit the spread of dopamine beyond the synapse [189]. Hence burst activity and tonic activity represent two relatively independent ways in which basal ganglia dopaminergic neurons act on the striatum [188]. Changes to dopamine availability in the striatum are associated with various movement and cognitive disorders, and there is evidence that burst firing plays a role in learning [693]. However, the SNc dopaminergic neurons also target the cortex, and the two firing modes interact in a complex fashion in both cortex and striatum [695].

The dorsal and ventral basal ganglia have very similar pattern of connectivity. One major difference is that the ventral striatum (nucleus accumbens) receives a higher volume of excitatory (glutamatergic) inputs from the basal amygdala [670].

As described earlier, at a more detailed level there are some significant differences between the nucleus accumbens and the dorsal striatum. Unlike the striatum, the nucleus accumbens is separated into two components, the core and the shell. The core physically resembles the striatum with a patch and matrix organisation and D1 and D2 projection neurons. In the shell there are complex arrangements of neurochemical sensitivities, input sources and output targets [670]. For example, different clusters of neurons in the shell receive different combinations of inputs from hippocampus, prefrontal cortex and amygdala, and project to different combinations of targets including the VP, the VTA and the hypothalamus. The outputs of the nucleus accumbens shell have a relatively direct influence on the core via both the VP and the VTA [696].

As defined by predominant connectivity, there are multiple parallel paths from the cortex through the striatum, GPi/SNr, the thalamus and back to the cortex. Each of these paths begins in a specific group of cortical areas, passes through different parts of the striatum, GPi/SNr, and thalamus, and returns to one of the initial cortical areas. The five such paths that have been identified [697] are illustrated in Fig. 6.13, but others may exist. However, it is important to note that these paths are not fully separated. As discussed earlier, one cortical area projects strongly to a limited striatal domain, but to a smaller degree projects to more extensive regions of the striatum.

The five paths can be associated with different behavioural functions, because damage to any point on a path results in similar deficits, whether to the cortex, basal ganglia or thalamus [698]. The motor path is associated with control of voluntary

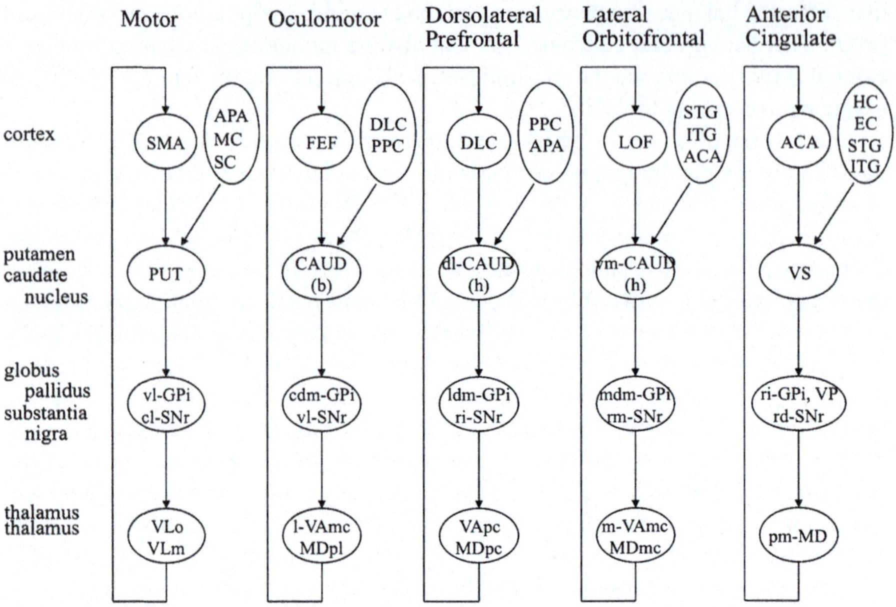

Fig. 6.13 Parallel cortex-basal ganglia-thalamus-cortex paths. In these connectivity paths, the thalamic nucleus heavily targets one cortical area. The same cortical area and a few others target a region of the striatum, that region of the striatum targets a region of the GPi/SNr/VTA, and that region of the GPi/SNr/VTA targets the thalamic nucleus. The different paths appear to correspond with different general types of behaviour. Abbreviations: *ACA* Anterior Cingulate Area, *APA* Arcuate Premotor Area, *DLC* Dorsolateral Prefrontal Cortex, *HC* Hippocampus CA fields, *EC* Entorhinal Cortex, *FEF* Frontal Eye Fields, *ITG* Inferior Temporal Gyrus, *LOF* Lateral Orbitofrontal Cortex, *MC* Motor Cortex, *PPC* Posterior Parietal Cortex, *SC* Somatosensory Cortex, *SMA* Supplementary Motor Area, *STG* Superior Temporal Gyrus, *VLo* Ventralis Lateralis pars oralis, *VLm* Ventralis Lateralis pars medialis, *l-VAmc* lateral Ventralis Anterior pars magnocellularis, *MDpl* Medialis Dorsalis pars paralamellaris, *VApc* Ventralis Anterior pars parvocellularis, *MDpc* Medialis Dorsalis pars parvocellularis, *m-VAmc* medial Ventralis Anterior pars magnocellularis, *MDmc* Medialis Dorsalis pars magnocellularis, *pm-MD* posteromedial Medialis Dorsalis, *PUT* Putamen, *CAUD* Caudate (b) body, (h) head, *dl-CAUD* dorsolateral Caudate, *vm-CAUD* ventromedial Caudate, *VS* Ventral Striatum (nucleus accumbens), *cdm-GPi* caudal dorsomedial Globus Pallidus internal, *ldm-GPi* lateral dorsomedial Globus Pallidus internal, *mdm-GPi* medial dorsomedial Globus Pallidus internal, *rl-GP* rostrolateral Globus Pallidus internal, *vl-GPi* ventrolateral Globus Pallidus internal, *cl-SNr* caudolateral Substantia Nigra pars reticulata, *vl-SNr* ventrolateral Substantia Nigra pars reticulata, *rl-SNr* rostrolateral Substantia Nigra pars reticulata, *rm-SNr* rostromedial Substantia Nigra pars reticulata, *rd-SNr* rostrodorsal Substantia Nigra pars reticulata, *VP* Ventral Palladium

movement, and the oculomotor path is concerned with the control of eye movements. The dorsolateral prefrontal path is associated with what are called executive cognitive functions, such as solving complex problems, planning etc. The lateral orbitofrontal path is associated with personality, including empathy and appropriate responses to social cues. The anterior cingulate path through the nucleus accumbens is associated with motivation, with damage resulting in apathy.

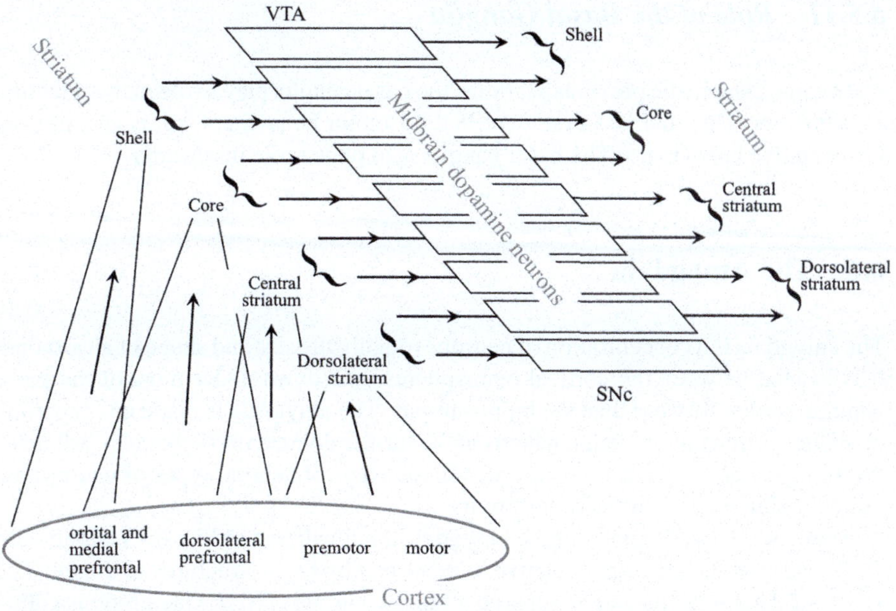

Fig. 6.14 The spiral of connectivity involving cortex, striatum and midbrain dopaminergic neurons (VTA and SNc). Inputs to the striatum come from cortical areas that vary continuously moving across the striatum from ventral to dorsal. Neurons in the striatum project to dopaminergic neuron areas which in turn project back to the striatum. The areas of the dopaminergic neurons that project to an area of the striatum overlap partially but not completely with the area of dopaminergic neurons that receive inputs from that striatal area

There are even more detailed parallel paths within these major parallel paths. For example, as described earlier, within the motor path there are different groups of neurons in the striatum that correspond to different behaviour related circumstances. One striatal group projects to a group of neurons in the GPi/SNr, which in turn project to a group of thalamic neurons, all of which correspond with the same type of behavioural circumstance. For example, the activity of some neurons correlates with the direction of arm movement, other groups with the use of specific muscles, and yet others with the location of a target object in space [699–701].

The reciprocal connectivity between the dorsal and ventral striatum and the midbrain dopamine neurons (i.e. the dopaminergic neurons of the SNc and VTA) is organized in a striking spiral pattern of connectivity [702] as illustrated in Fig. 6.14. Inputs to the striatum come from cortical areas that vary continuously moving across the striatum from ventral to dorsal. Neurons in the striatum project to dopaminergic neuron areas which in turn project back to the striatum. The areas of the dopaminergic neurons that project to an area of the striatum overlap partially but not completely with the dopaminergic neurons that receive inputs from that striatal area. As a result, via the midbrain dopamine neurons the shell influences the core, the core influences the central striatum, and the central striatum influences the dorsal striatum.

6.5.11 *Role of the Basal Ganglia*

As became clear in the previous sections, the basal ganglia play a role in the selection of behaviour. This role was first identified for motor behaviours, but is also present for cognitive behaviours. The basal ganglia also play a role in rewards.

6.6 The Amygdala

The amygdala is a concentration of neurons roughly the size and shape of an almond. It is located between the cortical hemispheres, low down under most of the basal ganglia, further forward than the hippocampus. The amygdala is made up of a group of about a dozen major nuclei, which can be further divided into subnuclei. All these divisions are on the basis of differences in the types of neuron, types of neurotransmitter, and input and output connectivity.

Some of the major nuclei of the amygdala are illustrated in Fig. 6.15, along with the location of the amygdala relative to other nearby brain structures in the rat. The nuclei of the amygdala can be separated into two groups [703]. One group is called the centromedial group, and includes the medial and central nuclei. These nuclei

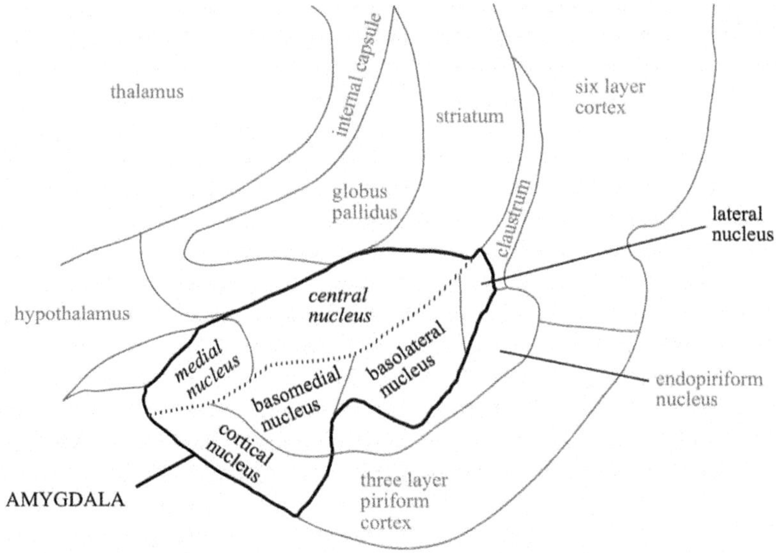

Fig. 6.15 Nuclei of the amygdala. The relative positions of the major nuclei of the amygdala and of some adjacent structures are shown for the rat brain. The projection neurons in the central and medial nuclei of the amygdala are GABAergic like the neighbouring striatum, while the projection neurons in the basolateral, basomedial and cortical nuclei are glutamatergic and resemble cortical pyramidal neurons

have some strong resemblances to the striatum, including the presence of large numbers of inhibitory (GABAergic) medium spiny projection neurons. The other group is called the corticomedial group, and includes the cortical, basomedial (alternatively called accessory basal) and basolateral (alternatively called basal) nuclei. These nuclei have some resemblances with the cortex, including the presence of large numbers of excitatory (glutamatergic) pyramidal projection neurons. In the cortical nucleus it is possible to see a cortex-like layered structure: a molecular layer I with mostly axons, a thin layer II with small multipolar neurons, and a layer III making up most of the thickness and containing pyramidal neurons [704]. It has been further argued that the corticomedial group should be subdivided into a group interacting with the autonomic nervous system and two groups dealing mainly with the olfactory system [705].

The amygdala is associated with a range of apparently disparate functions, many of which seem to fall under the general heading of influencing emotional behaviour. An important example is that it plays a major role in fear. Damage to the amygdala in monkeys results in reduced fear and aggression, increased submission, and excessive manual and oral exploration behaviours [706]. In rats, a cue which triggers fearful behaviour because it has become associated with an impending unpleasant event (such as an electrical shock) does not result in fearful behaviour if the amygdala (and in particular the basolateral group) is inactivated [707]. Neuroimaging of human beings reveals amygdala activation in response to threatening or fear provoking images [708], and patients with amygdala damage are unable to learn to associate cues with the typical physiological indications of fear [709]. These physiological indications of fear (increased heart and breathing rates, increased blood pressure, muscle changes, changes in skin conductance) are generated outside of conscious control by the autonomic nervous system. The amygdala must therefore have some influence over that system. Furthermore, the amygdala influences attention, and in particular the shifting of attention from where it is currently engaged to something unexpected that has occurred [710], particularly if the something unexpected could be a threat [708].

It has been demonstrated that certain facial expressions are recognized as indicating certain emotions, independent of cultural background. Expressions indicating happiness, anger, sadness, disgust, surprise and fear [711] and more recently pride [68] have been shown to be recognized across widely different cultures. Damage to the amygdala results in some impairment in the recognition of negative emotions like fear [64].

The amygdala appears to also play a role in emotions other than fear. Direct electrical stimulations of the amygdala in human patients induced both negative emotions (fear, anxiety or sadness but rarely anger) and positive emotions (happiness) [712].

The amygdala plays an important role in learning behaviours, both fearful behaviours and behaviours with positive rewards. For example, fear conditioning or learning to associate a cue with an impending unpleasant event is disrupted by damage to the amygdala [713]. Learning of positive associations (such as between a cue and a food reward) also involve the amygdala, although the role is more subtle than for fear [714]. If a monkey is offered a choice between two objects, and one of the objects is associated with a food reward, the monkey rapidly learns to choose the rewarded object. Removal of the amygdala does not affect this capability. If a monkey is trained to associate one food type with one object and a different food type with a different object, and then offered the choice of objects after being satiated with one food type, a normal monkey will avoid choosing the object associated with the food with which it has been satiated. However, this effect is removed by lesioning of the amygdala [715]. Different nuclei in the amygdala play different roles in rewards [714, 716].

The amygdala also plays a role in episodic memory. Normal subjects show an enhanced memory for the gist of a scene observed during emotional arousal (fear, alarm etc.) but no such enhancement (and perhaps a suppression) for visual details. Patients with amygdala damage show a memory impairment for the gist of a scene observed during emotional arousal [717]. The activity of the amygdala predicts the degree of subsequent memory for emotionally intense scenes [16].

Finally, the amygdala is intimately involved with the olfactory system. Amygdala damage impairs the ability to recognize and name odors [718].

6.6.1 *Neurons of the Amygdala*

The majority of neurons in the basolateral and cortical nuclei are pyramidal-like excitatory (glutamatergic) neurons. One difference from the cortex is that in the basolateral nuclei the dendrites are oriented randomly [719], not parallel to each other as they are in the cortex. Some glutamatergic neurons resemble the spiny stellates in the cortex. In addition there is a cortex like range of inhibitory (GABAergic) interneurons. In the centromedial group of nuclei, the majority of neurons are GABAergic projection neurons similar to those found in the striatum.

Some individual neurons in the basolateral group respond to specific biologically relevant sounds or objects, and responses change with reward type feedback. For example, a neuron might fire strongly in response to seeing a piece of watermelon, but with considerably reduced firing following a piece of salted watermelon [720].

6.6.2 *Synaptic Weight Changes in the Amygdala*

In the lateral nucleus of the amygdala, multiple forms of synaptic weight changes have been found in the same projection neurons [721]. Firstly, simultaneous 30 Hz stimulation of both cortical and thalamic inputs results in LTP of cortical but not thalamic synapses [722]. Stimulation of just one type of input does not produce weight change. This LTP depends upon presynaptic NMDA receptors and is independent of postsynaptic neuron activity. Secondly, at synapses derived from thalamic axons, LTP and LTD are observed depending on the relative timing of presynaptic action potential and postsynaptic neuron firing [723]. This is the standard postsynaptic NMDA dependent weight change observed in hippocampal and cortical pyramidal neurons. This standard type of LTP/LTD is not observed in synapses from cortical axons on to lateral nucleus projection neurons, unless a more sustained projection neuron depolarisation is induced.

There is also evidence for presynaptic, NMDA-dependent LTP at glutamatergic synapses of thalamic axons on to projection neurons in the central nucleus of the amygdala [724]. Timing dependent synaptic weight changes implemented presynaptically require a signalling mechanism from postsynaptic neuron to presynaptic neuron, which is generally implemented by diffusion of endocannabinoids. LTP/LTD in both the lateral and central nuclei appear to be important for fear conditioning, such as learning an association between a neutral stimulus (e.g. a light) and a painful stimulus (e.g. an electric shock) [725].

6.6.3 Connectivity of the Amygdala

As illustrated in Fig. 6.16, the pattern of connectivity between the amygdala and other brain regions is very complex. Different nuclei can receive inputs from and project to about 15 sites on average [726]. The amygdala receives inputs from the sensory cortices (auditory, visual, somatosensory, gustatory and olfactory). In general these inputs come from the monomodal association cortices not the primary sensory cortices. The amygdala also receives inputs from polymodal association cortices, the prefrontal cortex, and the hippocampal system. In addition, there are some inputs from the thalamus. The hypothalamus targets the amygdala. Inputs to the amygdala also come from the brainstem, including regions that receive pain, taste and intestinal information. Outputs from the amygdala target the hypothalamus, the dorsal striatum, the nucleus accumbens, the autonomic nervous system regions of the brainstem, and the cortical areas from which inputs are received, although outputs to the monomodal sensory association areas are light. Outputs from the amygdala that target the striatum also project to cortical regions that themselves project to the same striatal region [727, 728].

Within the amygdala there is also a complex pattern of connectivity between different nuclei and subnuclei [535]. For example, the pattern of connectivity believed to support the phenomenon of fear conditioning [729] is illustrated in Fig. 6.17. In the brainstem, the periaqueductal grey (PAG) nucleus drives the freezing behaviour fear response, and the dorsal vagal complex (DVC) drives the cardiovascular fear response. The two type of behaviour are initiated in these two nuclei by two populations of neurons in the medial subnucleus of the central nucleus of the amygdala (CeM) [730]. The two CeM populations are influenced by two populations of neurons in the lateral nucleus of the amygdala. The primary information flow is believed to be that information about the stimulus (e.g. a light) enters the lateral nucleus of the amygdala (LA) from the cortex. From there, information proceeds to the CeM by two paths. Firstly, glutamatergic outputs from the

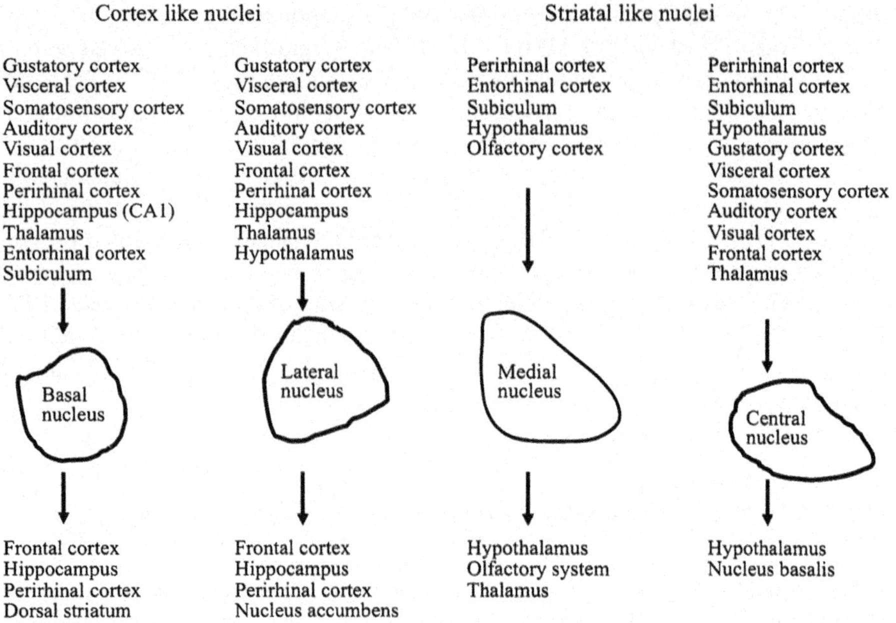

Fig. 6.16 Pattern of connectivity between various nuclei of the amygdala and other brain structures. All the cortex like nuclei tend to receive inputs from sensory cortices, frontal cortex, hippocampal formation and cortices associated with the hippocampus. The lateral nucleus also receives inputs from the hypothalamus. The inputs to the striatum-like nuclei come from similar sources. However, the outputs from the cortex-like nuclei go to the cortex, hippocampal formation and striatum, while the outputs from the striatum like nuclei go to the hypothalamus and other subcortical nuclei

LA target the basal nucleus of the amygdala (BA) which in turn targets CeM. Secondly, glutamatergic outputs from the LA target the lateral subnucleus of the central nucleus of the amygdala (CeL). In CeL there are spatially mixed but different populations of neurons corresponding with freezing and cardiovascular behaviours [731]. These two populations can mutually inhibit each other, and each target a corresponding population in CeM. Some additional circuitry also shown in Fig. 6.17 links oxytocin neurons in the paraventricular nucleus of the hypothalamus to just the population in CeL corresponding with freezing behaviour. Oxytocin excites the CeL neurons, resulting in increased inhibition of the CeM neurons corresponding with freezing behaviour, reducing the probability of such behaviour [732]. Oxytocin is associated with social behaviours such as protecting young, and this circuitry could therefore reduce the probability of freezing behaviour if young of the species are threatened by the same source of fear. Furthermore, when freezing behaviour is inhibited by oxytocin, the CeL neurons target the cortex (via the basal forebrain) to increase risk assessment and exploration type behaviours in response to the fearful stimulus [733].

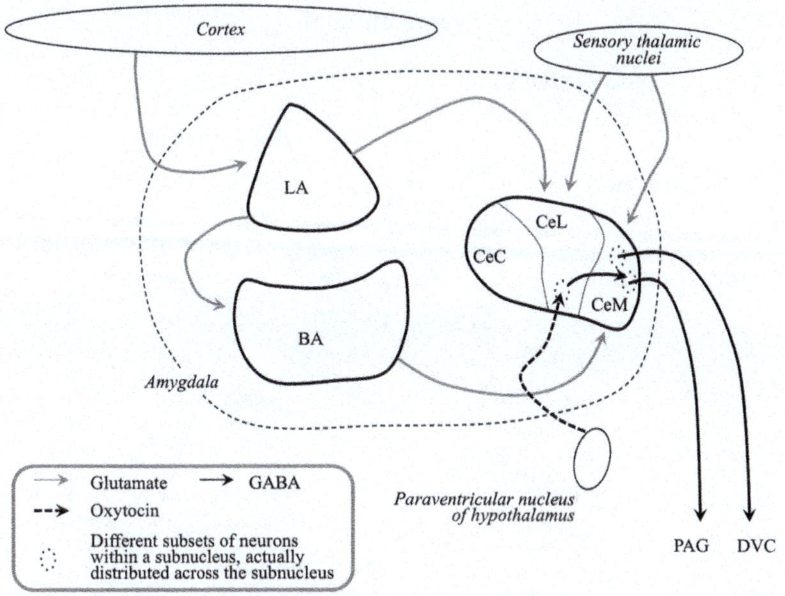

Fig. 6.17 Pattern of connectivity believed to support fear conditioning. The PAG is implicated in driving the freezing response to fear, the DVC affects cardiovascular fear responses. Activity of these two structures is controlled by two physically intermingled but separate sets of neurons in the CeM. The hypothalamic paraventricular nucleus excites neurons in the CeL which in turn inhibit the CeM neurons that control freezing behaviour, reducing the probability of that freezing behaviour. Abbreviations: *PAG* periaqueductal gray, *DVC* dorsal vagal complex, *CeM* medial central nucleus, *CeL* lateral central nucleus, *CeC* central nucleus, *LA* lateral amygdala nucleus, *BA* basal amygdala nucleus

6.6.4 Role of the Amygdala

It is clear from the above discussion that the amygdala plays a role in emotion, but that role is intertwined with the hypothalamus.

6.7 The Hypothalamus and Pituitary Gland

The hypothalamus is a relatively small but very complex concentration of neurons located centrally at the bottom of the brain. It is connected by a stem called the infundibulum to the pituitary gland. The infundibulum is a pair of routes by which signals can pass from the hypothalamus to the pituitary. One of these routes is a bundle of axons that target the posterior pituitary gland. The other route is a set of veins, called the portal veins, that carry blood from the hypothalamus to the anterior pituitary. Hormones secreted by the hypothalamus into

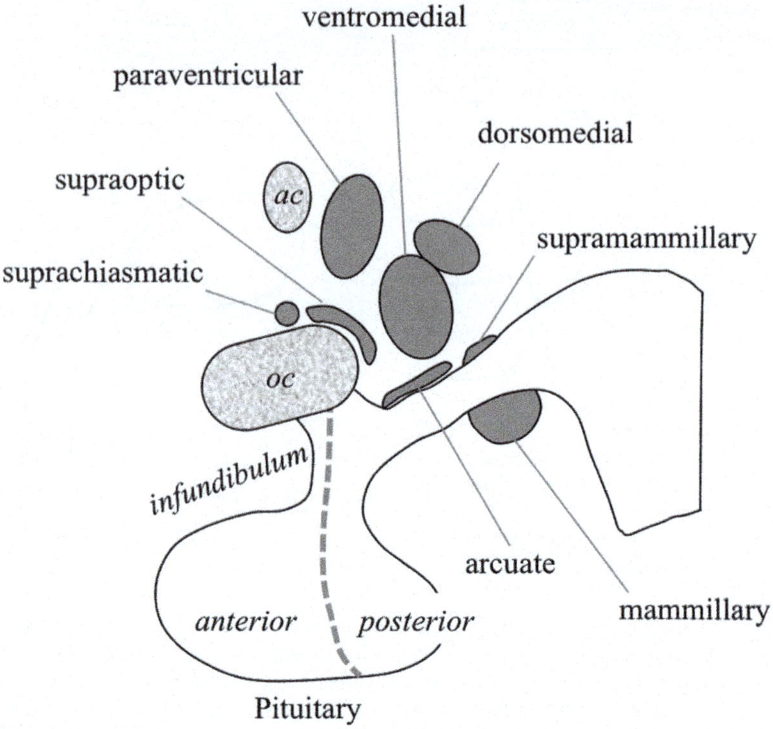

Fig. 6.18 Hypothalamic nuclei. Some of the many hypothalamic nuclei are illustrated. The hypothalamus is located under the thalamus. The pituitary gland is located beneath the hypothalamus, linked to it by the infundibulum, which contains both blood vein and axon bundle links between the two structures. The optic chiasm (*oc*) is the nerve bundle where the optic nerves from the left and right eyes cross over to go to the right and left hemispheres. The anterior commissure (*ac*) is also a nerve bundle

these veins regulate the anterior pituitary. Both parts of the pituitary release hormones into the blood stream that regulate different parts of the body, including in some cases the brain.

There are a large number of hypothalamic nuclei, some of which are illustrated in Fig. 6.18. On the basis of general appearance of neuron somas, three divisions are distinguished in the hypothalamus from anterior to posterior (the preoptic, tuberal and posterior) and each of these divisions are separated from medial to lateral into three zones (periventricular, medial and lateral) [734]. Some nuclei in each of these regions are listed in Table 6.3. However, at least 40 cell groups have been distinguished on the basis of neuron morphology, neurotransmitter type, and connectivity [735], and there could be significantly more. The hypothalamus is crossed by many diffuse and complex fibre systems, and the borders of some cell groups are poorly delineated. For example, the supramammillary area is clearly identifiable in rats and projects to the hippocampus, but in

Table 6.3 Hypothalamic divisions and zones, and some of the nuclei and areas

	Preoptic division	Anterior division	Tuberal division	Mammillary division	
Lateral zone	Lateral preoptic Magnocellular preoptic	Lateral hypothalamic Supraoptic	Lateral hypothalamic Tuberal	Lateral hypothalamic Supramammillary	*LAT*
Medial zone	Medial preoptic	Anterior hypothalamic	Dorsomedial Ventromedial	Medial mammillary Lateral mammillary Posterior hypothalamic	
Periventricular zone	Median preoptic	Suprachiasmic Paraventricular	Arcuate	Posterior periventricular	*MED*
ANT			*POST*		

Abbreviations: *LAT* more lateral location, *MED* more medial location, *ANT* more anterior location, *POST* more posterior location

humans it is much less clearly identifiable [734]. However there is nevertheless a cell group in humans at a corresponding location that similarly projects to the hippocampus [736].

The hypothalamus plays an important role in maintaining body homeostasis: keeping temperature, water absorption and discharge, electrolyte concentrations (e.g. sodium), glucose levels etc. within the appropriate range. It adjusts body parameters like temperature, blood pressure, arousal including the wake/sleep cycle for the time of day (circadian rhythms). It acts on the autonomic nervous system to affect heart rate, digestion, urination, salivation, perspiration, pupil diameter, and breathing. In addition, it makes appropriate body adjustments in response to external threats (preparing for fight or flight by increasing heart rate, blood pressure etc.). It makes adjustments in preparation for reproductive behaviours such as sexual arousal, labour, and lactation.

The hypothalamus also affects complex behaviours linked with these body requirements: for example, it influences initiation of food seeking behaviours. It also appears to be associates with pleasure, aggressive and aversion behaviours. The links with behaviours have often been established by the ways in which damage to the hypothalamus affects specific functions [737]. For example, sometimes a hypothalamic tumor results in fits of laughing, crying and more severe seizures, combined with severe aggressiveness, hyperactivity, and socially inappropriate behaviour [738]. Damage to the suprachiasmatic nucleus result in loss of daily rhythms of wake–sleep activity, feeding, body temperature, and a variety of hormones. Damage to the ventromedial area result in obesity, while damage to the lateral hypothalamic area dramatically reduced feeding. Damage to the medial preoptic nucleus results in loss of fine control of body temperature, although temperature regulation within a broader range is preserved.

There is strong connectivity between different hypothalamic nuclei, and in disorders of eating resulting from lesions of the hypothalamus there are often mood, sleep, and circadian and annual rhythm disorders. For example, tumors affecting the ventromedial hypothalamic area, in addition to obesity, often result in aggressive behaviour, emotional instability, and memory deficits [737]. However, it could be that some symptoms are caused by damage to axons that pass through an affected area but do not interact with it.

Damage to the mammillary bodies have a very specific effect on memory. A tumor strictly limited to the mammillary bodies resulted in loss of the ability to create new declarative memories (both semantic and episodic), while leaving past memories, speech, general intelligence and skills unaffected [739].

6.7.1 Hypothalamic Neurons

The range of neurons in the hypothalamus is very large, in terms of size, shape, and neurotransmitters generated. Many neurons in the hypothalamus generate neurohormones that are released into the blood and are carried by the portal veins into the pituitary gland.

Some neurons in the hypothalamus appear to respond to very specific behavioural circumstances. For example [740], in monkeys some individual neurons change their firing rate in response to seeing a favoured food item, provided the monkey is hungry. Other neurons respond just to the taste of food or just to the combination of sight and taste. A strictly visual neuron does not respond to the sight of non food items, or to grasping and eating food in the dark, but does respond to a good imitation food item. The changes in response can be an increase in firing rate or a decrease in firing rate.

6.7.2 Synaptic Weight Changes in the Hypothalamus

Activity related LTP has been observed in glutamatergic synapses in the hypothalamus. In the supraoptic nucleus, LTP is triggered by 100 Hz stimulation [741]. In the lateral nucleus, LTP is triggered by intensive activation caused by prolonged wakefulness [742].

6.7.3 Hypothalamic Connectivity

Synaptic inputs to the hypothalamus come from a very wide range of brain structures. There is very heavy input from a part of the hippocampal system called the subiculum, via an axon bundle called the fornix. This input targets the mammillary bodies. The mammillary bodies provide outputs to the anterior thalamus, which in turn targets area CA1, the subiculum and the entorhinal cortex in the hippocampal system (see below). There are inputs from the amygdala to the hypothalamus.

The orbital and medial prefrontal cortex areas project to the hypothalamus, but the hypothalamus projects back to all prefrontal areas [743]. In addition, the hypothalamus projects back to the entire cortex by means of a number of different populations of neurons using different neurotransmitters including oxytocin, melanin, orexin, histamine, GABA, acetylcholine and galanin, plus other currently unknown neurotransmitters [734]. One nucleus or group of neurons, such as the cholinergic neurons located in the supramammillary area, may project to almost every cortical area [744]. Such projections may also reach the hippocampus and amygdala. Cortical areas heavily involved in emotion receive very dense projections from the hypothalamus (and also from the amygdala) [745].

For example, oxytocin neurons mainly in the paraventricular nucleus project to the central nucleus of the amygdala and synapse on to selective neurons in the central nucleus [732]. Delivery of oxytocin by this connectivity appears to be responsible for reducing the probability of freeze behaviour in response to a fearful stimulus, for example if the need to protect young individuals is present [730] (see Fig. 6.17). Note that the oxytocin neurons also project back to the prefrontal cortex [746].

Most of the brain is isolated from a wide range of blood chemicals by the blood-brain barrier (a network of very fine capillaries) which prevent passive entry of hydrophilic chemicals unless they are actively transported in. There are some structures called the circumvetricular organs which are outside the blood-brain barrier and detect the concentrations of hormones, peptides and other chemicals in the blood, communicating this information to the hypothalamus, which in turn projects back to those organs [747].

In addition to the connectivity with the higher brain systems, there are a number of outputs from the hypothalamus which affect a wide range of body functions. Firstly, hypothalamic neurons project axons that target neurons of the autonomic nervous system, and also target neurons in the brain stem that control autonomic responses. Secondly, hypothalamic axon outputs control hormone release into the blood stream by the posterior pituitary. Thirdly, the hypothalamus releases hormones into the blood flowing to the anterior pituitary, regulating its release of hormones.

6.8 Habenula

The habenula is a small structure located above the posterior part of the thalamus. It has two nuclei, the lateral and medial habenula, and some subdivisions within these nuclei. Neurons in the medial habenula are acetylcholinergic or contain substance P [748], but acetylcholine and glutamate are co-released by the same neurons [749]. The lateral habenula contains glutamatergic neurons [750].

The connectivity of the habenula [751] is illustrated in Fig. 6.19. The habenula receives ATPergic [752], GABAergic and glutamatergic inputs from the basal forebrain [748], GABAergic inputs from the GPi nucleus in the basal ganglia, and inputs from the hypothalamus, some of which are GABAergic [753].

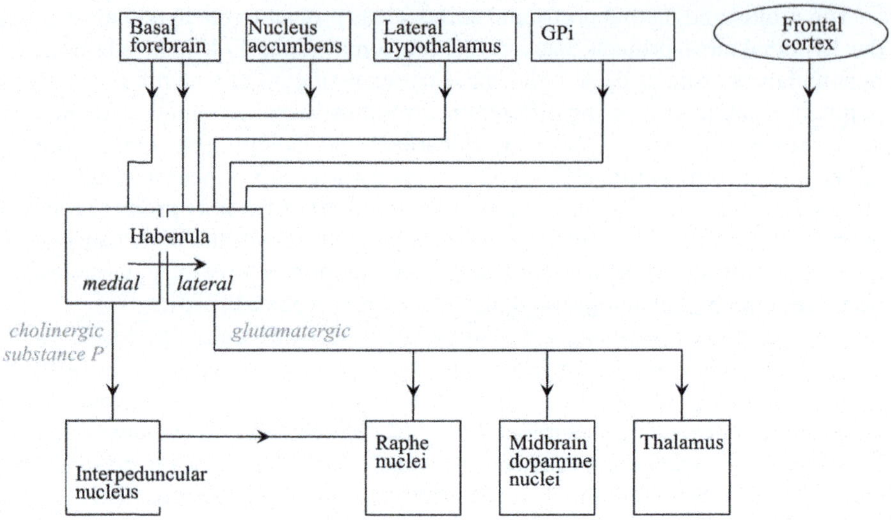

Fig. 6.19 The connectivity of the habenula. The habenula is believed to be involved in the detection of negative feedback following a behaviour

Two strongly targetted structures are the midbrain dopamine neurons and the raphe nuclei. This targetting may include glutamatergic synapses [754] but inhibits the two structures [755, 756], perhaps via inhibitory interneurons [757].

The habenula appears to be involved in detection of the absence of an expected reward and in negative feedback following a behaviour [758]. For example, in human subjects given the task of predicting the outcome of a race between two visual objects starting from different points and proceeding at different speeds, the habenula is activated if they are informed that their prediction was incorrect [681]. The habenula is hyperactive in depressed human patients [759].

6.9 Cerebellum

The cerebellum is a large structure at the back of the brain, underneath the cortical hemispheres and behind the spinal cord. The outside of the cerebellum resembles the cortex in that it is a thin folded sheet of tissue separated into layers, and is in fact called the cerebellar cortex. To avoid confusion, in this section the cortex will be referred to as the cerebral cortex, although elsewhere, the cerebral cortex is simply called the cortex and the cerebellar cortex is always referred to by the full term. Also like the cerebral cortex, within the cerebellar cortex is a volume of white matter made up of axons, and inside the white matter is a set of clusters of neurons called the cerebellar nuclei. Three large bundles of axons called the cerebellar peduncles are located at the front of the cerebellum and connect it to the rest of the brain.

The inferior peduncle mainly carries inputs from the inferior olive nucleus, the middle peduncle is the largest and mainly carries input from the pontine nuclei. The superior peduncle mainly carries outputs from the cerebellum to the thalamus and (much via the red nucleus) to the spinal cord [760]. The pontine nuclei derive most of their inputs from the cerebral cortex, and the pontine nuclei axons targetting the cerebellum are called mossy fibres. The inferior olive nucleus derives most of its inputs from the spinal cord but a significant proportion from the cerebral cortex, and the inferior olive axons targetting the cerebellum are called climbing fibres. The two types of axon are named after a major physical characteristic as discussed below.

The cerebellar cortex is intricately folded, resulting in a surface area which is about 70 % of the area of the cerebral cortex [761]. It can be divided into three major lobes: the anterior lobe (also called the paleocerebellum); the posterior lobe (also called the neocerebellum); and the flocculonodular lobe (also called the archicerebellum). There are also smaller subdivisions, called lobules, for which there are various naming conventions. In one approach, lobules I, II, III, IV and V are identified in the paleocerebellum, lobules VI, VII, VIII and IX form the neocerebellum, and lobule X forms the archicerebellum [762].

There are also three cerebellar nuclei, called the fastigial nucleus, the interpositus nucleus (which can be subdivided into the emboliform and globose nuclei) and the much larger dentate nucleus. The archicerebellum is connected to the fastigial nucleus, and the combination plays a role in balance and eye movement. The paleocerebellum is connected to the interpositus nucleus, and the combination plays a role in muscle tone and posture. The neocerebellum is connected to the dentate nucleus, and the combination plays a role in motor control, and also higher cognition as discussed below.

The electrical oscillations observed in the cortex, thalamus, basal ganglia etc. are also observed in the cerebellum. During performance of a motor task, a synchrony appears between the cerebral cortex and cerebellum in the 10–40 Hz bands [763].

The cerebellar cortex has three well defined layers. The innermost layer is called the granular layer, and contains very large numbers of very small neurons called granule cells. There are of the order of 100 billion granule cells in the cerebellum [761], 5–10 times as many as the number of pyramidal neurons in the cortex. The middle layer is called the Purkinje layer because it contains mainly the somas of ~30 million Purkinje neurons. The outer layer is called the molecular layer, and contains mainly axons from granule cells and the dendrites of Purkinje cells, plus a few GABAergic interneurons (stellate and basket cells).

The cerebellum plays a major role in motor control, and damage to the human cerebellum results in a range of movement related deficits [764]. One result of damage can be loss of muscle tone, or the constant slight tension that is maintained in healthy muscle and which results in resistance when a passive limb is moved by someone else. Another result of damage is loss of strength in a limb. The most prominent symptoms of damage are disorders in voluntary movements. For example, when asked to reach for an object, the start of a motion may be delayed, normal speed is reached more slowly, and there is a lack of uniformity in the motion. Towards the end of the movement, the speed of the limb may not be reduced until the target object has been passed, and the error is then corrected with a series of jerky movements. One patient was quoted as observing that "the movements of my left arm are done subconsciously, but I have to think out each movement of the right [affected] arm. I come to a dead stop in turning and have to think before I start again" [764]. The irregularities of movement are greater when multiple limbs and muscle groups are involved: abnormalities of gait are a common symptom. However, one striking characteristic of cerebellar damage is that symptoms decrease with time, and in some cases may disappear.

It has become clear that the cerebellum also plays many roles in higher cognition. Cerebellar damage results in some specific deficits in working memory [765], and imaging reveals that the cerebellum is active during working memory tasks [766]. The cerebellum is active during both generating and understanding speech [767]. There appear to be some specific cognitive deficits and a generalized tendency towards poorer cognitive performance in patients with damage strictly limited to the cerebellum, with, for example, no damage to the cerebral cortex [768].

6.9.1 Cerebellar Neurons

The outermost molecular layer of the cerebellar cortex contains axons from a very large number of glutamatergic granule cells in the granule layer. These axons come up from the deeper layer where granule cells are located, and form T-branches so that they flow parallel to each other across the molecular layer. The granule cell axons are therefore called the parallel fibres. The molecular layer also contains two types of inhibitory (GABAergic) interneurons: stellate cells and basket cells. These interneurons receive inputs from granule cells in the bottom layer and target Purkinje cells in the middle layer.

The middle Purkinje layer contains the somas of GABAergic Purkinje cells. The dendrites of these cells are flat and extend in parallel across the molecular layer, perpendicular to the parallel fibres. Each Purkinje cell receives a very large number (~200,000) of excitatory inputs from the parallel fibres. Only a few percent of the parallel fibre synapses have synaptic strength, but Purkinje cells nevertheless have very specific receptive fields [769].

The basket cells target the Purkinje soma and could therefore have a stronger effect than the stellates, that target more distal regions of the dendrite [770]. Purkinje cells are inhibitory (GABAergic) and target neurons in the cerebellar nuclei.

The bottom granule layer contains three types of neuron: granule cells, Golgi cells, and unipolar brush cells (UBCs). The most numerous are the granule cells. These are small excitatory (glutamatergic) neurons with about four short dendrites each with input from just one mossy fibre, and additional excitatory inputs from UBCs. They also receive inhibitory inputs from Golgi cells. Granule cell outputs target Purkinje cells, interneurons in the molecular layer, and also Golgi cells. UBCs are excitatory (glutamatergic [771]) interneurons found only in the flocculonodular lobe [772]. One UBC has a single dendrite with a giant synapse from just one mossy fibre. This giant synapse traps glutamate neurotransmitter molecules from the mossy fibre, and therefore results in prolonged output from the UBC [773]. A UBC output targets granule cells and other UBCs [774]. Their axons can project long distances across the granule layer (sometimes passing through intermediate white matter). The third type of neuron is the inhibitory (GABAergic) Golgi cell which targets granule cells and is itself targetted by granule cells.

None of the neurons in any cerebellar cortex layer project outside the cerebellum, but the cerebellar nuclei contain several types of projection neurons and also several types of interneuron [775]. The three types of projection neurons utilize three different neurotransmitters: glutamate; GABA; and glycine, but the third type is found only in the fastigial nucleus. Different locally projecting interneurons appear to use the same three neurotransmitters, and the glycinergic interneuron projects back to target Golgi cells in the cerebellar cortex.

6.9.2 Cerebellar Connectivity

The major connectivity of the cerebellum and associated nuclei is illustrated in Fig. 6.20. There are two major sources of input to the cerebellum: the mossy fibres and the climbing fibres. Most mossy fibres come from the pontine nuclei located in the brainstem, but some come from other nuclei in the brainstem and some direct from the spinal cord [776]. Mossy fibres target the cerebellar cortex granular layer with side branches to neurons in the cerebellar nuclei. In the granular layer, the mossy fibres branch extensively, and each branch has multiple swellings that visually resemble moss. Each swelling is a synapse on a granule cell, and one mossy fibre from the pontine nuclei therefore contacts many (about 600 on average) granule cells. Granule cells have an average of about four short dendrites that end in a claw shape that connects to just one mossy fibre. Inputs to two or three claws are required to fire the granule cell [777]. Excitatory UBCs in the granule layer are also targetted by mossy fibres, but one UBC is contacted by just one mossy fibre. A UBC targets a number of granule cells, sometimes at long distances across the cerebellar cortex [778]. Climbing fibres come from the inferior olive, also located in the brainstem, and target both Purkinje cells in the cerebellar cortex and neurons in the cerebellar nuclei. In the cerebellar cortex, one Purkinje is targetted by just one climbing fibre, which climbs through the Purkinje dendritic tree making multiple synaptic contacts. One climbing fibre targets a small group of (on average seven) Purkinje neurons [779].

Purkinje neurons provide the only outputs from the cerebellar cortex, and only target neurons in the cerebellar nuclei. The same Purkinje neuron can target both GABAergic and glutamatergic projection neurons in the cerebellar nuclei [780]. Cerebellar nuclei neurons fire tonically in vitro, and there is evidence of mechanisms that maintain total inhibition of cerebellar nuclei neurons above some defined level [781]. The cerebellar nuclei provide the outputs from the cerebellum.

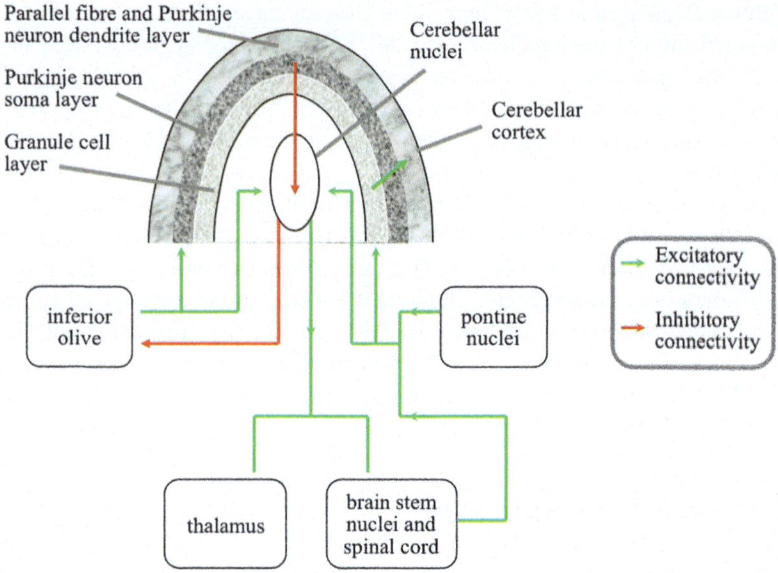

Fig. 6.20 Major connectivity paths of the cerebellum and associated nuclei. Inputs to the cerebellum come from the pontine nuclei and from the inferior olive. Pontine nuclei inputs are excitatory and are called mossy fibres and target granule cells in the cerebellar cortex with side branches to neurons in the cerebellar nuclei. The inferior olive inputs are excitatory and are called climbing fibres and make numerous synapses across the dendritic tree of one or a small number of Purkinje cells. Climbing fibres also have side branches to cerebellar nuclei neurons. The granule cells target Purkinje cells, with one Purkinje cell having one synapse from each of ~200,000 granule cells but synapses from just one climbing fibre. Purkinje cell outputs are inhibitory and target both inhibitory and excitatory cells in the cerebellar nuclei. Excitatory cerebellar nuclei neurons target the thalamus and the spinal cord, inhibitory neurons target the inferior olive

The pontine nuclei receive most of their inputs from the cortex [782], including both visual [783] and auditory [784] cortices and a high volume of input from the prefrontal cortex, especially in human beings [785]. Layer V pyramidal neurons provide most of the cerebral cortex input to the pontine nuclei [760]. A small number of pontine inputs come from other nuclei carrying somatosensory information, the spinal cord itself, and from a range of other brain structures including hypothalamus, thalamus and amygdala [786]. The most common neuron in the pontine nuclei has a large number of asymmetric synapses terminating on spines, probably from cortical inputs [787].

The inferior olive receives excitatory inputs directly and indirectly from the spinal cord [788]. It also receives excitatory inputs from the cerebral motor cortex [789], and from the prefrontal cortex [790]. The inferior olive receives inhibitory inputs back from the cerebellar nuclei [780]. Individual neurons in the inferior olive are excited by different specific somatosensory circumstances such as moving a single hair, stroking a limb surface, or pinching [791]. It is also striking that all the connectivity between the projection neurons in the inferior olive that are the sources

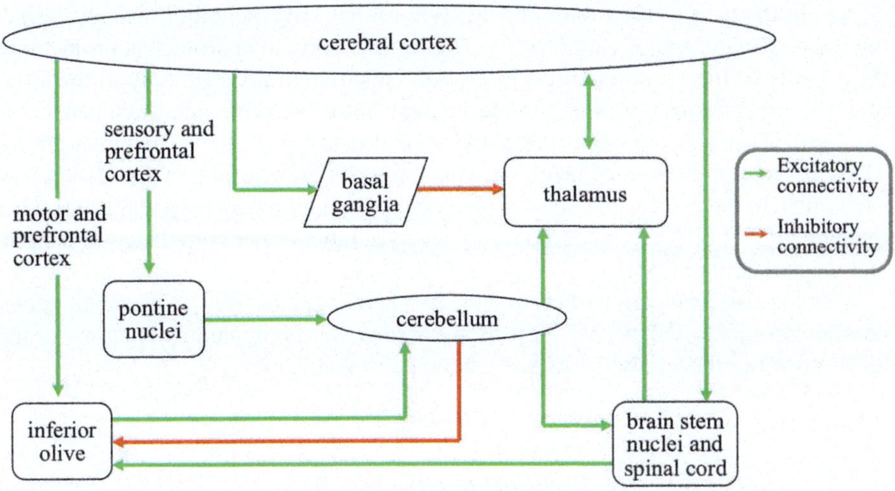

Fig. 6.21 Parallel processing paths for cortical outputs. Outputs from the same cortical region can proceed from the cortex by the basal ganglia-thalamus path back to the same cortical area, or by the pontine nucleus-cerebellum-thalamus path. In the case of motor control, the cerebellum can also directly target muscles via brain stem and spinal cord

of the climbing fibres is by electrical synapses, which are faster responding than chemical synapses and therefore allow closer coordination of timing [792].

Large numbers of combinations of cerebral cortex outputs contribute to the receptive field of a Purkinje neuron [761]. There are over 10 million cortical axons targetting the pontine nuclei, but 100 billion granule cells each with about four inputs. Hence granule cells detect a huge number of presumably different combinations of cortical outputs. Each of the 30 million Purkinje neurons has ~200,000 inputs from granule cells, so a Purkinje neuron has the potential to develop a receptive field that is an extremely specific combination of cortical outputs. There are about five million neurons in the dentate nucleus, which is much the largest of the human cerebellar nuclei. Each cerebellar nuclei neuron is targetted by many Purkinje cells [775].

The outputs from the cerebellum come from the cerebellar nuclei. The GABAergic outputs target the inferior olive, the glutamatergic outputs target both the spinal cord (often via the red nucleus) and the thalamus [793]. The outputs targetting the thalamus influence the cerebral cortex. The cerebellar nuclei neurons respond to a reduction in inhibition by producing a burst of action potentials called a post-inhibitory rebound burst [794]. A pause in Purkinje firing would be effective in triggering such burst firing [795].

Figure 6.21 shows how there are two parallel paths over which cerebral cortex outputs are processed. For example, from the frontal cortex one path goes from the cortex through the basal ganglia and thalamus back to the cortex [785]. The other goes from the cortex through the pontine nuclei and cerebellum and back to the cortex. In the case of the motor cortex [796] routes, the cerebellum can also target the brainstem and spinal cord directly. The layer V cortical neurons that project into these paths form two populations [797]. One population projects only to the striatum, the other projects to the pontine nucleus but often with side branches to the striatum. These side branches target the striatal neurons in the D2 pathway [636]. The separate populations of layer V cortical neurons are within less than 100 μm of each other in the cortex, but show some patchiness in their distribution [798]. One thalamocortical projection neuron rarely receives inputs from both the basal ganglia and the cerebellum [799].

There is also connectivity between the two parallel paths [800]. Cerebellar nuclei neurons project to D2 striatal projection neurons via the thalamus. Neurons in the STN project to the cerebellar cortex via the pontine nuclei.

6.9.3 Cerebellar Microcomponents

Groups of neurons have been identified in the cerebellum which form functional units. These microcomponents include neurons in the cerebellar cortex, the inferior olive and the cerebellar nuclei. A microcomponent is identified by a microzone on the cerebellar cortex within which all the Purkinje cells receive climbing fibre inputs from a set of inferior olive neurons with similar receptive fields [801]. These inferior olive neurons are often arranged in a column [802]. The Purkinje cells project to a limited group of cerebellar nuclei neurons. In cats, the microzone on the cerebellar cortex can be >100 mm long and <200 μm wide [801]. It is estimated that a microzone contains ~10,000 Purkinje cells, and that the human brain may contain ~5,000 microzones [803]. Interneurons in the molecular layer inhibit Purkinje neurons within the same zone [769]. However, granule cell inputs to Purkinjes in a zone which have non-zero synaptic weights mainly come from outside the zone [769].

A number of microzones may project to the same group of cerebellar nuclei neurons [804]. Different groups of cerebellar nuclei neurons generate different types of behaviour, such as different limb movements [805].

6.9.4 Cerebellar Synaptic Weight Changes

The most investigated cerebellar synaptic weight changes are in the glutamatergic parallel fibre synapse on to a Purkinje cell. It is found that LTD is induced in a glutamatergic parallel fibre synapse when an input to the synapse is paired with a strong Purkinje output resulting from a climbing fibre input, and LTP is induced when an input to the synapse is not paired with a climbing fibre input [806]. However, if very strong parallel fibre input occurs in the absence of climbing fibre input, there is also LTD, but in this case the LTD spreads to other parallel fibre synapses [807]. LTP is triggered by presynaptic Ca^{++} entry and is implemented by presynaptic changes, while LTD is triggered by Ca^{++} entry through postsynaptic voltage gated channels opened by climbing fibre activity, resulting in reduction of the number of AMPA receptor channels [808]. Thus, unlike pyramidal neurons in the cerebral cortex, LTP and LTD do not reverse each other.

Cerebellar nuclei neurons receive glutamatergic inputs from side branches of mossy fibre and climbing fibre inputs, and GABAergic inputs from Purkinje cells. LTD of the GABAergic synapses can be observed following 10 Hz stimulation of inputs from Purkinje cells while the neuron membrane is depolarised. This reduction in the weights of GABAergic synapses is not limited to the synapses targetted by the 10 Hz stimulation, and is triggered by entry of Ca^{++} ions through NMDA receptor channels [809]. LTP of GABAergic synapses can be observed following 100 Hz burst stimulation of inputs from Purkinje cells while the neuron membrane is depolarised, and again is triggered by Ca^{++} entry through NMDA channels and not limited to stimulated synapses [810].

Both LTP and LTD are observed in the glutamatergic synapses of mossy fibres on to granule cells, with induction and mechanisms generally similar to those in cortical pyramidal neurons [811]. LTD is induced by low frequency stimulation. LTP is induced by high frequency stimulation, and requires membrane depolarisation, Ca^{++} entry through NMDA receptor channels, and activation of mGlu receptors [812].

6.9.5 Role of the Cerebellum

Learning to associate a behaviour automatically triggered by one stimulus (the unconditioned stimulus) with an unrelated stimulus (the conditioned stimulus) is known as classical conditioning. The cerebellum is believed to play a critical role in many examples of classical conditioning. For example, when an eye experiences a puff of air, there is an automatic blinking response triggered shortly after the puff begins. In tone and puff eyeblink conditioning, an auditory tone is regularly presented, starting a fixed time before a puff and continuing beyond the puff. After about 100 such presentations the blink occurs at a point just before the start of the puff, a fixed time after the tone begins. Hence the blink is now "anticipating" the puff. Lesions to the cerebellum of a rabbit that had successfully learned the response, on the same side as the eye targetted for learning, eliminates the learning. New learning with the other eye is still possible [813]. Lesions to the cerebellum on one side also prevented future learning for the eye on that side [814]. Information on the conditioned stimulus (e.g. the tone) is carried to the cerebellum by mossy fibres, while information on the unconditioned stimulus (e.g. the puff) arrives via the climbing fibres, and cerebellar nuclei drive the response [815].

However, brain activity related to learning of eye blink conditioning is not limited to the cerebellum, but has been identified in a number of other structures including the thalamus, striatum and hippocampus [816]. In addition, if the tone ends before the puff occurs, lesions to the hippocampus also disrupt the learning [817], although the cerebellum remains important [818].

6.10 Basal Forebrain

The basal forebrain is a group of nuclei located centrally between the cortical hemispheres, a little further forward than the thalamus and hypothalamus. These nuclei are unusual in that they have significant numbers of three different types of projection neurons: cholinergic, GABAergic and glutamatergic. The nuclei can be divided into two major groups on the basis of projection targets. One group, which includes the septal nucleus and the vertical limb of the diagonal band of Broca, projects mainly to the hippocampus although some projections also go to the cingulate and occipital cortices. The other group, which includes the nucleus basalis and the horizontal limb of the diagonal band of Broca, projects extensively throughout the cortex and also to the amygdala and the thalamic reticular nucleus [819]. The basal forebrain is the primary source of cholinergic projection to all these targets.

There is deterioration of the cholinergic neurons in the basal nuclei in Alzheimer's disease, although there is also deterioration in neurons in other nuclei that project widely to the cortex, such as the locus coeruleus and to a lesser extent the raphe nucleus [820]. Lesions to the basal nuclei that destroy all the neurons result in learning deficits, especially when there are delays between different information which must be associated by learning [821]. When the damage is to the nucleus basalis, addition of acetylcholine to the cortex can ameliorate the deficit [822]. Lesions also result in loss of past memory, but in some cases these memories are gradually recovered over a period of a few weeks [823]. Some experiments indicate that deficits following damage to just cholinergic neurons do not produce memory deficits [824], but only deficits in the performance of attention tasks [825], especially when attention is required to multiple objects in quick succession [826]. Other experiments indicate that blocking the action of acetylcholine in the cortex produces deficits in recognition memory [827]. The nucleus basalis is essential for more complex cortical learning but not for simpler learning [828, 829]. It is known that acetylcholine enhances LTP in hippocampal [830] and cortical [831] neurons.

6.10.1 Neurons of the Basal Forebrain Nuclei

The basal forebrain nuclei have three different types of projection neurons. Most of the larger neurons are cholinergic, some of the larger neurons are glutamatergic, and there are also GABAergic projection neurons [832]. There are also GABAergic interneurons and interneurons containing a range of neuropeptides [833]. All three types of projection neuron participate in management of theta and also gamma band frequencies in the cortex [834, 832] and hippocampus [835, 836]. Burst firing of different cholinergic neurons can support slightly different theta band frequencies in different cortical areas [837].

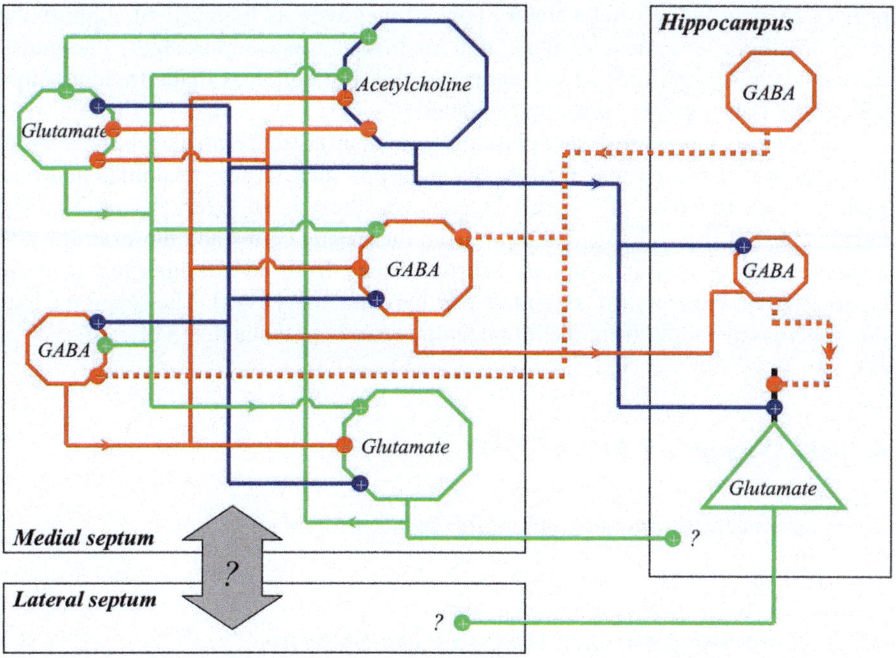

Fig. 6.22 Complex interconnectivity of the septal nuclei and the hippocampus. There are cholinergic, GABAergic and glutamatergic projections from the medial septal nucleus to the hippocampus proper

6.10.2 Connectivity of the Basal Forebrain Nuclei

The basal forebrain nuclei are the major source of cholinergic innovation of cortex, hippocampus and amygdala [832]. There are also significant numbers of GABAergic [838] and glutamatergic [839] projection neurons in the basal forebrain nuclei. Different nuclei within the basal forebrain project to different brain regions, with the nucleus basalis projecting to the cortex and the septal nuclei to the hippocampus, and within these nuclei different neurons project to different cortical areas [840]. In the case of the visual system, cholinergic neurons project to the visual cortex, while GABAergic neurons project to the visual sector of the thalamic reticular nucleus. A different type of cholinergic neuron projects to both the thalamic lateral geniculate nucleus and the visual sector of the thalamic reticular nucleus [841].

The connectivity between the septal nuclei and the hippocampus has been extensively investigated, and is illustrated in Fig. 6.22 [835]. Projection neurons in the medial septal nucleus target the hippocampus, and a return projection reaches the medial septal nucleus via the lateral septal nucleus. Septal cholinergic neurons target hippocampal pyramidal and interneurons, septal GABAergic projection neurons target only hippocampal interneurons, and the targets of septal glutamatergic projection neurons are not known. This connectivity is believed to support the septal nuclei driving theta band oscillations in the hippocampus [835]. The multi-neurotransmitter, multitargetting nature of this projection perhaps indicates that something is being very carefully regulated.

Cholinergic, GABAergic and glutamatergic neurons in the nucleus basalis target interneurons in the cortex, GABAergic neurons also target pyramidal neurons. Synapses are found in all cortical layers, most heavily in layers V and VI. The nucleus basalis receives massive input from the frontal cortex and the striatum, and a smaller degree of input from other cortical areas, from several other nuclei in the basal ganglia, and from the amygdala and hypothalamus [842]. The septal nucleus receives massive input from the hippocampus and hypothalamus [843].

6.10.3 Synaptic LTP and LTD

There have been few studies of LTP/LTD in basal forebrain neurons.

6.11 Neurotransmitter Distribution Systems

There are a number of nuclei that distribute one neurotransmitter widely across the pyramidal neuron containing regions (i.e. cortex, hippocampus and amygdala) and also other brain regions, and are almost the only source for that neurotransmitter. The two most typical examples are the locus coeruleus distributing norepinephrine and the raphe nucleus distributing serotonin. Another example is the tuberomammillary nucleus of the hypothalamus distributing histamine. The midbrain dopamine neurons, which include the SNc and VTA, are another example which has been discussed under the basal ganglia. The projections of the midbrain dopamine neurons to the pyramidal neuron containing regions (i.e. cortex, hippocampus and amygdala) will be discussed in more detail in this section. The other example sometimes included in this category is the acetylcholine projection of the basal forebrain discussed in the previous section.

One role of these distribution systems is to modulate the overall state of the brain, for example managing shifts between REM sleep, NREM sleep, quiet inattentive wake and aroused vigilant wake [844]. For example, stimulation of the locus coeruleus results in an immediate transition from sleep to wake [845]. Infusion of histamine results in higher motor activity, and lesion of histamine neurons induces a somnolent state [846]. Stimulation of serotonin and dopamine receptors increases wakefulness [847]. Thus in general terms, activity of neurons in the locus coeruleus, raphe and hypothalamic tuberomammillary and midbrain dopamine nuclei all enhance wakefulness. These effects on general state of alertness are implemented by neurotransmitter receptors that modulate K^+ ion channels [848].

However, it is also clear that these neurotransmitters also play a role in the management of synaptic weight changes in their target structures. Norepinephrine from the locus coeruleus is known to modulate such changes in the amygdala, cortex and hippocampus, with effects on emotional enhancement of memories [849, 850]. Serotonin from the raphe nuclei is known to modulate such changes in the hippocampus [851] and cortex [852, 853]. Histamine from the hypothalamic tuberomammillary nucleus also modulates synaptic weight changes in hippocampus [854] and cortex [855]. Dopamine also facilitates changes in both the cortex [856] and hippocampus [857].

These four neurotransmitter distribution systems are directly interconnected in a number of ways [858–861] and therefore influence each others activity.

6.11.1 Locus Coeruleus

The locus coeruleus nucleus contains only norepinephrinergic neurons [862] that project widely, to almost all brain regions including cortex, thalamus, hypothalamus, amygdala and cerebellum. The only major region without such projections is the basal ganglia [863]. There are relatively few neurons in the locus coeruleus, and a single neuron can target many different regions [864].

Outputs from the locus coeruleus to the cortex are widespread [865] and excitatory. However, some cortical areas receive more input than others. In general the sensory cortices receive a higher level of input, the degree of input varies across the polymodal parietal cortex, while input to prefrontal and temporal cortices is low but homogeneous across different cortical areas [866]. For example, in the primate cortex, visual areas concerned with spatial analysis and visuomotor response are more heavily targetted than areas with feature detection functions [864]. The locus coeruleus also sends excitatory outputs to the thalamus, with a tendency for one neuron to send outputs to the both a thalamic nucleus and the cortical area with which that nucleus interacts [867]. The amygdala and cerebellum are also targets for excitatory outputs. Locus coeruleus outputs target the autonomic nervous system, both directly to targets in the brainstem and indirectly via the hypothalamus. Some of these outputs inhibit the area of the hypothalamus that regulates sleep. There is strong input to the locus coeruleus from the prefrontal cortex, much less from the rest of the cortex. There are also inputs from the amygdala, hypothalamus and brainstem [198].

The effect of neuropinephrine on burst firing layer V cortical neurons is to modify the firing mode. The pattern of firing is changed from burst generation (3–6 spikes with interburst frequency ~1 Hz) to single spike outputs, and the average spike rate is increased by a factor of three [868].

The locus coeruleus is a major wakefulness promoting nucleus, and activation of the autonomic nervous system and inhibition of sleep are two of its effects [869]. An important cognitive capability is to be able to focus attention on a limited range of objects in the environment to carry out a complex behaviour, screening out irrelevant objects but retaining the ability to adjust to an unexpected event. It has been observed that at low levels of locus coeruleus activity, drowsiness interferes with performance of a complex task. At intermediate levels, performance is optimal, while at high levels, a task is constantly interrupted be scanning of the environment. These observations have been interpreted to indicate that the locus coeruleus manages the balance between task focus and general watchfulness [870].

Both overactivity and degeneration of the locus coeruleus are associated with various mental deficits. The treatment of depression by tricyclic antidepressants reduces the activity of the locus coeruleus, and there is evidence that excess activity can lead to suicide [871]. Degeneration of the locus coeruleus is present in a wide range of dementias, such as Parkinson's disease, Alzheimer's disease, and Down syndrome [872]. However, it is not clear whether the generation is a cause, or a result of degeneration elsewhere.

6.11.2 Raphe Nuclei

The raphe nuclei fall into two groups, the rostral group made up of the caudal linear, dorsal and medial nuclei, and the caudal group made up of the magnus, obscurus and pallidus nuclei. All these nuclei contain serotonergic neurons, but dopaminergic and GABAergic neurons are present in some of the rostral group nuclei, and neurons containing substance P and neurons containing different neuropeptides are present in many of the rostral and caudal nuclei [873].

The major source of inputs to the rostral group is the habenula. Other sources include the medial prefrontal cortex, the basal forebrain, the ventral pallidum, several hypothalamic nuclei and the central nucleus of the amygdala. Inputs to the caudal group include several hypothalamic nuclei and the central nucleus of the amygdala.

The outputs of the rostral group target all cortical areas but most heavily the frontal lobe [874]. Rostral group projections also target the hippocampus, amygdala, striatum and substantia nigra pars compacta, thalamus, basal forebrain, mammillary nucleus of the hypothalamus and cerebellum [873]. The outputs of the caudal group are mainly targetted to motor nuclei and the spinal cord [873].

Serotonin acts on target neurons by a number of different 5HT-receptors, making its possible effects very complex. For example, the effect of serotonin on pyramidal neurons in the hippocampus is to initially reduce the firing rate and then to increase the rate, acting on multiple types of receptor [875]. Serotonin can facilitate LTP and LTD in pyramidal neurons [853, 852].

The raphe nuclei play a role in the management of pain, modulating both motor and cognitive responses [876]. Serotonin plays a role in managing the sleep-wake cycle, generally suppressing REM sleep and encouraging wakefulness, but in some situations increasing the tendency towards sleep [877]. It has a role in mood that is not fully understood, but serotonin receptor blockers (antagonists) can be effective in treating schizophrenia [878].

6.11.3 Tuberomammillary Nucleus of the Hypothalamus

The tuberomammillary nucleus is located in the posterior hypothalamus. It contains large histaminergic neurons, and is the only brain structure that projects the neurotransmitter histamine. A small proportion of the neurons also contain substance P [879].

Inputs to the tuberomammillary nucleus come from many brain regions, with substantial inputs from the infralimbic cortex (Brodmann area 25), the lateral septal nucleus in the basal forebrain, the subicular complex of the hippocampus and the hypothalamic preoptic nucleus [880].

The ~64,000 histaminergic neurons target all regions of the brain [881], and one neuron can target multiple widely separated brain regions [879]. The types of receptor present in different brain regions can differ, for example H1-receptors are more common in the neocortex while H2 and H3 receptors are present in the cerebral cortex but more common in the basal ganglia [882]. H1 receptors generally increase firing rate of target neurons, H2 receptors increase the number of action potentials generated in response to a single depolarisation event, while H3 receptors inhibit by reducing glutamate release [881].

The histamine distributed by the tuberomammillary nucleus plays a role in learning and memory, consistent with its influence on synaptic plasticity [881]. It also plays a role in sleep, and in regulation of activity and body homeostasis [881]. Histamine may be a danger response, reflected in the experience of anxiety [883]. It is involved in the modulation of pain [861].

6.11.4 Midbrain Dopamine

The midbrain dopamine neuron region has been very extensively investigated because of its role in human neurological diseases like Parkinson's. Most of the dopaminergic neurons in the brain are located the three regions in the midbrain: the retrorubral nucleus, and the SNc and VTA in the basal ganglia [884]. These three regions are also designated A8, A9 and A10, respectively. The retrorubral nucleus is adjacent to the SNc and although physically distinguishable, also projects heavily to the striatum and could be regarded as part of the basal ganglia. There are a number of other smaller groups of dopaminergic neurons, including groups in several hypothalamic nuclei. The main focus in this section will be on the three larger groups and their projections to the cortex, hippocampus and amygdala.

The midbrain dopamine region contains dopaminergic, GABAergic and glutamatergic neurons. All three types project to the cortex and also locally [885]. Some neurons that project to the striatum contain both dopamine and GABA [667].

As described earlier, the major source of inputs to the midbrain dopamine neurons is from within the basal ganglia, especially from the striatum. The VTA also receives inputs from the frontal cortex. The SNc receives inputs from the cerebellum.

In primates including humans the midbrain dopamine neurons target the entire cortex [884], the hippocampus [886], the amygdala [887] and the thalamus [593]. In the primate cortex there are dopamine D1 and D2 type receptors in all areas, lowest in the primary visual areas, with more D1 type. D1-type receptors occur in all areas and all layers, but fewer in layer IV, while D2-type receptors occur most heavily in layer V [888]. Both D1 and D2 types are found in the amygdala [889]. D1 and D2 type are present in the hippocampus in comparable amounts, fewer D1 type but somewhat more D2 type than in the cortex [890]. The concentration of dopamine is highest in the basal ganglia, followed by the thalamus and then the cortex [891]. The midbrain dopamine neurons are similar to the other neurotransmitter distribution systems in that single dopaminergic neurons have hundreds of thousands of synapses, but different in that subregions of the midbrain dopamine neurons target different subregions of the target structures to a much greater degree [892].

The midbrain dopamine neurons are involved in a range of cognitive functions. Working memory depends on dopamine, and defective D1 receptors in the dorsolateral prefrontal cortex are associated with the working memory deficits observed in schizophrenia [893]. Dopamine plays a role in the recording of episodic memories [894] and in the management of rewards [895]. Dopamine regulates the release of other neurotransmitters [896], and facilitates long term changes to synaptic weights in cortex [856], hippocampus [857] and as discussed earlier, in the striatum.

6.12 The Hippocampal System

Under the lower posterior part of the cerebral hemispheres and curving up between the hemispheres are three cortical regions called the parahippocampal, perirhinal and entorhinal cortices. Further up between the hemispheres the cortical sheet thins down and curls over, and finally ends at a wedge of very small neurons called the dentate gyrus. This region is illustrated in cross section in Fig. 6.23.

The perirhinal, parahippocampal and entorhinal cortices have the standard six layers, but layer IV in the entorhinal cortex contains axons and no neurons [897], and hence only four layers of pyramidal neurons. The other two cortices have relatively few or sometimes no neurons in layer IV [898]. As the cortex thins down towards the dentate gyrus there is initially a region called the subicular complex, usually divided into three areas: parasubiculum, presubiculum and subiculum in order of distance from the entorhinal cortex [899]. Beyond the subicular complex are the CA fields, first CA1, then the small CA2 followed by CA3. Finally at the end of CA3 is the dentate gyrus. The region between the arms of the dentate gyrus wedge of neurons is called the hilus (or hilar region) which is regarded as part of the dentate gyrus. Across the subicular complex the number of distinguishable layers of pyramidal neurons reduces from four down to one, and there is just one layer of pyramidal neurons in the CA fields.

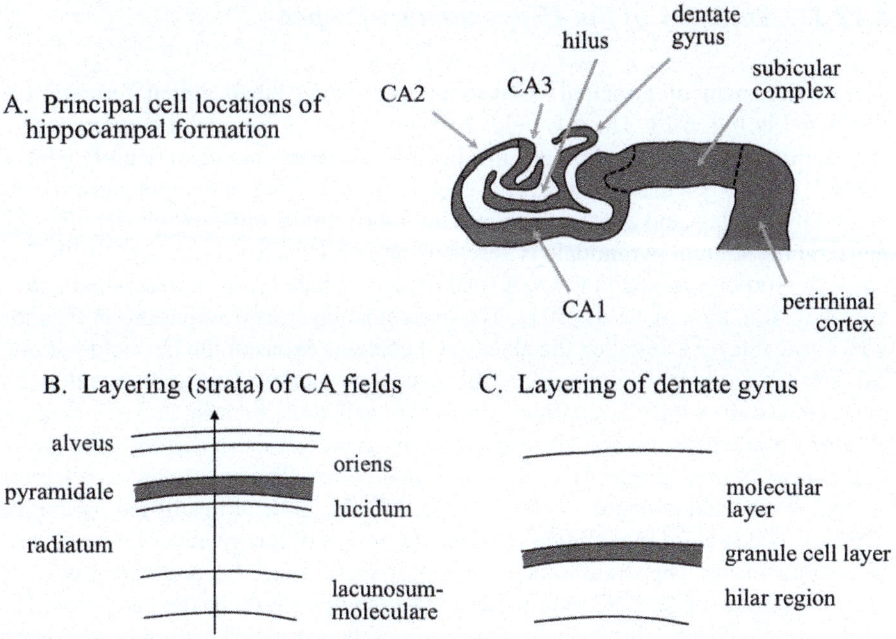

Fig. 6.23 The structure of the hippocampal formation. (**a**) Regions of principal neurons in the hippocampal formation. The formation is a continuation of the neighbouring cortex, which is sometimes the perirhinal and sometimes the entorhinal depending on the level of the section. In the CA fields there is just one layer of pyramidal neurons. In the human brain, CA1 has the thickest pyramidal layer and some signs of sublayering, CA2 has the thinnest pyramidal layer. In the rat brain, CA3 is the thickest layer. (**b**) In the CA fields, six or seven layers are distinguished. The alveus contains axons of CA pyramidal neurons. The stratum oriens contains the basal dendrites of the pyramidal neurons, and the stratum pyramidale contains their somas. The stratum lucidum is only present in CA3 and contains axons coming from granule cells in the dentate gyrus. The main body of the apical dendrites is in the stratum radiatum, and their tufts in the stratum lacunosum-moleculare, which is sometimes separated into two layers. (**c**) In the dentate gyrus, three layers are identified. The middle granule cell layer contains the granule cell somas, and their dendrites are located in the molecular layer. The hilus contains the dentate gyrus mossy cells

In this book the term *hippocampus proper* will be used for the CA fields and the dentate gyrus. The term *hippocampal formation* will be used for the hippocampus proper plus the subicular complex. The term *hippocampal system* will be used for the hippocampal formation plus the three associated cortices plus a number of additional subcortical structures that are strongly interconnected with the hippocampal formation.

Because of its relatively simple layering organisation, pattern of connectivity, and the similarity of its neurons to those in the cortex, the hippocampus proper has been very extensively investigated.

6.12.1 Anatomy of the Hippocampus Proper

The arrangement of principal neurons of the human hippocampal formation is illustrated in Fig. 6.23. The CA fields have one layer of pyramidal neurons, called the stratum pyramidale. In humans this layer is narrowest and most densely packed in CA2, in CA3 is wider and in CA1 is even wider but with smaller and more widely spaced pyramidals and a hint of sublayering [900]. In the more extensively investigated rat the stratum pyramidale is very narrow and densely packed in CA1, thicker and more loosely packed in CA3. A total of six separate layers (often called strata) are labelled in the CA fields [901]. The outermost layer, corresponding in location with cortical layer I, is called the alveus and contains axons of the CA field pyramidal neurons on their way to various subcortical structures. This axon bundle adds outputs from the subicular complex and is then called the fimbria, and renamed the fornix as it proceeds towards its target regions. The alveus also carries some inputs to the hippocampus proper from the entorhinal cortex. The next layer is called the stratum oriens and contains interneurons and the basal dendrites of the pyramidal neurons. The stratum pyramidale contains the somas of the pyramidal neurons and also interneurons. The stratum lucidum is only present in CA3, and contains many of the axons that target CA3 pyramidals from granule cells in the dentate gyrus. The stratum radiatum contains the apical dendrites of the pyramidal neurons, some interneurons, and axon fibres from CA3 to CA1, called the Schaeffer projection. The stratum lacunosum-moleculare (sometimes regarded as two layers) contains the tufts of the pyramidal apical dendrites, most of the input axons from the entorhinal cortex (called the penetrant path) and some Schaeffer collaterals.

The dentate gyrus has a simpler three layer structure. The middle layer contains the somas of principal cells called granule cells, plus interneurons. The outer molecular layer contains the dendrites of the granule cells and axons from the entorhinal cortex. The inner layer, called the hilus (also called the hilar region or the polymorphic layer), contains the somas of another type of principal cell called the mossy cell, and some interneurons.

6.12.2 Neurons of the Hippocampus Proper

The projection neurons of the subicular complex and CA fields are pyramidal neurons similar to those in the cortex. In addition, within the CA fields are many different types of GABAergic interneurons, in some cases with resemblances to cortical interneurons. About six types of interneuron have been identified in the CA fields, including axo-axonic cells in the stratum pyramidale, basket cells in the stratum pyramidale and stratum radiatum, and OLM cells in the stratum oriens [419]. Basket cells support gamma band oscillations (as in the cortex) and OLM cells support theta band [902]. The gamma band activity can be modulated at theta frequencies [903]. Cortical theta band activity appears to be controlled by hippocampal system pyramidal outputs [904]. Axo-axonic cells target pyramidal axon initial segments, basket cells the somas, and OLM cells the distal apical dendrites.

The dentate gyrus has two types of glutamatergic neurons: granule cells in the granule cell layer and mossy cells in the hilus. There are also about five types of GABAergic interneurons including basket cells and axo-axonic cells in or close to the granule cell layer, the MOPP interneuron type in the molecular layer, and the HIPP type in the hilus [419].

6.12.3 Synaptic Weight Changes in the Hippocampus Proper

In general, weight changes to hippocampal glutamatergic synapses are similar to those found in cortical pyramidal neurons. The relatively simple morphology and connectivity of the hippocampus proper has made it relatively easy to investigate synaptic weight changes, and LTP was in fact first demonstrated in vivo in glutamatergic synapses on to granule cells in the dentate gyrus [905]. Synapses in different locations on hippocampal neurons have been found to have different LTP properties. For example, for the same level of input activity the threshold for LTP is higher at apical dendrite synapses than at basal dendrite synapses in CA1 pyramidal neurons, partly because of interneurons preferentially synapsing on the apical dendrites [906]. Differences have also been found for CA3 pyramidals and dentate gyrus granule cells. Synapses of dentate gyrus mossy cells on to CA3 pyramidals show LTP, but mossy cell synapses on to CA3 interneurons do not [907]. This difference is associated with the very different physical form of the two types of synapse from the same axon [908].

6.12.4 Hippocampal System Connectivity

The entire cortex with the exception of the primary sensory areas is reciprocally connected with the hippocampal formation via the associated cortices [909]. This connectivity is illustrated in Fig. 6.24. The monomodal sensory cortices and some polymodal cortices are connected with the perirhinal and parahippocampal cortices. The entorhinal cortex is connected with other polymodal cortices and with the perirhinal and parahippocampal cortices. The entorhinal cortex has strong connectivity with the hippocampal formation.

As shown in Fig. 6.25, the interconnectivity between the entorhinal cortex and the hippocampal formation follows some specific paths. The inputs to the entorhinal cortex from the perirhinal and parahippocampal cortices arrive in layers II and III. Entorhinal cortex layer II provides outputs to CA3 and the dentate gyrus, layer III to CA1. This entorhinal output is called the perforant path because it is a very visible axon bundle that penetrates through the subiculum and the CA fields. There is then a return connectivity path that goes one way from the dentate

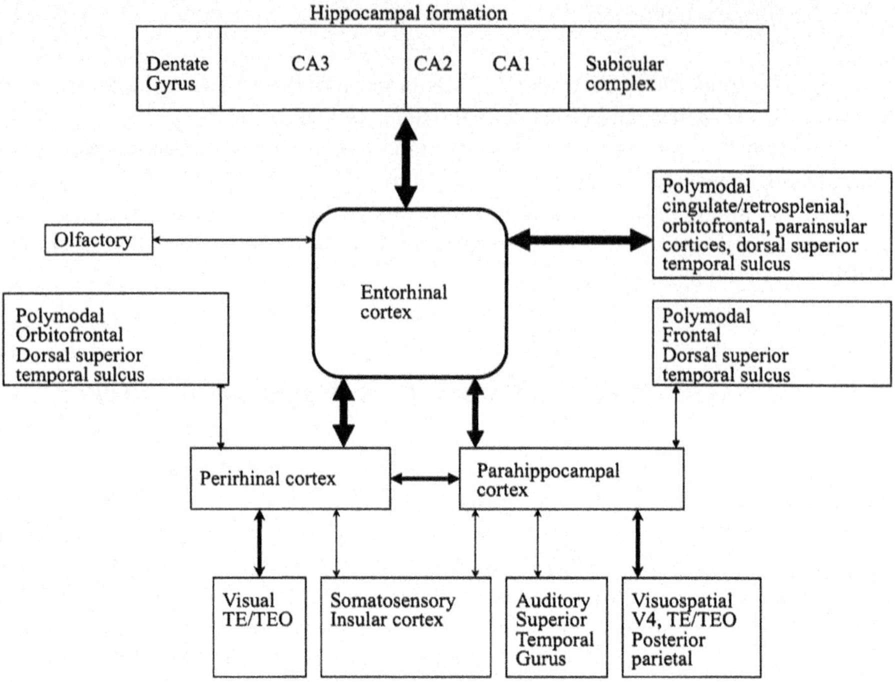

Fig. 6.24 Connectivity of the hippocampal formation with the cortex. There is reciprocal connectivity between the hippocampal formation and the entire cortex with the exception of the primary sensory areas. Most of this connectivity is funnelled through the parahippocampal and perirhinal cortices and then the entorhinal cortex

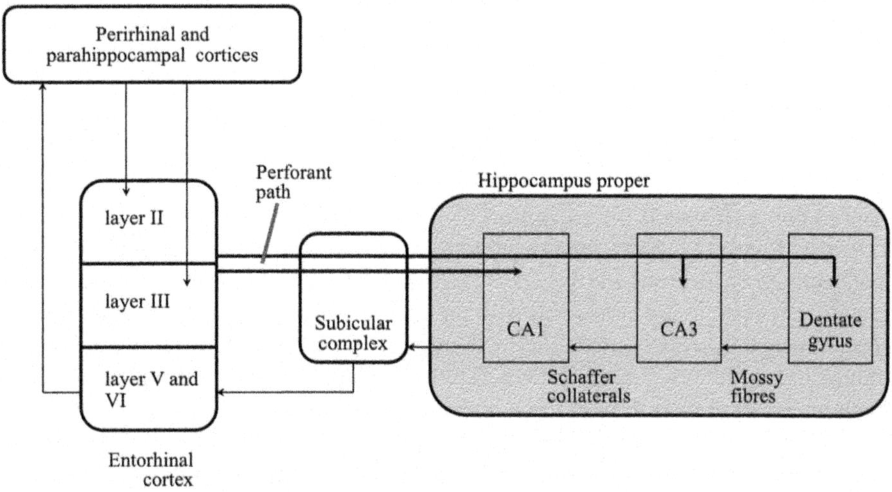

Fig. 6.25 Connectivity of the entorhinal cortex and the hippocampal formation. Inputs from the rest of the cortex arrive in entorhinal layers II/III and outputs back to the cortex go from layers V/VI. There is a one way direction of flow from entorhinal cortex layer II directly to CA3 and the dentate gyrus via the perforant path, from the dentate gyrus to CA3, from CA3 to CA1, and from CA1 (generally via the subicular complex) to layers V/VI of the entorhinal cortex. Layer III of the entorhinal cortex projects directly to CA1, also via the perforant path

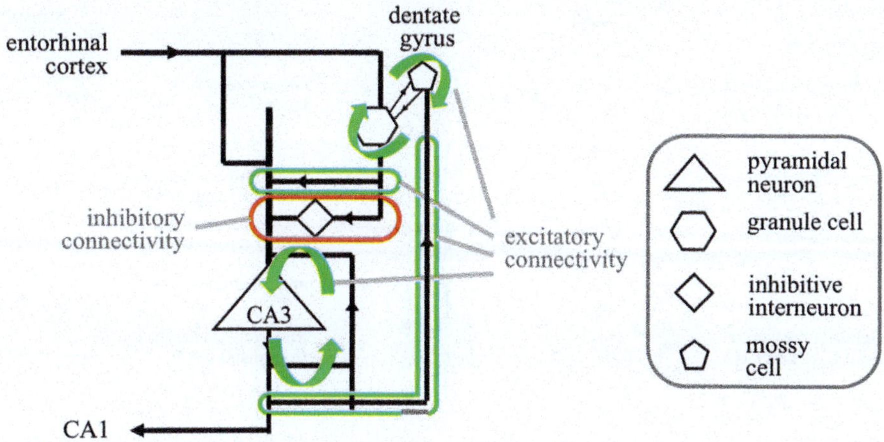

Fig. 6.26 Feedback loops within the hippocampus proper. Both pyramidal neurons in CA3 and granule cells in the dentate gyrus receive excitatory inputs from the entorhinal cortex. In the dentate gyrus, granule cells excite mossy cells and mossy cells excite granule cells. In CA3, one pyramidal neuron excites many other pyramidal neurons. Between these two positive feedback loops there are two excitatory links. CA3 pyramidals excite mossy cells in the dentate gyrus, and each granule cell in the dentate gyrus excites a small number of CA3 pyramidals. There is also an inhibitory link in which dentate gyrus granule cells inhibit a wide range of CA3 pyramidals via CA3 interneurons

gyrus to CA3 (called the mossy fibres), from CA3 to CA1 (called the Schaeffer collaterals), from CA1 to the subicular complex and from there to layers V and VI of the entorhinal cortex.

The connectivity within the dentate gyrus and CA3 is shown conceptually in Fig. 6.26. There are two positive feedback loops, one within CA3, the other within the dentate gyrus. Within CA3, one pyramidal targets large numbers of other pyramidals. Typically, the number of synapses on to one CA3 pyramidal from other CA3 pyramidals is several times larger than the number of synapses on to that pyramidal from the entorhinal cortex [910]. Within the dentate gyrus, granule cells target mossy cells [911] and mossy cells target granule cells [912]. There are also links between the two feedback loops: each CA3 pyramidal receives a small number of inputs directly from granule cells, but a much larger number of inputs from inhibitory interneurons that are themselves targetted by granule cells [908]. Hence a dentate gyrus granule cell is "communicating with a handful of CA3 pyramidal cells while silencing most others" [913]. These positive feedback loops are also damped down by other interneuron connectivity routes, for example granule cells establish many synapses on to GABAergic interneurons in the hilus [908]. These HIPP interneurons project back to the molecular layer [119], probably targetting granule cell dendrites.

The CA2 region can be distinguished on a number of grounds. There are a number of neurochemical and morphological differences between CA2 neurons and those in CA1 and CA3. CA2 is similar to CA3 in that the pyramidal neurons receive inputs from entorhinal cortex layer II, but similar to CA1 in having inputs from CA3

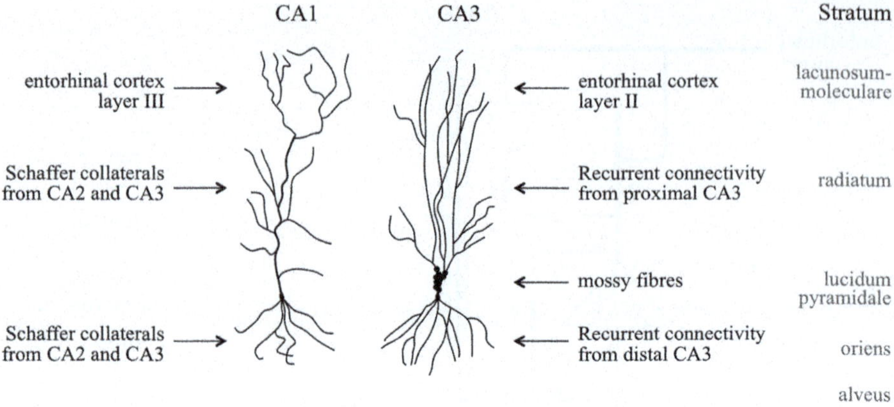

Fig. 6.27 Segregation of input connectivity on dendrites of CA field pyramidal neurons. On a CA3 pyramidal neuron, inputs from layer II of the entorhinal cortex arrive on the distal tufts of the apical dendrite, and inputs from the dentate gyrus arrive on the most proximal regions of the apical dendrite. Inputs from nearby CA3 pyramidals arrive on the main part of the apical dendrite, while inputs from more distal CA3 pyramidals arrive on the basal dendrites. On a CA1 pyramidal, inputs from layer III of the entorhinal cortex arrive on the distal tufts of apical dendrites, while inputs from CA2 and CA3 arrive on the apical dendrites and basal dendrites

but no inputs from the dentate gyrus [901]. Another CA2 difference is that it is the only CA field that receives strong input from the supramammillary area [914] as discussed below. Neurons in the CA2 region are relatively spared in epileptic seizures that damage neurons in CA1 and CA3 [915].

CA1 pyramidal neurons do not target other CA1 pyramidals, but form the only source of hippocampus proper outputs targetting the cortex directly or via the subicular complex [900].

The different input sources to the hippocampus proper tend to target different regions of the principal neurons [901]. For example, as shown in Fig. 6.27, in CA3 the pyramidal neurons receive inputs from the dentate gyrus on the most proximal parts of their apical dendrites, inputs from layer II of the entorhinal cortex on the distal tufts of those dendrites, and inputs from relatively nearby CA3 pyramidals in the central regions of the apical dendrites. Inputs from more distant CA3 pyramidals arrive on the basal dendrites. In CA1 the pyramidal neurons receive inputs from layer III of the entorhinal cortex on the distal tufts of their apical dendrites, and inputs from CA3 pyramidals on the more proximal parts of their apical dendrites and on the basal dendrites [916]. Information from different sources can thus be integrated separately, before overall integration.

There is also strong connectivity between specific subcortical structures and specific regions of the hippocampal formation, and in some cases the entorhinal cortex. This connectivity is illustrated in Fig. 6.28. Four key subcortical structures are the mammillary bodies of the hypothalamus, the anterior thalamic nucleus, the basolateral nuclei of the amygdala, and the septal nuclei in the basal forebrain. The basolateral amygdala targets layer III of the entorhinal cortex and various parts of the

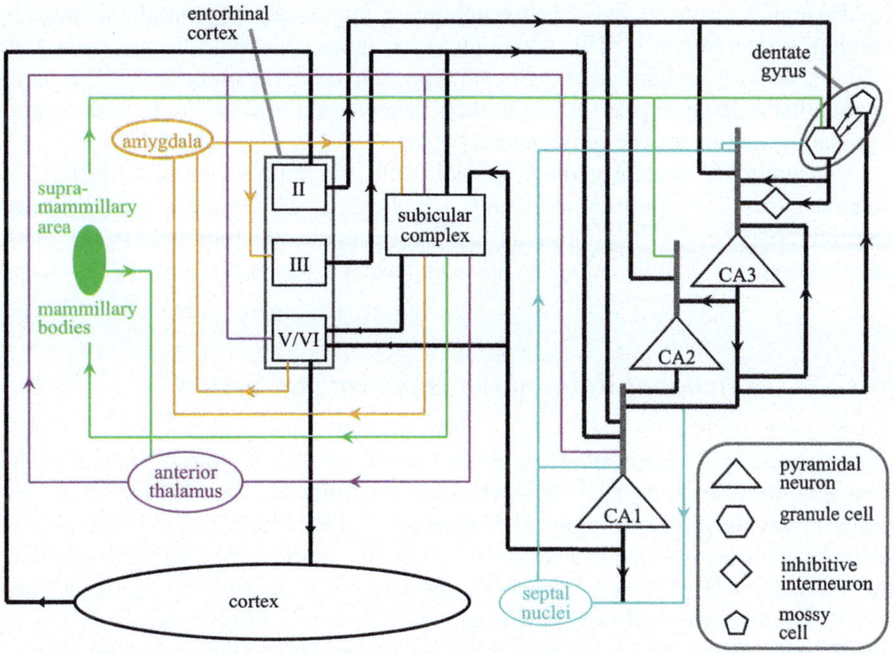

Fig. 6.28 Connectivity of the hippocampal system. In addition to the reciprocal connectivity with the cortex, a number of subcortical structures are connected with the hippocampal formation and entorhinal cortex. Much of this subcortical connectivity travels in the axon bundle called the fornix. The septal nuclei receive substantial inputs from CA3, and project to CA1, CA3 and the dentate gyrus. The mammillary bodies receive inputs from the subicular complex, and the supramammillary area projects to CA2 and also the dentate gyrus. The anterior thalamic nucleus receives inputs from the subicular complex and also from the mammillary nucleus. It projects to CA1, the subicular complex and layers V/VI of the entorhinal cortex. The amygdala receives inputs from the subicular complex and from layers V/VI of the entorhinal cortex and projects back to the subicular complex and layer II of the entorhinal cortex. In a human being, damage to any of these hippocampal or subcortical structures can result in declarative memory deficits

subicular complex, with projection back originating in the subicular complex and layers V/VI of the entorhinal cortex [900]. The subicular complex projects to the anterior thalamic nucleus, which projects back to CA1 [917, 918], and also to the subicular complex and layer V/VI of the entorhinal cortex [919]. The mammillary bodies receive strong inputs from the subicular complex [920], and project strongly to the anterior thalamic nuclei. The associated supramammillary area projects to CA2 pyramidals and also to dentate gyrus granule cells [921]. The septal nuclei project to CA1, CA3 and the dentate gyrus utilizing cholinergic, GABAergic and glutamatergic projection neurons [835], terminating in all regions of the hippocampus proper [900]. There is a large projection to the septal nuclei from CA3 [901], and some projection from CA1 [922]. The septal projection to the hippocampus proper is heavily involved in generation of theta band activity. Septal glutamatergic neurons influence the frequency and cholinergic neurons the magnitude of the waves, while

GABAergic neurons increase the excitability of hippocampal pyramidal neurons by inhibiting interneurons [819]. Although septal input appears to trigger theta band activity, activity is maintained by local hippocampal interneurons [903]. The supramammillary area also plays a role in theta generation, projecting to both the hippocampal formation and the septal nuclei [914].

Unusually, the dentate gyrus projects to only one brain region, CA3 [901]. The dentate gyrus is also unusual in that it is one of only two brain regions in which new neurons are generated in adulthood (the other being the olfactory bulb) [923]. These new neurons are granule cells, and show enhanced synaptic plasticity [924].

6.12.5 Functional Roles of the Hippocampal System

The hippocampal system plays an important role in certain types of memory. This was first understood in the 1950s, when an experimental treatment for severe epilepsy involving surgery to removal of large parts of the hippocampal formation and entorhinal cortex resulted in a striking combination of memory deficits. The deficits in one patient (HM) were extensively investigated over a period of many years [925]. The ability to learn any new events, facts or words was lost, although general intelligence, conversation skills, perception and reasoning ability were unaffected. Working memory was unaffected [926]. Memory for events that had occurred in the decade preceding the surgery were impaired [27] but memory for words learned in the same period was unaffected [28]. Skills acquired prior to surgery were unaffected, and new sensorimotor skills like mirror tracing could even be acquired. However, memory of the acquisition process was lost: HM improved over a number of sessions, but in each session could not remember attempting mirror tracing before [927]. Priming memory was also unaffected [30].

It is also observed that similar groups of deficits can follow damage to the subcortical structures discussed earlier. Damage limited to the mammillary bodies can result in loss of ability to create new fact or event memories, but preservation of other cognitive capabilities [739]. Damage to the anterior thalamic nucleus can result similar deficits [597]. The amygdala plays a role in enhancing the memory of emotional events [928]. The evidence is that these structures have these effects through their action on the hippocampal formation [929, 930], and should therefore be regarded as an integral part of that system. Damage to the septal nuclei also produce similar memory deficits [931], although not necessarily when only the cholinergic neurons are damaged [819].

6.13 Relating Structure and Function

A general conclusion that can be drawn from this overview of anatomy is that in general no one anatomical structure corresponds with a specific type of cognitive task. Any task requires many different structures, and any structure supports many different types of task. Even for functions like rewarding, determining behaviour, pattern detection and declarative memory the same lack of one-to-one correlation between brain structures and functions applies. For example, reward behaviours are associated with the nucleus accumbens and VTA in the basal ganglia, but also with the amygdala, the habenula, and various cortical areas. Each of these structures is also associated with non-reward related functions.

The implication is that the driving force for the existence of different anatomical structures cannot be found in specialisation for types of cognitive tasks, or even specialisation for types of function. Rather, the role of a structure needs to be found in the type of information processing performed by the structure, with any one type of information processing being required by many different types of function and cognition. Such information processing roles are likely to be very abstract relative to commonly understood cognitive functions.

In the next chapter we will describe some constraints on physical architecture that exist for any sufficiently complex learning system. These constraints arise as a result of practical considerations including the need to limit physical information processing resources and the need to limit the interference between earlier and later learning. These constraints include the requirement to separate the physical resources of the system into subsystems that perform different types of information processes. In Chap. 8 we will show how the anatomical structures discussed in this chapter correspond with these subsystems. This correspondence provides the basis for understanding the information processes performed by each anatomical structure.

Chapter 7
Constraints on the Physical Architecture of the Brain

A number of practical requirements constrain the physical architecture of any system which must learn to perform a complex combination of behaviours with resources that are not unlimited [932]. In this chapter we will review the nature of these constraints and the way in which they arise. Analogous practical requirements constrain the architecture of a complex system in which behaviours are defined under external intellectual control (i.e. designed). These constraints will be considered briefly for comparison. Learning and design result in analogous but qualitatively different architectures. The constraints impose limits on the range of possible architectural forms, and the limits are tighter if the ratio of behaviours to resources is larger.

7.1 Conditions and Behaviours

A complex control system must act on its environment in a manner that is appropriate to achieve its objectives. Such a system receives a large number of inputs carrying information about the state of the environment in which it operates and also about its own state. These inputs are constantly changing in time. The system must detect conditions within that input information, and associate different combinations of conditions with different behaviours. A behaviour can be an action on the environment or on the system itself.

The meaning of the *condition* concept is important for the rest of the discussion, and there are different levels of precision with which *condition* can be defined. Inputs from the environment come from sensors, and a sensor is in different states depending on the current environment. A basic definition of a condition is a group of inputs plus a specified state for each input. If all of the inputs in the group are in their specified state, the condition is detected. A condition can also be defined by a group of simpler conditions, where if all of the simpler conditions in the group are present, the combination condition is detected.

L.A. Coward, *Towards a Theoretical Neuroscience: from Cell Chemistry to Cognition*, Springer Series in Cognitive and Neural Systems 8, DOI 10.1007/978-94-007-7107-9_7,

A condition could also be defined to be present if a specified proportion of its components are in their specified state. A yet more complex definition would include a weighting algorithm indicating how the detection of each component of the condition combines with the detections of other components to determine whether or not the condition itself is detected. The basic definition will be useful for higher, more approximate levels of description as discussed in Chap. 1, and the more complex definitions will be relevant for more detailed, more accurate levels.

One major difference between a designed system and a learning system is that in a designed system the conditions are generally predefined under external intellectual control. For a learning system many of the conditions will be defined heuristically from the system experience. In a designed system condition definitions are therefore relatively fixed, but in a learning system the definition of a condition may constantly change over time.

Conditions must be associated with behaviours. In a designed system this association is relatively fixed, and the behavioural meaning of a condition detection is generally an unambiguous command. For example, a common line of software is

$$\text{if } (\text{x} = \text{a}) \text{ do: } [\text{instruction}]$$

where x is some variable, a is some fixed value, and (x = a) therefore indicates a condition detection. The detection of the condition results in a command to carry out a specified instruction. In a learning system, the situation is much more complex because the associations between conditions and behaviours must be defined heuristically, and condition definitions are changing over time. A simple association between a condition and a command is therefore generally impractical, and as discussed more fully below, for learning systems condition detections can only be associated with behavioural recommendations.

7.1.1 Similar Conditions

Two conditions are similar if there is significant overlap in the two groups of components that define them. If conditions are similar in this sense, there is a higher probability that they can be associated with the same behaviours than if they are completely dissimilar. For example, a visual condition corresponding with something approximately spherical might be detected in a ball and be associated with kicking. The same condition might be detected in many different balls, but could also be present in some apples, some balloons or a globe of the earth. A different condition corresponding with something approximately cubical would be much less likely to be appropriately associated with kicking. Hence the more similar condition is more likely to be associated with the same behaviour, but this is not guaranteed.

The definition of similarity can be expanded to include conditions with less overlap in their groups of components but with a tendency to be present at similar times. Conditions with a strong tendency to occur at the same time will also have a higher probability that they can be associated with the same behaviours, but again not a guarantee.

7.1.2 *Conditions on Different Levels of Complexity*

The complexity of a condition can be defined as the total number of raw sensory inputs that contribute to its definition. Conditions on different levels of complexity will often be appropriate for association with different types of behaviour. To give a conceptual example, again consider conditions detected within visual inputs. Relatively small numbers of inputs will be needed to define a condition relevant to the presence or absence of a visual feature like a leg. Rather more inputs will be needed to define conditions relevant to the presence of an object like a dog. Even more inputs will be required to define conditions relevant to the presence of a group of objects such as a chase situation (e.g. a dog chasing a cat up a tree).

Behaviourally useful complex conditions will include many subsets corresponding with behaviourally useful simpler conditions. The more complex conditions could repeat the processing required to detect the simpler conditions, or detection of the simpler conditions could be communicated to resources that detect the more complex conditions. The latter approach will tend to require fewer processing resources.

7.2 Practical Requirements

Practical requirements place significant constraints on the physical architecture of any sufficiently complex system. One general requirement that interacts with all the others is the need to limit the total physical information handling resources needed by the system. A second requirement is the need to be able to modify the behaviours of the system in an appropriate fashion, in a way that does not introduce undesirable side effects on system behaviours that should remain unchanged. In a learning system, this means being able to learn without undesirable changes to previous learning. A third requirement is to be able to build a system from "blueprints" in such a way that the building process is efficient and not prone to errors. In the case of a brain, the "blueprints" are the DNA. A fourth requirement is to be able to recover to some degree from damage, such as damage caused by strokes. The fifth requirement is called synchronicity. At each point in time a system will be receiving inputs from its sensors. The combination of inputs from all sensors at one point in time is called an input state. Conditions must be detected in the input states at each point in time, without confusion of information across different states. This requirement for separate detection of conditions in different input states is called synchronicity.

These requirements impose a wide range of constraints on the physical architecture. In addition they often conflict with each other, and the need to find an adequate compromise places further constraints on the architecture. In the following sections each requirement will first be discussed in turn, then the wide range of architectural constraints resulting from interaction between the constraints will be considered.

7.2.1 Resource Limitations and Modular Hierarchy

Any information process requires some physical resources on which it can be performed, which could for example be transistors or neurons. A brain that required more neurons to learn the same behaviours would have a greater weight and a need for more food intake, giving it a disadvantage in natural selection terms.

One task that must be performed by the information processing resources is definition and detection of conditions. A very large number of conditions could be required to adequately guide the behaviour of a complex system. A system without resource constraints could use different physical resources for every condition. In the extreme, every condition that was used to directly influence behaviour would have separate inputs from sensors and separate implementation of algorithms to determine its presence or absence. This would be very expensive in both physical connectivity and physical processing resources.

If resource constraints exist, it will be necessary to take advantage of condition similarities. In other words, conditions with significant overlap in their definitions will be collected into groups, and connectivity and processing resources can be shared across a group. Such a group will be called a module. The resources of a module can be customized to make performance of its collection of processes as efficient as possible. As shown in Fig. 7.1, groups of modules with a some degree of similarity between conditions in different modules can be collected into higher level modules. Yet higher level modules can be defined on the basis of somewhat less similarity. Separate modules will also interact with each other. If a component condition required by one module is also required by another module, it will be more efficient for just one module to detect the component condition, and communicate it to the other module. Such communications between separate modules will be called information exchanges.

A primary result of the resource limitations that will be present in any complex control system is therefore organisation of physical resources into a hierarchy of modules. Essentially, a module is defined by a collection of similar information processes, where similarity means that they can share some of the resources across different processes. The resources of any one module are customized to make them as efficient as possible in performing their type of information process. As discussed earlier, a condition may be relevant for many different behaviours, and similar conditions do not necessarily carry the same behavioural implications. Hence modules will not correspond on any level with groups of related cognitive behaviours. Such a group of related cognitive behaviours will be called a feature.

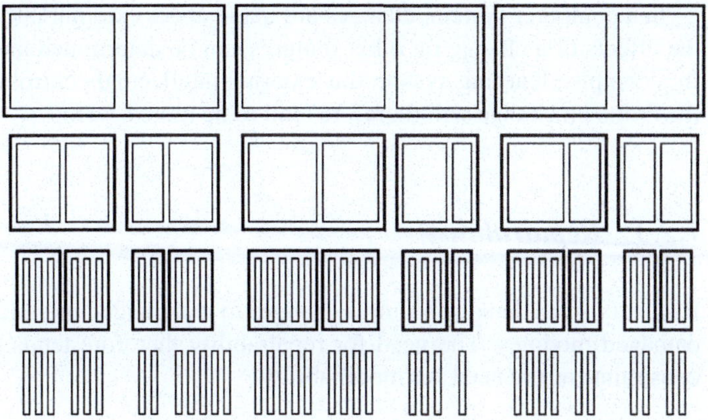

Fig. 7.1 Physical resources for condition definition and detection organized into a modular hierarchy. Resource constraints tend to result in similar conditions being collected into groups, with the conditions in a group sharing the resources needed for their definition and detection. Such a group is called a module. Groups of modules detecting conditions with a somewhat smaller degree of similarity can be collected into higher level modules sharing a smaller proportion of resources, and so on to yet higher modules as long as the degree of similarity is sufficient to provide an adequate degree of resource economy

Any type of cognitive feature will require information processes performed by many different modules, and any one module will perform processes in support of many different features.

7.2.2 *Modifiability*

If the physical information handling resources to support one type of behaviour were completely separate from the resources to support another type, then changes to implement a modification to one behaviour would have no effect on any other. However, resource limitations require the existence of resource modules shared across behaviour types. There is therefore a risk that a change to a module required to implement a change to one feature will have undesirable side effects on the other features supported by the same module. Furthermore, a change to a module could affect conditions that are communicated to other modules, potentially spreading the undesirable side effects to features supported by those other modules.

Modifiability therefore means that limits must be placed on changes to modules and on the information exchanges between modules. However, changes are required for learning, and information exchanges are important ways to reduce resource requirements. There is therefore a conflict between resource economy and modifiability. A critical problem for the architecture of any complex control system is therefore to find an adequate compromise between these two conflicting pressures.

In a computer system, changes are made under external intellectual control, and the effects of a change on other features can be determined and corrections made. In a complex learning system this external intellectual control is not available, and the limits on change will therefore tend to be more severe.

7.2.3 Repairability

Recovery from damage requires changes to undamaged modules to compensate for damaged modules. The need for repairability therefore tends to impose the same constraints as the need for modifiability.

7.2.4 Constructability

If every module on one level was identical to every other module, then the amount of information needed to guide the construction process would be minimized, and the construction errors would be relatively easily corrected. However, the resource requirement means that every module must be customized to perform a particular type of information process as efficiently as possible. Hence a compromise will be required, with each module on any level being physically similar to the other modules, but with some detailed differences.

7.2.5 Synchronicity

The problem of synchronicity arises because physical resources require a certain amount of time to carry out an information process, and this time depends on the detailed nature of the process. Consider how information derived from two successive input states can be handled by the arrangement of modules illustrated in Fig. 7.2. Any of the modules in the figure could receive inputs directly from the sensors, and because of resource constraints, some conditions are detected by one module and communicated by information exchange to other modules. Chains of such information exchanges could be present across multiple modules, two such possible chains between modules M1 and M2 are illustrated. Each module in each chain could take a different amount of time from receipt of inputs to completion of condition detections. Suppose all the modules receive some sensory inputs from one input state at the same time, and shortly afterwards receives inputs from the next input state and so on. In each chain, each module could take a different amount of time to perform its condition detections, including the waiting time required for any input condition detections from other modules. If delays in one chain are longer that the other, condition detections within the first input state via one chain could arrive after condition detections within the second input state via the other chain. The final

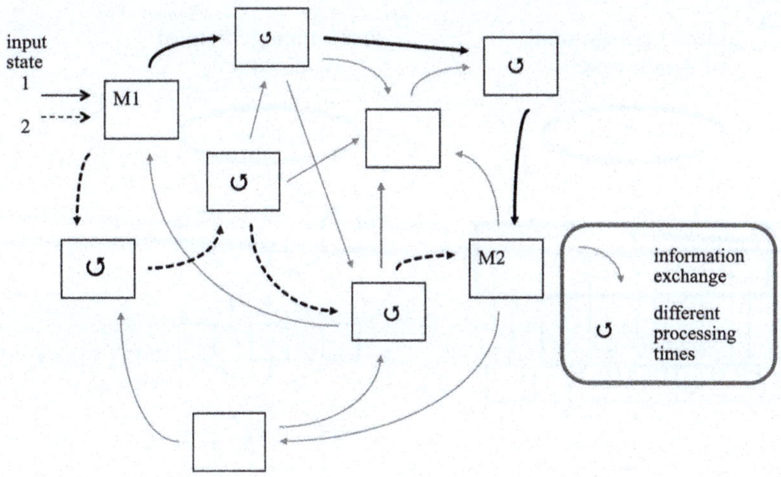

Fig. 7.2 Maintaining synchronicity in an arrangement of modules. A sequence of input states are received by module M1. Each input state is processed to detect conditions, and some condition detections are passed to other modules. These modules perform processing to detect conditions within the input state, including the detections from M1, and pass some condition detections to yet other modules and so on. Information derived from the same input state can therefore reach a module like M2 by multiple routes. Along the route, different modules take different amounts of time to perform their processing. Because of the different routes and different processing times, asynchronicities can develop in the inputs to a module like M1. Information derived from one input state by one route could even arrive after information derived from a later input state by a different route. The system must have mechanisms to maintain the integrity of the links between condition detections and corresponding input states. Maintaining such integrity is a key aspect of the synchronicity requirement

module must be able to distinguish between information derived from different input states. This is one example of the synchronicity problem in an arrangement of modules that share condition detection resources to some degree.

In an electronic system, the global solution to the synchronicity problem involves use of an external memory as illustrated in Fig. 7.3a. Inputs are recorded in a reference memory along with an indication of the input state from which they derive. A module gets inputs from a consistent input state by accessing the memory, and records any condition detections back in the memory, along with an indication of their corresponding input state. Each input state and associated condition detections is therefore separately recorded in the memory. This global solution to synchronicity carries with it the resource and processing costs of the reference memory. An alternative is the local solution illustrated in Fig. 7.3b. In the local solution, modules are arranged in layers. Raw sensory inputs only go to the first layer, and condition detection is completed in that layer before detections are released to the next layer and so on. This local solution is not completely accurate, because for example there may be differences in the amount of time required by modules in the same layer to complete their condition detections. If some degree of synchronisation errors can be tolerated, this solution avoids the cost of the reference memory.

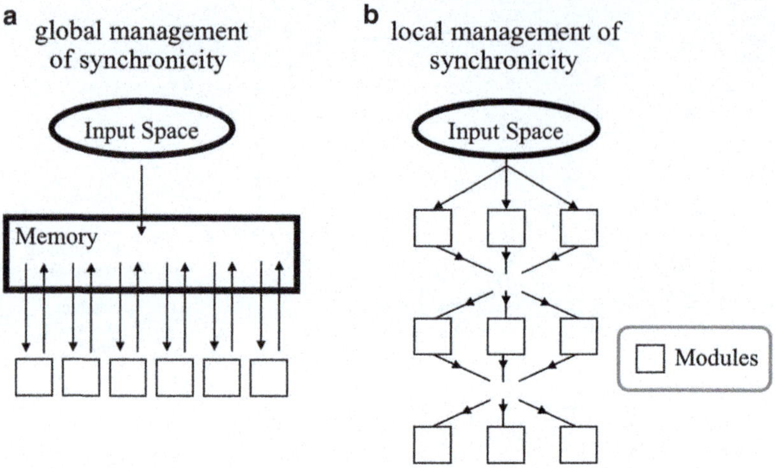

Fig. 7.3 Two alternative approaches to achieving synchronicity, global and local. (**a**) In the global approach, all the information from the input state is recorded in a reference memory. Modules can go to the memory and obtain the information relevant to their conditions, and record in the memory whether or not their conditions are present in the input state. If one module incorporates conditions from another module in its own conditions, the other module must determine the presence or absence of its conditions first i.e. condition detection must be performed in a specific sequence. The separate memory and sequential processing is the von Neumann architectural form ubiquitous in electronic systems. (**b**) In the local approach, modules are arranged in layers, with condition detection information proceeding only from one layer to the next. Conditions at a given level of complexity are all detected at the same time in one layer. However, this approach assumes that every module in a layer takes exactly the same amount of time to detect its conditions, and to communicate detections to the next layer. Such precision is unlikely in practice. If all information exchanges are behaviourally unambiguous, errors introduced by the layering approach will be unacceptable. However, if behavioural meanings are partially ambiguous, some level of errors can be tolerated and the resource cost of a reference memory avoided. Unambiguous information exchanges therefore require the von Neumann memory/processing form, but partially ambiguous information exchanges can tolerate the errors of the layering approach and avoid the resource costs of the global approach

7.3 Behavioural Meanings of Condition Detections and Exchanges

Condition detections must be associated with behaviours. For condition to be useful, its detection must indicate that the range of all possible system behaviours can be divided into two groups. One group contains behaviours that could be appropriate given the condition detection. The second group contains behaviours that are not appropriate given the condition detection. Detection of enough conditions will narrow the appropriate group to a single behaviour. The detection of a very complex condition defined by some combination of conditions could therefore indicate that one and only one behaviour is appropriate. The detection of a simpler condition will only limit the range of possibly appropriate behaviours to a particular subset.

There are two qualitatively different types of association between condition detections and behaviours, which will be labelled unambiguous and partially ambiguous. In the unambiguous case, the detection of a condition indicates that the currently appropriate behaviour is limited to a specific subset of system behaviours with 100 % confidence. In this case, a condition detection can be interpreted as a command. In the partially ambiguous case, a detection indicates that the currently appropriate behaviour is probably in a specific subset of system behaviours. In this case, a condition detection can only be interpreted as a behavioural recommendation.

If condition detections are only associated with recommendations, many more condition detections will be required to achieve a high integrity behavioural determination. Hence the physical information processing resources required will be greater if partially ambiguous meanings are utilized. As indicated earlier, electronic systems maintain an association between condition detections and commands, but there is a problem with this approach in complex learning systems.

7.3.1 Learning Systems and Partially Ambiguous Information

The requirement to modify features without undesirable side effects on other features has implications for the type of association between conditions and behaviours in a complex learning system. To understand the implications, consider the information exchanges between modules illustrated in Fig. 7.4. Ten modules are illustrated, and the information exchange required to reduce the use of resources has been conceptually limited in that one module receives information from just two other modules, and passes information to just two modules.

To modify an existing behaviour or add a new behaviour, often some changes to conditions will be required. Suppose that the changes needed to implement some behaviour modification apply just to module M in Fig. 7.4. Resource constraints mean that any condition supports many different behaviours. So the first problem is that changes to module M could affect not just the behaviour to be changed but many other behaviours also supported by M. The second problem is that module M passes information to a couple of other modules (M4 and M5 in Fig. 7.4), and changes to module M could therefore affect conditions detected by these other modules. Furthermore, these other modules pass information to yet other modules, and so on. Hence the conditions detected by these other modules could be affected, and the effect of a change could even come back to introduce condition changes in module M itself. If the behavioural meaning of every condition detection is a command, the information exchanges between modules mean that a change to one module to implement a behaviour modification could lead to an exponentially increasing wave of undesirable side effects on other behaviours.

In the case of an electronic system, a designer can follow through all the effects on other modules, testing for and correcting all the side effects. The use of unambiguous behavioural meanings for condition detections is therefore feasible. For a learning

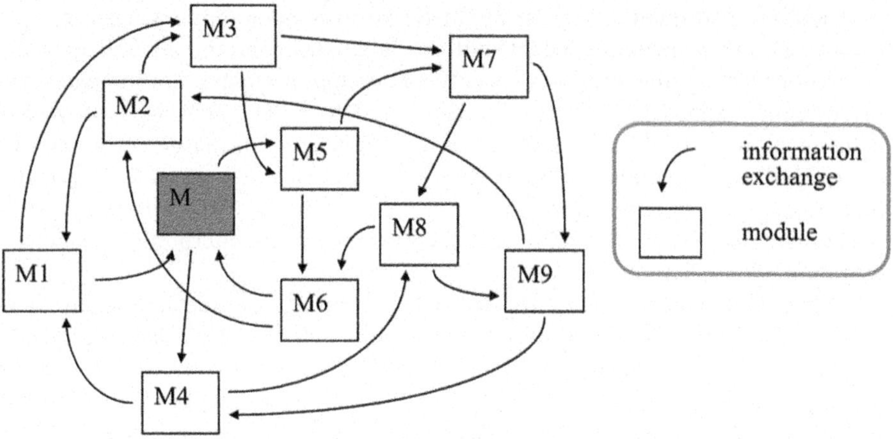

Fig. 7.4 The information integrity problem in a learning system. A set of modules with limited information exchange is illustrated. If learning is required, the system must make experimental changes to at least one module such as M, and perform the changed behaviour. The only immediate feedback is the consequences of the one changed behaviour. However, the change could affect other behaviours also influenced by the module M. In addition, module M passes information to other modules, and the change could affect that information. Those other modules pass potentially changed information to yet other modules. Hence a wide range of behaviours influenced by all those downstream modules could also be affected. If every such information transfer carries an unambiguous behavioural meaning (i.e. behavioural commands), in later circumstances appropriate to many of these behaviours the behaviour would no longer be selected. Consequence feedback is not specific enough to correct this situation and system behaviour would become less and less appropriate. However, if behavioural meanings are partially ambiguous (i.e. behavioural recommendations), in later circumstances there could still be a predominant recommendation strength in favour of the appropriate behaviour. Consequence feedback following that behaviour could then adjust to changes since the previous occasion when the behaviour was appropriate

system, there is no external intellectual guidance for the selection of condition changes, and such changes cannot be determined in advance to be appropriate. Furthermore, consequence feedback (i.e. positive or negative rewards) is the only information available to correct the associations between conditions and behaviours. Suppose now that some experimental change to a condition definition is made in module M, and also suppose that the resultant behavioural modification is successful and followed by positive consequence feedback. If another, previously learned behaviour depends upon some of the conditions before they were changed and every condition detection is a command, the behaviour may well not occur next time it is appropriate.

However, if the behavioural meaning of a condition detection is only a recommendation, any learned behaviour will be carried out only in response to a large number of condition detections, each recommending that behaviour. If a behaviour has been learned, a large number of conditions will have acquired recommendation strengths in favour of that behaviour. After past occasions when the behaviour was performed these recommendation strengths will have been increased by positive consequence feedback. If changes to some of the relevant condition definitions are made to implement learning of other behaviours, there could still be enough recommendation strength the next time the behaviour is appropriate to result in that behaviour. Consequence feedback could then adjust recommendation strengths to compensate for the changes in condition definitions.

Thus the use of unambiguous behavioural meanings for condition detections in a complex learning system is not practical, because it will lead to major interference between new learning and prior learning. Use of partially ambiguous behavioural meanings requires more information handling resources, but makes it feasible to limit interference between new and prior learning.

Evolution under natural selection pressures provides another reason for complex learning systems to utilise partially ambiguous information exchanges. The evolution process involves random mutations which will tend to persist in descendants if they are functionally valuable. If information processes are recommendations, there is a lower probability that a mutation will have fatal consequences, because any behaviour must be supported by recommendations from multiple sources.

7.4 Pressures on the Definition of Modules

Because of resource limitations, the physical resources of a complex control system will be organized into a hierarchy of modules. Each module will be made up of information handling resources customized to perform a particular group of similar information processes. Higher level modules will be made up of more detailed modules. Because modules are defined on the basis of similar information processes, they will not correspond with features as perceived by a system user (see Box 7.1).

Resource limitation implies that there will be information exchanges between modules, but such information exchanges make feature modifications more likely to introduce undesirable side effects. Hence the degree of information exchange will be a compromise between the conflicting demands of resource conservation and modifiability. In general there will be much more information exchange within a module on any level (i.e. between its submodules) than between peer modules. Constructability is better if all modules on one level are identical, but resource limitation means that each module must be customized to perform a different type of information process very efficiently. Hence modules on one level will tend to be generally similar but with detailed differences.

In a complex learning system, the behavioural meaning of a condition detection by a module will be a recommendation in favour of a range of different behaviours. Because this recommendation meaning is more error tolerant than the instruction meaning, synchronisation can be achieved by module layering and the resource cost of an external reference memory can be avoided. Conditions on different levels of complexity are appropriate for recommending different types of behaviour. The layering of modules means that more complex conditions will be detected in deeper layers. The modular hierarchy will therefore be imposed on a layered structure as illustrated in Fig. 7.5.

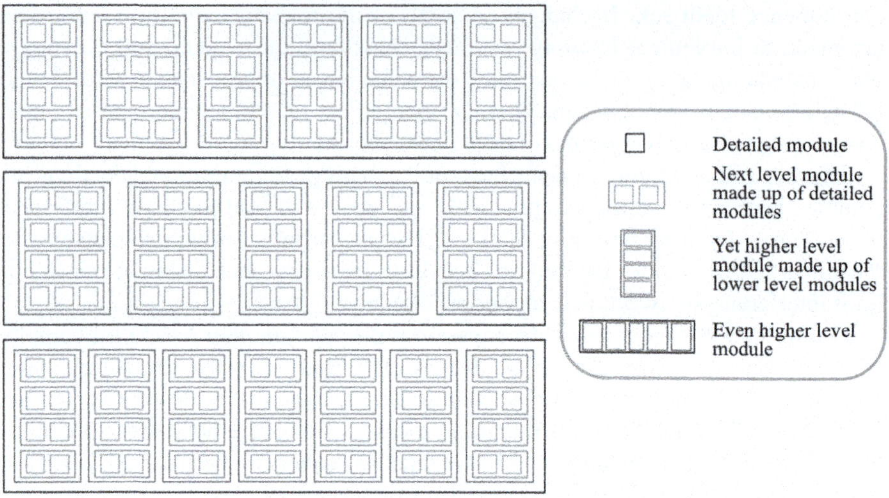

Fig. 7.5 Layering to support synchronicity superimposed on modular hierarchy. The local solution to synchronicity requires that conditions at one level of complexity are detected within all modules at that level, and then passed to modules detecting conditions at the next higher level of complexity. Hence the modular hierarchy must be consistent with a layered arrangement of resources

Box 7.1: Terminology: Sensory Inputs, Conditions, Condition Complexity, Modules, Higher Level Modules and Receptive Fields

The terms "sensory input", "condition", "condition complexity", "module" and "receptive field" have been defined in many different ways. In this book they have the specific meanings defined in this box.

A complex system can acquire information about its environment, including about its own internal state. Such acquisition uses a large number of different sensors, which provide information about the state of an aspect of the environment at frequent intervals in time. A *sensory input* is the information provided by one sensor at one point in time. At each point in time this information could have different magnitudes depending on the state of the environment.

A *condition* is defined by a set of sensors, an integrating algorithm for combining their information, and a threshold. The condition is detected if the information from the set of sensors at one point in time, when combined using the algorithm, exceeds the threshold. A condition can also be defined via a set of simpler conditions, using the simpler condition detections as inputs, an integrating algorithm, and a threshold. The rough concept of *condition complexity* can be defined as the total number of sensory inputs that contribute to the condition definition, directly or via intermediate conditions, and including a separate count of all duplications.

(continued)

Box 7.1 (continued)

A *module* is defined as some information processing resources, optimized to perform the detection of a group of similar conditions very efficiently. Different modules detect different groups of conditions. A module has an integrating algorithm defining the module outputs generated in response to different combinations of condition detections.

The *receptive field* of a module is the set of circumstances in which the module produces outputs, as defined by its group of conditions and its integrating algorithm.

A group of modules detecting somewhat similar receptive fields can share resources in proportion to the degree of similarity. Such a group defines a *higher level module*. A receptive field for the higher level module can be defined in the same way as for a simpler module.

Note that by these definitions, receptive fields are equivalent to very complex conditions detected by modules. Receptive field complexity can be defined in the same way as condition complexity.

In cognitive science, there is a long history of controversy over the existence of modules in the mind and brain, where these "cognitive modules" correspond to varying degrees with cognitive features, and/or different types of restrictions on access to information [933]. In this book, the primary role of a module is to achieve resource economies. Such modules will not in general correspond with cognitive features of any kind to any useful degree. However, as discussed in this chapter, the need to modify features without excessive side effects on other features does result in a pressure for modules to share condition detections as little as possible, in conflict with the need to achieve resource economies. This minimisation of information exchange is analogous with the concept of restrictions on access to information postulated for cognitive modules.

7.5 Hierarchy of Modules and Hierarchy of Descriptions

The requirement for a hierarchy of modules thus emerged from the practical needs of a complex control system. However, this hierarchy of modules has properties that mean that it can support the hierarchy of descriptions required for understanding of a complex system as discussed in Chap. 1.

A process in a system can be described approximately by referring only to the operations of the highest level modules and their interactions, and neglecting their internal operations. Because of the requirement that information exchange between modules be minimized as far as possible, these interactions are relatively limited and therefore comprehensible. A more detailed and more accurate description will be in terms of the submodules of the highest level modules, a yet more accurate description will be in terms of the submodules of those submodules and so on. Only a small part of a comprehensible description at high level will be comprehensible at one time at a more detailed level, but there are clear ways to shift between levels as required.

For example, as described in Chap. 1, in a complex electronic system the physical resources are organized in a modular hierarchy. The use of the instruction and data read/write information models makes it possible to describe processes on different levels of detail, with different degrees of approximation. In an electronic system this hierarchy has been imposed by design. However, the need for a modular hierarchy is created by various practical requirements, and these also exist for a system that has evolved under natural selection pressures. It is therefore possible that such a modular hierarchy, and therefore a hierarchy of descriptions, also exists for brains. In the case of brains, because of the need for learning of complex combinations of behaviours, we can expect the condition define/detect and recommendation information models to be utilized in the hierarchies.

7.6 Sharing of Resources Across the Processing of Different Input States

As discussed earlier, synchronicity includes the requirement that sensory inputs derived from input states at different times are not confused. However, the need to detect very complex conditions composed of conditions detected within multiple input states combines with the need to conserve resources to add another architectural constraint.

Consider the conceptual example of the need to develop an appropriate behavioural response to a group of objects. Such an appropriate response requires detection of appropriate conditions. For example, seeing a dog chase a cat up a tree may require a behavioural response. This response must be associated with detection of complex visual conditions containing sensory information derived from all three objects. However, conditions must first be detected within the three objects separately, then these conditions combined into more complex conditions correlating with the presence of a particular type of group. In other words, mixing information derived from the body of the cat and the leaves of the tree would result in detection of behaviourally misleading conditions, but once conditions combining just information derived from the cat alone have been detected, it will be necessary to combine these conditions with the conditions detected within just the tree and just the dog. If there is no reference memory where condition detections can be stored, then in order to combine simpler conditions into more complex conditions all the detections of simpler conditions must be present simultaneously.

Resource limitations mean that in general there will be overlaps in the groups of modules detecting conditions in different objects. If a module is simultaneously detecting simpler conditions in more than one object there is a high risk that

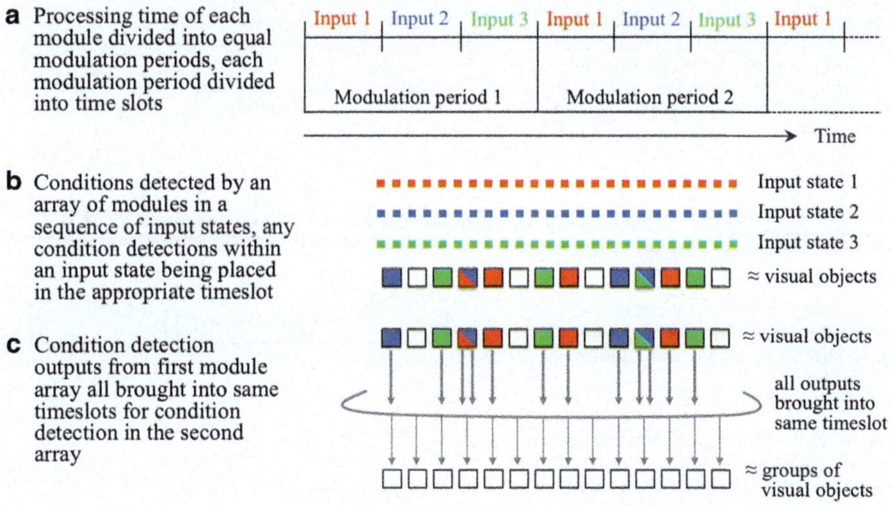

Fig. 7.6 Time division multiplexing of the processing resources of a module across multiple input states. (**a**) The processing capabilities of a module are applied to the first input state for a short period of time, then to the second input state for the next such period, then to the third input state for the third such period. Processing resources in the next time period are applied to the first state again, and so on. (**b**) With this modulation plan, conditions are detected in the same modules in response to three different input states, conditions corresponding with different input states detected at different times. For example, conditions detected within three different visual objects could be maintained active simultaneously. (**c**) If higher complexity conditions need to be detected within the three visual objects, the lower complexity condition detections could be brought into the same timeslot at the input level to the higher complexity modules. Thus, for example, conditions at the higher level corresponding with a dog chasing a cat up a tree could be detected, while avoiding the problem of invalid conditions corresponding with, say, a cat with leaves being detected at the lower complexity level

information derived from multiple objects will be inappropriately mixed. There is therefore a conflict between resource conservation and synchronicity.

One way of handling separate detection of object conditions followed by detection of group of object conditions would be to maintain separate modules to handle input states at different times. This spatial separation of information would be very resource intensive. A less resource intensive alternative is temporal separation of information as conceptually illustrated in Fig. 7.6. In this approach, the information handling capability of a module is divided up in time into a constant series of equal modulation periods. Each modulation period is divided up into a number of equal timeslots. Suppose, as in Fig. 7.6a, there are three such timeslots in each modulation period. If conditions must be detected simultaneously in three different input states in order to be combined at a higher condition complexity, then conditions within the first input state are detected in the first timeslot and the detection activity prolonged just in that timeslot. Conditions within the second input state are detected in the second timeslot and so on (Fig. 7.6b). Once condition detections

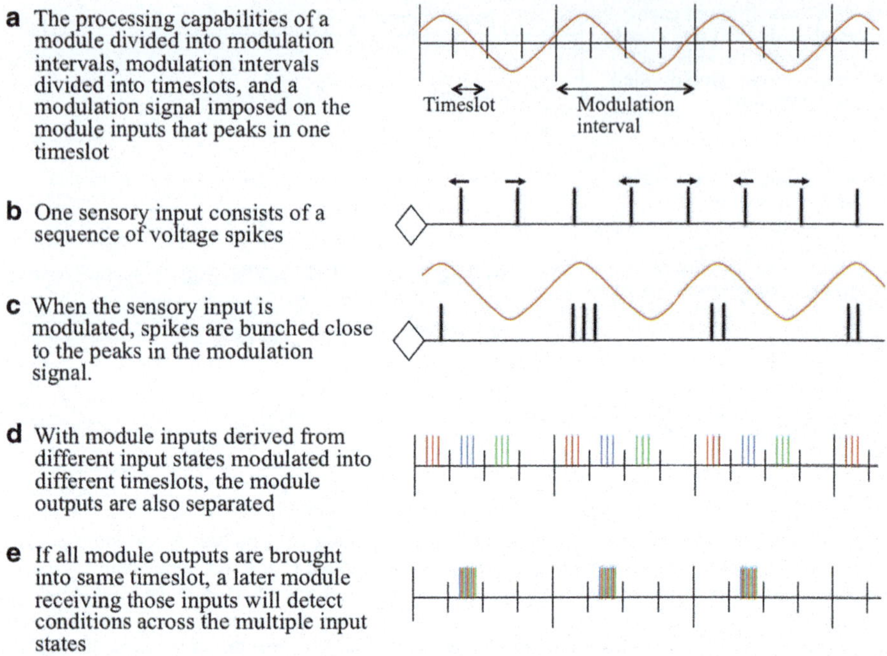

Fig. 7.7 Possible implementation of the separation of input states by modulation, using brief voltage spikes to indicate condition detections. (**a**) As in Fig. 7.6, module processing resources are divided up into modulation intervals, each interval divided up into timeslots. (**b**) The raw inputs from a sensory state are voltage spikes distributed randomly in time. (**c**) The modulation signal shifts each spike towards the nearest peak in the signal. (**d**) If sensory inputs derived from different input states are modulated by the same frequency at different phases. The voltage spikes from the different states are located in different timeslots. (**e**) If output spikes for the three states are brought into the same timeslot, higher complexity conditions can be detected within the combination of states

are present in all three timeslots, the module outputs can be brought into the same timeslot, i.e. synchronized, and passed to the next modules for detection of more complex conditions (Fig. 7.6c).

One way in which such an approach could be implemented is shown in Fig. 7.7. In this example, there are four timeslots per modulation interval, and a modulation signal can be applied with a frequency that is the reciprocal of the modulation interval. The peak of the modulation signal will therefore occur at the same point in the modulation interval in each cycle, and the phase of the modulation signal can be varied so that the peak occurs in the middle of any one timeslot. Sensory inputs and module outputs are sequences of voltage spikes. If a modulation signal is applied to an input state, the effect will be to shift the spikes so that they tend to occupy the timeslot within which the modulation signal peaks. If all the sensory inputs for one input state are modulated in this fashion, they will tend to arrive bunched in that timeslot (Fig. 7.7c). If the activity generated by an input spike lasts no longer that about the width of a timeslot, spikes will only interact with other spikes in the same timeslot. If sensory inputs from a later input state are modulated with a different phase, they will tend to arrive in a different timeslot. Hence if the condition detection activity resulting from the first input state is prolonged in one timeslot, conditions can be independently detected in the same module within a second input state by modulating the inputs at a different phase. Once condition detection has taken place within all three input states, the output detection spikes can be brought into the same timeslot and communicated to a later module. In the later module, conditions that are combinations across the three input states will be detected.

7.7 Integration of Module Inputs on Different Levels of Detail

A module receives a set of inputs indicating the presence or absence of a set of relatively simple conditions. The module integrates these inputs to determine the presence or absence of more complex conditions, and indicates detections by generating outputs. The set of circumstances in which a module produces outputs will be called the receptive field of the module. These outputs may be communicated to other modules as inputs and/or interpreted as behavioural recommendations.

If two modules detected exactly the same receptive fields, one could be eliminated to save the resources with no loss of behavioural effectiveness. If there was a large overlap in the receptive fields detected by two modules, they would tend to be active in similar circumstances and their value for discriminating between circumstances in which different behaviours were appropriate would be limited. Hence there will be a tendency for the receptive fields of modules to be relatively orthogonal.

The algorithm used to detect the module receptive field from inputs could vary considerably in complexity. At its simplest, the module receptive field is detected only if all its input conditions are detected. Somewhat more complex would be if the receptive field is detected when more than some defined proportion of its input conditions are detected. In this case there are a number of different circumstances in which the receptive field is detected. A yet more complex algorithm would be if each input condition has an individual weight, and the module receptive field is detected if the total weight of all the currently detected input conditions exceeds a threshold.

All three of these types of algorithm are relatively linear. An even more complex type of algorithm is illustrated in Fig. 7.8. In this type of algorithm, intermediate conditions are detected. An intermediate condition is defined by a different subset of input conditions, each input with a different weight. An intermediate condition is detected if the total weight of its currently detected inputs exceeds a threshold. There are then more complex conditions defined by different subsets of intermediate conditions, again with different weights. Any one more complex condition detection is an output from the module indicating detection of its receptive field. The multiple module outputs communicate information about where the current input state is located within the

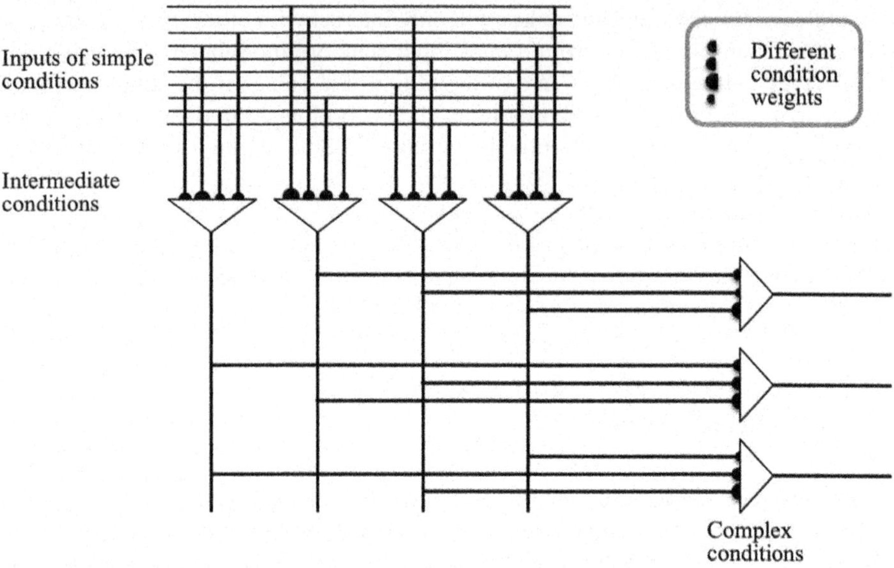

Fig. 7.8 Nonlinear algorithm for constructing more complex conditions. Intermediate conditions can be defined by combinations of simple conditions in which each separate simpler condition has a different weight. Complex conditions can similarly be defined by weighted combinations of intermediate conditions

circumstances that define the module receptive field. Even more complex algorithms could have multiple levels of intermediate conditions.

Modules in real systems will generally utilize more complex algorithms. However, often a simpler algorithm can be used as an approximation for some levels of description in order to keep the level of detail at a comprehensible level. At the highest level of description, a module detects complex conditions within inputs detecting simpler conditions. At a more detailed level, one or more intermediate levels of condition detection can be considered. At another detailed level, the relative weights of different inputs can be considered, and so on.

7.8 Limitations to Receptive Field Changes

In a complex learning system, many of the receptive fields needed to guide behaviour must be defined heuristically. There are very large numbers of possible receptive fields, and few criteria by which appropriate fields can be specified in advance. However, once a receptive field has been initially defined, it may acquire multiple different behavioural meanings. Any future change to the field will therefore jeopardize those meanings.

If a receptive field enlarges the range of circumstances in which it is detected, it will still always be detected if a circumstance identical to one in which it was detected in the past reoccurs. The behavioural meanings acquired during past receptive field detections are therefore preserved. Reductions or qualitative changes to a receptive field carry more behavioural risk. However, there are a couple of situations in which such receptive field changes are reasonable. Firstly, if a receptive field is detected in every or many input states, it has little value for discriminating between circumstances in which different behaviours are appropriate. Frequent detection therefore justifies some tightening. For example, in an integration algorithm in which the receptive field is detected if more than a given proportion of inputs are active, an increase in the proportion would be appropriate if the field was detected too frequently. Secondly, if an input is rarely present when the receptive field is detected, deletion of that input will have little impact on the integrity of the behavioural meanings associated with the receptive field. Thirdly, if one of the inputs to a receptive field is rarely detected at all, deletion of the input will have little impact.

The concept that a receptive field can expand by enlarging the range of circumstances in which it is detected, but cannot otherwise change those circumstances, is a good first approximation and can be used for higher level descriptions. Occasionally it will be necessary to shift to a more detailed level to understand some phenomena.

7.8.1 Management of Receptive Field Changes

When should receptive fields be changed, and which fields specifically should change? Given the dependence of behavioural meanings on receptive field definitions, the general principles must be only to change receptive fields if necessary, and change receptive fields by as small amount as possible.

Receptive field detections are the sources for behavioural recommendations. In order to achieve a high integrity behaviour, a reasonable range of recommendations is needed. Hence the primary indication that receptive field expansions are required is that few or no receptive fields are being detected in response to some input state. If less than some minimum are being detected, receptive field expansions must be triggered until the minimum is reached.

When some receptive field expansions are required, in order to minimize the behavioural impact, those receptive fields which require very little change to be detected in the current circumstances must be identified. The way such identification could occur can be understood by considering the module integration algorithm illustrated in Fig. 7.8. In the illustrated module, an output from one or more of the complex conditions indicates a receptive field detection. Suppose that some receptive field expansions are required, and the illustrated module has no complex condition detections, but some detections of intermediate conditions. The implication is that much less receptive field expansion would be needed for this module than for a module in which no intermediate conditions were being detected. The requirement is therefore to identify the module(s) with the most intermediate condition detection as illustrated in Fig. 7.9a, and drive receptive field expansion in those modules.

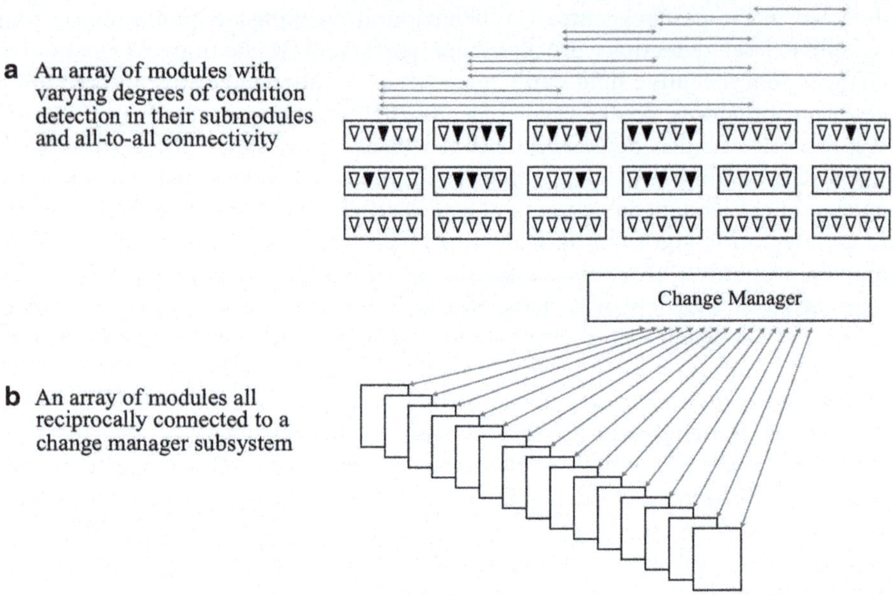

Fig. 7.9 Alternative approaches to managing the selection of appropriate modules for receptive field expansion. (**a**) All-to-all connectivity of modules, with internal activity of each module inhibiting receptive field expansions in all other modules. (**b**) Internal activity of all modules reciprocally connected to central change manager which determines the most appropriate columns and drives expansions of their receptive fields. The change manager approach requires fewer connectivity resources and allows creation of information on past temporally correlated expansions to improve selection of columns for new expansions

The requirement to identify appropriate modules for receptive field expansion therefore implies that the primary module level at which such expansions are managed must be organized as a sequence of submodules. Each submodule in the sequence detects conditions within the same overall input space at a gradually increasing level of complexity. Higher levels of complexity are the sources of outputs with general behavioural recommendation strengths, intermediate levels of complexity are the source of information that guides receptive field expansions. Hence there will be a tendency for submodules to be arranged in a column, with a predominance of connectivity between column layers.

7.8.2 Implementation of Expansions

In the module illustrated in Fig. 7.8, receptive field expansion could be implemented by adding inputs to intermediate conditions and/or changing input weights. Expansion could also be implemented by adding inputs to complex conditions and/or changing input weights. To be relevant, the changed inputs must actually be active in response to the input state that is driving the need for expansions. To minimize the risk of undesirable side effects, if inputs are being added they must be similar to other inputs at the same level. As discussed earlier, "similar" means that there must be overlap in the information that defines the conditions corresponding with the new and existing inputs and/or the new input must often be active at the same time as the existing inputs.

If the condition detections are implemented by physical connectivity, there could be difficulty in creating appropriate connectivity at the moment the expansion is required. Firstly there is the problem of making new physical connections in real time. Secondly, there is the problem of ensuring that new connections are from inputs similar to existing inputs. In general it will therefore be necessary to create provisional new inputs in advance of experience. These provisional inputs would be selected to be similar to existing inputs, but would have zero weight in defining conditions. They would only acquire weight if they were active at a point in time when receptive field expansion was needed in the module in which they were located.

7.9 Entanglement of Condition Definition and Detection

Condition definition is triggered by inadequate condition detection. Hence condition definition only occurs in the course of condition detection, and condition detection only occurs in a context where the requirement for condition detection is being determined. Hence it is misleading to regard definition and detection as separate processes, in a complex learning system there is one condition definition/detection process.

7.10 Corollaries of Receptive Field Consistency

Because receptive field definitions change relatively slightly over time, and their associated recommendation strengths are also fairly stable, there is an additional source of information which can be used to guide behaviour, in addition to current sensory inputs and resultant receptive field detections.

If a receptive field is not being detected in current sensory inputs, but has often been detected at similar times in the past as many of the receptive fields that are being currently detected, there is a reasonable probability that the undetected receptive field has associated recommendation strengths that are relevant to the current situation. There could therefore be behavioural value in indirectly activating such receptive fields, adding their recommendation strengths even though they are not being detected directly. Such indirect activations must of course be carefully managed to avoid confusion.

7.11 Modulation of Total Receptive Field Detection

If a very large number of receptive fields are detected, the range of available recommendations would also become excessive and selection of the most appropriate behaviour made more difficult. Hence there will be value in limiting total receptive field detection, for example by cross inhibition between modules, or between peer submodules of a group of modules etc. Excessive activity in one submodule layer could also be addressed by inhibition of a submodule that is the main source of its inputs.

7.12 Development of a Modular Hierarchy

To a considerable degree the modular hierarchy of a complex learning system will need to be defined from experience. The overall structure including numbers of modules, hierarchical organisation and initial biases on connectivity could be defined a priori (for example under genetic control) but the detailed module definitions must be learned. To illustrate the process, consider the conceptual array of modules illustrated in Fig. 7.10. In Fig. 7.10a there are seven modules, but these modules detect no conditions. The requirement placed on the array is that at least two modules must detect their receptive fields in any input state. When this array is exposed to a visual object for the first time (Fig. 7.10b), enough conditions present in the inputs are added to two modules to create a receptive field in those modules. These conditions define the initial receptive fields. Within a somewhat similar object seen soon afterwards (Fig. 7.10c), neither receptive field is initially detected. However, enough existing conditions are detected in one of the modules that the criterion for receptive field expansion is met, and conditions are added until detection occurs. To meet the requirement for at least two module receptive field detections, another "empty" module is initiated. After much more visual experience, all the modules have receptive fields, and a new visual object may result in simple detection of one receptive field and some condition detection short of receptive field detection in several other modules. To meet the requirement for at least two receptive field detections, the module with the most such condition detections adds conditions present in the input state to its portfolio until receptive field detection occurs. Note that at a greater level of detail, each of the modules in Fig. 7.10 could be of the type illustrated in Fig. 7.8, with intermediate conditions corresponding with the individual conditions illustrated in Fig. 7.10.

As discussed earlier, this algorithm could be modified to allow removal of a condition from a module if the condition was rarely detected at a time that contributed to receptive field detection, or if it was very rarely detected at all. This modifications are still based only on temporal correlation and have no direct dependence on consequence feedback. Although direct dependence on feedback could lead to excessive undesirable side effects, an indirect use of such feedback

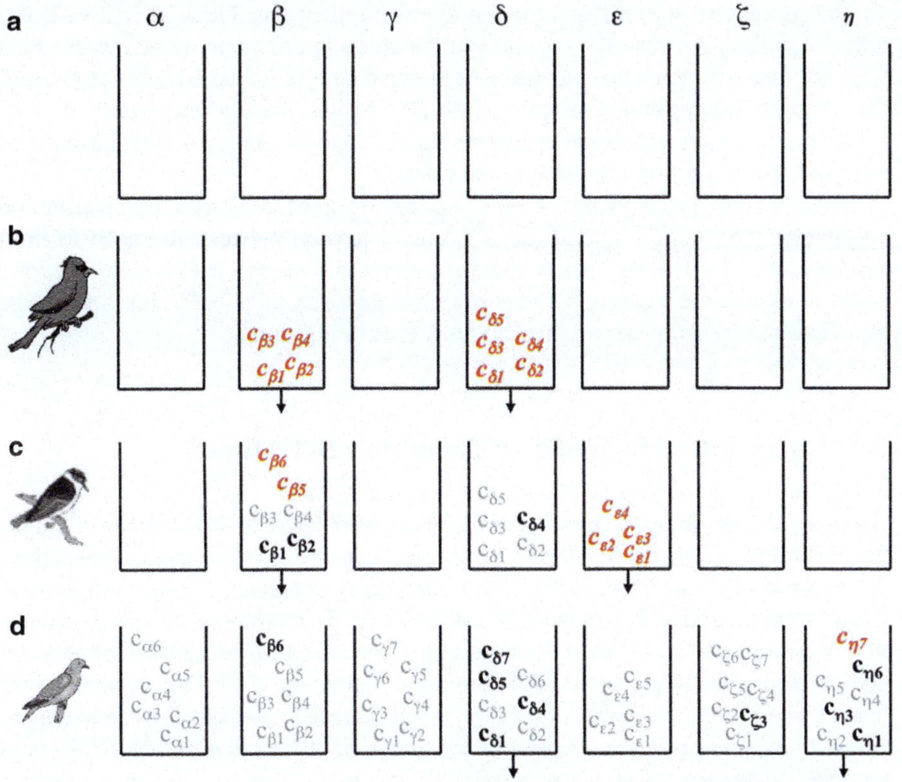

Fig. 7.10 Simple conceptual mechanism for initiation and evolution of receptive fields. The same array of seven modules $\alpha - \eta$, all at the same level of complexity, is shown at different times in **a** through **d**, along with the sensory experiences at those times. Modules can become programmed with conditions. The learning criteria are: (i). At least two modules must detect their receptive fields in response to every sensory experience; (ii). A module detects its receptive field if four or more of its programmed conditions are detected; (iii). When not detecting its receptive field, a module can add conditions currently present in the sensory input provided that two or more of its programmed conditions are being detected; (iv). If fewer than two receptive fields are being detected and the minimum cannot be reached by expansions, "empty" columns can be forced to record conditions to reach the minimum. Such "empty" columns will generally only be relevant in early learning. (**a**) Initially the modules are not programmed with any conditions. (**b**) In the first experience, condition recording is forced in two modules. (**c**) In early experiences, generally modules do not detect their receptive fields, but expansions and empty column initiations achieve the minimum detections. (**d**) In later experiences, detections and occasional expansions are adequate to reach the minimum number of module receptive field detections

could be possible. Suppose that the same group of receptive fields were detected a number of times, and resulted in predominant recommendation and acceptance of the same behaviour. However, also suppose that this behaviour was sometimes followed by positive consequences, sometimes negative. The implication is that the current set of receptive fields do not adequately discriminate between two similar situations with different behavioural implications. Other receptive fields detecting this contradictory consequence feedback could recommend that the next time the group is detected, receptive field expansion is forced in additional modules even if the minimum number of detections has already been reached. This forcing means that additional receptive fields could be detected which would be detected in one situation and not in the other.

Note that this type of receptive field expansion algorithm based only on temporal correlation information emerges as a result of tension between the need to limit resources and the need to learn without interference with past learning. As the ratio of behaviours to be learned to resources increases, receptive field changes will be more and more tightly constrained into this type of algorithm.

7.13 Cascades of Condition Definition/Detection

One way of visualizing the process of receptive field detection is a cascade of condition detection as illustrated in Fig. 7.11. Information derived from input states arrives at the top level array of modules and in that array generates condition detections within some modules and receptive field detections in modules with enough condition detections. Receptive field detections go to the next array and generate condition detections and some receptive field detections and so on. If initially in some array fewer than the required number of receptive fields are detected, receptive field expansions will occur to reach the minimum. Receptive field detections in any array could recommend behaviours of different types.

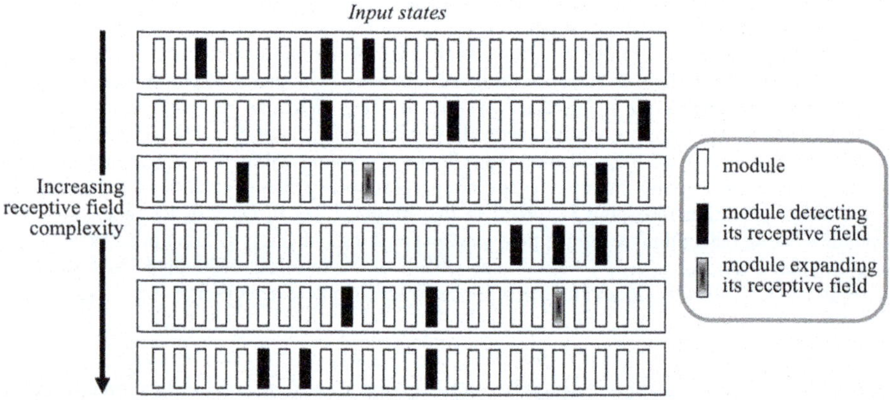

Fig. 7.11 A cascade of condition detection. Conditions are detected at increasing levels of complexity through a sequence of arrays of high level modules, each made up of an array of more detailed modules. In this example, at least three detailed modules must detect their receptive fields in each high level module, with receptive field expansions as required to reach the minimum

7.14 Associations Between Receptive Fields and Behaviours

As discussed earlier, receptive fields must be associated with behavioural recommendations, with each receptive field detection corresponding with a range of different recommendations. In a complex learning system, the primary information available to create associations between receptive field detections and behavioural recommendations is consequence feedback following a behaviour. Such consequence feedback cannot be used directly to modify receptive fields, because such modification would introduce undesirable side effects on all the other behaviours associated with the receptive field. It is therefore necessary for a complex learning system to have two separate subsystems as illustrated in Fig. 7.12.

One system is the modular hierarchy, which we will call clustering. This modular hierarchy defines and detects conditions within the information available to the system, and groups conditions into receptive fields. Information exchanges within clustering have very complex behavioural meanings, and changes can only use information about condition similarity and temporal correlations as described earlier. Clustering defines receptive fields, and communicates any detections to the second subsystem, called competition. Each receptive field detection is interpreted in competition as a set of recommendations in favour of a range of different behaviours, each recommendation having an individual weight. Competition determines

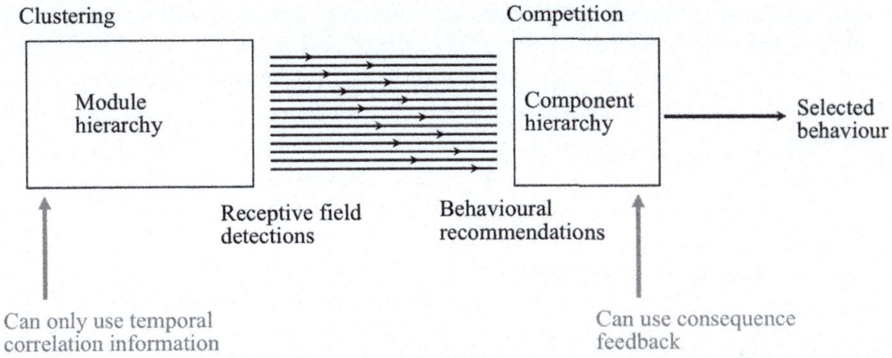

Fig. 7.12 The primary recommendation architecture separation in the physical resource architecture of a complex learning system. This separation is between clustering and competition, and is analogous with but qualitatively different from the separation between memory and processing in electronic systems. Clustering is a modular hierarchy that defines and detects conditions in the information available to the system. Detections of groups of conditions, called receptive fields, are communicated to clustering. In clustering, each detection is interpreted as a range of recommendations in favour of different behaviours, each recommendation having a different weight. Competition is a component hierarchy that determines and implements the behaviour most strongly recommended across all currently detected receptive fields. Consequence feedback is received following each behaviour and can be used to change the recommendation weights in favour of that behaviour, but cannot be used to directly change receptive field definitions. Such receptive field changes can only be on the basis of temporal correlation information

the most strongly recommended behaviour across all currently detected receptive fields, and implements it. Consequence feedback following a behaviour adjusts all the recommendation weights in favour of the behaviour that were employed in generating it, but does not change the associated receptive field definitions or the recommendation weights in favour of other behaviours associated with the same receptive fields.

Practical considerations will also tend to constrain the organisation of resources in competition into a hierarchy. However, because of the use of consequence feedback within this hierarchy, information exchanges cannot carry complex behavioural meanings. An information exchange can only be a recommendation in favour of one behaviour and/or a recommendation against any other behaviour. Because of this restriction on information exchange, the resource units in competition must correspond with behaviours, or groups of related behaviours. These resource units will be called components, and will be organized in a hierarchy with more detailed components corresponding with specific behaviours, higher level components corresponding with types of behaviour.

The separation between clustering and competition is fundamental for any complex learning system. The separation emerges as a result of practical requirements, and is analogous with but qualitatively different from the memory/processing separation that exists in complex electronic systems. As a result of the memory/processing separation, two key information models exist in complex electronic systems: *data read/write* and *instruction*, and these models are the basis for hierarchies of description that make understanding of such systems possible. As a result of the clustering/competition separation, two key information models will tend to exist in complex learning systems: *condition definition and detection* and *behavioural recommendation*. The clustering/competition separation can therefore also be called the recommendation architecture, in contrast with the instruction architecture in electronic systems.

7.15 Types of Behaviour

Many behaviours of a complex learning system directly act on the external environment. Other important behaviours act on the system itself, ultimately as an indirect influence on the environment. Both types of behaviour must be recommended by receptive field detection and accepted by the competition subsystem. Motor behaviours are an example of often direct action on the external environment. Some important examples of actions on the learning system itself include changes to receptive fields, reward behaviours that change recommendation weights, indirect activation behaviours that make recommendation weights of undetected receptive fields available to competition, and attention behaviours that manage the flow of information within clustering. We will consider each of these behaviour types in turn.

7.15.1 Change Management Behaviours

Changes to receptive fields can have major effects on system behaviour, and must therefore be carefully managed. Such management involves detection of special purpose receptive fields with recommendation strengths is in favour of expansions to other receptive fields or groups of receptive fields. As discussed earlier, the criteria for expanding a receptive field is that too few receptive fields are being detected at some level of complexity, and that the receptive fields to be changed have higher levels of internal condition detection than other fields. Such criteria could be implemented by all-to-all connectivity between modules. However, this approach requires very high levels of connectivity resources. A more effective approach is to make use of a central change manager as illustrated in Fig. 7.9b. This change manager receives inputs from all modules indicating the degree of internal activity, determines if changes are required, identifies appropriate modules to be changed, and send signals to the selected modules implementing the change. A further benefit of a central change manager is that it can take advantage of a possible additional criterion for identifying appropriate modules for change. If a group of modules have often expanded their receptive fields at similar times in the past, and if a significant proportion of those modules are currently detecting their receptive fields, then the remaining modules in the group may be good candidates for receptive field expansion in the current situation if required. Hence a combination of strong internal activity and past expansion at the same time as a number of currently detected receptive fields makes an inactive receptive field a particularly good candidate for expansion. A central change manager could develop receptive fields corresponding with such groups of modules with correlated past receptive field expansions. Such a receptive field could recommend expansion in the modules in its group.

7.15.2 Reward Behaviours

Changes to recommendation weights are also critical behaviours with major effects on a complex learning system. Such behaviours must therefore be recommended by detection of appropriate receptive fields and implemented if there is sufficient total recommendation weight. Because reward behaviours act upon the competition system itself, a subsystem of the competition system will be required to manage them effectively.

Another consideration in reward behaviours is the requirement to support strategic, tactical, and detailed rewards. To understand these concepts, consider the requirement that various behaviours of a musician are rewarded. Consequence feedback modulates the probability that recent behaviours will be repeated in similar circumstances in the future. Such consequence feedback is needed, for example, to modulate the detailed muscle movements manipulating an instrument. However, applause at the end of a concert cannot discriminate between muscle movements employed at different points, some parts of the concert may have been very good, others less so. Such applause would definitely be relevant to reward strategic behaviours such as the selection of the musical programme, concert location, partner musicians etc. Applause at the end of each piece of music would be relevant to the tactical behaviours such as selection of the particular piece. Rewarding of general limb movements and detailed finger movements requires some internal mental models against which the results of such movements are compared, with appropriate consequence feedback depending on the result of the comparisons. In receptive field detection terms, if receptive fields corresponding with applause are detected, they can recommend increasing recommendation weights in favour of strategic behaviours, and if receptive fields corresponding with certain combinations of sounds are detected, they can recommend increasing recommendation weights in favour of detailed movements. The receptive fields recommending detailed rewards will initially be acquired as a result of teachers rewarding certain kinds of muscle behaviour by providing, for example, comments. However, for an advanced musician, strategic rewards such as applause can be used to reward the more detailed reward behaviours. In other words, receptive field detections following applause can also have recommendation strengths in favour of modulating the recommendation strengths of the receptive fields that recommend modulation of recommendation strengths in favour of detailed muscle movements.

Receptive fields with recommendation strengths in favour of rewarding strategic behaviours can thus also recommend changes to the recommendation strengths of receptive fields recommending more detailed rewards. There will therefore be a reward hierarchy, with strategic rewards recommending changes to tactical rewards, and tactical rewards recommending changes to more detailed rewards.

7.15.3 Indirect Activation Behaviours

As mentioned earlier, some receptive fields that are not detected in current sensory inputs may nevertheless have relevant recommendation strengths. For example, if a receptive field is not being detected, but has often been detected in the past when a number of currently detected receptive fields were also being detected, that undetected receptive field may have some recommendation strengths relevant to the current situation. If a receptive field is not being detected, but expanded its receptive field in the past at the same time as a number of currently detected receptive fields, that undetected receptive field may also have currently relevant recommendation strengths. If a receptive field is not being detected, but was recently detected at the same time as a number of currently detected receptive fields, again that undetected receptive field may have relevant recommendation strengths.

There could therefore be behavioural value in "indirectly activating" receptive fields on the basis of such past temporally correlated activity. The correlations could be simultaneous past activity with a group of currently detected receptive fields, or activity just before or just after such a group. Furthermore, a group of indirectly activated receptive fields could be the basis for secondary indirect activation of another group and so on. Such indirect activations could lead to large, incoherent populations of receptive field detections. Indirect activations must therefore be behaviours which are recommended by receptive field detections and accepted or rejected by the competition subsystem.

Management of indirect activations means that a module receptive field acquires recommendation strengths in favour of any other module that detects its receptive field at the same time. Immediately after simultaneous activity, the probability of relevance is fairly high, so the initial recommendation strengths should be high but decay with time. Repetition of simultaneous activity should block or reverse the decay. Simultaneous receptive field expansion will generally be a stronger indicator of relevance that just simultaneous activity, and the recommendation weights will therefore decay more slowly. Positive consequence feedback following an indirect activation will tend to increase the weights that recommended that activation.

Indirect activations add a layer of complication to receptive field definition. Conditions are groups of currently active inputs, and receptive fields are groups of conditions. Conditions which mixed information derived from both direct and indirect activations could lead to undesirable activations. Hence there is a need within a module to separate conditions and receptive fields based on directly activated information from those based on indirect information.

Indirect activations can allow introduction of a much more extensive range of information derived from past experience. Suppose for example that a set of receptive fields detected within auditory inputs have often been active at the same time as a set of receptive fields detected within visual inputs. Such a situation would occur if a word was often heard at the same time as a type of object was viewed. Indirect activations on the basis of frequent simultaneous past activity would then lead to a situation in which hearing the word would tend to activate a "pseudoimage" of the type of visual object Such an indirect activation would be a semantic memory. Indirect activations on the basis of simultaneous past receptive field expansion would tend to activate all the receptive fields that were active in the past during some novel event or at some novel location. Such an indirect activation would be an episodic memory. Indirect activation on the basis of recent simultaneous activity would make it possible to access more information about a recently perceived object on the basis of limited current information, in other words a priming memory.

As discussed in more detail later, different types of indirect activation thus support different types of memory. One point to note is that the change manager discussed earlier, as part of its regular function, will tend to develop receptive fields correlating with groups of other receptive fields that expanded at similar times in the past. Such receptive fields would also be relevant to indirect activations on the basis of simultaneous past receptive field expansion. To support semantic memory, special purpose receptive fields corresponding with groups of modules often active at the same time will need to be developed.

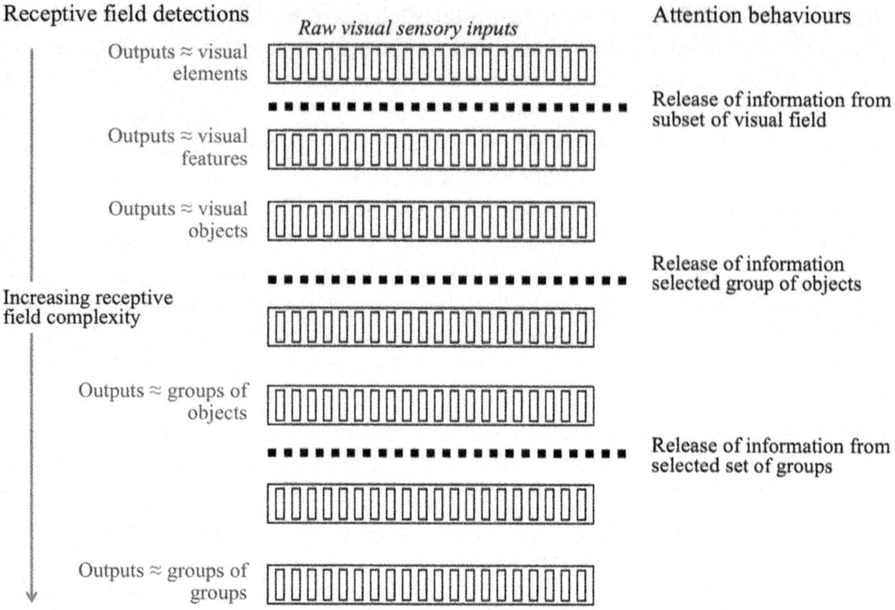

Fig. 7.13 Attention: managing visual receptive field detection information to avoid excessive processing requirements. The number of receptive fields detected across a retina at any one point in time will be extremely large. To reduce the volume of information within which receptive fields must be detected at a higher level of complexity, only a subset of the retinal receptive fields will be allowed to proceed at any one time to the modules detecting receptive fields discriminating between (≈) different visual objects. This selection is the process of visual attention. Similar selections limit the number of ≈visual objects receptive field detections proceeding to receptive field detections at the ≈groups of objects to those detected within a small number of objects. Similar selections must occur at higher levels. Without such selections, the processing resources for receptive field detections and for determining the most strongly recommended behaviour would be excessive

7.15.4 Information Release or Attention Behaviours

As conceptually illustrated in Fig. 7.13 for the processing of visual information, a modular hierarchy can detect receptive fields on many different levels of complexity. Individual receptive fields do not correspond exactly with visual features, objects etc. as they would be defined cognitively, but receptive fields on a given level of complexity tend to be more effective for discriminating between different features, or objects etc. This property of discrimination but not correspondence is indicated using the symbol ≈.

Relatively simple modules receive raw sensory inputs and develop receptive fields able to discriminate between different types of visual feature (*≈visual features*). Somewhat more complex modules receive detections of *≈features* receptive fields and generate *≈visual objects* receptive fields. Yet more complex modules detect ≈*groups of objects* receptive fields and so on. Detections on different levels of complexity are most effective for recommending different types of behaviours. For example, ≈objects receptive fields have complexities particularly effective for appropriate behaviours in response to objects, such as naming them etc. Note that receptive fields at other levels of complexity may have some recommendation strengths appropriate for such behaviours. Thus the presence of a particular feature or a particular group of objects may be relevant to appropriate naming of an object.

A problem with the hierarchy is that information derived from the entire visual field is available to the ≈*visual features* modules, but handling all this information simultaneously would be impractical. Information derived from many visual objects could be available in the ≈*visual objects* modules, but the ≈*groups of objects* needs to have inputs corresponding with just a few such objects. The same problem exists at higher levels. Hence there is a requirement that limited information from the visual field be released at any point in time to the ≈*objects* modules, limited information from the ≈*objects* modules be released at any point in time to the ≈*groups* modules and so on. This requirement interacts with limits to the number of different sets of inputs that can be independently processed at the same time within one module, as discussed earlier.

There must therefore be selections of the subsets of information available at one level of complexity that will be released to a higher level of complexity. These selections are of critical importance in determining appropriate system responses, and must therefore by behaviours that are recommended by receptive field detections and accepted or rejected by a competition subsystem. Such behaviours manage the internal flows of information within the modular hierarchy. For example, the selection of which domain in the visual field will be allowed to release information deeper into the modular hierarchy to generate ≈*object* receptive fields is the visual attention function. Another part of his function is to select the ≈*object* receptive fields detected within different objects that will be combined to drive detection of ≈*group of objects* receptive fields. If different ≈*objects* populations are maintained simultaneously in the same array of modules using the frequency modulation mechanism discussed earlier, the release behaviour to the ≈*group of objects* array would be implemented by bringing the different outputs from the ≈*objects* array into the same phase of modulation.

7.16 Structure of Competition

The various types of behaviour will correspond with components on various levels of detail in Fig. 7.14. General types of behaviour could be motor, speech, attention, reward, receptive field expansion and indirect activation. More specific types within the general indirect activation type could be indirect activation on the basis of temporally correlated recent activity, frequent past activity and past receptive field expansion. Yet more specific behaviours within the indirect activation on the basis of frequent past activity type would be indirect activation of a particular module or

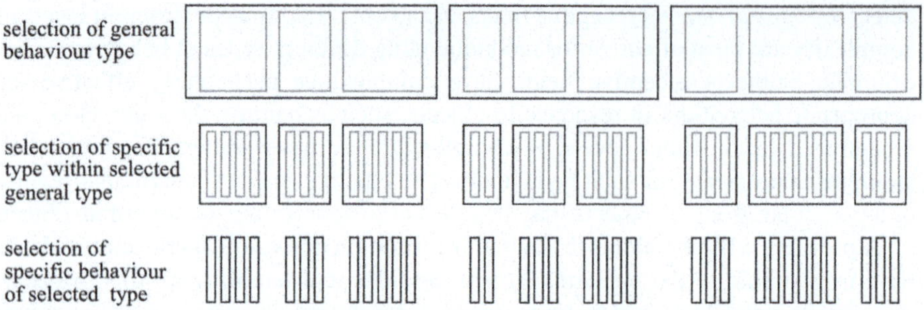

Fig. 7.14 The structure of competition. Components corresponding with different behaviours are organized in a hierarchy. There are components corresponding with different general types of behaviours, different more specific types within each general types, and different individual behaviours of each specific type. Because of the use of consequence feedback, components within competition must correspond with behaviours. However, by first selecting a general type of behaviour, then limiting further competition to more specific behaviours of the general type, the information processing load can be reduced

group of modules. The implementation of a general type of behaviour could often be release of the receptive field detections that recommended the behaviour to the components corresponding with the more specific behaviours of the general type. Releases could also be directed to other arrays of modules detecting receptive fields of higher complexity, more appropriate to recommending more specific types of behaviour.

For example, as illustrated in Fig. 7.15, an array of modules detecting receptive fields within auditory sensory information could have connectivity to an array of intermediate modules with receptive fields corresponding with groups of auditory modules often active at the same time. An intermediate module would target visual modules that were often active when it was active. These visual modules would also receive visual sensory inputs via various other arrays. Receptive field detections in the auditory array would have recommendation strengths (among many others) in favour of indirect activations. These recommendation strengths are implemented by input weights into the general indirect activation component. This component will receive inputs from other arrays with relevant recommendation strengths. If the general behaviour is accepted, it is implemented by release of receptive field detections from the auditory array to the intermediate array, where receptive field detections will occur. These receptive field detections have recommendation strengths in favour of indirect activations in the visual array. These recommendation strengths are implemented by input weights into a detailed release component. If this behaviour has the predominant weight, it is implemented by release of intermediate module receptive field detections to the visual array. The modules activated in this array are some of the same as would be activated by direct visual inputs, so a pseudovisual experience is created, but only at ≈visual objects levels of complexity, not in the arrays closer to visual sensory inputs. Hence the semantic memory is not a visual hallucination.

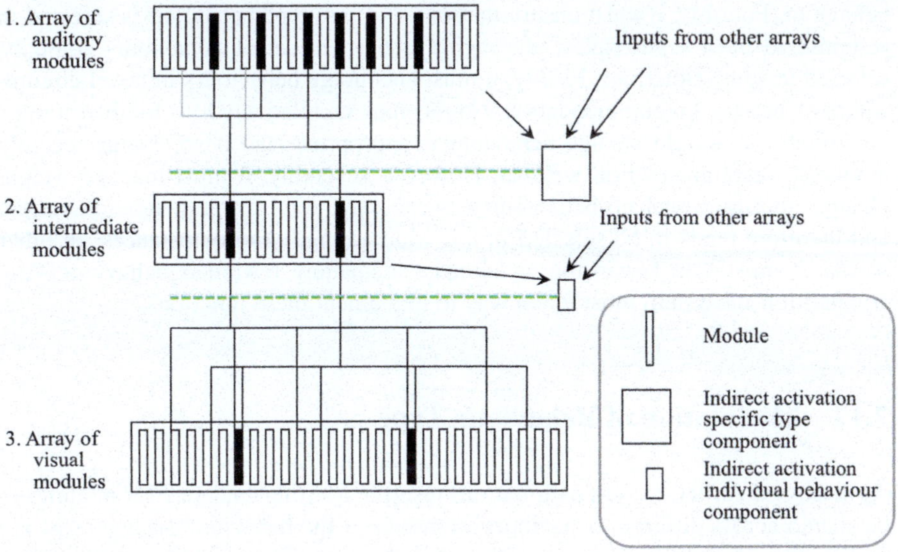

Fig. 7.15 Management of indirect activation on the basis of frequent past simultaneous activity. In this possible architecture, there are three array modules that detect receptive fields at different levels of complexity. Array 1 detects receptive fields at a level of complexity able to discriminate between different complex sounds (such as words). Array 3 detects receptive fields at a level of complexity able to discriminate between different categories of visual object. Array 2 detects receptive fields corresponding with groups of array 1 modules that have often been detected at the same time in the past. A module in array 2 targets modules in array 3 that have often been active at the same time as the array 2 module. Hence if for example a word has often been heard when a type of object has been seen, this arrangement will result in a pseudovisual image of the object when the word is heard in the future. Such indirect activations could interfere with processing of direct sensory inputs. Hence indirect activation must be a behaviour that is recommended by receptive field detections and accepted or rejected by behavioural components. Selection can occur in two stages. First the general behaviour of indirect activation on the basis of frequent past simultaneous activity must be selected. Then the behaviour of activating specific modules in array 3 must be selected

The component hierarchy reduces the degree of information processing required, by allowing receptive field detection only in module arrays appropriate to the already selected type of behaviour.

7.16.1 Roles of Competition: The Behaviour Selection Subsystem

There are a number of information functions which must be performed by the competition subsystem. Firstly, it must interpret each receptive field detection that it receives as a recommendation in favour of many different behaviours, each with an individual weight. Secondly, it must determine the total weights in favour of each behaviour. Thirdly, it must select and implement the most strongly recommended behaviour. Fourthly, it must ensure that one and only one behaviour is selected in response to most input states, although a consistent set of behaviours could be selected in some situations. Fifthly, it must encourage behaviours in novel circumstances when total recommendation weights may be relatively low. Sixthly, it must determine if a weight change behaviour is appropriate and if so change recently employed recommendation weights. However, seventhly, it must manage weight changes to ensure that no behaviour gets excessive total recommendation weight and therefore tends to be selected in too wide a range of circumstances. In other words, it must stop rewarding an already adequately rewarded behaviour. Any competition subsystem must find a way to implement these functions.

7.17 Modulation of Behaviour Type

As discussed earlier, resource limitations result in a strong tendency for modules to be shared across different behaviours. However, if the behavioural advantages are sufficiently great, some behaviourally specialized modules could be justified. For example, suppose that there are a number of different general types of behaviour such as aggressive, fearful, food seeking etc. Suppose further that there are circumstances in which different general types of behaviour are more likely to be appropriate. Circumstances in which important possessions are threatened could indicate that aggressive behaviour is likely to be appropriate. Circumstances in which self is weak relative to a threat could indicate that fearful (concealment or flight) behaviour is appropriate. Low blood sugar could indicate that food seeking behaviour is appropriate.

If there were modules that tended to recommend just one of these types of behaviour, then receptive fields detected within the different circumstances could recommend reducing the thresholds for detection of receptive fields in the appropriate set of modules, increasing the probability of an accepted recommendation of the general type. Furthermore, detection of circumstances indicating appropriateness of one type of behaviour could also act on the competition subsystem, recommending the behaviour type and reinforcing recommendations of that type from other module sources.

Resource limitations will tend to result in just a limited range of critical general behavioural types being supported in this fashion, reflecting a compromise between resource costs and behavioural effectiveness gains. One type of behaviour which could be modulated in a complementary fashion is receptive field expansion. Such expansion records information within current circumstances that could be relevant to guiding future behaviour. If the general circumstances strongly indicate a critical behaviour type, there could be value in increasing the degree of receptive field expansion because of a higher probability that the expansions will be relevant to future critical behaviour selections. Receptive field expansion behaviours could also be modulated by detection of strong consequence feedback, since information recorded in such circumstances could also be particularly relevant to future behaviour selections.

7.18 Management of Frequent Action Sequences

There are sometimes behaviours that are sequences of detailed actions always performed in the same order. One example could be a human being climbing stairs. The same sequence of muscle movements need to be performed repeatedly for each pair of steps. The learning process involves defining receptive fields in the combination of visual and somatosensory (including propioceptic) sensory inputs. These receptive fields must acquire recommendation strengths in competition in favour of the appropriate muscle movements.

If each action in a sequence needs to be determined by the full receptive field detection and behaviour selection process, the performance of the full behaviour will be relatively slow, and the risk of error at some point in the sequence relatively high. One way to achieve greater speed and accuracy once a sequence has been learned is for it to be recorded in a behaviour sequence manager. Such a manager is activated by detection of a set of receptive fields recommending initiation of the sequence, and the sequence is then driven to completion by the manager without going through the full receptive field detection and behaviour selection at each point. The manager could of course be interrupted if another behaviour is strongly recommended through the regular process.

As will be discussed later, action sequences may also be sequences of indirect activations and information releases that implement previously learned cognitive processes. Such processes could also be efficiently supported by a behaviour sequence manager.

7.19 Identification of Provisional Connections

A complex learning system does not have full a priori information on which sensory inputs need to be connected to which modules, which modules need to be connected to which other modules, or which modules need to be connected to which behavioural components and acquire recommendation weights. In addition, the information that a connection is required will not generally be available until the moment it is needed. There if limited information available at that point to select the most useful connection, and establishing the connection by physical growth at that point may not be practical.

One indication that a connection between two modules has a higher probability of being useful is if the two modules have often been active at the same time in the past. Similarly, a connection between a module and a component is more likely to be useful if they have often been active in the past at the same time. There will also be a tendency on average for more recent simultaneous activity to be more relevant that activity more remote in time. One way to use such information would be to periodically perform a partial rerun of past activity, with a bias in favour of more recent, and establish provisional connections with zero weights between modules and components that are often active at the same time during the rerun. Such provisional connections would then be available for future learning, selected by a process that increases the probability of their being relevant.

Note that a priori information, for example derived from genetic programming, could be used to guide initial connectivity. Such guidance could improve the speed and effectiveness with which receptive fields and recommendation weights were bootstrapped from experience.

7.20 Practical Requirements and Architectural Constraints

This chapter has demonstrated that there are some practical requirements that tend to constrain the physical information handling resources of a complex learning system into some specific forms. The practical requirements sometimes conflict, and the need to find an adequate compromise is one of the factors driving the architectural forms. The way in which the practical requirements lead to a modular hierarchy of condition definition and detection and a component hierarchy of behaviour selection from recommendations is illustrated in Fig. 7.16.

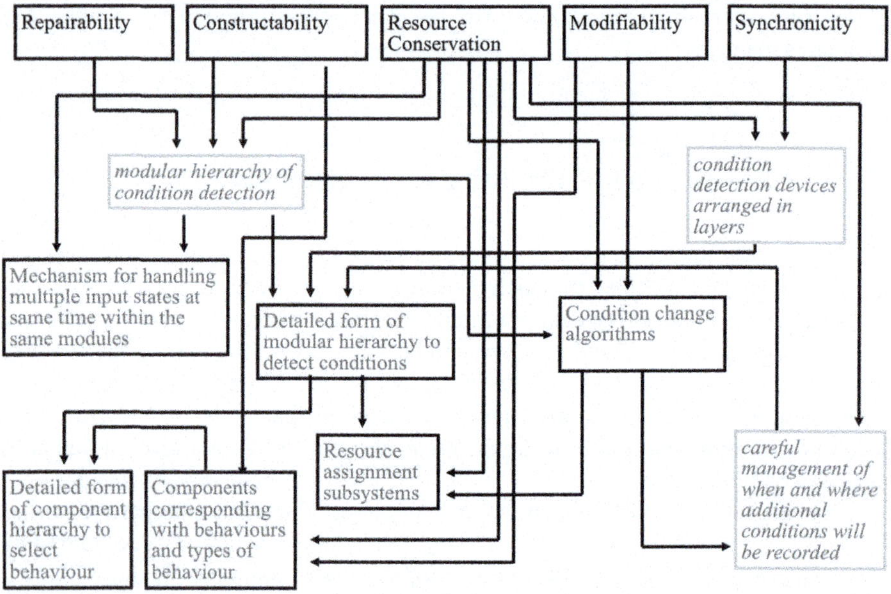

Fig. 7.16 Five practical needs drive architectural forms. The requirements for resource limitation, modifiability, synchronicity, constructability and repairability place pressures on the physical architecture of a complex learning system. The architectural forms are a compromise between sometimes conflicting requirements. The key requirement is resource limitation, and as the ratio of number of behaviours that must be learned to available information handling resources increases, the resource architecture will be more tightly constrained into the indicated forms

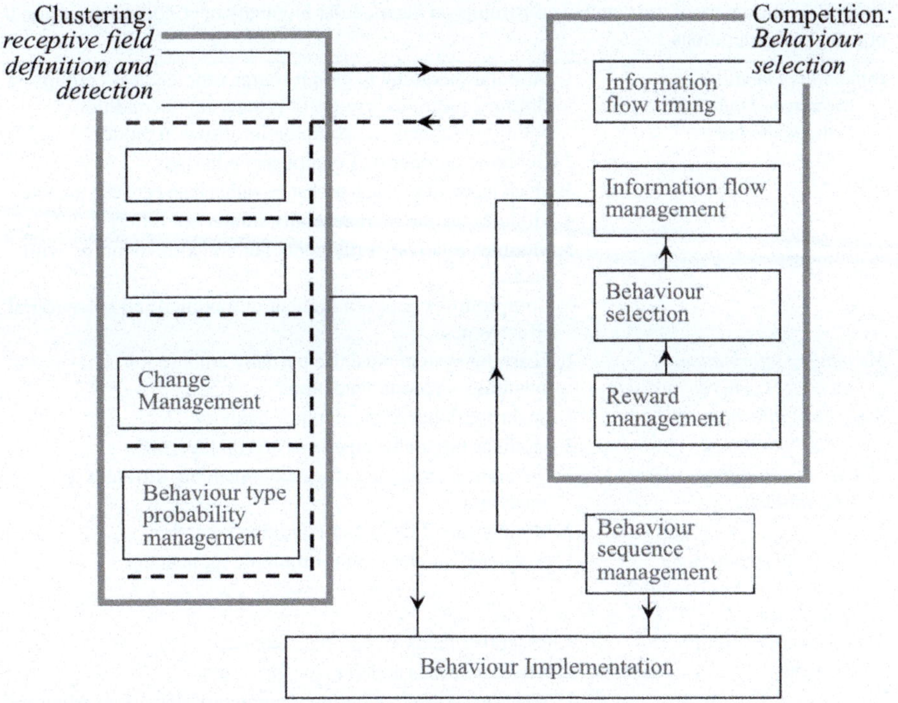

Fig. 7.17 Recommendation architecture major subsystems driven by practical requirements. Three modules within the major clustering subsystem are required. One is the primary receptive field definition and detection module. The second is a module detecting receptive fields that are effective for recommending changes to other receptive fields. The third is a module detecting receptive fields that are effective for recommending different general types of behaviour. Four components within the competition subsystem are required. Behaviour selection determines and implements the currently most strongly recommended behaviours. Information flow management implements behaviours by release of information flows into, within or out of clustering. Information flow timing ensures that the relative timing of information releases is appropriate. Reward management determines and implements the currently most strongly recommended reward behaviours. Finally, a special purpose subsystem records frequently utilized sequences of behaviours to increase performance speed and accuracy

At a more detailed level, the same practical requirements tend to result in a range of subsystems as illustrated in Fig. 7.17. The major separation between a receptive field definition and detection subsystem (called clustering) and a behaviour selection subsystem (called competition) is called the recommendation architecture. Clustering is a hierarchy of modules that define and detect conditions on different levels of complexity. Competition is a hierarchy of components where components correspond with different behaviours or types of behaviour. The existence of these hierarchies and the associated common information models make it possible to create hierarchies of description linking low information content, more approximate high level descriptions of complete system processes to high information content, more precise detailed descriptions of segments of system processes. Such hierarchies of description are essential for scientific system understanding.

Table 7.1 Summary of the detailed constraints on the modular hierarchy of condition detection by practical considerations

Definition of modules and exchange of information between modules	1. Modular hierarchy to gain resource efficiencies – groups of similar conditions, groups of groups etc. – condition definition/detection a common information model
	2. Heuristic definition of complex conditions
	3. Much more information exchange within than between modules
	4. Modules on one level generally similar but different in detail
	5. Modules relatively orthogonal but not corresponding with features
	6. Compromises between number of modules and behavioural effectiveness
Meaning of information exchanges between modules	1. Learning system must use partially ambiguous behavioural meanings – recommendations
Support of information synchronicity	1. Layering rather than reference memory
	2. Modular hierarchy superimposed on layering
Conditions on different levels of complexity	1. Different complexities for recommending different types of behaviour
	2. Modulation of degree of condition detection
Independent populations of simultaneous condition detections	1. Frequency modulation for multiple input states
Management of condition changes	1. Receptive fields tend only to expand
	2. New conditions added to existing set
	3. Need for at least a minimum number of detections triggers receptive field expansions
	4. Strong internal module activity indicates change appropriateness
Implications of limits on receptive field changes	1. Value of indirect activation
	2. Segregation of direct and indirect receptive fields

The practical requirements also drive the existence of some subsystems within clustering and competition. Because of the critical need for management of changes to receptive fields, a change manager containing receptive fields particularly effective for change management will be needed. Because of the behavioural advantages of management of the relative priorities of a range of critical behaviours, a behaviour type probability manager containing receptive fields particularly effective for this purpose will be needed. Within the behaviour selection subsystem and special purpose component managing reward behaviours is needed. Because many implementations of behaviours are releases of sensory inputs into clustering, or releases of receptive field detections between arrays in clustering or out of clustering to drive externally directed behaviours, special purpose subsystems to manage the selection and timing of these information flows will be needed. Finally, there will be a need for a behaviour sequence manager to rapidly and accurately implement previously learned behaviour sequences. There is then a range of more detailed constraints placed on each of these subsystems by the practical considerations. The more detailed constraints discussed in this chapter are summarized in Tables 7.1, 7.2, and 7.3.

Table 7.2 Summary of the detailed constraints on the behaviour selection subsystem by practical considerations

Use of consequence feedback	1. Consequence feedback cannot change condition definitions
	2. Separate subsystem (competition) to interpret receptive field detections into behavioural recommendations
	3. Components in competition correspond with behaviours
	4. One receptive field detection recommends wide range of behaviours
	5. Each behavioural recommendation has individual weight
Management of connectivity resources and information flow	1. Limits on information releases between condition complexity levels
General and specific behaviour types	1. Hierarchy of components corresponding with general and more specific behaviour types
	2. General behaviour selection followed by more specific behaviour selection
Behaviour selection and rewards	1. Ensure that one and only one behaviour selected at each point in time
	2. Determine if weight change is appropriate
	3. Change weights, but ensure individual weights do not become excessive
	4. Strategic and tactical rewards
Information release within modular hierarchy	1. Coordinated management of information flows within modular hierarchy
	2. Ensure information releases occur at the correct, coordinated time

Table 7.3 Summary of other detailed constraints

Resource management subsystem	1. Extensive reciprocal connectivity with modular hierarchy
	2. Inputs organized by groups of modules that have changed at the same time in the past
	3. Competition to determine appropriate modules for expansion
	4. Outputs across modular hierarchy to drive expansions, released after completion of competition
	5. Management of provisional connectivity in modular hierarchy
	6. Support of some types of indirect activation
Behavioural effectiveness vs. resource economy compromises	1. Modulation of general behaviour type
	2. Modulation of rate of condition recording
	3. Modulation of general system activity state
Optimisation of action sequence execution	1. Inputs from behaviour drivers
	2. Very specific circumstances for each action
	3. Relative timing critical
	4. Outputs to behaviours

These constraints are architectural limits that will tend to be approached as the ratio of behaviours that must be learned to resources increases. Natural selection will favour brains that can learn more behaviours with fewer resources. The architectural forms will therefore tend to appear in the brains of different species. Hence much of the experimental information obtained for other mammal species will be applicable to understanding the human brain. Chapter 8 will review the evidence that the limits as discussed in this chapter are indeed visible in the mammal and human brains

Chapter 8
Appearance of Architectural Constraints in the Brain

8.1 Brain Structures and Information Processes

As discussed in Chap. 7, a number of practical requirements tend to constrain the physical resources of a complex learning system into specific forms called the recommendation architecture. These forms include the existence of hierarchies of modules and components, with modules and components on different levels using the same two types of information process, called *condition definition and detection* and *behavioural recommendation.* This common use of information models means that the hierarchies of modules and components can be used as hierarchies of description to support scientific understanding.

Natural selection will tend to result in these practical requirements applying to brains. This chapter will discuss how the human brain conforms with these recommendation architecture forms. The correspondence between these modules and components and major anatomical structures of the brain is illustrated in Fig. 8.1. The cortex defines and detects receptive fields within the input information available to the brain, including sensory inputs. The hippocampus manages changes to cortical receptive fields, defining and detecting its own receptive fields appropriate to guide such changes. The amygdala and the hypothalamus detect receptive fields with recommendation strengths in favour of different very general types of behaviour. Such receptive fields in the amygdala tend to be defined heuristically, other receptive fields are specified genetically to a significant degree. The interpretation of cortical receptive field detections into behavioural recommendations and selection of the most strongly recommended behaviours is performed by the basal ganglia. The ventral basal ganglia selects more strategic behaviours including strategic rewards, the dorsal basal ganglia more tactical behaviours. The thalamus manages implementation of the behaviour selected by the basal ganglia. Such an implementation is an information flow in and out of the cortex and/or between different cortical regions. The basal forebrain manages the precise timing of such information flows. The cerebellum records previously learned sequences of behaviours, and implements those sequences rapidly and accurately when required.

L.A. Coward, *Towards a Theoretical Neuroscience: from Cell Chemistry to Cognition*, Springer Series in Cognitive and Neural Systems 8, DOI 10.1007/978-94-007-7107-9_8,

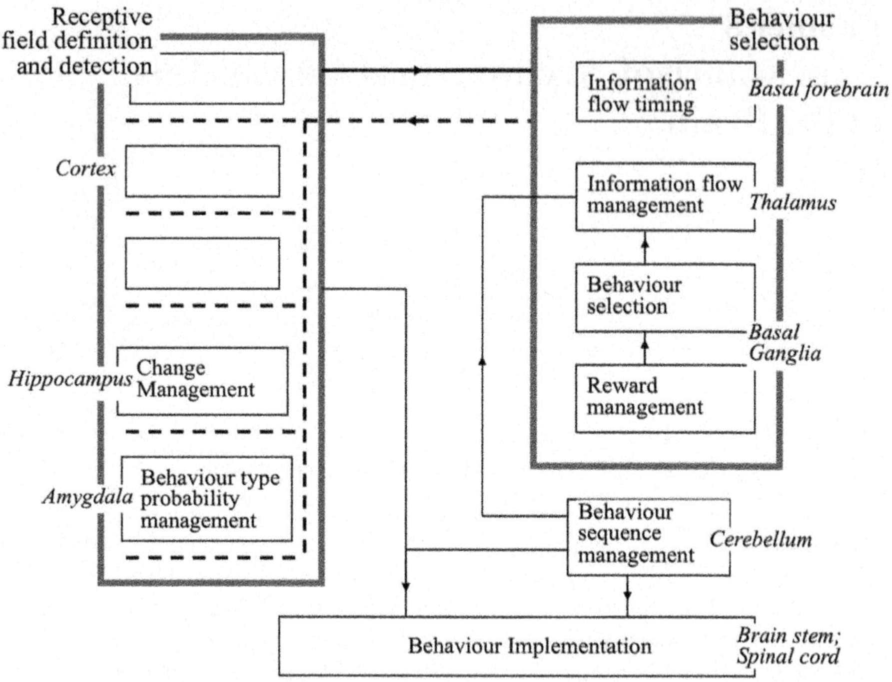

Fig. 8.1 The recommendation architecture in the mammal brain. Mapping between subsystems, modules and components required theoretically on the basis of resource constraints plus other practical requirements and the major anatomical structures of the brain

In this chapter, firstly there will be a quick overview of the evidence for these correspondences for each major anatomical structure. The evidence comes from the psychological, anatomical, physiological and neurochemical material described in earlier chapters. Then the way in which each major anatomical structure performs its corresponding information processes will be described at relatively high level. Examples of how the information models make it possible to map segments of these higher level descriptions to more detailed physiological and neurochemical levels will be provided.

8.1.1 Information Process Based Architecture in the Brain

The central constraint is that in the physical architecture, separate resource subsystems will be optimized for different types of information process, not different types of cognitive process. One resource subsystem will be shared by many cognitive processes, and one cognitive process will require information processes performed by many different subsystems. In the brain, multiple cortical areas and other anatomical structures are active in any cognitive feature, and no one area supports just one feature. For example, a successful episodic memory retrieval involves participation of multiple cortical areas, the thalamus, basal ganglia, hippocampus, amygdala and cerebellum. However, each of these cortical areas and brain regions is also active in other types of cognitive features. Some areas participate strongly in one type of feature, but in general that area will also participate in other feature types.

8.1.2 General Evidence for the Existence of the Recommendation Architecture in the Brain

In the recommendation architecture modular hierarchy, detailed modules define and detect groups of similar conditions and higher modules made up of more detailed modules define and detect larger groups of somewhat less similar conditions. The physical structure of the cortex is as would be expected for such a modular hierarchy. Groups of inputs to one branch of a pyramidal neuron define very detailed conditions. The group of conditions defined by the branches making up a dendrite define a higher level condition. The group of conditions made up of different dendrites define a neuron receptive field. The group of neurons in one layer of a column define a more complex receptive field, and the layers of a column together define the column receptive field. The group of columns make up the information space within which a cortical area detects receptive fields.

There is thus a modular hierarchy of dendrite branches, dendrites, pyramidal neurons, column-layers, columns, and areas. As required by constructability, there is a general physical similarity between peer modules on one level, but differences in detail. A module on one level detects conditions that are combinations of the conditions detected by its submodules. There is much more information exchange between submodules than between modules. For example, there is massive interaction within a neuron, but the neuron produces a single output. There is much more connectivity within a column than between columns. Cortical areas have connectivity with a limited range of other areas, indicating that information exchange is minimized as far as possible. Conditions and receptive fields are largely defined heuristically, and as demonstrated by the LTP/LTD algorithms the definition process is based upon temporal correlations. As demonstrated by the receptive fields found in area TE, fields are relatively orthogonal but do not correspond with cognitive features. In other words, information exchanges are partially behaviourally ambiguous.

The cortex is arranged in layers, as expected for synchronisation in a system utilizing partially ambiguous information, and the modular hierarchy is imposed on the layers. Areas detect receptive fields on different levels of complexity, and these different levels of complexity are more appropriate for different types of behaviour. The degree of receptive field detection is modulated on different modular levels by different targetting of pyramidal neurons by interneurons.

The existence of different frequencies in the electrical activity of the cortex is consistent with the use of frequency modulation both to allow simultaneous processing of different input states in the same physical resources and to select the subsets of inputs that will be processed.

The predominance of LTP in the cortex is consistent with a tendency for receptive fields to expand, and contract only when a contributing condition is irrelevant on the basis of temporal correlation. The value given to indirect activation by receptive field consistency is shown by the existence of phenomena like semantic, episodic and priming memory. The need for segregation of inputs to support direct and indirect activations is consistent with the segregation of inputs from different sources into different regions of pyramidal neuron dendritic trees.

As expected for a change manager, the hippocampal system receives inputs from all across the cortex, and sends outputs back to those cortical areas. Within the hippocampus proper are multiple feedback loops consistent with the competition to determine appropriate targets for receptive field expansion. At a psychological level, damage to the hippocampal system results in loss of ability to create new declarative memories, consistent with loss of ability to achieve cortical receptive field expansions. The loss of some episodic memory is consistent with the use of hippocampal system receptive fields to indirectly activate cortical receptive fields on the basis of past simultaneous receptive field expansion. The activation of the hippocampal system observed in response to novelty is also consistent with its role in driving receptive field expansions in such circumstances.

The action selection subsystem receives many inputs from the modular hierarchy, each input recommending a range of different behaviours. The subsystem determines and implements the most strongly recommended behaviour. In the behaviour selection subsystem, components correspond with individual behaviours or groups of behaviours. In the basal ganglia, and in particular the striatum and GPi/SNr, small groups of neurons correspond with different behaviours. The striatum receives massive input from the cortex. Any one striatal projection neuron receives large numbers of inputs from the cortex, but only a few from the thousands of synapses made by any one cortical neuron, and a striatal neuron requires simultaneous inputs from many cortical neurons to produce an output. This connectivity is consistent with the interpretation that a behaviour requires recommendation by many receptive field detections to be accepted, and that one receptive field detection recommends many different behaviours. The synaptic weights of cortical inputs can be individually adjusted, consistent with their correspondence with different behavioural recommendations. As discussed in more detail later, the direct and indirect paths can be interpreted as managing the competition between different behaviours, with one role of the dopamine feedback path being to ensure the selection of one and only one behaviour in most circumstances. The outputs from the GPi/SNr can be interpreted as implementing the selected behaviour by acting on the thalamus to release information into, out of or between cortical areas. The observations that the ventral basal ganglia plays a role in rewards and its similar connectivity suggests that it manages the selection of strategic reward behaviours.

As expected for a subsystem that manages information flow in the modular hierarchy, there is strong reciprocal connectivity between different thalamic nuclei and different cortical areas, and the basal ganglia action selection subsystem manages the activity of the thalamus. Damage to a thalamic nucleus results in cognitive deficits similar to those resulting from damage to the corresponding cortical area. Consistent with the use of frequency modulation to manage information flows, the thalamus is intimately involved in various cortical modulation frequencies, such as the gamma band associated with the thalamic reticular nucleus.

The basal forebrain plays a role in the generation of theta band oscillations in the cortex and hippocampal system, with theta and gamma oscillations being brought into phase with each other. One indicator of a role in timing is the connectivity of the hippocampus proper with the septal nuclei, consistent with the restraining of outputs to drive cortical receptive field expansions until the competition to determine the appropriate fields has resolved.

Emotions can be regarded as mechanisms to adjust the relative probability of different types of behaviour. For example, anger increases the probability of aggressive behaviours, fear increases the probability of avoidance and flight behaviours. The association of the amygdala and hypothalamus with emotions therefore indicates that they are associated with the behaviour type probability management subsystem.

Regulation of general behavioural priorities is not limited to these subcortical structures. A range of emotions have been defined by cognitive psychology, including hunger, anger, fear, disgust, surprise, sadness, happiness and pride. Disgust increases the probability of rejection behaviours, surprise puts priority on delaying behavioral response and recording information. Sadness puts priority on avoiding behaviour when radical changes in the social environment are detected. Happiness puts priority on recently selected behaviours. Pride puts priority on behaviours that assert social dominance. In each case, receptive fields are detected which recommend the behavioural priority, and the priority is implemented in response to sufficient total recommendation weight. Implementation can be by direct action on cortical regions tending to recommend specific behaviours of the general type, or by general release of neurotransmitters in the cortex with differential effects. "Primitive" emotions like hunger, anger, fear and disgust tend to be more recommended by receptive fields located in the hypothalamus and amygdala that have a strong degree of genetic predefinition. "Higher level" emotions like surprise, sadness, happiness and pride tend to be more recommended by more complex receptive fields in the prefrontal cortex. However, complex receptive fields will also be relevant to primitive emotions and some genetic guidance relevant to higher level emotions.

Damage to the cerebellum results in loss of ability to perform smooth movements, but the ability to perform the movements in a jerky fashion remains. The existence of this type of deficit supports the view that the cerebellum manages frequently utilized behaviour sequences.

We will now consider the information models for each structure in turn.

8.2 Information Model for the Cortex

The combined pressures of resource economy and modifiability result in a hierarchy of physical condition detection modules in which information exchange between modules is minimized as far as possible. The levels in this hierarchy are areas, columns, layers, pyramidal neurons, dendrites, terminal dendritic branches and synapses. A synapse detects the presence of an input condition. A terminal dendritic branch detects a condition that is a combination of synaptic branch conditions. A dendrite detects conditions that are combinations of branch conditions. A neuron detects conditions that are combinations of its dendrite conditions and so on.

8.2.1 Cortical Areas

At the highest level, different cortical areas detect receptive fields at different levels of complexity. As discussed earlier, complexity is defined by a combination of input sources and the number of inputs that must be present to trigger receptive field detection (i.e. thresholds). Resource considerations limit the number of different areas, and the number of different levels of complexity is therefore limited. Natural selection pressures determine the complexity levels, in such a way that the combination of different levels is adequate for discrimination between circumstances in which different behaviours are appropriate. A group of receptive fields in various areas is activated in response to every circumstance. "Adequate discrimination" means that the group activated when a specific behaviour is appropriate is sufficiently similar to the groups activated in other circumstances in which the same behaviour is appropriate, and sufficiently different from the groups activated in circumstances in which other behaviours are appropriate. "Sufficiently" means that it is possible to assign behavioural recommendation weights to each receptive field detection in such a way that the predominant weight is always in favour of the appropriate behaviour.

Genetic information therefore places biases on the connectivity between areas. Inputs to an area come from other areas and/or from the senses. In order to limit undesirable side effects of learning, any one area exchanges information with a limited number of other areas as conceptually illustrated in Fig. 8.2. One area typically receives minimal inputs from most areas, a very small number of inputs from a significant proportion of areas, a moderate level of inputs from a smaller proportion, and a high volume of inputs from a very small number of other areas. Hence information exchange between areas is minimised as far as is consistent with resource economy. Information exchange with another area is often but not always reciprocal, and the volume of connections may be larger one way than the other.

Each area detects receptive fields in a particular range of complexity and each type of receptive field is effective for discriminating between circumstances in which different behaviours of a range of types are appropriate. Thus each area contributes recommendations in favour of a range of behaviour types. The total number

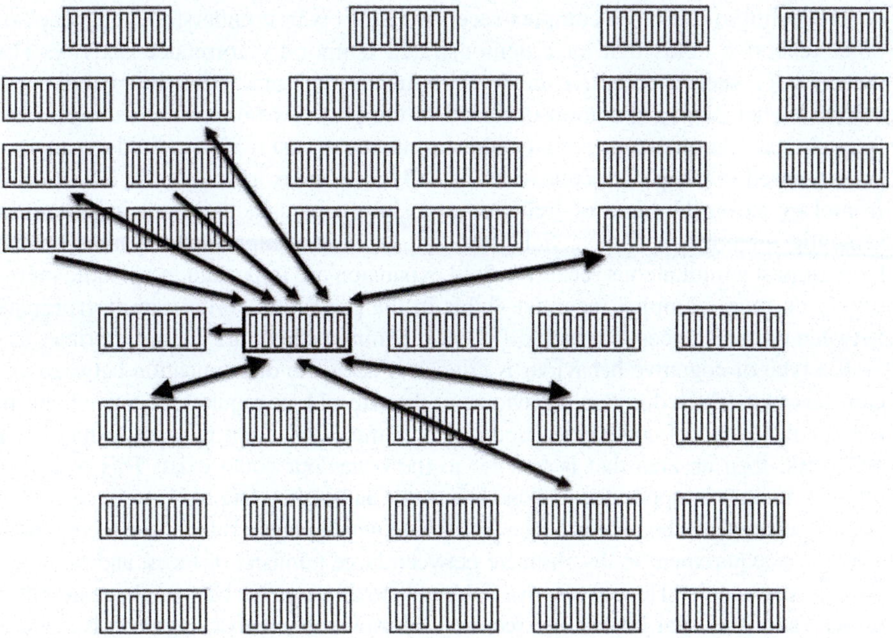

Fig. 8.2 Information model for interconnectivity of cortical areas. As in this conceptual illustration, in the cortex one area is strongly interconnected with a small number of areas, and more weakly interconnected with a slightly larger number of areas. A small proportion of areas receive sensory inputs directly. Interconnectivity is usually reciprocal, but can be much stronger one way than the other (indicated by size of *arrows*). The receptive field complexity in an area is defined by the group of areas from which it receives inputs. These groups reflect natural selection pressures in favour of receptive field complexities with high discrimination between circumstances in which different behaviours are appropriate, and pressures in favour of minimizing information exchange as far as possible to reduce the undesirable side effects which new learning can have on prior learning

of areas is a compromise between better discrimination achieved by more areas and the resource cost of extra areas. This compromise is defined genetically.

In the case of visual areas, relatively simple complexities are effective for discriminating between circumstances in which different objects should be the focus of attention, and more complex receptive fields discriminate between different types of objects and different object positions. In the auditory areas, relatively simple receptive fields discriminate between different sounds, more complex fields between sequences of sounds such as words.

In the motor cortices, receptive fields discriminate between circumstances in which different motor behaviours are appropriate. These circumstances include current body position and movements, and receptive fields able to discriminate between different body positions and movements defined in the somatosensory cortices must therefore contribute to motor cortex receptive field definitions.

The definition of more complex receptive fields is best understood as groups of other receptive fields that have demonstrated temporally correlated activity. The detection of such a receptive field recommends the same behaviours that were selected when such correlation occurred in the past. Two important examples are frequent past simultaneous activity and simultaneous past receptive field expansion. As discussed in Chap. 9, receptive fields defined in areas including the left inferior frontal gyrus on the basis of frequent past simultaneous activity are important in semantic memory. Receptive fields defined in various hippocampal areas on the basis of past simultaneous receptive field expansion are important in episodic memory. Even more complex receptive fields in the prefrontal cortex can discriminate between circumstances in which different planning behaviours are appropriate.

If a type of cognitive behaviour is critical, if adequate discrimination between circumstances in which different behaviours of the critical type requires receptive fields in a particular range of complexity, and if this complexity is not relevant to any other behaviour, then an area that is specific to the behaviour could exist. This is a very resource intensive support for just one behaviour and is therefore likely to be rare. One possible example is the fusiform face area (FFA) which plays a role in face recognition. There is a requirement to discriminate between large numbers of faces, and faces are visually very similar. Hence receptive fields able to discriminate between large numbers of very similar visual objects are required. However, if another cognitive behaviour requires a similar information capability, the area will be utilized. Thus experts in bird recognition able to discriminate between very large numbers of different birds make use of the FFA, although non-experts make no use of the area for bird recognition.

To achieve an adequate range of behavioural recommendations, a cortical area detects at least a minimum number of receptive fields within every input state to the area. Different receptive fields at the area level of complexity are defined by cortical columns within the area. At a somewhat more detailed level, a column actually detects a group of very similar receptive fields instantiated by pyramidal neurons in its output layers. These receptive fields may have slightly different behavioural implications. One column receptive field is relatively orthogonal to other column receptive fields in the same area. A column-layer module detects a set of receptive fields at a complexity level that varies slightly depending on the layer, but similar to other receptive fields in the column.

8.2.2 Cortical Columns

An area module is made up of a set of column submodules, each detecting a different receptive field within the information available to the area. The area manages its submodules to ensure that at least a minimum but not an excessive number of submodule receptive fields are detected whenever the area receives a set of inputs. Receptive fields are expanded when necessary to achieve the minimum number.

Outputs of area receptive field detections directly to subcortical regions have explicit weights in favour of a range of different behaviours, outputs to other cortical areas contribute recommendations via intermediate receptive fields.

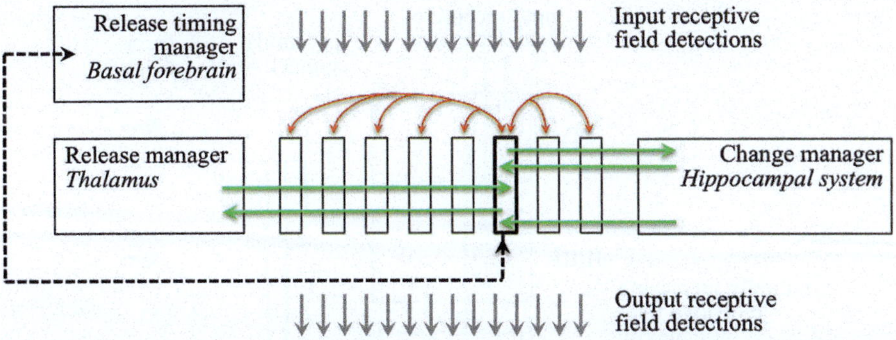

Fig. 8.3 Information model for a cortical area. An area is made up of a large number of columns with different receptive fields. Key connectivity of one column is illustrated, all columns are similarly connected. An area detects a number of receptive fields in all input states to the area. This number is managed by the area to be above a minimum but not excessive. If the number is below the minimum, this condition is detected by the central change manager which determines which columns are most appropriate for receptive field expansion and drives those expansions. Receptive fields are expanded until the required minimum number is reached. Inhibitive cross connection between columns limits the total number detected. This cross inhibition is implemented by interneurons that target pyramidals within a column but receive inputs from pyramidals in other columns. The release manager implements behaviour selections that are releases of selected receptive field detections to other cortical areas or brain regions. Releases are implemented by imposing a gamma band frequency modulation on output signals. The release timing manager ensures that releases have an appropriate timing relative to other signals. Timing is implemented by modulating the gamma band at a theta band frequency. Cortex – hippocampus and cortex – thalamus connectivity is glutamatergic in both directions. Connectivity from cortex to basal forebrain is glutamatergic, the return projection is cholinergic, GABAergic and glutamatergic. Reciprocal inhibitory connections between columns are implemented by glutamatergic outputs from a column activating GABAergic neurons that only act within the target column

Figure 8.3 illustrates the information model for management of columns within an area. Each column submodule has potential access to all the inputs to the area and defines its receptive field within a subset of those inputs. Every column is also reciprocally connected with the hippocampal system change manager. The hippocampal system receives inputs indicating the degree of internal activity in the column. The hippocampal system determines whether sufficient receptive fields are being detected, and if not it sends signals back to columns with the highest degree of internal activity to drive receptive field expansions. Too many receptive field detections raises the processing cost of behaviour selection, and inhibitive connectivity between columns therefore ensures that the degree of receptive field detection across the area does not become excessive. Output receptive field detections are released to other areas or subcortical structures. Such releases are themselves behaviours which must be recommended by receptive field detections in the area and in other areas, and accepted by the basal ganglia and thalamus. If such a release behaviour is accepted it is implemented by signals from the thalamus release manager, which places an appropriate frequency modulation on the outputs to increase their effect on their targets. The exact timing of receptive field releases can also be

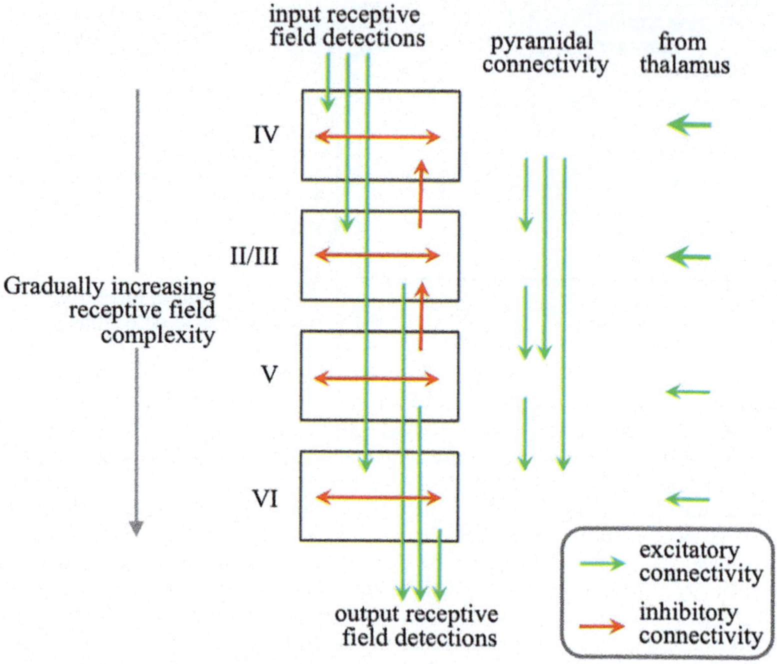

Fig. 8.4 Information model for a cortical column. The column receives input receptive field detections and within those inputs it detects receptive fields at several slightly different levels of complexity. The output receptive field detections are at different complexities for different targets. Some of these receptive fields are interpreted as behavioural recommendations in subcortical structures, some carry detections to other cortical areas where they are part of more complex receptive fields, and some contribute to receptive field expansion management. Inhibition within and between layers ensures that activity does not become excessive. Inputs from the thalamus regulate release of information from the column to its targets, and especially in primary sensory areas also contribute to receptive field definitions. The gradual increase in receptive field complexity is supported by pyramidal neuron connectivity in the general direction layer IV → layer II/III → layer V → layer VI. The hippocampal system receives outputs from layer II indicating internal activity of the column. If the hippocampal system determines that receptive field expansion is appropriate it returns activity to layer III and perhaps other layers including VI

critical, and such timing is triggered by the basal forebrain and implemented by a further modulation imposed on the release modulation. These modulations will be discussed later.

8.2.3 Column-Layers

Figure 8.4 illustrates the information model for one column module in more detail. The column is made up of a sequence of column-layer submodules. In the layer sequence IV → II/III → V → VI the forward connectivity of the pyramidal

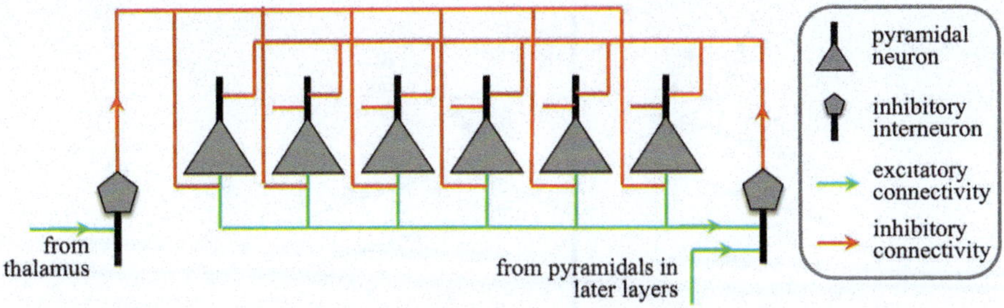

Fig. 8.5 Information model for regulation of column-layer module activity by interneurons. In the illustration, the pyramidal neurons in a column-layer target one interneuron, which in turn targets the pyramidals. This connectivity limits total pyramidal neuron receptive field detections in a column-layer module. Inputs to such an interneuron from other layers of the column regulate total detections within their column module. Different interneurons with substantial inputs from the thalamus regulate the timing of spike outputs indicating receptive field detections. These different interneurons may target the pyramidal neuron axon initial segment

neurons results in a slightly increasing receptive field complexity, although the receptive fields are all detected within a very similar information space. These slightly different complexities mean that different column-layers are most effective for different behavioural purposes. Layer II/III is effective for providing the internal activity information on which column receptive field changes are based. Layer V provides receptive field detections that are interpreted by the basal ganglia and thalamus as behavioural recommendations. Layer VI provides receptive field detections that are components in the receptive fields of other areas. To prevent excessive column output, each column-layer limits its own total activity and in addition high activity in one column-layer inhibits activity in earlier column-layers. Additional layers may provide better management of different column processes, but at a resource cost. Hence the number of layers may differ between different areas depending on the necessary compromises between these two considerations.

Inputs from the thalamus result in frequency modulation of output receptive field detections, managing the effect of those detections on their targets as discussed later.

At a more detailed level as illustrated in Fig. 8.5, a column layer model is made up of pyramidal neuron submodules and interneurons. Regulation of total activity within the column layer is regulated by interneurons that are excited by the pyramidal neurons and inhibit those neurons. Interneurons in one layer can also be targeted by pyramidals in a different layer, providing another limit on layer activity. Primary regulation of frequency modulation to control column-layer outputs is performed by inputs to interneurons from the thalamus. These interneurons tend to produce outputs at the modulation frequency. Thalamic axons can also target pyramidals directly, where they may contribute to receptive field definition or modulation.

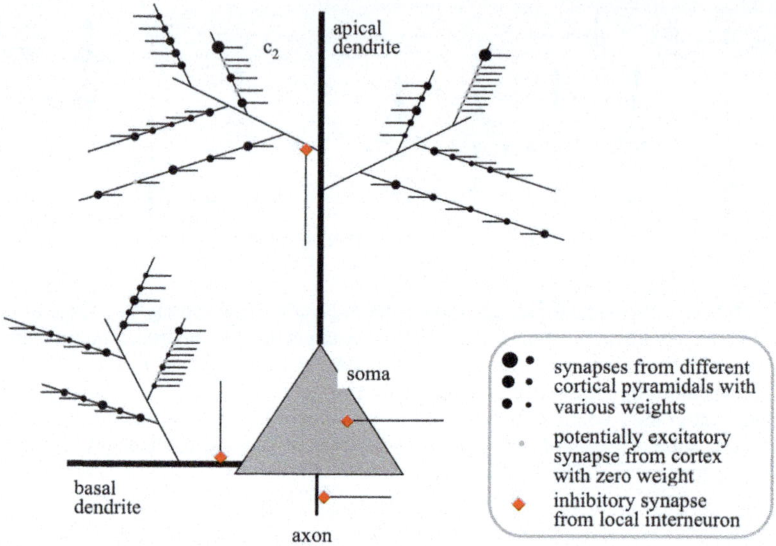

Fig. 8.6 Organisation of fast excitatory and inhibitory inputs to a pyramidal neuron. Glutamatergic inputs from other pyramidal neurons define the receptive field. Conditions are defined by groups of glutamatergic synapses on a dendritic branch and the neuron fires if a large enough subset of the conditions are detected. Inputs derived from different sources arrive on different types of dendrite. The primary receptive field is defined on basal dendrites. The receptive field defined on the apical dendrite defines the circumstances in which the neuron will be indirectly activated on the basis of past temporal correlations in activity with neurons in other areas. Inhibitory synapses from local interneurons regulate the probability of neuron receptive field detection in response to a given degree of condition detection, the relative influence of direct and indirect receptive fields, and the timing of spikes indicating detections

8.2.4 Pyramidal Neurons

Figure 8.6 illustrates the information model for one pyramidal neuron. The neuron has different dendrites and regions of dendrites that detect groups of conditions within different sources of input. Basal dendrites detect conditions within inputs from areas with simpler receptive fields, apical dendrites within inputs from areas with more complex receptive fields (higher areas). Hence basal dendrites instantiate the direct receptive field, while apical dendrites instantiate the receptive field corresponding with the circumstances in which the neuron should be indirectly activated on the basis of past temporal correlations of its activity with the activity of other neurons. Indirect activation of layer IV neurons in the primary visual cortex that are the entry point for inputs from the retina, would result in activations that would be difficult to distinguish from actual sensory inputs. Such "false" inputs could be behaviourally hazardous and layer IV excitatory neurons in the primary visual cortex do not have apical dendrites. The terminal branches on a dendritic tree detect individual conditions. The degree of branching of pyramidal neurons

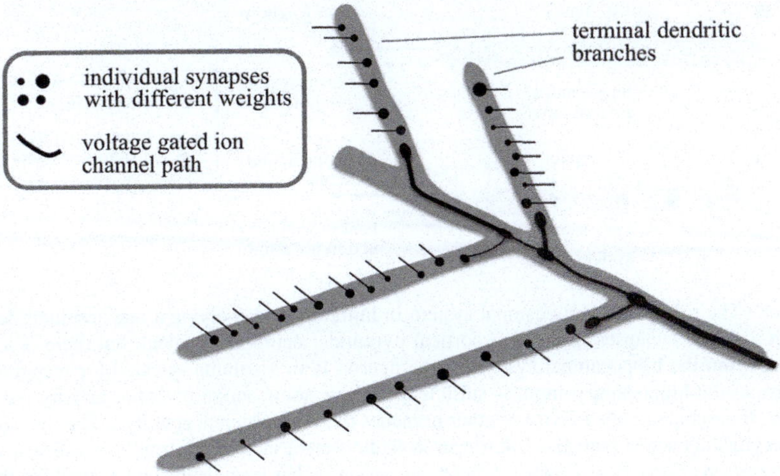

Fig. 8.7 Integration of terminal dendritic branch conditions. Synapses on a terminal branch define conditions, and detection of a condition is indicated by total postsynaptic potential in the branch being sufficient to initiate a dendritic action potential along a path defined by voltage gated ion channels and leading towards the soma. Detections of branch conditions are integrated at intersections in voltage gated ion channel paths, the contribution of all the currently detected branch conditions determining further propagation towards the soma

in different areas can differ significantly, reflecting different genetic biases on receptive field complexities.

Interneurons can target different parts of the neuron to limit the effect of inputs from different sources or to place frequency modulation on outputs. The information model for a neuron is that it defines and detects a receptive field made up of a group of similar conditions. The rate of firing of the pyramidal neuron indicates the degree to which the receptive field is present, i.e. the proportion of the conditions that define the field that is currently present. The detection of a branch condition can be viewed as a recommendation in favour of the detection of the dendritic receptive field, and the detection of a dendritic receptive field as a recommendation in favour of detection of the neuron receptive field. Interneuron inputs can be regarded as recommendations in favour or against receptive field detections at different points in time.

As illustrated in Fig. 8.7, individual conditions instantiated by different branches of a dendrite are integrated to determine whether the dendrite will inject potential into the soma. If the total postsynaptic potential injected by all the synapses on a branch reaches a threshold, a calcium action potential is launched down a path of voltage gated calcium channels and carried towards the soma. This action potential decays unless it is reinforced by action potentials entering the path from other branches. The concentration of voltage gated calcium channels along the path and at nodes where the influence of branches enters the path specify the integration algorithm. Voltage gated potassium channels inhibit passage of action potentials

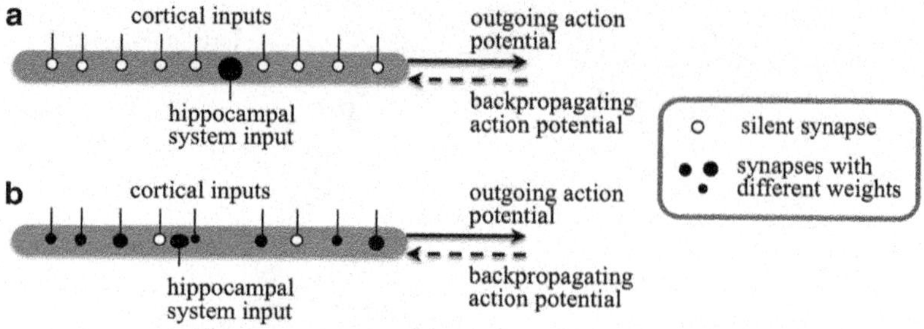

Fig. 8.8 The role of the hippocampal system in initiating and evolving a new terminal dendritic branch. (**a**) The synapses from other cortical pyramidal neurons are silent, but there is a strong synapse from the hippocampal system. If the neuron is in a column selected for receptive field expansion, the hippocampal input is sufficient to drive some action potential activity out of the branch. If combined with activity of other branches there is sufficient activity to fire the neuron, a backpropagating action potential will trigger synaptic weight increases in any of the silent synapses that recently receive inputs. Hence a new branch condition has been defined by a condition that was present in recent inputs. The branch condition can continue to expand by synaptic weight increases with the aid of hippocampal inputs. (**b**) Silent synapses that do not acquire weight in some period of time are disconnected, and silent synapses from other cortical neurons added if required

and therefore regulate the integration algorithm. Release of neurotransmitters like serotonin by the raphe nucleus and norepinephrine by the locus coeruleus regulate the opening probabilities of voltage gated ion channels and can therefore be viewed as recommendations in favour of or against receptive field detections in the targetted areas. Interneuron activity can also be viewed as recommendation strengths against receptive field detections at the time of action potential generation, although in some cases it can also provide recommendation strength in favour of receptive field detection out of phase with the GABAergic spikes. Activity of the pyramidal inputs to an interneuron can be regarded as recommendations against receptive field detections by the pyramidal neurons targetted by the interneuron.

8.2.5 Terminal Dendrite Branches

The role of the hippocampal system in the definition of receptive fields is illustrated for one branch in Fig. 8.8. A new condition can be initiated from a branch which has a number of silent synapses from cortical pyramidal neurons, plus strong synapses from the hippocampal system (Fig. 8.8a). The hippocampal system input is present only if receptive field expansion is required. In the absence of hippocampal system input the branch can make no contribution to firing the neuron. However, if the hippocampal system input is present, a calcium action potential may be initiated in the branch and propagate down the dendrite. If shortly afterwards the neuron fires, a backpropagating action potential from the soma will increase the weights of the silent synapses. Such an increase means that the branch in the future could contribute to the firing of the neuron independent of the presence of the hippocampal system inputs. In information terms a new condition has been added to the neuron receptive field, slightly expanding it. The same branch could continue to contribute to receptive field expansion with the aid of the hippocampal system inputs (Fig. 8.8b). One additional mechanism could be that hippocampal inputs target the base of a branch, increasing the chance that a relatively small branch output contributes to neuron firing. A further mechanism could be that hippocampal inputs target the proximal dendrite, increasing the probability that the dendrite contributes to neuron firing.

From information point of view an active hippocampal system input can therefore be regarded as a recommendation in favour of receptive field expansion. Such a recommendation will not be accepted unless supported by the recommendation strengths implicit in the presence of inputs to the cortical synapses on the same branch. Furthermore, for long-term synaptic weight changes there must also be recommendation strengths derived from the activity of other brain structures. For example, as discussed below the presence of activity at synapses derived from dopamine neurons is required for long-term synaptic weight changes and can be regarded as recommendation weights in favour of such changes.

8.2.6 Synapses and Neurochemistry

At an even more detailed level (see Fig. 4.7), an incoming action potential to a synapse injects a postsynaptic potential. The arrival of the action potential triggers opening of voltage gated calcium channels in the presynaptic region. Influx of calcium triggers release of glutamate within a vesicle into the synaptic cleft. This glutamate binds to AMPA channels in the postsynaptic neuron and these channels open to allow entry of sodium ions, implementing a postsynaptic depolarisation.

Glutamate also binds to NMDA channels in the postsynaptic neuron. However, these channels are blocked by magnesium ions and do not open just in response to glutamate binding. If while the glutamate is still bound, a backpropagating action potential arrives at the synapse, the potential removes the magnesium block permitting entry of calcium ions. These calcium ions trigger creation of additional AMPA channels, increasing the synaptic weight. To make the feedback mechanism more specific, the strength of backpropagation to a dendrite or branch needs to depend on the presence of an earlier forward propagating calcium action potential initiated by branch or dendrite activity.

Activation of NMDA receptors can therefore be regarded as recommendations in favour of a relatively short term increase in synaptic weight. Longer term increases need to be recommended by other inputs, leading to activation of genetic information. For example, as illustrated in Fig. 8.9, the addition of AMPA channels is implemented at a more detailed level by calcium entry triggering the activation of local mRNA that codes for AMPA channels. For long-term changes, input is required from other brain regions. A dopamine synapse may be located close to a

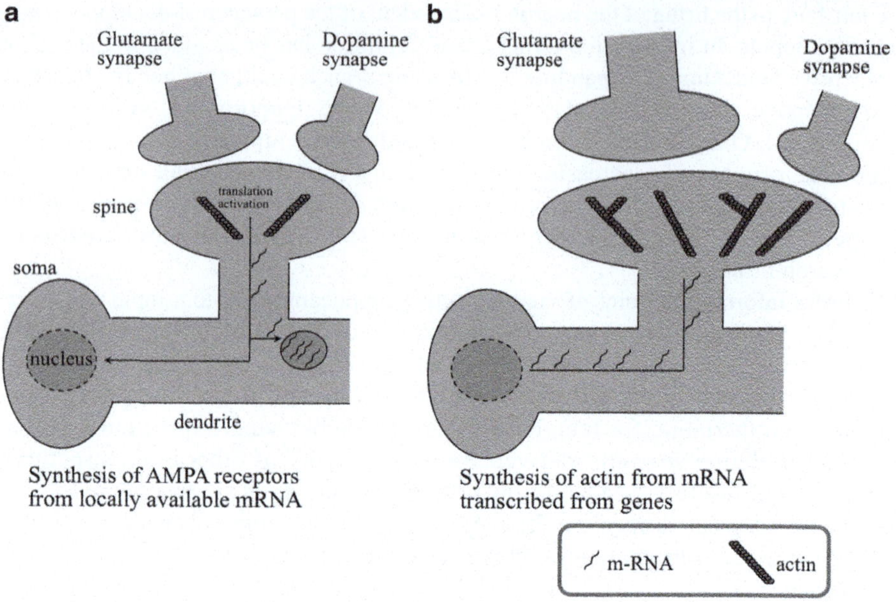

Fig. 8.9 Size increase of a synaptic spine to implement long term synaptic weight increase. (**a**) Moderate term synaptic weight increases are triggered by a combination of glutamate receptors and implemented by synthesis of AMPA receptors from mRNA available close to the spine. Such additional AMPA receptors will be recycled unless the spine size increases. (**b**) Longer term synaptic weight increases require recommendation by a later dopamine signal, which triggers transcription of mRNA from the *actin* gene, transport of the mRNA to the region of the spine and synthesis of actin that drive physical expansion of the spine. For more detail see Figs. 5.8 and 5.9

glutamate synapse. Action potentials arriving at the dopamine synapse result in synthesis in the nucleus of new mRNA that codes for actin molecules. This mRNA is transported to the synapse where the resultant actin synthesis causes expansion in synapse size supporting long-term weight change. In information terms the dopaminergic input from another brain region is a recommendation strength in favour of such long-term change.

Glutamate, dopamine, GABA, acetylcholine, serotonin and norepinephrine all influence LTP. Glycine, ATP, zinc, neurotrophins and neuropeptides also affect LTP. Entry of Ca^{++} ions by voltage gated channels can also influence LTP. Hence many other brain regions acting via other neurotransmitters may contribute recommendation strengths for or against synaptic weight changes and duration of such changes in one neuron, or even one synapse of one neuron.

At an even more detailed chemical process level each recommendation is transmitted by interacting kinase cascades as illustrated in Fig. 8.10. Each kinase cascade can be regarded as composed of more detailed recommendations in favour of activation of genetic information. Recommendations at this detailed chemical level add together to determine if such activation will take place.

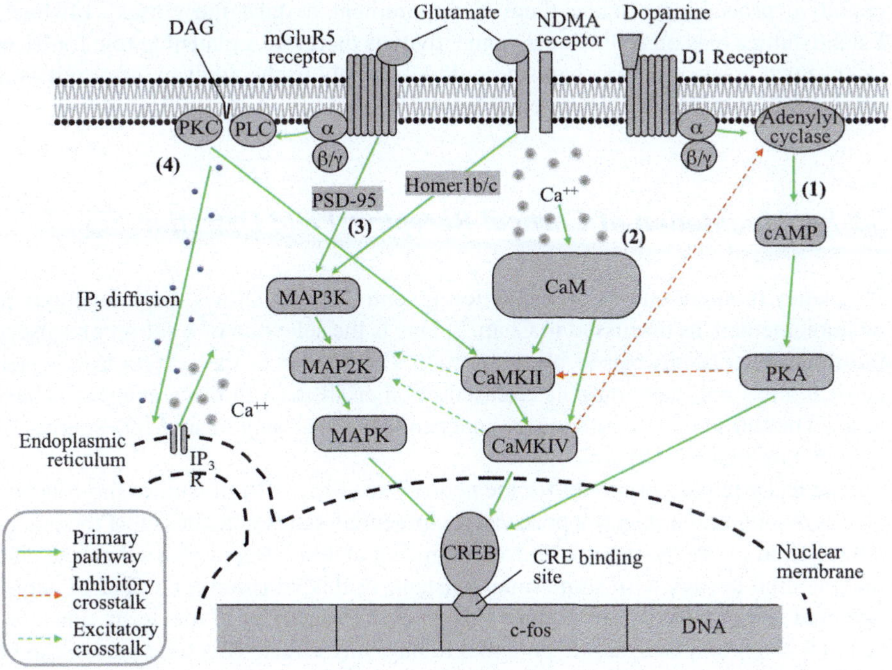

Fig. 8.10 Different kinase cascades carry recommendation strengths in favour and against transcription of different genetic information. Transcription occurs in response to sufficient recommendation strength, and in some cases transcription rate can be higher with greater total recommendation strength

8.2.7 REM Sleep and Configuration of Provisional Connectivity

One issue with the receptive field expansion mechanism is the initial configuration of dendritic branches and addition of new inputs to ongoing branches. Creation of a new branch with a range of silent or very weak provisional synapses from input pyramidal neurons requires biological resources. Random selection of the sources of these synapses could result in many that are irrelevant. A way to reduce the resource cost would be to select the sources of the silent synapses on the basis of past simultaneous activity by the source and target neuron. Recent simultaneous activity would indicate a somewhat higher chance of being relevant than simultaneous activity in the more distant past, but such more remote activity would still have some chance of being relevant. A way to achieve such a selection bias would be to perform a fast rerun of past cortical activity with a bias in favour of the more recent, and establish provisional connections on the basis of correlated activity. Such a rerun would avoid the need to continuously record separate information about simultaneous activity. However the rerun would severely interfere with current cognitive processing. Hence there is requirement to take the cortex "off-line". Establishing a bias on provisional connectivity is therefore a plausible role for REM sleep. Such a selection mechanism could also apply to the addition of provisional synapses to already established branches.

8.2.8 Modulation of Cortical Receptive Field Detection Signals

Frequency modulation of cortical action potential spikes allows three functions to be implemented, as discussed in Chap. 7. One is the selection of a subset of sensory information to be processed by the cortex. The second is the selection of which cortical areas will have outputs released to other areas or to drive external behaviours. The third is the simultaneous independent processing of multiple sources of information in the same area.

The leaky integrator model for the neuron (see Fig. 3.10) illustrates the concept. One action potential injects a postsynaptic potential that is much less that the potential required to fire the neuron or even to inject potential deeper into a dendrite, and the potential decays with some time constant. In this situation a number of action potential spikes must occur within a time period of the order of the decay time constant for the total postsynaptic potential to have any effect. This time period can be called an integration window. If, as shown in Fig. 8.11, the source of inputs to a pyramidal neuron are unmodulated and occur at a relatively low frequency, the threshold for producing further activity in their target may not be reached. However, if a frequency modulation is imposed on those input sources, input spikes tend to occur close to the peak signal. As a result, there is a much higher probability of a target response. Hence imposing a frequency modulation on a subset of sensory inputs, such as a specific region of the retina, means that those inputs will generate firing in the target cortical area. Inputs derived from other retinal regions will generate a much smaller response.

An additional frequency modulation at a different frequency but phase locked to the first frequency can further concentrate the action potentials in time, and allow an additional criterion for selection of the inputs. In the cortex, the gamma band reflects the presence of an imposed modulation managing information release, and the theta band reflects imposition of an additional modulation managing receptive field expansions.

The cortex performs very behaviourally complex processing, but the results of cortical processing can only be interpreted into actual behaviours by subcortical structures that directly drive behaviours. These structures use consequence feedback to manage the recommendation strengths of cortical outputs. The presence of consequence feedback means that subcortical structures cannot perform behaviourally complex processing. Hence subcortical structures in general can perform processing to select the cortical information that will be allowed to drive behaviour, but cannot change that information in complex ways. These structures must interpret cortical outputs as predominantly recommending one behaviour, and implement that behaviour by releasing a subset of cortical outputs to drive the behaviour.

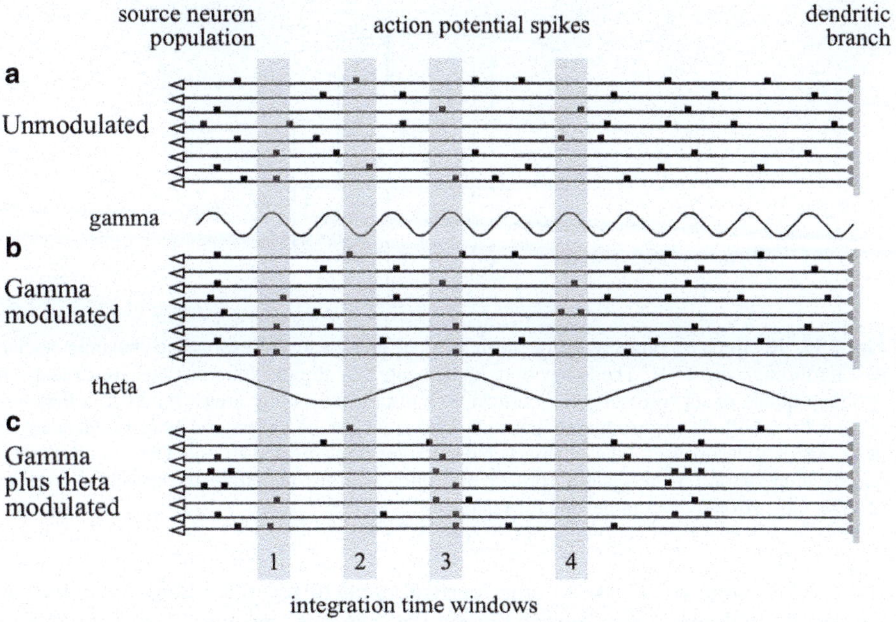

Fig. 8.11 Different modulations of an input population of action potential spikes to different synapses on one dendritic branch. Action potentials generated over a period of about 250 ms by a source neuron population that targets one dendritic branch are shown. Integration windows at different points are *shaded*. (**a**) The output action potentials from the source neuron population are randomly distributed in time. As a result, the number of action potentials in any one integration window is small, and the total postsynaptic potential injected into the branch is less likely to reach the threshold for dendritic action potential initiation towards the soma of the target neuron. (**b**) The action potentials are modulated at the gamma frequency (40 Hz). Action potentials tend to be concentrated close to the peaks in the modulation frequency. Integration windows near these peaks have above average action potentials generation of a branch output is more likely. (**c**) The action potentials are modulated at the gamma frequency (40 Hz) and the theta frequency (10 Hz). Action potentials tend to be concentrated where the peaks in the two modulation frequencies coincide. Action potentials are very concentrated in integration windows near these combined peaks and a strong branch reaction is likely

8.3 Information Model for the Thalamus

The thalamus performs the detailed selection of the exact combination of cortical outputs to be released and implements the selected releases by imposing frequency modulations. As illustrated in Fig. 8.12 for two example cortical areas, there is reciprocal excitatory connectivity between a thalamic nucleus and an area. Figure 8.13 illustrates this connectivity in more detail. One thalamocortical projection neuron targets many cortical pyramidal neurons and interneurons, and receives inputs from many cortical pyramidal neurons. The synaptic strength of a cortical pyramidal neuron on to a projection neuron can be interpreted as a recommendation

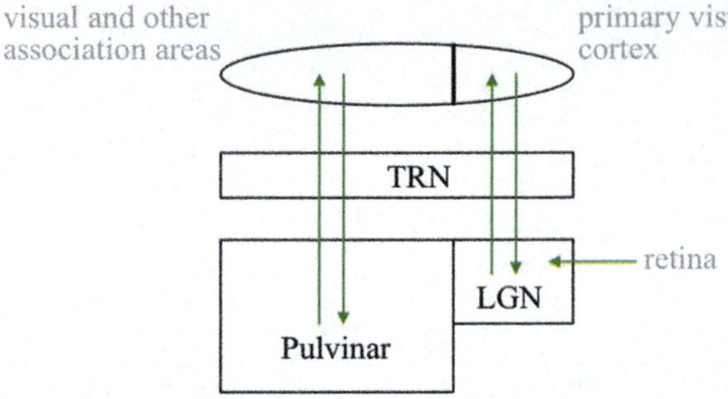

Fig. 8.12 Reciprocal connectivity between cortical areas and corresponding thalamic nuclei, passing through the TRN. The behaviours implemented by the thalamic nucleus are releases of specific outputs of its corresponding cortical area to targets in other areas. Inputs to a thalamic nucleus from both its corresponding area and other areas can be understood as recommendations in favour of such releases. The release behaviours are implemented by the TRN placing a frequency modulation on the thalamocortical projection neurons targetting the groups of cortical neurons most strongly recommended for release

strength in favour of release to other cortical areas of receptive field detections by the specific columns containing the pyramidal neurons targetted by that projection neuron. As discussed below, such release recommendations are blocked by inhibitory input from the basal ganglia unless a general behaviour of release from the cortical area containing the columns has been selected by the basal ganglia.

The thalamic synaptic weights that were active on recently active thalamocortical neurons correspond with the recommendation weights in favour of the recently accepted release behaviour. If a release recommendation is accepted and followed by a positive reward recommendation, these synaptic weights are increased, increasing the probability of acceptance of similar behaviours in similar circumstances in the future. As discussed in the basal ganglia section, reward behaviours are implemented by activation of dopaminergic neurons. These dopaminergic neurons target the thalamus, triggering long term LTP in the appropriate synapses.

If the release behaviour is selected by the basal ganglia, a group of thalamocortical neurons will be disinhibited and there will be strong activity in the axons passing both ways between the cortex and the thalamus. These axons also activate interneurons in the TRN, which tend to fire at frequencies used for modulation. These TRN interneurons target thalamic projection neurons, imposing a frequency modulation on the action potential outputs of those neurons. Basket cell interneurons in layer IV of the cortex tend to fire at the modulation frequencies, and these basket cells are interconnected in a network in a way that tends to synchronize their firing into the same phase of modulation. Thalamic projection neurons excite these interneuron networks. The TRN imposed modulation is therefore enhanced in the interneuron network, and the enhanced modulation imposed on the active pyramidal neurons.

To prevent excessive release of information, there is feedback from the cortex to the thalamic nucleus, targetting inhibitory microneurons. These microneurons target projection neurons in the same thalamic nucleus, limiting the thalamocortical

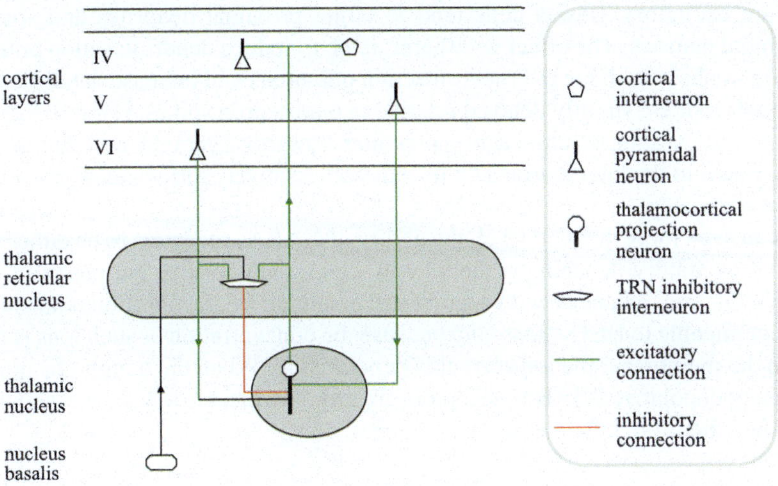

Fig. 8.13 Management of gamma and theta frequency modulations. Outputs from pyramidal neurons in cortical layers V/VI that are received by thalamocortical projection neurons can be interpreted as recommendations in favour of releases of different groups of layer V/VI outputs to other areas. Additional recommendations for such releases come from other cortical areas. Cross inhibition between thalamocortical projection neurons in one nucleus via interneurons determines the strongest recommendations. A thalamocortical projection neuron targets a group of layer IV pyramidal neurons and if its outputs are frequency modulated, the modulation will tend to propagate through the cortical layers to the outputs. Thalamic reticular nucleus (TRN) neurons tend to fire at the gamma modulation frequency, and are excited by axons passing between the nucleus and the cortex in both directions. Hence the thalamocortical projection neurons with the strongest activity will be modulated at the gamma band frequency, implementing the selected release behaviour. The nucleus basilis in the basal forebrain via a combination of excitatory and inhibitory outputs superimposes a theta band frequency, generating even stronger activity. Recommendations in favour of this theta band behaviour are received by the nucleus basilis from the frontal cortex

activity. This prevention of excessive release can equivalently be viewed as a competition to narrow the releases to those most strongly recommended.

The basal forebrain imposes a theta modulation at a frequency that is an integral fraction of the current gamma frequency (see Fig. 8.11). Part of the way in which this is implemented is by the basal forebrain targetting the TRN interneurons. As discussed below, the information model for this theta modulation is to ensure the correct relative timing of releases with respect to other brain events.

At a more detailed level, the ion channel properties of interneurons are such that they tend to burst fire at the gamma frequency under steady excitatory stimulus. The behaviourally important information is carried by pyramidal neuron outputs, and the inhibitory signals must not distort that information by excessive reduction of pyramidal outputs. It could be that one way in which the distortion is reduced is that the $GABA_A$ receptors supporting frequency modulation can be either inhibitive or excitatory depending on local membrane potential. When the membrane is depolarised from resting potential these receptors are inhibitory, but at resting potential they are excitatory. Basket cells tend to target proximal dendrites and somas of pyramidal neurons. The effect will therefore be to reduce dendritic action potentials that are in phase with the interneuron action potentials and increase dendritic action potentials that are slightly delayed relative to interneuron spikes. Overall, there will be a shift of pyramidal output action potentials towards a gamma modulation, opposite in phase to the interneuron activity, but with relatively little reduction in average output frequency.

At an even more detailed level, the ability of $GABA_A$ receptors to be either inhibitory or excitatory depends on the way in which their relative permeability to Cl^- and HCO_3^- ions changes with membrane potential. At resting potential they are more permeable to HCO_3^- ions, and because the concentration of such ions is higher inside the membrane, the ion current is depolarising. When the membrane potential is already depolarised $GABA_A$ receptors are more permeable to Cl^- ions, and the ion current is therefore hyperpolarising.

8.4 Information Model for the Basal Ganglia

As illustrated in Fig. 8.14, the connectivity of the basal ganglia implements the selection of the most appropriate type of behaviour at each point in time. Its selection is communicated to the thalamus which then determines the very specific behaviour within the selected type. For example, the basal ganglia might select an arm movement and the thalamus the particular combination of muscle movements to implement the arm movement. Each current receptive field detection by the cortex is interpreted in the striatum as a recommendation in favour of a range of different behaviours. D1 population neurons in the striatum determine the total recommendation weights in favour of each behaviour, and communicated those totals to the GPi/SNr/vVP by the direct path. D2 population neurons determine the total recommendation weights against all but each one behaviour, and also communicate those totals to the GPi/SNr/vVP by the indirect path. There is a competition within the GPi/SNr/vVP to determine the most strongly recommended behaviour, and that behaviour is communicated from the GPi/SNr/vVP to the thalamus, and the most strongly recommended specific behaviour of the type implemented by the thalamus into the cortex. One path from the midbrain dopamine neurons to the striatum regulates the balance between the direct and indirect paths to prevent the selection of multiple incompatible behaviours and to ensure that some behaviour is selected in most circumstances. In novel situations there may be weaker total recommendation weights, and the hippocampal system detects the presence of novelty, recommends an increase on the probability of a behaviour selection, which is implemented via the striatum and the midbrain dopamine neurons. The second path from the midbrain dopamine neurons to the striatum adjusts recently accepted recommendation weights on the basis of consequence feedback. These adjustments increase the probability of the same behaviour being accepted in similar circumstances in the future if the consequence feedback was positive, and decrease the

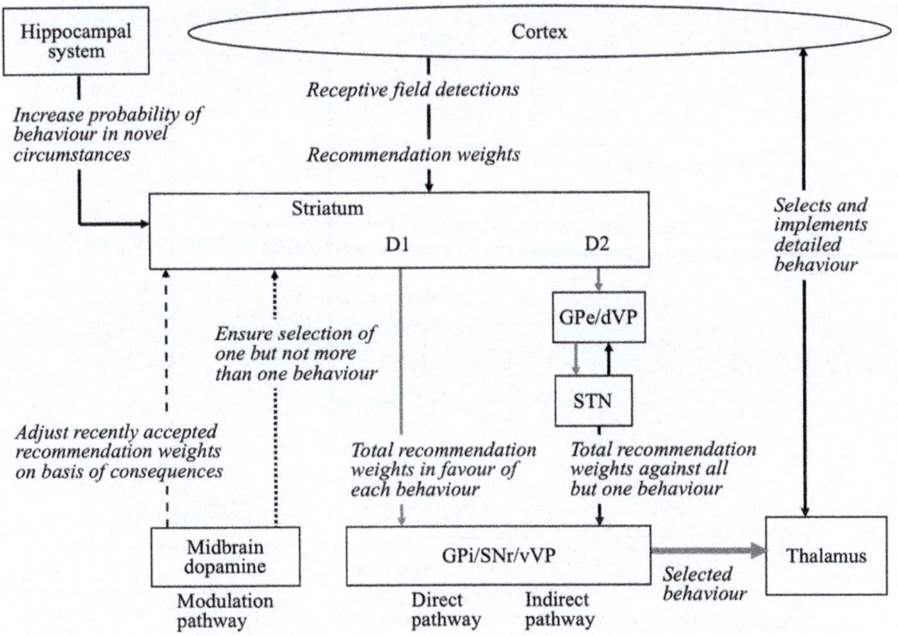

Fig. 8.14 Information interpretation of basal ganglia connectivity. The striatum receives recommendations from the cortex in favour of a wide range of different behaviours, with recommendation weights instantiated by the weights of corticostriatal synapses. A striatal neuron output indicates the total current recommendation weight in favour of the behaviour or type of behaviour corresponding with that neuron. A competition over the direct and indirect pathways into the GPi/SNr/vVP identifies the most strongly recommended behaviour type, corresponding with the GPi/SNr/vVP neurons that have their activity reduced. This reduced activity reduces inhibition of a region of the thalamus, implementing the most strongly recommended detailed behaviour of the type selected. A feedback loop from the midbrain dopamine neurons to the striatum adjusts the background dopamine level in the striatum to ensure that there are no selections of multiple inconsistent behaviours. Inputs from the hippocampal system to the striatum indicate the degree of novelty in the current environment, and recommend exploratory behaviours. Some striatal neurons correspond with positive reward behaviours, which increase the recommendation weights of recently accepted behaviours. If there is sufficient cortical recommendation weight in favour of such a reward behaviour, it is implemented by the striatal neuron causing burst firing of dopamine neurons targetting the striatum. Acronyms: *GPi* globus pallidus internal segment, *GPe* globus pallidus external segment, *SNr* substantia nigra pars reticulata, *SNc* substantia nigra pars compacta, *STN* subthalamic nucleus

probability following negative feedback. There are also limits on increase to ensure that the behaviour of one behaviour does not become excessive.

The implementation of these information processes is shown in more detail for the dorsal striatum in Fig. 8.15. In both the striatum and the GPi/SNr, neurons and small groups of neurons correspond with behaviours. One cortical pyramidal neuron targets a large number of striatal projection neurons, making one or perhaps a very small number of synapses on each projection neuron. This connectivity reflects

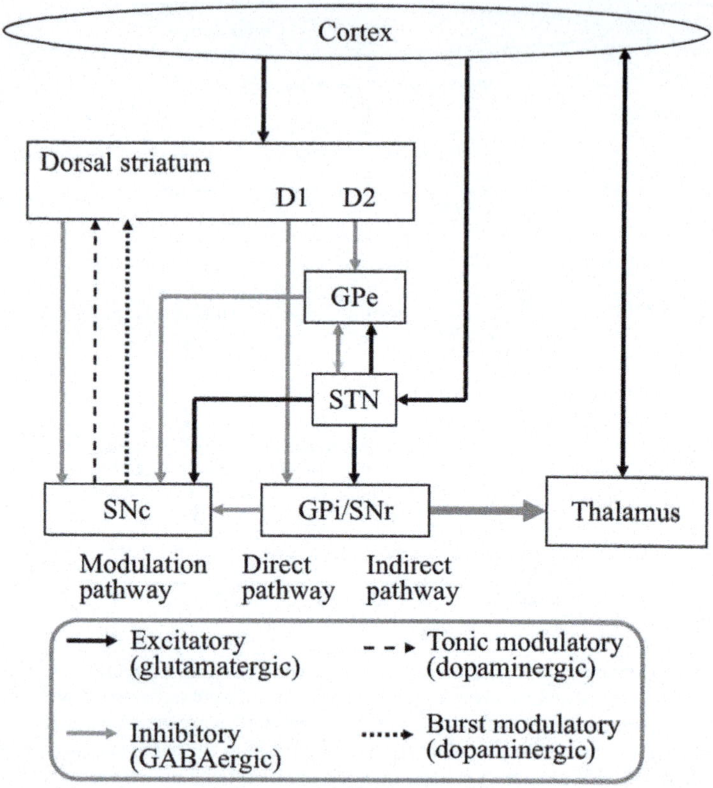

Fig. 8.15 Implementation of behaviour selection information processes in the basal ganglia. Excitatory cortical inputs to the striatum result in activity of striatal neurons corresponding with strongly recommended behaviours. Activity of a D1 neuron is interpreted in the GPi/SNr as the total recommendation strength in favour of the corresponding behaviour. Activity of a D2 neuron is interpreted in the GPi/SNr as a total recommendation strength against any behaviour except the corresponding behaviour. If the ratio of direct/indirect pathway activity is too high, no behaviour will be selected. If the ratio is too low, multiple behaviours will be selected. The ratio is adjusted by inputs to the SNc from the GPe, STN and GPi/SNr. These inputs regulate the number of SNc dopamine neurons contributing to the background dopamine concentration in the striatum, thus adjusting the relative activity of D1 and D2 neuron populations

the recommendation strengths possessed by one cortical output in favour of many different behaviours. For one type of behaviour, some cortical areas have receptive field complexities that are most effective for recommending the type, but many areas may have some relevant recommendation strengths. Hence one region of the striatum receives a high volume of inputs from a small number of cortical areas, and some inputs from a rather larger number of areas.

Corticostriatal synapses are glutamatergic and are located on spines with weights that are independently adjustable. These synaptic weights implement the recommendation weight of the cortical receptive field detection in favour of the behaviour corresponding with the projection neuron. Hence the output firing of the striatal neuron corresponds with the total weight in favour of that behaviour. GPi/SNr neurons are inhibitory and fire tonally targetting the thalamus. Hence they constantly inhibit the release of cortical information by the thalamus. A direct path D1 striatal neuron inhibits the GPi/SNr neuron corresponding with the same behaviour, reducing its activity and therefore tending to result in implementation of its behaviour via the thalamus. In the indirect path, STN neurons are glutamatergic and fire tonically. They therefore provide excitatory input to GPi/SNr neurons which inhibits the implementation of behaviour. STN neurons are also connected reciprocally with inhibitory GPe neurons. An indirect path D2 striatal neuron corresponding with one behaviour inhibits GPe neurons and therefore increases the excitatory activity of the STN neurons targetting GPi/SNr neurons which correspond with different behaviours. The output of a striatal D2 neuron thus corresponds with a total recommendation weight against any behaviour except its corresponding behaviour. Thus a D1 striatal output targets a small group of GPi/SNr neurons, while an STN output targets a large number of GPi/SNr neurons.

In the absence of input, dopaminergic SNc neurons fire tonically at low frequency. The proportion of the population of neurons firing in this mode is regulated by inputs from the indirect path including the STN and the GPe. The tonic firing results in a background concentration of dopamine in the striatum, which predominantly inhibits D2 neurons and to a small degree excites D1 neurons. Excess activity in the indirect path tends to prevent any behaviour, and the feedback loop to the background dopamine level in the striatum tends to ensure that a behaviour will be selected in most circumstances.

Some striatal projection neurons correspond with reward behaviours. Cortical inputs to such a neuron recommend the reward behaviour, which is to adjust the weights of other recommendation weights that were recently utilized. If there is sufficient total recommendation weight the striatal neuron fires, and inhibits interneurons or other sources of inhibition to dopaminergic neurons, which as a result produce burst firing. When a striatal neuron fires strongly, there is a short term LTP in its recent synaptic weights that decays in a period ~ minutes. The type of dopamine concentration generated by the burst firing results in longer term persistence of the synaptic weight changes, implementing longer term recommendation weight changes. The habenula detects negative consequences and absence of an expected reward. This information is provided to the midbrain dopamine neurons which manage recommendation strengths in the striatum.

Dopaminergic neurons regulate the competition between different behaviours, and reward recently selected behaviours. One dopaminergic neuron therefore targets very large numbers of striatal neurons corresponding with different behaviours.

In the ventral basal ganglia there is analogous circuitry supporting similar information processes. However, the recommended behaviours tend to be more strategic. This includes selection of general types of behaviour, and selection of rewards for such behaviours. As discussed below, the amygdala detects receptive fields appropriate for recommending different general types of behaviour, and therefore provides input to the ventral striatum. Also as discussed below, the hippocampal system detects novelty in current circumstances. The hippocampal receptive field detections corresponding with novelty have recommendation strengths into the striatum to increase the probability that general behaviours of some types (such as exploratory) will occur. This increase in probability is itself a behaviour corresponding with ventral striatal neurons, but in this case the behaviour is implemented by these striatal neurons targetting interneurons in the SNc, increasing the proportion of dopaminergic neurons that are firing tonically and therefore increasing behaviour probability.

Selection of some general types of behaviour may be compatible with simultaneous selection of other behaviour types. For example, a body movement, head movement reflecting attention, thinking behaviour and reward behaviour may all occur at the same time. This capability is reflected in the parallel cortex-basal ganglia-thalamus-cortex paths through different subnuclei illustrated in Fig. 6.13 in Chap. 6. In that figure, a motor behaviour could be selected by the motor path, an attention behaviour by the oculomotor path, different thinking behaviours by the dorsolateral prefrontal and lateral orbitofrontal paths, and a reward behaviour by the anterior cingulate path.

As discussed in Chap. 7, an important constraint is imposed by the need to manage detailed rewards that affect detailed behaviours, tactical rewards that affect tactical behaviours and detailed rewards, and strategic rewards that affect strategic behaviours and tactical rewards. The aspect of the basal ganglia circuitry addressing this management is illustrated in Fig. 8.16.

Striatal outputs corresponding with reward behaviours originate in the striosomes and target the SNc, while outputs into both the direct and indirect pathways corresponding with other behaviours are located in the matrisomes. Thus as described in Chap. 6, different matrisomes correspond with different regular behaviour types, while damage to striosomes can result in general behavioural disturbances such as compulsive behaviour. If dopaminergic neuron damage results in failure to maintain an adequate level of background dopamine in the striatum, the result will be the D2 pathway overwhelming the D1 pathway. Such a result will mean loss of ability to initiate any behaviours, as observed in Parkinson's disease. On the other hand, if the indirect path is damaged, the result will be a failure to reduce selections to a single behaviour. Outputs corresponding with multiple incompatible behaviours such as different motions of the same muscles will be released from the cortex. Such incompatible behaviours cannot be implemented in the muscles, so an actual movement will be initiated in the spinal cord corresponding with a momentary signal predominance. The continual presence of signals corresponding with other movements means that the current movement may be constantly interrupted by random shifts in the mix of signals. The result will be intrusion of extraneous body movements as observed in Hemiballism. Similarly, excess dopamine production in the striatum would lead to the behavioural tics and jumps observed in Tourette's Syndrome.

The mapping of reward behaviours from higher to more detailed levels in illustrated in Fig. 8.17. As illustrated in the figure, a monkey can be trained to interpret a visual image as an instruction to perform a specific task in the near future, and to perform the task when a trigger stimulus is heard. During training a successful

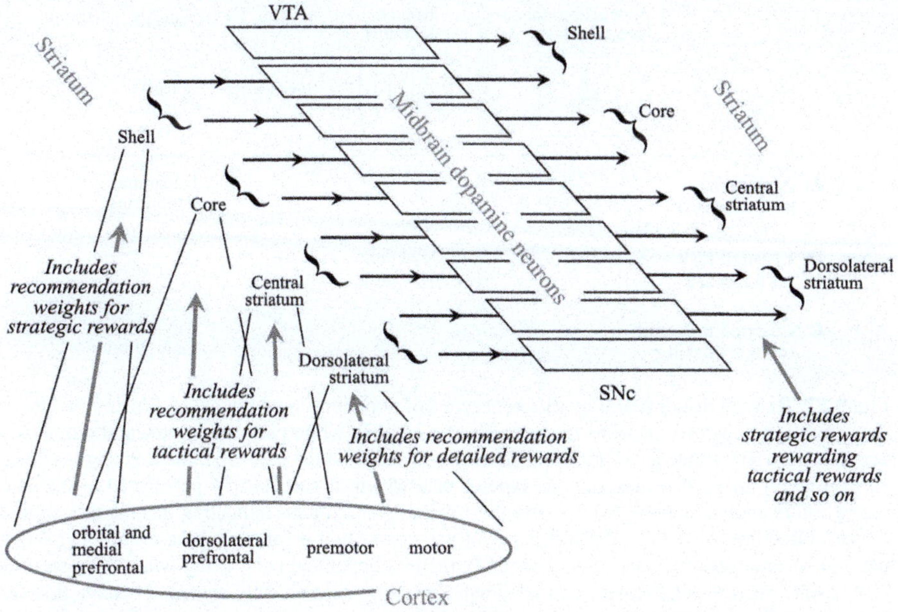

Fig. 8.16 Management the rewarding of strategic, tactical and detailed behaviours. A positive reward following a strategic behaviour is not a reliable guide to rewarding all the tactical behaviours that were performed in the course of the strategic behaviour. However, it is an appropriate guide to rewarding the reward behaviours that rewarded the tactical behaviours. Different parts of the cortex tend to recommend strategic, tactical or detailed behaviours, including rewards. Different parts of the striatum determine the predominant recommendation strengths for these different levels of behaviour. Reward behaviours are implemented by the striatal neuron activating dopaminergic neurons that in turn target the striatum with burst firing. Outputs from dopaminergic regions implementing more strategic rewards therefore tend to target striatal regions corresponding with somewhat less strategic behaviours

performance of the task is followed by a food reward. After training, absence of the reward reduces the chance of future successful performance. It is observed that during learning, a burst of dopamine firing follows the reward. However, once learning is achieved, there is a burst of dopamine firing following the stimulus but not following the reward. If the reward is not given at the appropriate time, there is a dip in dopamine activity.

During learning, cortical receptive fields are detected within the instruction image and trigger stimulus. These receptive fields have various recommendation strengths in the striatum in favour of different behaviours including the target behaviour. The most strongly recommended behaviour is selected, corresponding with the most strongly activated striatal neuron(s). LTP lasting ~ an hour occurs to synaptic strengths of recently active synapses on this neuron. If the behaviour is followed by a reward, cortical receptive fields correlating with a reward are detected. These receptive fields have recommendation strengths in favour of burst firing dopamine neurons. This burst firing prolongs the LTP induced earlier. Hence the recommendation strengths

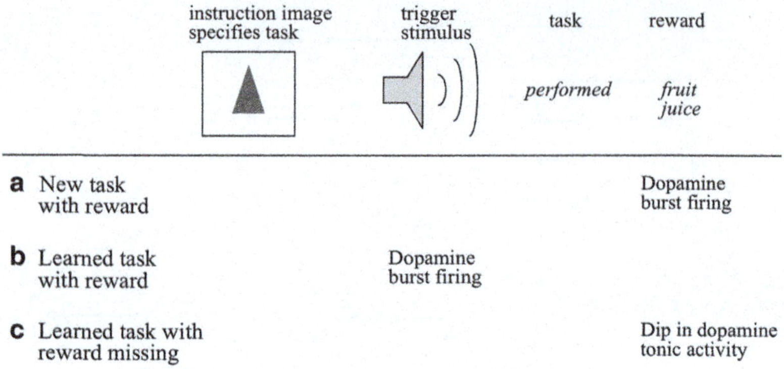

Fig. 8.17 Reward management by different types of dopamine signalling. The illustration shows the activity of dopamine neurons in the brain of a monkey while performing new and previously learned tasks. The sequence of task related events is shown at the *top*. First, the monkey is shown an instruction image that indicates the type of task which, if successfully performed in the near future will result in a reward. A little later there is a trigger stimulus indicating that the task should be performed immediately. If the task is performed correctly it is followed by a reward. (**a**) When the task is unfamiliar (i.e. learning is needed), there is dopamine burst firing when the reward is given which increases the corticostriatal synaptic weights into recently accepted behaviours. (**b**) When the task is familiar, there is a burst of dopamine firing at the time of the trigger stimulus, but no changes at the time of the reward. The burst firing does not follow a behaviour, so no behaviour will be rewarded. However, the burst firing will slightly increases the background dopamine, increasing the chance that the most strongly recommended behaviour in the near future will be implemented (rather than no behaviour). (**c**) When the task is familiar, but no reward is provided at the appropriate time, the tonic firing of dopamine neurons decreases briefly. This decreased dopamine results in reduction of the corticostriatal synaptic weights into the recently accepted behaviour

of the receptive fields detected within instruction image and stimulus in favour of the target behaviour have increased long term, increasing the probability of the same behaviour in similar circumstances in the future.

After learning, when the reward continues to follow the behaviour, further increases in recommendation weights would generally be inappropriate. The reason for this is that all the receptive fields in the group also recommend many other behaviours which are appropriate following detection of different groups of receptive fields including some from this group. An excessive growth in the recommendation strengths of these fields in favour of the one behaviour could result in the behaviour being selected in inappropriate circumstances.

No behaviour is performed prior to the trigger stimulus, and the burst firing following that stimulus will therefore have no effect on recommendation weights. However, such burst firing will slightly increase background dopamine. The behavioural value of this increase is that it increases the chance of a behaviour being selected in the near future by adjusting the balance between the direct and indirect paths. The target behaviour is being recommended by receptive field detections, and the background dopamine increase triggers immediate acceptance of that behaviour.

If the reward is missing, detection of this circumstance results in activity in the habenula, resulting in a decrease in background dopamine. Such a decrease triggers LTD in recently active synapses on the recently active striatal neuron. Recommendation strengths of recently detected receptive fields in favour of the behaviour are therefore reduced.

8.5 Information Model for the Amygdala and Hypothalamus

The information model for the amygdala and hypothalamus is management of the selection of types of behaviour. For example, in some circumstances food seeking behaviours are urgently needed, and a bias needs to be placed to favour such behaviours. In other circumstances, aggressive or avoidance behaviours may be urgent. Emotions are thus the manifestation of current behavioural priorities, placing a bias upon the selection of general behaviour type as indicated in Table 8.1. In some cases the external communication of the current bias can have behavioural value, and the existence of the bias influences facial expression and body language. If there is bias on more than one behaviour type at the same time, complex emotions that are combinations of other emotions will occur, such as contempt resulting from simultaneous bias in favour of disgust and anger.

In some cases, circumstances and/or behaviours are very strongly influenced by genetic programming, in which case there is a tendency for processing to occur in the hypothalamus. When there is a larger learning from experience component, there is a tendency for processing to occur in the amygdala. There is therefore a need for close coordination between these two structures. Furthermore, receptive fields discriminating between very complex circumstances in which different biases are appropriate can be defined in the prefrontal cortex. These receptive fields acquire recommendation strengths in favour of different types either directly into the basal ganglia or via the amygdala.

Table 8.1 Emotions and corresponding behavioural biases

Emotion	Favoured behaviour type
Hunger	Food seeking
Thirst	Drink seeking
Sexual desire	Mate seeking
Panic	Random behaviour selection
Anger	Aggressive
Disgust	Rejection
Fear	Avoidance
Happiness	Continuation of recent types
Pride	Assertion of social dominance
Sadness	Delay behaviours because of radical social change
Surprise	Delay behaviours because of novel environment

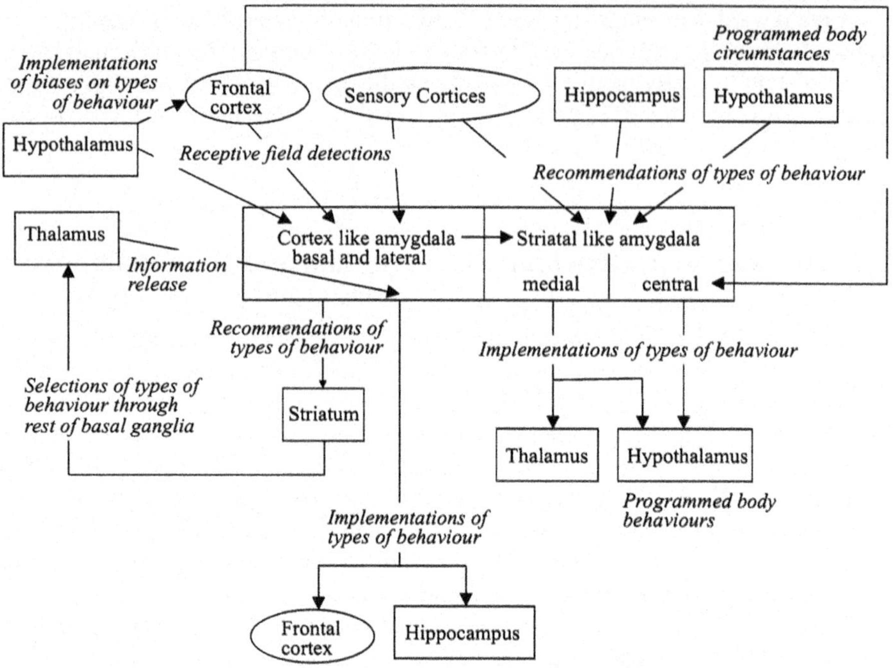

Fig. 8.18 Information model for amygdala and hypothalamus. The cortex-like amygdala receives inputs from sensory cortices, frontal cortex and amygdala, and within these inputs detects receptive fields which can be interpreted as recommendations in favour of different general types of behaviour. The striatal-like amygdala receives receptive field detections from the cortex-like amygdala, the sensory cortices, and the hippocampus, and detections of body circumstances from the hypothalamus. The striatal-like amygdala interprets these inputs as recommendations in favour of different general types of behaviour, determines the most strongly recommended, and implements that behaviour by outputs targetting the thalamus and hypothalamus. Outputs from the cortex like amygdala can also target the striatum, influencing behaviour selections through the thalamus. The thalamus can implement behaviours as releases of cortex-like amygdala outputs to the frontal cortex and hippocampus

Figure 8.18 illustrates the high level information model. The cortex-like amygdala receives receptive field detections from the sensory cortices and the frontal cortex. These receptive field detections are at levels of complexity useful as components in amygdala receptive fields able to discriminate between circumstances in which different general types of behaviour are appropriate. Receptive field detections by the cortex-like amygdala are interpreted in the ventral striatum as recommendations in favour of general types of behaviour. Such general types could be aggressive, fearful etc. but also increases in the degree of cortical information recording or strategic reward behaviours. The dorsal striatum interprets amygdala outputs as recommendations in favour of somewhat more specific types of behaviour. The striatum implements the selection of a behaviour type by increasing the probability of detection of frontal cortex receptive fields with recommendation strengths of the selected behaviour type. This implementation is achieved by release of amygdala receptive field detections to the frontal cortex, which reduce the proportion of regular components required for detection of the frontal cortex receptive fields. The behaviour of increasing the degree of information recording in the cortex is implemented by release of amygdala receptive field detections to the hippocampal system as discussed further below.

For general types of behaviour triggered by more genetically programmed circumstances or circumstances defined largely by information derived from internal body status, the striatal-like amygdala receives information from the hypothalamus, but also from the sensory cortices and hippocampal system. This information is interpreted as recommendations in favour of different general types of behaviour, and the striatal-like amygdala determines and implements the most strongly recommended behaviour. Implementation may be by triggering a programmed group of behaviours in the hypothalamus (such as the fear or anger response), or by releasing cortical information via the thalamus. Note that for the limited range of more genetically programmed behaviours, the more complex basal ganglia structure in not used. However, the separation between condition definition and detection and behaviour selection is maintained.

As an example of a more detailed level of description, consider the interactions in the rat between cortex, amygdala and hypothalamus in modulating fear behaviours by maternal behaviours as discussed in Chap. 6 (see Fig. 6.18). Receptive fields detected in the cortex correlate with circumstances in which fearful behaviours are appropriate. These receptive fields are communicated to the cortex-like amygdala, and contribute to detections of receptive fields in the amygdala. These amygdala receptive fields recommend fearful behaviours into the striatal-like amygdala. Receptive fields in the sensory thalamic nuclei correlate with the presence of fearful objects and have similar recommendation strengths into the striatal-like amygdala. If there is sufficient total recommendation strength, appropriate behaviours (such as cardiovascular and freezing responses) are implemented into the brainstem. However, if receptive fields correlating with circumstances in which behaviours protective of offspring (maternal behaviours) are detected in the prefrontal cortex, these receptive fields recommend those type of behaviours into the paraventricular nucleus of the hypothalamus. If there is sufficient such recommendation strength, excitatory oxytocin outputs from the paraventricular nucleus to the striatal-like amygdala inhibit the fear responses, and outputs to the prefrontal cortex increase the probability of more detailed recommendations in favour of maternal behaviours. Hence freezing behaviour in response to a fearful object becomes less likely and some protective behaviour more likely.

At a somewhat deeper level of description, oxytocin acts upon G_q type G-protein receptors in target neurons, and G_q receptors can increase activation of their target neurons by opening voltage dependent calcium channels and closing potassium channels. Hence oxytocin neuron activity reduces the thresholds for receptive field detection by neurons in the prefrontal cortex recommending maternal behaviours, making more recommendation strengths in favour of this type of behaviour available. Oxytocin neuron activity in the paraventricular nucleus of the hypothalamus.

Oxytocin neuron activity increases inhibitive neuron activity in the CeL nucleus of the striatal-like amygdala. This increased inhibitive activity reduces outputs from the CeM nucleus of the striatal-like amygdala that drive fear responses.

8.6 Information Model for the Hippocampal System

The primary role of the hippocampal system is to manage cortical receptive field expansions. Such expansions ensure that at least a minimum number of receptive field detections occur in response to every cortical input state, in turn ensuring that an adequate range of behavioural recommendations is always available. Because there is a degree of novelty in almost every situation, there will generally be some degree of receptive field expansion, although in familiar situations it will be small. Receptive field expansions are behaviours, and these behaviours must be recommended and selected. Unlike motor behaviours, novelty will always result in a receptive field expansion behaviour, and the competition process in the basal ganglia is therefore not required. However, the selection of information releases by the thalamus, the timing of releases by the basal forebrain, and placing general behavioural priorities on the degree of expansion by the amygdala and hypothalamus are required.

Different receptive field detections in the hippocampal system can be understood as recommendations in favour of receptive field expansions by different groups of cortical columns. The total recommendation strengths are determined in CA3 and the dentate gyrus, modulated by current behavioural priorities in CA2 on the basis of recommendation strengths from the supramammillary nucleus. Once the selection of the strongest recommendation strengths has been determined, the results are released by the septal nuclei to CA1. The CA1 outputs recommend receptive field expansions in different large groups of cortical columns, and are translated back into more specific groups of columns through the subicular complex, the entorhinal cortex and then the perirhinal and parahippocampal cortices. Release of outputs at each stage are recommended by different receptive field detections and acceptance determined by the anterior thalamic nucleus. Receptive field detections by the amygdala recommend increases in the degree of receptive field expansion in specific cortical areas on the basis of very high behavioural priorities (i.e. strong emotion). Such strong emotion indicates that information about the current experience has a higher than average probability of being relevant to future behaviour selections.

The more detailed information model for the connectivity between the cortex and the hippocampus proper is illustrated in Fig. 8.19. Pyramidal neurons in the middle layers of cortical columns provide the inputs to columns in the perirhinal and parahippocampal cortices. In these cortices, receptive fields correspond with groups of columns elsewhere in the cortex that have tended to expand their receptive fields at the same time in the past. Middle layers of these two cortices provide inputs to the entorhinal cortex, where receptive fields therefore correspond with groups of

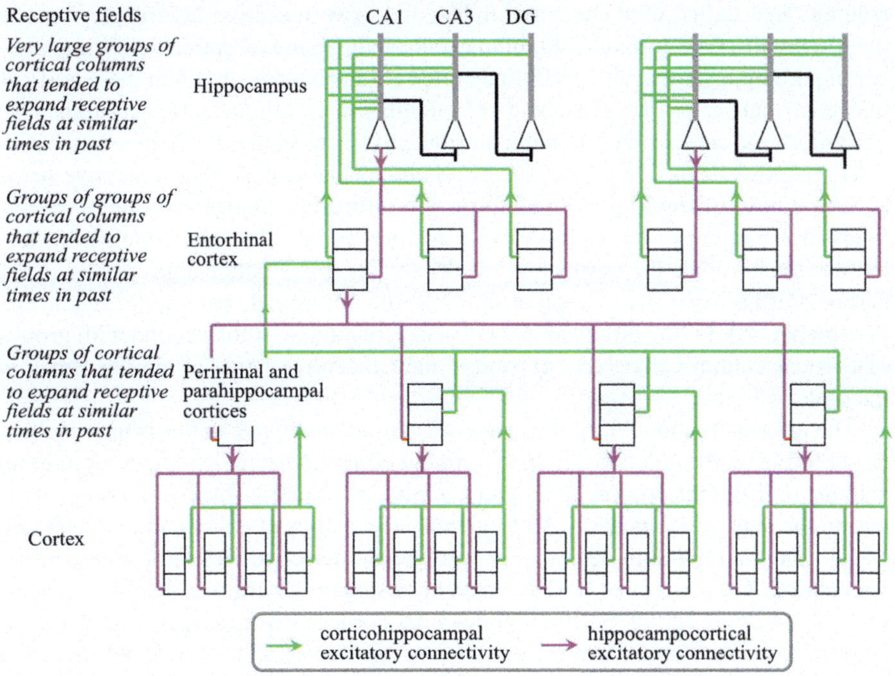

Fig. 8.19 Information model for reciprocal connectivity between hippocampus proper and cortex. In the associated cortices and hippocampus proper, receptive fields correspond with groups of cortical columns that have tended to expand their receptive fields at the same time in the past. The receptive fields are detected if there is strong internal activity across the corresponding group. Within the hippocampus proper there is a competition to determine the groups with the highest internal activity, and if there is inadequate column receptive field detections in some cortical areas, the columns in those areas with the strongest internal activity but no outputs are targetted for receptive field expansions. This targetting proceeds back through the associated cortices

groups of columns across the cortex that have tended to expand their receptive fields at the same time in the past. Outputs from the middle layers of the entorhinal cortex provide inputs to CA1, CA3 and the dentate gyrus. Neurons in all three structures therefore have receptive fields corresponding with the internal activity of very large groups of cortical columns that have tended to expand their receptive fields at the same time. Within the hippocampus proper there is a process to determine if receptive field expansion is required, and the specific very large groups of receptive fields for which expansion is most appropriate. The CA1 pyramidals that produce outputs have receptive fields corresponding with the selected groups. A CA1 pyramidal neuron is connected back to layer VI of the entorhinal cortex columns that define its output receptive field. This connectivity drives output activity from the entorhinal cortex, with receptive field expansion if required. Entorhinal cortex outputs target the perirhinal and parahippocampal columns that define its receptive field, and therefore drive output activity from those cortices, again with receptive field expansion if required. The perirhinal and parahippocampal columns target the cortical columns that define their receptive fields, and drive receptive field expansions in those columns. One cortical column may contribute to the receptive fields of a number of parahippocampal and perirhinal cortex columns, one column in these cortices may contribute to the receptive fields of a number of entorhinal cortex columns, and one entorhinal cortex column may contribute to the receptive fields of a number of CA1 receptive fields. Hence the cortical columns that expand their receptive fields at one point in time will tend to be those that ultimately contribute to the receptive fields of a number of currently active CA1 pyramidals. The hippocampal system outputs thus provide the cortical inputs driving receptive field expansion discussed earlier. Furthermore, the receptive fields in the entorhinal, parahippocampal and perirhinal cortices (the hippocampal cortices) are guided to correspond with groups of cortical columns that have expanded their receptive fields at similar times in the past.

The information model for the processes within the hippocampus proper is illustrated in Fig. 8.20. Inputs from the entorhinal cortex indicate the degree of internal activity in different groups of columns across the cortex. There is a competition within the dentate gyrus and CA3 to identify the groups of columns in which there is less than the minimum degree of receptive field detection. Once this competition is complete, CA3 activates CA1 neurons corresponding with the identified groups, and CA1 outputs acting via the hippocampal cortices drive receptive field expansions in cortical columns forming parts of those groups. The CA1 outputs also drive receptive field expansions as required in the entorhinal, parahippocampal and perirhinal cortices, updating their receptive fields to reflect the current receptive field expansions.

At a more detailed level within the hippocampus proper, receptive fields of granule cells, CA3 pyramidals and CA1 pyramidals are groups of entorhinal cortex columns that tend to have past temporally correlated activity. Detection of such a receptive field reflects current simultaneous activity in a large group of cortical columns. However, layer II entorhinal cortex activity reflects internal activity in the group, while layer III activity reflects internal activity discounted if there is column output. This discounting occurs by inhibitive interneurons with inputs from layers V/VI pyramidals and outputs to layer III pyramidals. Thus in information terms, the degree of activity of layer II entorhinal cortex neurons reflects the degree of familiarity of current cortical inputs, while activity of layer III reflects the columns available for receptive field expansion. Layer II targets CA3 and the dentate gyrus, while layer III targets CA1.

Granule cells excite CA3 pyramidals with similar receptive fields, but via interneurons inhibit CA3 pyramidals with similar receptive fields. Hence moderate granule cell activity will result in activity of CA3 pyramidals with similar receptive fields, but strong granule cell activity will suppress such CA3 activity. Dentate gyrus receptive fields are defined on the basis of simultaneous activity with no guidance, while CA3 receptive fields are defined with guidance from granule cells. Hence the CA3 receptive fields are somewhat more sharply focussed on appropriate groups of cortical columns. CA1 receptive fields are groups of entorhinal cortex columns that tend to have strong layer III activity at the same time. Layer III activity reflects the

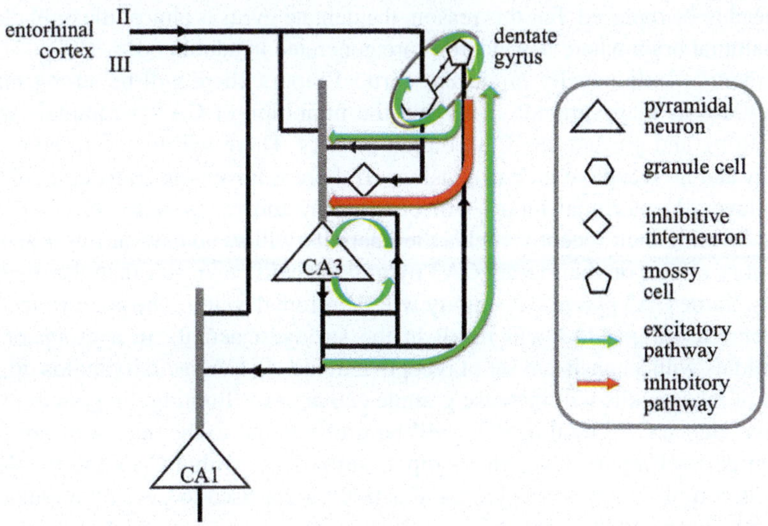

Fig. 8.20 Information model for the competitive process within the hippocampus proper. Entorhinal cortex inputs correlating with internal activity of groups of cortical columns target CA1 pyramidals, CA3 pyramidals, and dentate gyrus granule cells. Similar neuron receptive fields exist in all three regions, corresponding with groups of cortical columns that have tended to expand their receptive fields at similar times in the past. In the dentate gyrus and CA3 there are two positive feedback loops. In the dentate gyrus, granule cells excite mossy cells and mossy cells excite granule cells. Within CA3 there is massive feedback from pyramidal neurons to pyramidal neurons. The two positive feedback loops are linked. Pyramidal neurons in CA3 excite mossy cells in the dentate gyrus. Granule cells in the dentate gyrus excite pyramidal neurons in CA3, and inhibit CA3 pyramidals via interneurons, with inhibition much stronger than excitation for significant granule cell activity. CA3 pyramidals target CA1 pyramidals, and CA1 pyramidal activity drives receptive field expansions across the cortex. If there is little novelty in the input states across all the cortical areas, there will be strong activity in all dentate gyrus granule cells. This strong activity will result in strong inhibition of CA3 pyramidal activity. With no CA3 activity, there will be no CA1 activity and no cortical receptive field expansions. If novelty is present in the inputs to some cortical areas, the granule cells that include such areas in their receptive fields will have lower activity, allowing activity to develop in CA3 pyramidals that include these areas in their receptive fields. The development of CA3 activity is limited because such activity also excites the dentate gyrus. Hence the activity of CA3 pyramidals will be proportional to the degree of novelty in current inputs to cortical areas, and the active CA3 pyramidals will have receptive fields that include those areas. This CA3 activity drives activity of CA1 pyramidals, and the CA1 activity drives appropriate receptive field expansions in the cortex

presence of strong internal activity but little output activity in the cortical columns making up the corresponding group. CA3 pyramidals target CA1 pyramidals with receptive fields corresponding with similar groups of cortical columns. Because CA3 activity guides receptive field expansions of CA1 pyramidals, the CA1 receptive fields tend to be even more sharply focussed than CA3 receptive fields on groups of cortical columns that have expanded their receptive fields at the same time in the past. The lack of guidance to dentate gyrus granule cell receptive fields means that these fields may eventually expand excessively, become functionally useless, and need to be replaced. For this reason, the dentate gyrus is almost the only place in the mammal brain where new neurons are generated in adulthood.

If there is little novelty in current cortical inputs, there will be strong internal activity across many granule cells, and the inhibition of CA3 pyramidals will be very strong and prevent any CA3 output activity. There will therefore be no CA1 outputs and no receptive field expansions. If there is novelty in some cortical areas, some layer II entorhinal inputs will be smaller, and for granule cells with these inputs forming their receptive fields their activity will be somewhat lower and their excitatory effects on their target CA3 pyramidals will be greater than the inhibitory effects. Some CA3 pyramidal activity will therefore develop. The positive feedback competition within CA3 will result in the strongest activity in a group of CA3 pyramidals with a consistent set of receptive fields. As CA3 activity builds up, excitation of mossy cells will increase granule cell activity, limiting the growth of CA3 activity. Hence CA3 total activity will be proportional to the degree of novelty in current cortical inputs. Once the competition process within CA3 and the dentate gyrus is complete, any CA3 pyramidal activity is released to CA1 pyramidals with similar receptive fields and CA1 activity is generated.

The effectiveness of the process for identifying the appropriate cortical columns for receptive field expansion depends upon receptive fields corresponding with groups of columns that have expanded their receptive fields at the same time in the past. The group receptive fields are constantly updated by the process for driving expansion in the appropriate columns. As discussed later, these receptive fields can also be used for indirect activation of cortical information, provided that undesirable side effect receptive field expansions are avoided. Such avoidance will depend upon using neurons that do not directly drive receptive field expansions, but have similar receptive fields.

The preceding description has some simplifications. Information on column outputs in addition to internal activity would make the process more effective. One way of achieving this would be if some of the layer II/III outputs were inhibited by layer V/VI pyramidal activity via interneurons. Some interneurons in layer II/III are observed to receive inputs from layers V/VI pyramidals as required by this model.

The hippocampal system as a whole includes a number of subcortical structures that carry out other supporting information processes, as illustrated in Fig. 8.21. The anterior thalamic nucleus interprets subicular complex outputs as recommendations for release of hippocampal outputs to appropriate cortical areas. This nucleus targets the hippocampal system output structures (CA1, subicular complex and entorhinal cortex) to trigger the appropriate releases by imposing a gamma band modulation. The hypothalamus and amygdala manage the relative priority of different general types of behaviour. In the context of receptive field expansion behaviours, the supramammillary nucleus of the hypothalamus gives priority to receptive field expansions in cortical regions that tend to produce recommendations of the currently prioritized general type. It performs this function by acting on CA2, a special purpose region in the hippocampus proper, to modulate the groups of cortical columns selected for receptive field expansion. The supramammillary nucleus also acts directly on the cortex to encourage receptive field expansions in

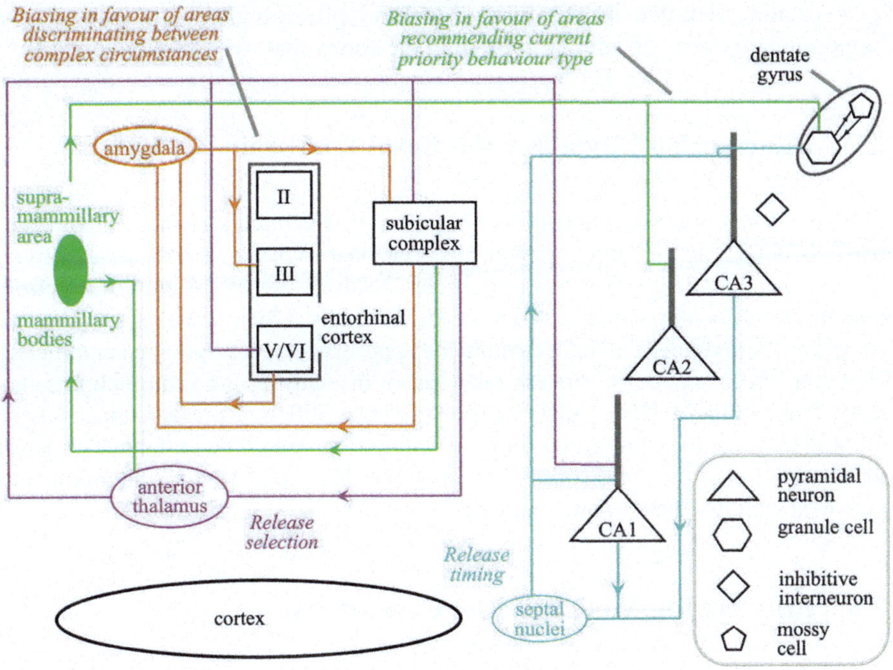

Fig. 8.21 Information model for subcortical structures of the hippocampal system. The anterior thalamus receives outputs from the subicular complex that recommend releases of hippocampal outputs to different cortical areas, selects the most strongly recommended releases and implements them by imposing a gamma band modulation. The septal nuclei receive inputs from CA3 correlating with the state of the competition process in CA3 and the dentate gyrus. When those inputs indicate that the competition process is complete, the septal nuclei trigger the full release of CA1 outputs by imposing a theta band modulation. The mammillary region of the hypothalamus biases the competition process in favour of receptive field expansions in cortical areas that tend to recommend the current priority general behavior type. The amygdala imposes enhanced receptive field expansions on cortical areas that discriminate between different general situations when very high behavioural priorities (i.e. high emotion) exist

appropriate areas. In situations in which a very high priority is placed on one behaviour type, information about the general situation may be particularly important for determining behaviour in future situations. There is therefore behavioural value in increasing the degree of receptive field expansion in such situations. This function is performed by the amygdala, acting somewhat later in the output generation process so that its action can be specific to outputs affecting specific cortical areas. If outputs from CA3 to CA1 drove operational CA1 outputs while the competition process is still occurring within CA3 and the dentate gyrus, inappropriate receptive field expansions would occur. The septal nuclei in the basal forebrain therefore receive inputs indicating the status of this competition, and when it resolves the nuclei impose a theta band modulation on CA1 outputs which combines with the gamma modulation to generate the higher temporal concentrations of action potentials required to generate receptive field expansions in their targets.

At a more detailed level, the neuron level physiological processes in the hippocampal system are similar to those in the cortex that were discussed earlier.

8.7 Information Model for the Basal Forebrain

The basal forebrain manages the relative timing between releases of cortical receptive field detections and other brain events. The clearest example is the management of hippocampal releases by the septal nuclei. As discussed in the previous section, outputs from the hippocampal system to drive receptive field expansions must not occur until the competition to determine the appropriate targets has been completed. However, the competition process takes place in many of the same neurons that drive those outputs. Hence those neuron outputs, although present, must not be allowed to influence their targets until the appropriate time. The septal nuclei detect the completion of the competition, and impose a theta band modulation on the outputs which allows them to influence their targets.

8.8 Information Model for the Cerebellum

In many cases, the appropriate response to an environmental situation is a sequence of actions that has often been utilized in the same order in the past. To give a simple example, consider climbing stairs. Stair climbing requires a sequence of body movements: the left leg is lifted and shifted forwards; weight is moved on to the left foot; the right leg is lifted and shifted forwards; weight is moved onto the right foot; then the sequence repeats. At a more detailed level, this sequence involves a sequence of ankle, leg and thigh muscle movements, which must be driven by outputs from the brain. These brain outputs must be generated, taking into account sensory information. The full process of detecting receptive fields in sensory inputs, interpreting receptive field detections as behavioural recommendations in the basal ganglia, determining the most heavily recommended behaviour, and implementing that behaviour by the thalamus releasing the appropriate motor cortex outputs requires time. If after the completion of one muscle movement the full process was required to determine the next muscle movement, the action selection time would be required between each muscle movement. A way to accelerate the implementation of previously learned muscle movement sequences would be to record such sequences as a single behaviour, and given an initial set of receptive field detections interpreted in the basal ganglia as recommending the behaviour sequence, to execute the sequence without further reference to the basal ganglia. Such an independent execution requires a sequence of very specific receptive fields corresponding with the situations at the initiation of each action in the sequence, with detection of a receptive field triggering the execution of the next action and perhaps preparatory activity for the subsequent actions. Recording sequences in this way could lead to faster and more error free execution of such frequently used sequence behaviours.

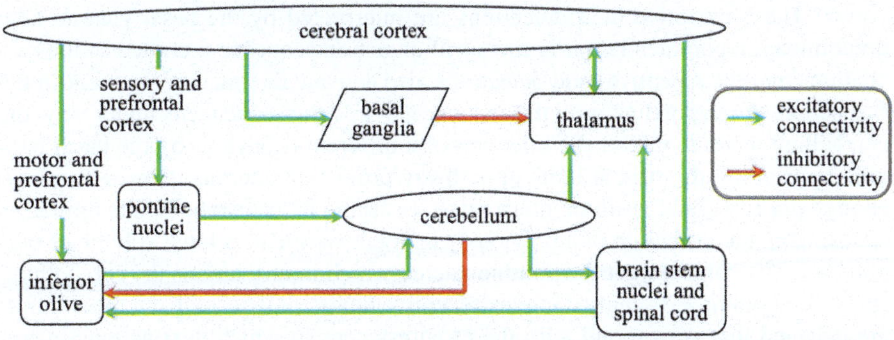

Fig. 8.22 Information model for management of frequently utilized behaviour sequences by the cerebellum. Initially, individual behaviours are learned by acquisition of appropriate synaptic recommendation weights in the basal ganglia and thalamus. Individual behaviours are implemented by thalamic release of cortical outputs, to the brainstem nuclei to drive muscle movements in the case of motor behaviours, and between cortical areas in the case of cognitive behaviours. If a sequence of behaviours occurs frequently, different neurons in a cerebellar component begin to target the different thalamic or brainstem nuclei activated to implement each behavior. Other cerebellar neurons develop receptive fields within cortical inputs through the pontine nuclei that correspond with the exact circumstances in which each behavior is initiated. The group of cerebellar neurons corresponding with the behaviour sequence can then be activated in turn without reference back to the basal ganglia after each behaviour. Timing errors in the implementation of each behaviour are detected by neurons in the inferior olive, triggering adjustments to the cerebellar receptive field that initiates the behaviour

Cognitive processing also may include frequently utilized sequences of actions, but in this case the actions will be information releases between cortical areas. To give a simple example, consider finding an answer to the written equation 6+7. As will be discussed in more detail later, the sequence of information process actions could include focussing attention on the three visual objects in turn, for each object detecting receptive fields at a specific level of complexity and holding the detections active but not interacting. Next, releasing all three sets of receptive field detections to other cortical areas and detecting higher complexity receptive fields. Finally, releasing the higher complexity detections to behaviour selection, biasing the selection process in favour of a verbal behaviour. If arithmetic has been learned, the higher complexity receptive fields will have previously acquired recommendation strengths in favour of speaking the correct answer. Such sequences of attention, cortical information release, and behaviour biasing behaviours could also be recorded and therefore made faster and more accurate.

The high level information model for the cerebellum performing the information processes associated with recording and implementing action sequence behaviours is illustrated in Fig. 8.22. The basal ganglia is the subsystem in which learning of the circumstances and order of the actions in a sequence take place. Once such learning has occurred, and if the same sequence is utilized repeatedly, the cerebellum takes over routine execution of the sequence. The circumstances in which a behaviour sequence is appropriate still result in receptive field detections by the cortex. These receptive field detections are interpreted by the basal ganglia as a predominant recommendation in favour of an action sequence recorded in the cerebellum, and the receptive field detections also activate the microcomponent in the cerebellum corresponding with the sequence. While the sequence is under way, the cerebellum tends to inhibit the selection of a different behaviour by the basal ganglia, by exciting the striatal indirect pathway projection neurons. However, a high enough priority could result in interruption. Learning in the cerebellum is limited to speeding up the execution of a previously learned sequence. A huge volume of outputs from the cortex target the pontine nuclei, which in turn targets the cerebellum. In the cerebellum, extremely complex receptive fields constructed from these inputs are detected that correspond with the circumstances in which next actions in frequently used sequences must be initiated. The actions are driven directly by cerebellar outputs to the spinal cord, or to the thalamus if the action is a cortical information release. In a sequence, the exact timing of the next action is adjusted by the inferior olive. This nucleus receives inputs from the spinal cord and cortex carrying information on the completion of actions. It also receives inputs from the cerebellum when an action is initiated. If the timing of the action is incorrect, it sends signals back to the cerebellum to modify the receptive field that triggered initiation of the action.

The information model at a more detailed level for one action sequence behaviour is illustrated in Fig. 8.23. A column of neurons in the inferior olive is connected to a group of Purkinje cells, and the Purkinje cells project to a limited number of cerebellar nuclei neurons. This set of neurons is called a microcomponent, and instantiates the record of one action sequence. A Purkinje neuron corresponds with one action in a sequence, and its receptive field corresponds with the exact circumstances in which that action must be initiated. These circumstances are very precisely defined: one Purkinje can have inputs from 200,000 granule cells, one granule cell has inputs from several pontine nuclei neurons. Each pontine nucleus projection neuron receives a large number of inputs from the cortex, and ~10 million cortical neurons provide inputs to the pontine nuclei. Hence the receptive field of a Purkinje is a combination of a very large number of cortical receptive fields. When a sequence of motor actions is being learned, spinal-cord neurons that trigger the required muscle movements are targeted by outputs from the motor cortex. Once initial learning of the sequence has been achieved, different excitatory cerebellar nuclei neurons begin to target the spinal-cord neurons active at different points in the sequence and therefore acquire the ability to drive the different actions. Both excitatory and inhibitory cerebellar nuclei neurons acquire receptive fields made up of pontine nuclei inputs corresponding with the general circumstances in which the action sequence is appropriate. A Purkinje neuron acquires a receptive field corresponding with whatever exact combination of cortical receptive field detections happens to be present at the instant one action in the sequence is initiated. This Purkinje neuron also establishes connectivity to all the cerebellar nuclei neurons except the ones corresponding with the same action. Inferior olive neurons receive inputs from the spinal-cord that are activated when the circumstances appropriate for the completion of one action are present.

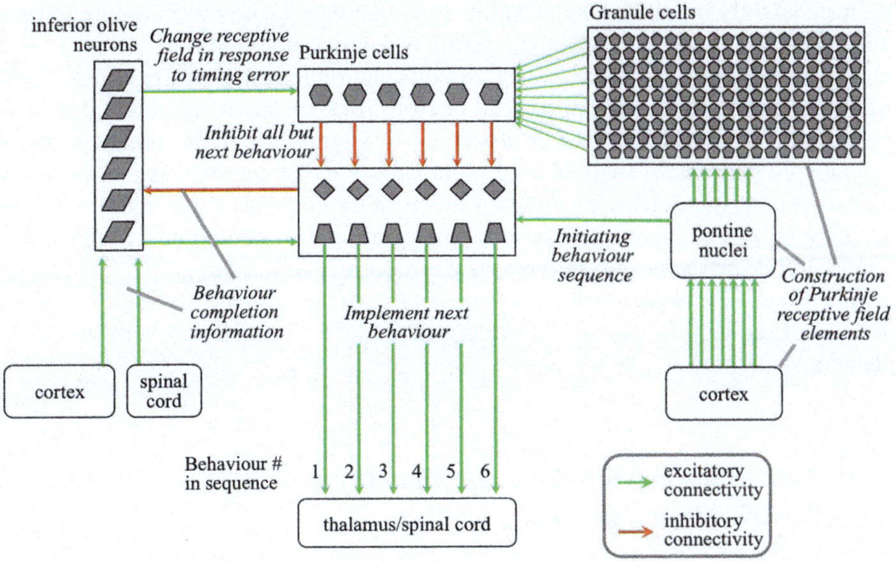

Fig. 8.23 Information model for cerebellar management of one action sequence by a microzone. Different inhibitory Purkinje cells in the cerebellar cortex develop very complex receptive fields corresponding with the very specific circumstances in which the next behaviour in a sequence should be initiated. Different excitatory cerebellar nuclei neurons target the neurons in the brainstem or thalamus that drive each behaviour. Sufficient recommendation strength in favour of the sequence into the basal ganglia releases cortical inputs via the pontine nuclei to the excitatory cerebellar nuclei neurons, sufficient to activate them in the absence of Purkinje inputs. The first Purkinje is also activated, inhibiting all except the first behaviour. Then the second Purkinje is activated, inhibiting all except the second behaviour and so on. Information about the actual implementation of each behaviour excites the inferior olive neuron targetting the Purkinje neuron corresponding with that behaviour. The inferior olive neuron is also inhibited by a cerebellar nucleus neuron that fires when the cerebellar nucleus triggers the behaviour. The behaviour occurred at the correct time, the inferior olive neuron does not fire. It there is a timing error, the inferior olive neuron fires, triggering change to the Purkinje neuron receptive field

When the circumstances in which the action sequence is appropriate are present, there will be strong activity in the inputs to both the cerebellar cortex granule cells and the inhibitory and excitatory projection neurons in the cerebellar nuclei. Purkinje cells inhibit all the cerebellar nucleus excitatory neurons in the behavioural microcomponent except the neuron corresponding with the same action as the Purkinje. Hence the action sequence is implemented by activation of a sequence of Purkinje cells which each inhibits all but the currently required action in the sequence. The inferior olive neuron corresponding with one action receives excitatory inputs from the spinal cord when the correct moment for implementation of its corresponding action is reached. If the timing of the action is appropriate, it will simultaneously receive inhibitory inputs from the cerebellar nucleus indicating implementation of the action. If the timing is incorrect, activation of the inferior olive neuron results in LTD of the recent inputs to the corresponding Purkinje neuron, slightly modifying its receptive field, guiding it towards an optimum field for the precise moment in which the action is appropriate.

For some action sequences, one individual action may not always involve activation of exactly the same spinal cord neurons. In such cases, the cerebellum acts upon brainstem nuclei to trigger activation of slightly different combinations of spinal cord neurons. If there is even more variation, the cerebellum acts upon the thalamus to trigger release of a limited subset of the currently active motor neurons in the cortex. The identity of these motor neurons then determines which spinal cord neurons are activated. In the case of frequently utilized cognitive behavioral action sequences, the cerebellum will generally action upon the thalamus to trigger release of the appropriate subset of currently active frontal cortex receptive field detections.

8.9 Information Model for the Neurotransmitter Distribution Systems

The information model for the locus coeruleus, raphe nuclei, tuberomammillary nucleus, midbrain dopamine etc. is that they implement general arousal changes that are recommended mainly by receptive field detections in the frontal cortex and amygdala. Increased arousal results in stronger cortical outputs recommending behaviours. Such stronger outputs can be increased neuron activity in response to the same input information and/or reductions in the thresholds for cortical receptive field detections resulting in more receptive field detections within the same input information. These systems can also regulate the degree and permanence of cortical receptive field expansions.

8.10 Application of Information Models

The information models described for the different brain structures are all based on the basic condition definition/detection and the behavioural recommendation information processing types. These types can be used to interpret the functions of anatomical structures, and we have outlined how neurons and neuron connectivity implement those functions. In the next chapters we will discuss in more detail how this approach can link psychological phenomena to neuron activity via intermediate levels of description.

Chapter 9
Memory and the Organisation of Experience

Memory involves the recording of information about experiences, and accessing that information to guide behaviour during later experiences that have some elements of similarity with earlier experiences. One new experience may have many different, potentially relevant elements of similarity with many past experiences.

As discussed in Chap. 2, there are a number of different types of memory at the psychological level of description that can be distinguished from each other on the basis of behavioural differences, and observations that damage to certain brain structures affects some types of memory but not others. These memory types are priming memory, semantic memory, episodic memory, procedural memory, working memory and the memory for complex sequences of actions. The differences between these memory types can be understood in terms of the differences between the information recording and accessing mechanisms utilized.

There are three primary physiological mechanisms by which information about experiences is recorded in the brain. One is to change (generally expand) receptive fields instantiated in cortical columns. The second is to modify behavioural recommendation weights instantiated as the synaptic weights of inputs into the striatum and thalamus. The third is to record a frequently used sequence of actions, instantiated by a group of Purkinje cell receptive fields in the cerebellum that correspond with the precise circumstances in which each action in the sequence must be initiated.

Information recorded in cortical column receptive fields can be accessed in various ways. In addition to direct detection of a cortical receptive field within current sensory inputs, receptive fields can be indirectly activated on the basis of various types of temporal correlations in the past activity of groups of columns. A column with a receptive field that is not directly activated by being present in current sensory inputs can be indirectly activated on the basis that it was recently active at the same time as a number of currently activated receptive fields. Columns can also be indirectly activated on the basis of frequent past simultaneous activity or simultaneous past receptive field expansion. Each type of indirect activation can also be based on past activity of a currently active group of columns just before or just after activity of the columns to be activated. There are limits to the number of populations of

L.A. Coward, *Towards a Theoretical Neuroscience: from Cell Chemistry to Cognition*, Springer Series in Cognitive and Neural Systems 8, DOI 10.1007/978-94-007-7107-9_9,

receptive field detections that can be independently active at the same time in the same cortical area.

To understand the relationships between these memory phenomena and detailed physiology, hierarchies of description are required. In this chapter, such hierarchies will first be described for organizing information derived from experience into similarity circumstances. Then the process for detecting such circumstances in current experience will be described. In these descriptions, elements of higher level processes will be described at a more detailed level and so on down to neurochemistry. Note that in reality the organisation and detection processes are entangled. Hence the two descriptions are emphasizing different aspects of one integrated process.

The key information processes supporting priming memory are information recording in cortical receptive fields and accessing that information on the basis of recent simultaneous activity. Semantic memory is supported by information recording in cortical receptive fields and accessing that information on the basis of frequent past simultaneous activity. Episodic memory is supported by information recording in cortical receptive fields and accessing that information on the basis of past simultaneous receptive field expansion. Procedural memory depends upon weight changes to corticostriatal synapses. Working memory relates to limits to the number of simultaneously active but independent populations of cortical columns in the same area, the limits deriving from the use of gamma band frequencies in the attention function as discussed in this chapter. Action sequence memories are recorded and accessed using Purkinje cell receptive fields in the cerebellum.

Hierarchies of description for semantic, priming, episodic and procedural and sequence memory types will be provided in this chapter. The memory phenomena will be described fairly completely at high level, with key narrow segments of a phenomenon being described at more detailed levels. Descriptions at the very detailed levels would be similar to those provided in the organizing and detecting similarity circumstances hierarchies. Hence the more detailed levels will be omitted. A hierarchy of description for working memory will be discussed in Chap. 10, after discussion of attention with which it shares some critical mechanisms.

Although all these types of memory can to some degree be investigated separately under laboratory conditions, real world cognitive behaviours are determined by current sensory inputs combined with selections from these recorded information sources, accessed in various ways. In Chap. 11 we will discuss how to understand such real world behaviours.

9.1 Organisation of Experience into Similarity Circumstances

In order to use past experience to guide current behaviour, that experience must be organised into similarity circumstances. If some circumstance in current experience is similar to a circumstance that occurred in past experience, that similarity implies that the most appropriate current behaviour may be similar to successful behaviours when the circumstance occurred in the past. If a circumstance has often occurred in the past and a specific behaviour was often selected and successful in that circumstance, the probability of that behaviour being successful in future occurrences of that circumstance is relatively high. Cortical column receptive fields correspond with circumstances that have tended to occur in many past experiences, and each receptive field can therefore recommend all the behaviours which have been successful in such past circumstances. In general, the more often a behaviour has been successful the stronger the recommendation. Appropriate behaviour can then be determined on the basis of the range of past circumstances present in current experience, and the predominant behaviour recommended across these circumstances.

Cortical column receptive fields correspond with different circumstances that have occurred a number of times in past experience. Such a receptive field is specified by the inputs to the column and a complex integration process by which the inputs are combined to determine if neurons in the output layers are activated. Different output neurons indicate detection of similar but slightly different circumstances.

9.1.1 Information Basis of Receptive Field Definition

As discussed in Chap. 7 and illustrated in Fig. 7.10, a high level description approximation that is very useful for understanding cognitive processes is to view a cortical column receptive field as a large group of conditions, each condition instantiated by synapses on one branch of a pyramidal neuron. If a sufficiently high proportion of the conditions is detected, the receptive field of the column is detected. Expansions to the receptive field can be viewed as additions of conditions to the group. Additions are triggered in a column by an external signal from the hippocampus, provided that a significant proportion of the existing conditions are being detected. The added conditions must have a degree of similarity to the existing conditions.

A complex learning system must heuristically define a range of receptive fields. One receptive field cannot correspond with all the experiences in which one behaviour is appropriate, and no other experiences. In any experience, a subset of receptive fields will be detected corresponding with the different circumstances present in the experience. The requirement for the set of receptive fields as a whole is that for any two experiences in which different behaviours are appropriate, the two subsets of detected receptive fields must be significantly different, although they may overlap partially. It is also valuable, but not as critical, that for two experiences in which the same behaviour is appropriate, the subsets of detected receptive fields have significant overlap. If these requirements are met, it is feasible to assign recommendation weights in the striatum to receptive field detections in such a way that high integrity behavioural selections are possible.

9.1.2 Description of Receptive Field Definition at Cortical Area Level

Genetic information can specify the range of complexity of the receptive fields in an area, and natural selection will result in different areas with good discrimination for all the different types of behaviours required. One area will tend to support multiple types of behaviour, but for any one type a small group of areas will tend to be particularly effective. These areas will have most of the recommendation strengths in favour of the type, although other areas may also contribute.

An area with receptive field complexities particularly effective for discriminating between different types of visual object is illustrated in Fig. 9.1. Cortical area TE is an example of such an area. The process for definition of receptive fields must result in the required type of discrimination. However, the achievement of discrimination can only utilize information that is readily available within the brain either from natural selection favouring certain types of connectivity or from utilisation of very general patterns in receptive field detection. Information only knowable by an external designer is not available. There are a number of factors which can be managed to increase the probability of achieving high discrimination within these information restrictions. *Firstly*, certain ranges of receptive field complexities may be particularly effective in discriminating between different experiences in which different behaviours of a particular type are appropriate. Receptive field complexity includes the sources of inputs, the total number of inputs that contribute to one receptive field, and the number of inputs that must be present for receptive field detection. Natural selection can result in each cortical area being genetically programmed for a behaviourally useful range of receptive field complexities. This programming constrains the sources of input to each area, the numbers of synapses on neurons in the area, and the thresholds for receptive field detection. Genetic bias can also constrain the outputs of a cortical area so that they target striatal regions corresponding with the behaviours effectively discriminated by that area. *Secondly*, natural selection can result in the initial sets of inputs to each column within an area being constrained to be different, so that the initial receptive fields are relatively orthogonal. The requirements that receptive field expansions only occur if less than a minimum number of fields are being detected, and that any one field only expands if a slight expansion is sufficient for detection, contribute to maintaining this orthogonality. *Thirdly*, if a column detects its receptive field too often, its threshold for detection can be raised. In addition, if two columns are generally active at the same time, one could be reprogrammed. Reprogramming would involve ensuring that the striatal targets of the reprogrammed column are eliminated after first being targetted by the remaining column, and then inputs to the reprogrammed column initialized. *Fourthly*, as illustrated in Fig. 9.1c, if the same group of columns is often active when the same behaviour is selected, but that behaviour is sometimes followed by positive consequences, sometimes negative, the implication is that column discrimination is inadequate: the same group of columns is being detected in situations in which different behaviours are appropriate. If this situation occurs, additional receptive

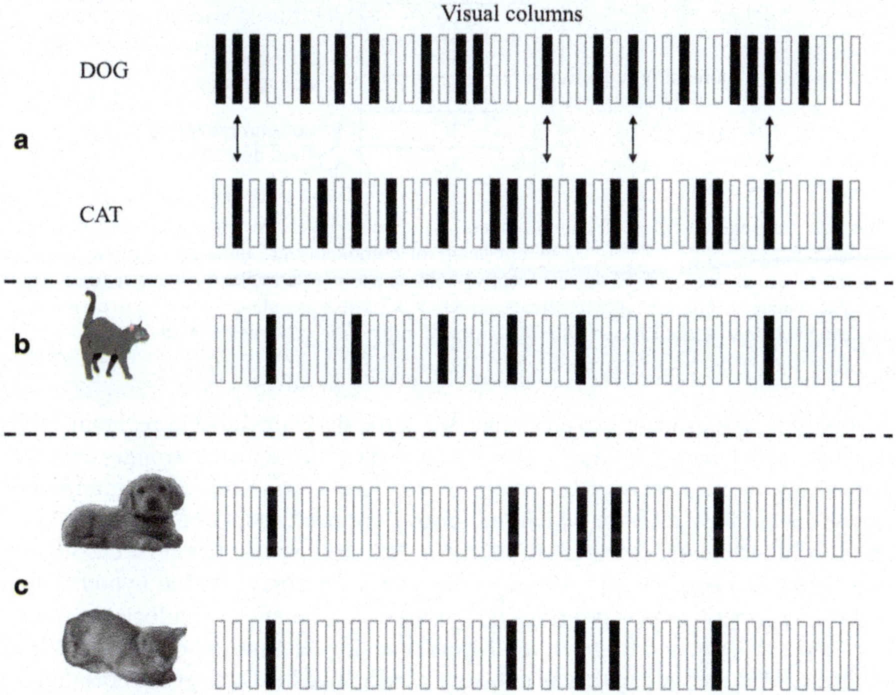

Fig. 9.1 Columns activated in the same cortical area in response to different visual circumstances. Receptive fields do not correspond with cognitive categories, but must discriminate between such categories. (a) Cortical columns in an area such as TE may detect their receptive fields within visual objects. A large number of columns may detect their receptive fields in response to different visual experiences of dogs. Each of these columns will have recommendation strengths in favour of dog-appropriate behaviours, but only a small proportion will be detected in response to any specific visual experience of a dog. Similarly, a large number of columns may detect their receptive fields in response to cats. In some cases a column may detect its receptive field in response to some cats and some dogs, in which case it will have recommendation strengths in favour of both cat- and dog-appropriate behaviours. (b) In response to a specific cat, the predominant recommendation strength across all detected receptive fields will be in favour of cat-appropriate behaviours. (c) If some instances of cats resulted in the same columns detecting their receptive fields as some instances of dogs, discrimination is inadequate. In this situation, detection of the same group of receptive fields followed by selection of the same behaviour sometimes results in positive consequences, sometimes negative. Detection of this situation triggers receptive field expansion whenever the same group is active again in the future, leading to greater discrimination between the similar circumstances

field expansion in the future when the same group is active will add receptive field detections. Detections will be added until adequate discrimination is achieved, as demonstrated by the conflicting consequence feedback no longer occurring.

Receptive field expansion in a cortical area is critical for selection of future behaviours but consumes resources. Hence decisions on such expansions must be carefully regulated. For example, if a general type of behaviour has been given priority,

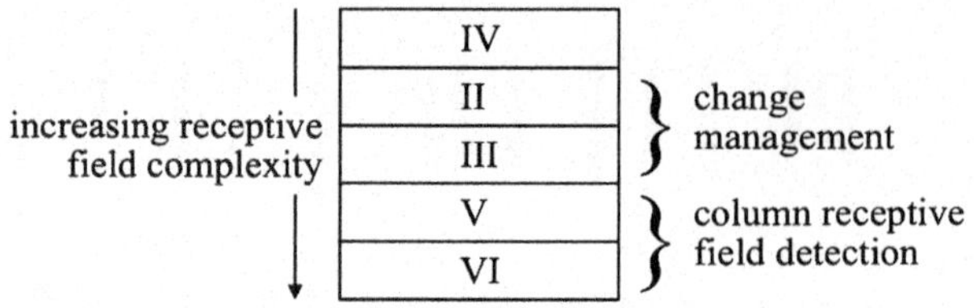

Fig. 9.2 Information roles of column-layers. Successive layers receptive fields with slightly increasing levels of complexity. Different levels of complexity are most effective for different information roles. Layers II/III detect receptive fields at a level of complexity effective for management of changes to column receptive fields. Layers V/VI detect receptive fields effective for recommending behaviours and for contributing to more complex receptive fields in other areas

receptive field expansions in areas that are particularly effective in recommending that type of behaviour should be favoured. If a recent behaviour is strongly rewarded (positively or negatively) then receptive field expansions recording information about the circumstances in which the reward occurred should be favoured. If the brain is in a state of high arousal, receptive field expansion should be encouraged more than in a passive state. Hence outputs from the amygdala and hypothalamus indicating general behavioral priorities, from the raphe nuclei and locus coeruleus indicating arousal, and from the midbrain dopamine neurons indicating reward can be regarded as recommendations in favour of receptive field changes. Depending on the information content of these outputs, the recommendations may be targetted directly to the cortex, or via different points in the hippocampal system, or both.

The length of time for which a change should persist also needs to be carefully regulated. Information recording resources are not unlimited, and sometimes there can be a high probability that some receptive field changes will not be needed beyond a certain period of time. Receptive fields that can discriminate between circumstances in which long term recording is needed and circumstances in which there is a time limit on the value of the information must recommend the persistence of any changes. If some receptive fields have not been detected for a long period, larger changes to such fields could be allowed to increase their behavioural value.

9.1.3 Description of Receptive Field Definition at Cortical Column Level

As illustrated in Fig. 9.2, the layering of a cortical column supports receptive field detections at slightly different levels of complexity but within the same set of column inputs, with the different levels being most effective for different information purposes. Receptive field detections by neurons in layer V are column receptive field detections that have behavioural recommendation weights into subcortical structures like the striatum. Receptive field detections by neurons in layer VI are column receptive field detections released to other areas where they may be incorporated into more complex receptive fields. They also carry recommendation weights in the thalamus in favour of such releases. The slightly different receptive fields detected by individual neurons in layers V/VI provide more detailed discrimination within the column receptive field.

Layers II/III are programmed with somewhat simpler receptive fields which may sometimes be detected in the absence of any layer V/VI detections. Such a pattern of activity indicates that only a relatively slight expansion in the column receptive field would be required to achieve detection. Hence layers II/III are used to manage receptive field expansions via communication with the hippocampal system.

Inputs to one cortical column recommending receptive field expansion can come from different sources. Such inputs can come from different columns in the cortices of the hippocampal system, corresponding with different groups of cortical columns that have expanded their receptive fields at the same time in the past. Each such input can be regarded as a recommendation in favour of receptive field expansion, with a weight proportional to the strength of the input. The strengths of these inputs are modulated by hypothalamus and amygdala inputs to the hippocampal system. These hypothalamic and amygdalar inputs can also be regarded as weighted recommendations. Inputs can also come directly to a column from the general neurotransmitter distribution systems, also recommending receptive field expansions. Hence the occurrence and degree and duration of receptive field expansion in a column depends upon the total weight of all the input recommendations in favour of such expansion.

9.1.4 *Description of Receptive Field Definition at Neuron Level*

The receptive field of a column is the relatively small range of receptive fields detected by the different pyramidal neurons in its output layers. The receptive field of a pyramidal neuron is defined by a group of conditions, each condition instantiated by a terminal branch on its dendritic tree. The elements defining a condition are synapses from pyramidal neurons in preceding layers of the column, and neurons that are the external inputs to the column. Each synapse has an individual weight, and the condition is detected if the total weight of the synapses currently receiving action potential inputs exceeds a threshold. Different combinations of active synapses may reach this threshold, and the condition corresponding with a branch is therefore in fact a combination of very similar conditions.

As illustrated in Fig. 9.3, the receptive field of a neuron can be expanded by increases in the weights of the synapses on a branch, addition of zero weight synapses to a branch followed by weight increases, or configuration of a new branch with zero weight synapses followed by weight increases. Weight increases are made more likely by the activity of inputs from the hippocampal system, and such activity is essential for new branches.

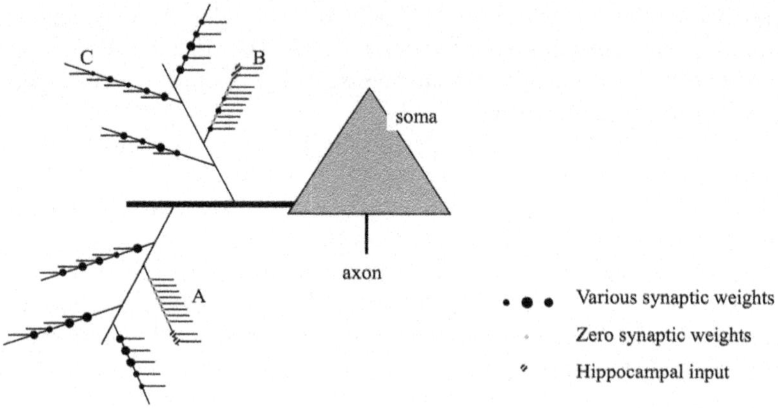

Fig. 9.3 Changes to the conditions detected by a dendrite. (*A*) Branches with zero weight synapses from other cortical neurons and strong synapses from the hippocampal system can define a new condition by increases in the zero weights. (*B*) A branch with small weight synapses from other cortical neurons and strong synapses from the hippocampal system can broaden the range of circumstances in which the condition is detected. (*C*) Branches with substantial synaptic weights from other cortical neurons may adjust relative weights on the basis of temporally correlated activity of the different source neurons

Increases in synaptic weights only occur if branch activity is followed immediately by neuron firing, and neuron firing only occurs if a number of branches are active. Hence receptive field expansion can only occur if a significant number of existing branch conditions are being detected. Hippocampal inputs are active only if three conditions are met: firstly, column receptive field expansion is needed; secondly, the column is not already producing an output; and thirdly, there is already significant activity in layers II/III of the column. Hence a new condition will be added only if that condition just tips the balance for receptive field detection.

In addition, pyramidal neurons have inputs from other structures including the raphe nucleus, the locus coeruleus and the midbrain dopamine neurons. These inputs can be regarded as recommending that synaptic weight changes are retained longer term and not allowed to decay.

9.1.5 Description of Receptive Field Definition at Dendritic Branch and Synapse Levels

The process for adding a new condition is illustrated in Fig. 9.4. A new branch has a group of synapses derived from cortical pyramidal neurons in preceding layers. These synapses are silent, in other words they have zero synaptic weight. The branch also has synapses from the hippocampal system, and these synapses alone have sufficient weight to activate the branch in the absence of any other inputs. If many or all of the cortical inputs are active, the branch will still not be active because the

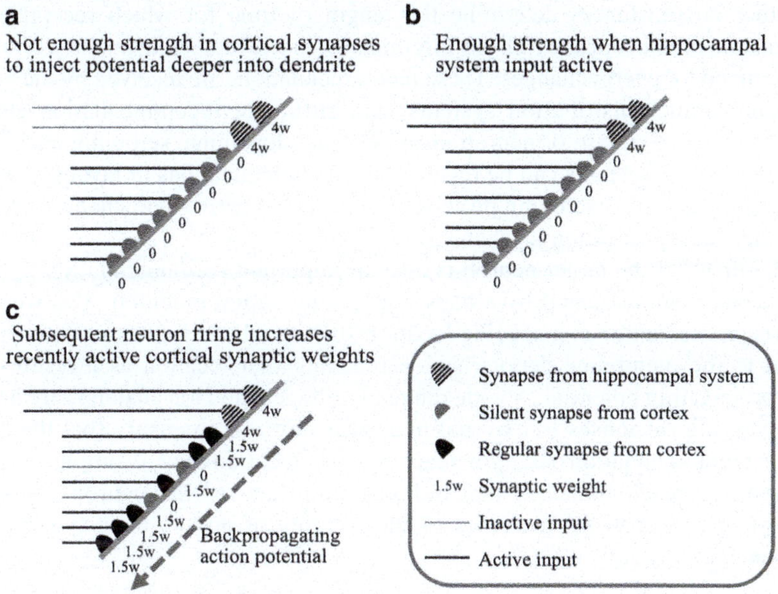

Fig. 9.4 Process for definition of a new condition on a dendritic branch. (**a**) The total weight of all synapses from other cortical neurons is too low to initiate a dendritic action potential deeper into the dendrite, even if all the synapses received action potentials at the same time. (**b**) If the strong synapses from the hippocampal system receive action potentials there is enough total input to initiate a dendritic action potential. (**c**) If some of the cortical inputs were active at the same time as the hippocampal inputs, and if the target neuron fires shortly afterwards, the backpropagating action potential results in increases in the weights of those cortical inputs. These increases mean that the branch could contribute to future neuron firings in the absence of the hippocampal input. In information terms, a new condition has been added to the neuron receptive field

synapses are silent. However, suppose that some of those cortical inputs are active while the hippocampal inputs are also active. If the neuron fires shortly afterwards, the backpropagating action potential will increase the weights of those recently active cortical synapses. The weight increases mean that the branch could be activated in the future by these cortical inputs, in the absence of active hippocampal inputs. In information terms, a new condition has been programmed on the neuron. The weights of the hippocampal inputs do not increase, and once a condition is well established they are disconnected. The weights of cortical synapses can continue to increase by the same backpropagation based mechanism. New silent synapses from additional sources can be added.

To ensure that the addition of new inputs does not risk too major a change to the receptive field, such inputs are selected on the basis that they have often been active in the past when the neuron receptive field has been detected. This is achieved by REM sleep conducting approximate partial reruns of cortical activity in past experiences, with silent synapses established on the basis of frequent simultaneous activity of presynaptic and postsynaptic neurons during the rerun.

Other circumstances determine the length of time for which receptive field expansion is appropriate. These circumstances are detected in the cortex and recommend long term changes. These recommendations are received by the general neurotransmitter distribution systems, and sufficient recommendation strength triggers cortical release of neurotransmitters like dopamine, serotonin and norepinephrine. These releases can be regarded as recommendations in favour of a range of neuron responses including prolongation of the receptive field changes resulting from the backpropagation mechanism.

The firing of the target neuron is also an important recommendation in favour of change, communicated by a backpropagating action potential. The weight of this recommendation at a specific location such as a branch or synapse is proportional to the membrane depolarisation reaching that location as a result of the backpropagating potential. A wide range of voltage gated ion channels are needed, with varying responses to external neurotransmitters, to ensure that the appropriate recommendation strength reaches each location. For example, backpropagation to reach branches that did not contribute to the neuron firing may be suppressed to avoid increases to the weights of any of their synapses that happened to be active.

9.1.6 Description of Receptive Field Definition at Receptor Level

Silent synapses have NMDA receptors but not AMPA receptors. A provisional branch has silent synapses from the cortex plus hippocampal system synapses with AMPA receptors but no NMDA receptors. Opening of AMPA receptors results in an Na^{+} current which depolarises the membrane, and with enough total current across the synapses on the branch results in potential injection deeper into the dendrite. Opening of NMDA receptors requires both glutamate binding and membrane depolarisation. Glutamate binding persists for a short time after glutamate release by an incoming action potential, and if a backpropagating action potential occurs during that time, the Ca^{++} current resulting from NMDA opening triggers addition of AMPA receptors to the synapse. Such addition increases the strength of the synapse, but decays over a period of about an hour. Releases of dopamine, serotonin, norepinephrine and other neurotransmitters in the vicinity of a synapse and/or neuron recommend long term extension of synaptic weight changes. For example, sufficient such recommendation weight results in further chemical processes that enlarge the synapses that have been active, making synaptic weight changes more permanent. Hence a weight change is determined by a complex combination of inputs from many sources, requiring a wide range of neurotransmitters and receptors. Each such input is functionally a recommendation in favour of change, and change will occur if there is sufficient total recommendation weight.

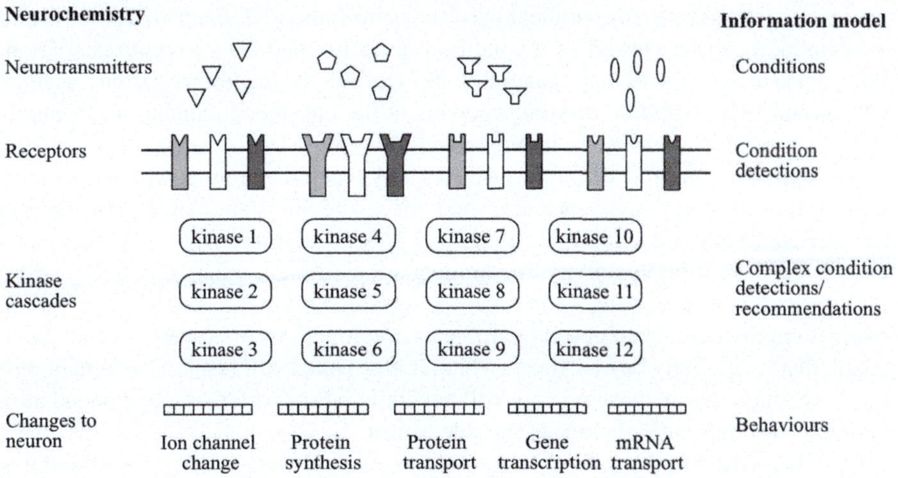

Fig. 9.5 Conceptual process for determining synaptic weight changes at neurochemical level. External neurotransmitters detected by different membrane receptors can be viewed as different condition detections that are also recommendations in favour of different chemical changes in the neuron. These condition detections initiate different kinase cascades reflecting the strengths of the receptor recommendations. These recommendations interact a multiple levels, and result in some predominant behavioural recommendations for specific chemical changes at the levels where these changes are initiated. Chemical changes can be initiated by various chemical behaviours including ion channel modifications, protein synthesis at the synapse from existing RNA, gene transcription, or transport of mRNA to a specific synapse. The chemical behaviours implemented depend upon the total recommendation strengths across current kinase cascades

9.1.7 Description of Receptive Field Definition at Neurochemical Level

As illustrated in Fig. 9.5, at a more detailed level a behaviour of changing a receptive field is made up of sequences and combinations of neurochemical behaviours. These neurochemical behaviours can include transcription of specific genes, transport of mRNA to specific locations, synthesis of specific proteins and transport to specific locations, insertion of new or pre-existing proteins into the membrane, and modifications to new or pre-existing proteins.

Releases of neurotransmitters into the extracellular environment can be viewed in information terms as conditions that are also recommendations in favour of a range of different types of neurochemical behaviours by the neuron. Activations of receptors are condition detections that are recommendations in favour of more specific behaviours of general types. The recommendations are more specific because they refer to the locations on the neuron where the receptors are located. At any point in time, there are very large numbers of recommendations at the receptor level, and these recommendations are integrated into predominant recommendations by many parallel kinase activation cascades. Each kinase activation in a cascade can be viewed as a condition detection and also a recommendation. Phosphatase activations can generally be viewed as recommendations against behaviours. The specific behaviours with sufficient recommendation strengths across all current cascades are implemented.

For example, multiple different cascades may reach the point at which transcription factors or coactivators are activated for a specific gene. Each cascade can be viewed as recommendation strength in favour of transcription of the gene. For strong transcription activity, recommendation strength must be contributed by multiple cascades activating a combination of transcription factors and coactivators. Such strength could be achieved by different combinations of cascade activity. As a result, there will rarely be one kinase which if inactivated will result in termination of memory, since any memory process will generally be supported by recommendation strengths through multiple kinase cascade routes.

Another aspect of the natural selection pressures relevant at this level of detail is that "learning" must be possible through random mutation. A mutation could have benefits to a species, but is also likely to have undesirable side effects. If there are multiple paths recommending any neurochemical behaviour, then the undesirable side effects of a beneficial mutation are less likely to be fatal.

9.2 Detection of Similarity Circumstances

A receptive field corresponds with a similarity circumstance and is associated with a range of behavioural recommendations. The presence of the similarity circumstance in current sensory inputs results in the detection of the receptive field and activates those recommendations. Depending on behavioural priorities, the threshold for detection of a receptive field (i.e. the proportion of component conditions required for detection) and/or the strength of the outputs indicating detection can be adjusted. For example, if the brain is at rest, the general threshold for detection of a receptive field may be set relatively high. Depending on arousal the threshold can then be lowered. The threshold for a receptive field detection can also be modulated more specifically. If some receptive field has recommendation strengths in favour of a general type of behaviour, such a receptive field may have recommendation strengths in favour of lowering the threshold for detection of other receptive fields that have recommendation strengths in favour of specific behaviours within the general type. These threshold modulations can either be viewed as lowering the threshold for detection of the receptive field or as additional conditions.

In addition, there are a number of situations in which the recommendation strengths associated with a receptive field may be useful even if it is not present within current sensory inputs. *Firstly*, if a receptive field has just been detected, there may be behavioural value in prolonging the activity indicating the detection, making the recommendation strengths available for a little longer. For example, to generate an appropriate behaviour in response to a group of objects, detections while one object is the focus of attention may need to be prolonged while the other objects become the focus of attention. *Secondly*, if a receptive field is not present, but was recently present at the same time as a number of currently detected receptive fields, there may be behavioural value in its indirect activation to make its recommendation strengths available. *Thirdly*, if a receptive field is not present, but has often been present in the past at the same time as a number of currently detected receptive fields, there can again be behavioural value in its indirect activation to make its recommendation strengths available. *Fourthly*, if a receptive field is not present, but has changed at the same time in the past as a number of receptive fields that are currently present, its indirect activation may have behavioural value.

Activity prolongations and indirect activations would lead to an excessive degree of activity if unregulated. They must therefore be behaviours which are recommended with various recommendation strengths by currently active receptive fields, and implemented if there is sufficient total strength. These recommendation strengths must themselves be managed. If two receptive fields have been recently active at the same time, their mutual relevance may be high for a short while, but declines rapidly with time. Recommendation strengths in favour of indirect activation could therefore be created reciprocally whenever two circumstances are present at the same time, but such recommendation strengths should decay relatively rapidly. However, if the simultaneous activity is repeated on a number of occasions, the mutual relevance is likely to be higher and recommendation strength decay will be slower. The highest probability of mutual relevance will be if the receptive fields change at the same time, and the associated indirect activation recommendation strengths will decay most slowly. If a recommendation strength contributes to an accepted indirect activation behaviour which is then followed by positive consequences, the strength will be increased and made less susceptible to decay.

Any one receptive field will therefore have varying recommendation strengths in favour of indirect activation of a wide range of other receptive fields on the basis of various temporal correlations in past activity, and consequence feedback following actual indirect activation behaviours. Behavioural recommendation strengths in favour of activity prolongation will also be modulated by consequence feedback.

9.2.1 Detection of Receptive Fields at Cortical Area Level

Receptive fields within a given level of complexity are implemented by the receptive fields of cortical columns within an area. There is an optimum number for the columns detecting their receptive fields at one point in time: below the optimum there is not enough range of recommendation strengths for high integrity selection; above the optimum the processing resources to determine the most strongly recommended behaviour will be excessive and the discrimination poor.

Input to an area can be regarded as recommendation strength in favour of receptive field detection by the area. If total input reaches a minimum level, receptive field detection will occur. If the number of receptive fields detected is below the minimum, receptive field expansion will take place to reach that minimum. Cross inhibition between columns via interneurons prevents levels of detection above the optimum by preventing activity in too many layers V/VI. The cross inhibition must select the receptive fields that most closely reflect current inputs to the area. Hence information about activity in all layers may be relevant, and inhibition between multiple layers will be required. Prevention of activity in layers II/III also avoids unnecessary activity in the hippocampal system.

Thresholds for receptive field detections and/or the strengths of the signals indicating detections can be modulated in two ways. One is relatively general modulation of arousal across wide regions of the cortex by neurotransmitter distribution systems such as the locus coeruleus, raphe nucleus, and the tuberomammillary nucleus of the hypothalamus. A second type of modulation is by the amygdala or other hypothalamic nuclei. This modulation is more specific, affecting the relative probability of detection of receptive fields with recommendation strengths in favour of different general types of behaviour. Different general types include aggressive, fearful, food-seeking, and sexual types.

9.2.2 Detection of Receptive Fields at Pyramidal Neuron Level

The neuron receptive field is defined by the group of conditions instantiated on terminal dendritic branches. One condition is defined by the group of inputs to the branch and their associated weights. A condition is detected if the synapses receiving action potentials within a short period of time inject enough postsynaptic potential for the branch to inject potential (generally via a calcium action potential) deeper into the dendrite. The neuron detects its receptive field if enough branches detect their conditions. In information terms, an input to a branch can be interpreted as a recommendation in favour of detection of the condition, and potential injection by a condition as a recommendation in favour of detection of the neuron receptive field.

A column can detect its receptive field directly, or indirectly on the basis of past temporally correlated activity. If dendritic branches mixed inputs recommending both direct and indirect activation, the branch condition could be detected in inappropriate situations. If branches recommending direct detection of the receptive field could combine with branches recommending indirect activation, the receptive field could again be detected in inappropriate circumstances. Hence direct and indirect recommendations must be segregated on the neuron.

This segregation is achieved by inputs recommending direct detection being located on different dendrites from inputs recommending indirect detection. Activity will only enter the soma if the total within the dendrite exceeds a threshold. For example, inputs from neurons with simpler receptive fields contribute to direct detection and arrive on basal dendrites. Inputs from neurons with more complex receptive fields contribute to indirect detection and arrive on apical dendrites. Different basal dendrites may add together. Alternatively, some basal dendrites could receive inputs from peer receptive fields and detect receptive fields supporting indirect activation on the basis of recent simultaneous activity, as discussed below under priming.

9.2.3 Detection of Receptive Fields at Dendritic Tree and Branch Level

A synapse on a branch injects a postsynaptic potential when it receives an incoming action potential, this postsynaptic potential being proportional to the weight of the synapse. Postsynaptic potentials injected by different synapses diffuse laterally within the branch, and combine to reinforce their individual depolarisations of the membrane potential. If the total depolarisation at the junction between the branch and the dendrite exceeds a threshold, a calcium action potential is initiated in the dendrite that propagates towards the soma. This calcium action potential decays as it propagates but can combine with the calcium action potentials generated by other branches, and if sufficient branches initiate calcium action potentials within a short period of time the calcium action potential reaches the soma and potential propagates towards the axon hillock.

General neurotransmitter distribution systems act on cortical pyramidal neurons by release of neurotransmitters that modulate the ability of calcium action potentials to propagate across the neuron and the ability to generate a second action potential soon after another. Hence release of these neurotransmitters reduces the threshold for receptive field detection and increases the spike generation rate in response to the same degree of condition detection. The amygdala targets dendritic branches with additional glutamatergic synapses, and increases the probability of detection of individual branch conditions.

9.2.4 Detection of Receptive Fields at the Neurochemical Level

Injection of postsynaptic potential occurs by opening of AMPA Na^+ channels triggered by binding to the glutamate released by the incoming action potential. Diffusion of the Na^+ ions laterally carries the membrane depolarisation towards other synapses. A concentration of voltage gated Ca^{++} channels is located at the junction of the branch with the dendrite, and if the membrane depolarisation reaches the threshold for opening these channels, a calcium action potential is initiated and propagates along a path defined by a somewhat lower concentration of Ca^{++} channels. The lower concentration means that the action potential decays before reaching the soma unless reinforced by action potentials initiated by other branches. The concentrations of Ca^{++} voltage gated ion channels thus defines the algorithm by which branch conditions are integrated into the neuron receptive field. This integration algorithm could also be evolved heuristically.

One action of neurotransmitters like norepinephrine (released by the locus coeruleus) and serotonin (released by the raphe nuclei) is to change the opening probability of some types of voltage gated potassium channels. These channels are responsible for the hyperpolarisation following an action potential that reduces the probability of an immediately following action potential. These neurotransmitters can therefore increase their target neuron firing rates. In addition, these neurotransmitters can act on voltage gated calcium channels, and may therefore affect the probability of propagation of calcium action potentials that integrate branch condition detections.

Note that the calcium ions released to propagate the action potentials must be segregated from the calcium ions released by NMDA channels that recommend synaptic weight changes.

The wide range of different types of neurotransmitter receptors and voltage gated ion channels makes very complex integration algorithms possible. However, each detailed chemical process can be regarded as detecting a condition that recommends a range of other chemical processes. At a higher level, combinations of such processes can be effectively approximated as detections of conditions that are complex combinations of the detailed conditions. These higher level conditions recommend neuron level condition detections and changes.

9.3 Semantic Memory

A characteristic example of semantic memory is that when the name of an object is heard, it is possible to describe the visual and other characteristics of the object. Thus hearing the word "bird" makes it possible to describe the appearance of a bird, including wings, two legs, beak and feathers. Furthermore, when an object is seen, it is possible to speak the name of the category to which the object belongs. Thus seeing a typical bird makes it possible to say "that is a bird".

An interesting semantic memory phenomenon is that when subjects are shown visual images of objects, and simultaneously asked if the object belongs to a specific category, the time taken to generate a response varies depending on how typical the object is of the category. Thus if a subject is shown a typical bird, and asked if it is a bird, the correct positive response comes within a characteristic response time. If shown a cat image, and asked if it is a bird, the correct negative response comes within the same response time. However, if shown an atypical bird (such as an ostrich or penguin), the correct positive response requires a slightly longer time.

In the following sections key processes in the cortex for semantic memory will be described at increasing levels of detail. Semantic memory also requires processes in the basal ganglia, thalamus, cerebellum etc., but these processes will not be discussed. Under episodic memory the approach to developing descriptions of detailed processes in these other structures will be provided.

9.3.1 Description of Semantic Memory at the Level of Cortical Areas

As illustrated in Fig. 9.6, visual inputs are processed through a succession of areas detecting receptive fields at increasing levels of complexity. Receptive fields in area TE, and less so in V4, are effective for discrimination between types of visual

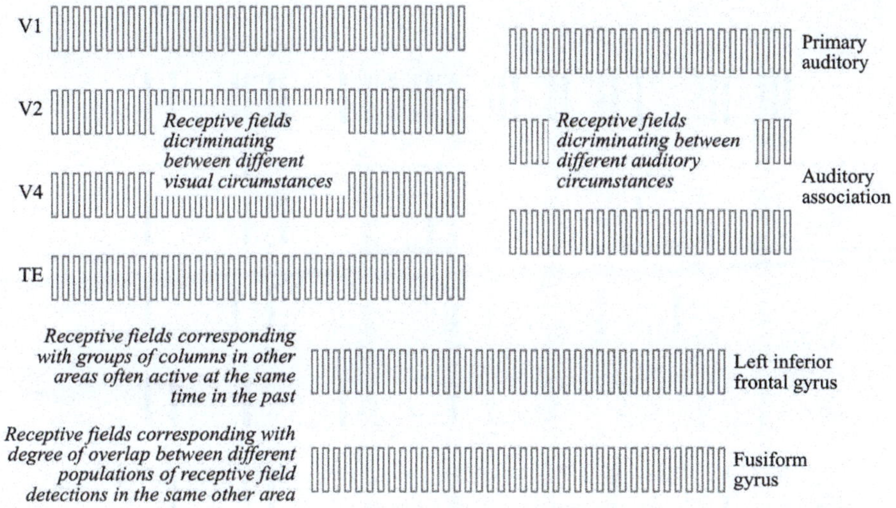

Fig. 9.6 Information model for cortical areas supporting some semantic memory tasks. Receptive fields in visual areas like TE discriminate between visual perceptions of different categories of objects. Receptive fields in auditory association areas discriminate between hearings of different words. Receptive fields in the left inferior frontal gyrus correspond with groups of columns in visual areas or in auditory areas that are often active at the same time, and recommend indirect activation of columns in the other sensory mode that are often active when they are active. These columns perform processes needed for semantic linking of visual and auditory experiences. When two populations of columns are simultaneously active in one area, receptive fields within the fusiform gyrus discriminate between situations in which the two populations contain many of the same columns and situations in which the populations are quite different. These columns perform processes needed for typicality judgements

object. Receptive fields in V4, and less so in TE, are effective for discrimination between features of visual objects. Auditory inputs are processed through primary and association areas, and receptive fields in some association areas are effective for discriminating between different spoken words. Cortical column receptive fields corresponding with groups of columns in visual areas that are often active at the same time are located in areas in the left inferior frontal gyrus, along with receptive fields corresponding with groups of auditory columns often active at the same time. Other receptive fields located in the fusiform gyrus detect the relative activity of different populations of columns within one visual area.

Experience results in the definition of receptive fields that discriminate between different circumstances. Consider the receptive fields defined as a result of visual experiences including different birds. A cortical area such as TE with receptive field complexities effective for discriminating between different types of visual objects is illustrated in Fig. 9.7a. A cortical area able to discriminate between auditory experiences of different words is also illustrated. Many different columns detect their receptive fields in visual experiences of birds. However, as illustrated in Fig. 9.7b, only a small subset of these columns is detected in response to any one

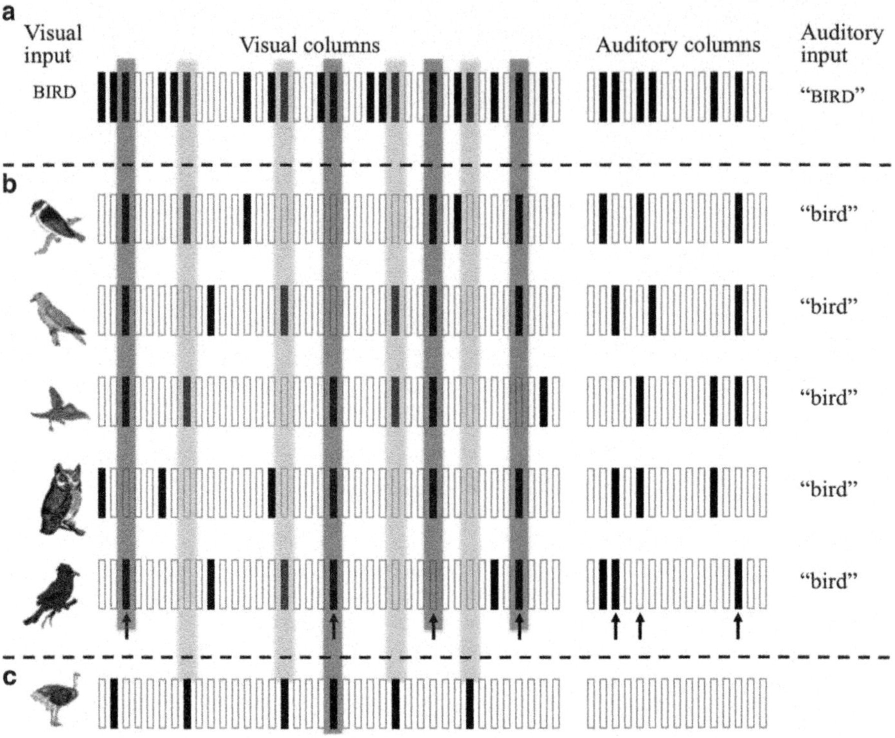

Fig. 9.7 Experiences and column activations in one visual and one auditory cortical area that result in acquisition and expression of semantic memory. (**a**) In a visual area like TE that is able to discriminate between different categories of visual objects, many different columns can detect their receptive fields in an object of any one category. The columns that could sometimes detect their receptive fields in response to seeing instances of the category BIRD are *shaded*. However, only a small subset of these columns will detect their receptive fields in any one instance of the category. Similarly, many different columns in an auditory association area can detect their receptive fields in a hearing of a particular word. The columns that could sometimes detect their receptive fields in response to hearing different instances of the word "BIRD" are *shaded*. Only a small subset of these columns will detect their receptive fields in any one hearing of the word. (**b**) When different instances of birds are seen, different subsets of visual columns detect their receptive fields. When different instances of the word "bird" are heard at the same time, different subsets of auditory columns detect their receptive fields. Because there are some visual similarities between birds, some visual columns are activated relatively frequently, but not always. Because there are some auditory similarities between different hearings of the word "bird", some auditory columns are activated relatively frequently, but not always. These columns are indicated by the *dark shaded rectangles*. Hence there is a subset of the auditory columns that is often active at the same time as a subset of the visual columns, indicated by the *arrows*. Typical visual bird experiences will tend to result in detection of a significant proportion of the visual subset, and typical hearings of the word "bird" a significant proportion of the auditory subset. (**c**) An atypical bird will activate a small proportion of the frequently active subset. However, some of the other columns activated in response to the atypical bird will have often been detected in viewings of more typical birds, as indicated by the *light shaded rectangles*. These atypical bird columns will therefore tend to have been active in the past at the same time as the typical bird columns. Second order information is therefore available to tie the atypical bird to the category

bird experience, and the subset is different for different birds. As also illustrated in Fig. 9.7b, because of the visual similarities between birds, some columns tend to be detected relatively often (but not always) across different bird instances. Similarly, across different experiences of hearing the word "bird" there is a set of columns that are sometimes activated: a different subset is detected in response to each experience of hearing the word, and because of auditory similarities there are some auditory columns detected relatively frequently in response to hearing the word "bird".

If the word "bird" is often heard at the same time as a visual bird is seen, there will be a group of visual columns often active at the same time as a group of auditory columns. On the basis of frequent past simultaneous activity, the auditory columns will therefore acquire recommendation strengths in favour of indirect activation of the visual columns, and vice versa. Hence hearing the word "bird" will activate auditory columns, and these columns will indirectly activate visual columns often active in the past at the same time as the auditory columns, creating a pseudo-visual image of a bird that is a weighted average of past visual bird experiences in which birds have been seen and the word has been heard. Conversely, the frequently active visual columns will acquire recommendation strengths in favour of behaviours like saying "that is a bird". When a typical bird is seen, a significant proportion of these visual columns will be activated, and such behaviours will be strongly recommended.

When different instances of the same category are seen, or the same instance is seen on different occasions, in different postures, or viewed from different angles, the visual columns activated in areas like V1 and V2 will be completely different, and only similar in higher areas like TE. Hence visual column groups defined in the left inferior frontal gyrus will tend to include TE columns but fewer V4 columns and few V1 and V2 columns. This tendency will be reinforced by a genetic bias on connectivity, for example favouring indirect activation connectivity from the left inferior frontal gyrus to TE. As a result, the pseudoimage activated by hearing a word will include few of the columns that are activated in the earlier visual areas in response to a direct visual experience. In other words, the semantic memory will not be a visual hallucination.

Once visual columns in TE have been activated in response to the word these columns can, via the left inferior frontal gyrus, activate columns in V4 and TE on the basis of frequent past simultaneous activity. Because bird features like wings, feathers etc. have often been present at the same time as birds, the secondary indirect activation contains columns often active in the past when different features have been present. The secondary activation population will therefore have recommendation strengths in favour, for example, of naming those features.

In the response time task described earlier, subjects are shown an image and asked "is this a bird?" and the time to come up with a response measured. The image directly activates a population of columns in the cortical area that discriminates between types of object. However, the columns activated in response to hearing the word "bird" indirectly activates columns in the same area. As illustrated in Fig. 9.8, if the image is a typical member of the category, there will substantial overlap between the directly and indirectly activated columns in TE. Receptive

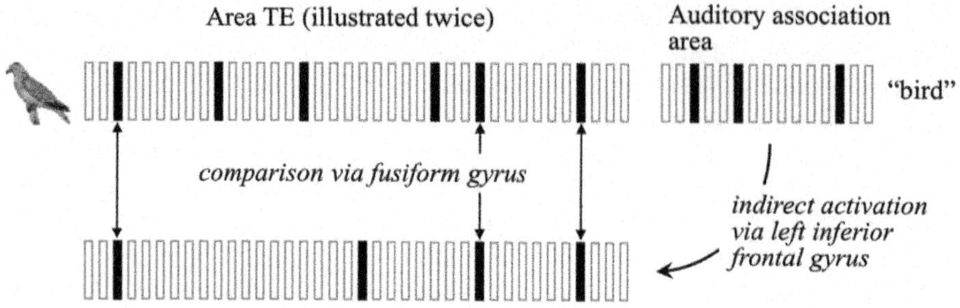

Fig. 9.8 Generating a response to the question "Is this a bird" by column activation typicality determination. A population of columns is directly activated at some phase of frequency modulation in area TE in response to the visual input. Another population is indirectly activated in TE at a different phase of frequency modulation in response to hearing the word bird. Outputs from both populations are synchronized and released to an area of the fusiform gyrus with receptive fields able to discriminate between situations in which there is high overlap between the columns in the two populations, and situations with low overlap. These fusiform gyrus columns can therefore recommend different behaviours such as saying "yes" or "no"

fields perhaps located in the fusiform gyrus can detect the degree of overlap and recommend saying "yes". If the image is a clear non-member of the category, receptive fields detecting the very small overlap can recommend saying "no".

If an atypical bird such as an ostrich is seen, the overlap may be intermediate and the total recommendation strength in favour of either verbal response may be too small to generate an immediate speech behaviour. However, as illustrated in Fig. 9.7c, some of the columns activated in response to seeing the ostrich that are not among the columns often active in the past in response to birds may nevertheless have been active in the past in response to some more typical birds. In information terms, these columns correspond with circumstances that occur less frequently when birds are seen. Because birds have been seen many times, these less frequently active columns have been active in the past relatively often when the more frequently active columns were also active. Hence these less frequently active columns have some recommendation strengths in favour of indirect activation of the more frequently active columns. Acceptance of these recommendation strengths will expand the overlap between the directly and indirectly activated visual column populations, generating adequate recommendation strength in favour of the appropriate response, but with a slight time delay to allow the extra indirect activation step.

An interesting sub-phenomenon of semantic memory is the "tip-of-the tongue" experience. Here, the active column population does not have quite enough total recommendation strength to achieve speaking a word. However, a significant subset of the relevant population and its recommendation strengths are activated. This subset may have sufficient total recommendation strength in favour of speaking the first syllable, or an approximate synonym of the target word.

Another sub-phenomenon is the generation of words from a target, such as thinking of five letter words beginning with the letter M. In this case, a column population is activated on the basis of frequent past simultaneous activity from the inputs "M" and "five letters". This population then drives other populations on the basis of frequent past simultaneous activity, some of which are sufficiently close to the populations activated in response to such a word that they have a predominant recommendation strength in favour of speaking the word.

9.3.2 *Description of Semantic Memory at the Level of Cortical Column Receptive Fields*

The receptive fields in the left inferior frontal gyrus are not cognitively unambiguous. In other words, one such receptive field will not correspond with the group of visual columns often active at the same time when one type of visual object is seen. Rather, as illustrated in Fig. 7.15, different groups of auditory association columns often active at the same time will define column receptive fields in the left inferior frontal gyrus. A column in the left inferior frontal gyrus will target a range of visual association columns that have tended to be active when it has also been active.

The intermediate receptive fields provide a number of advantages. Firstly, the volume of connectivity is less than would be required for direct connectivity from the auditory association to the visual association area. Secondly, connectivity via the left inferior frontal gyrus allows two management stages in the selection of an indirect activation behaviour. This two stage management is also illustrated in Fig. 7.15. Column outputs from the auditory association area have recommendation strengths into a component in the striatum corresponding with the behaviour of releasing auditory association outputs to the left inferior frontal gyrus. Other cortical areas may also have recommendation strengths of this type. If the total of such recommendation strengths is predominant, the outputs are released and columns in the left inferior frontal gyrus detect receptive fields. These left inferior frontal gyrus fields have recommendation strengths in favour of their own release to the visual association area. Again, other cortical areas may have such recommendation strengths. Hence the semantic memory behaviour must compete with alternative behaviours that are also appropriate in current circumstances.

Receptive fields required to support discrimination between category members and non-members are illustrated in Fig. 9.9. If an instance is a member of the category, the columns directly activated in TE by viewing that instance overlap with the columns indirectly activated by hearing the category name. For a non-member the overlap is small. Columns located in the fusiform gyrus have some receptive fields that are detected in response to strong average activity by a small group of TE columns, other receptive fields that are detected in response to weaker average activity by a larger group of TE columns. These columns can therefore discriminate between category members and non-members, and acquire appropriate recommendation strengths such as saying "yes" or "no" in response to category membership questions.

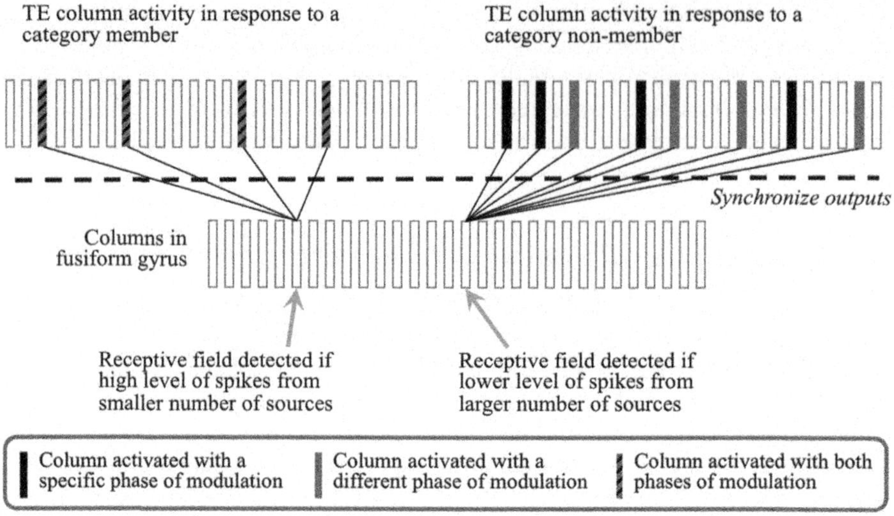

Fig. 9.9 Receptive field definitions to allow category typicality discriminations. Some columns in the fusiform gyrus have receptive fields that are detected in response to inputs from many different TE columns. Others have receptive fields detected in response to stronger inputs from fewer different TE columns. Two populations of receptive field detections are established in TE at different phases of frequency modulation, in response to direct visual inputs and indirect activation in response to auditory inputs. TE outputs from the two populations are brought into the same phase of frequency modulation for release to the fusiform gyrus area. Different fusiform gyrus columns will tend to be activated if the populations overlap or do not overlap, corresponding with category members and non members

9.3.3 Description of Semantic Memory at the Level of Neuron Receptive Fields

A neuron within a visual association area cortical column may be activated either by inputs from an earlier visual area or by inputs from the left inferior frontal gyrus. However, if the neuron was receiving a relatively small number of active inputs from each of the two sources, detection of either receptive field would be inappropriate. Hence inputs from the two sources must be segregated and integrated separately as illustrated in Fig. 9.10. Inputs derived from earlier visual areas target the basal dendrites, while inputs from the left inferior frontal gyrus target the apical dendrite.

The apical dendrite receptive field is defined by a set of conditions instantiated by terminal branches of the dendrite. A condition is made up of a group of inputs from the left inferior frontal gyrus, each input having an individual synaptic weight. A branch may instantiate a range of conditions if activity by different subsets of the inputs can all result in injection of potential from the branch deeper into the dendrite. This range can be extended by individual synaptic weight increases. Alternatively, a new condition can be added if the weights of all recently active synapses to a branch with very low or zero weight inputs are increased. Such changes to branches expand the apical dendrite receptive field.

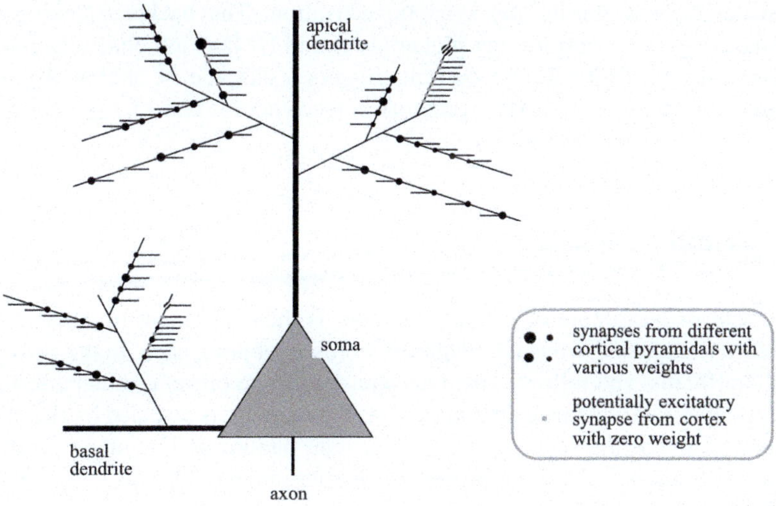

Fig. 9.10 Separate receptive fields for direct and indirect detections. Inputs from senses or from simpler monomodal or polymodal receptive fields are received on the basal dendrites. Inputs from (generally higher complexity) receptive fields corresponding with circumstances in which the direct receptive field should be indirectly activated on the basis of temporal correlations in past activity are received on the apical dendrite. Separate integration means that the chance of inappropriate neuron outputs based on partial receptive field detections is reduced

When left inferior frontal gyrus outputs are released to the visual association area, they drive receptive field detections. If there is less than a minimum number of such detections, the hippocampus drives receptive field expansions. Receptive field expansion can only occur if a significant proportion of existing conditions are already being detected. Hence expansion proceeds by increases in the synaptic weights inputs that are active at the same time as a significant number of other inputs. The synaptic weight of an input will be greater the more often it is active when the neuron fires. Hence the apical dendrite receptive field tends towards a group of left inferior frontal gyrus columns that are often active at the same time. Limits to synaptic weights prevent one input from becoming too predominant in the receptive field definition.

Similarly, typicality discrimination receptive fields in the fusiform gyrus develop on the basis of genetically defined bias on their initial connectivity and development on the basis of correlated activity.

9.3.4 Description of Semantic Memory at the Level of Neuron Chemistry

The firing of a neuron generates a backpropagating action potential which can increase the weights of recently active synapses in recently active branches. If a branch has recently been active, the backpropagating action potential reaches the branch. Recently active synapses on the branch have NMDA channels bound with glutamate released during the synaptic activation. The backpropagating action potential triggers opening of any glutamate bound NMDA ion channels, passing a calcium current. This calcium current triggers a cascade of chemical reactions leading to an increase in AMPA receptors in the synapse. This increase instantiates an increase in synaptic weight.

9.4 Priming Memory

A characteristic example of priming memory is the experiment in which subjects are shown a sequence of visual images of words or objects, each image shown very briefly (<100 ms) and followed by a masking pattern to prevent retinal afterimages. Subjects are asked to name each image. Priming memory is revealed by the observation that although the success rate for accurate naming is low, it is significantly increased for a repetition of an image seen earlier in the sequence.

9.4.1 Description of Priming Memory at the Level of Cortical Areas

The mechanism supporting priming memory is illustrated at the cortical area level in Fig. 9.11. The column activation for a normal viewing of an image is shown in Fig. 9.11a. These columns have a strong total recommendation strength in favour of correctly naming the object. The column activations for the first and second brief viewings are shown in Fig. 9.11b. Because of its brevity, the first viewing directly activates only a subset of the columns activated by a normal viewing. This subset typically does not have an adequate total recommendation strength in favour of correct naming. For a period of time after the first viewing, the activated columns have recommendation strengths in favour of indirectly activating each other on the basis of recent simultaneous activity. The second brief viewing activates another subset which could be different from but partially overlap with the first. If there is overlap, the columns activated in both the first and second viewings have recommendation strengths during the second viewing in favour of indirectly activating the rest of the columns activated in the first viewing. The combination of directly and indirectly activated columns following the second viewing is more likely to have sufficient recommendation strength to correctly name the image.

9.4.2 Description of Priming Memory at the Level of Neuron Receptive Fields

One behavioural value of priming memory is to construct an appropriate behavioural response to an object that is glimpsed partially several times in a brief period

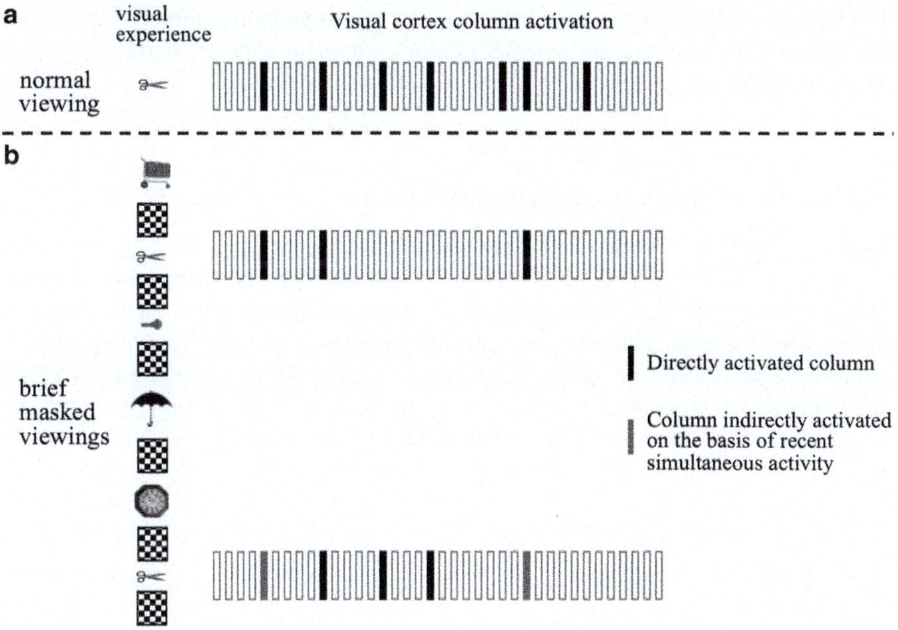

Fig. 9.11 Column activations in an area such as TE in response to the different possible visual experiences relevant to priming memory. (**a**) In a normal viewing of an object, a set of columns detect their receptive fields, and this set has a predominant recommendation strength in favour of behaviours appropriate to the object, such as naming it. (**b**) When the same object is viewed briefly with masking, the small amount of relevant visual input means that the group of activated columns is only a small proportion of the columns activated by a regular viewing. These columns will generally have too little total recommendation strength to speak the name. When the same object is viewed shortly afterwards, there may be some overlap between the two small groups. If such an overlap is present, active columns in the second group that were also active in the first group will have recommendation strengths in favour of indirect activation of the other columns in the first group. These recommendation strengths will increase the proportion of normal viewing columns that are active, increasing the available appropriate speech recommendation strength, and increasing the probability of a correct response

of time. Indirect activation on the basis of recent simultaneous activity is therefore valuable within visual processing areas. Because the value lies in rapid enhancement to immediate perception, such indirect activations can be applied immediately without the need for selection through the basal ganglia. Because the recommendation strengths are required rapidly and are only relevant short term, there is little value in creation of column receptive fields in a separate area. Connectivity to support such indirect activation can therefore be expected to be within visual areas such as V2, V4 and TE.

Priming memory can then be implemented by lateral connectivity between neurons in different columns of the same area. Such lateral connectivity could target basal dendrites. This lateral connectivity has very low synaptic weights which are briefly increased following correlated activity. Inputs supporting indirect activation on the basis of recent simultaneous activity could be widely distributed on the basal dendrites.

9.4.3 Description of Priming Memory at the Level of Neuron Chemistry

In their long term state, synapses supporting indirect activation on the basis of recent simultaneous activity have NMDA receptors but few AMPA receptors. Simultaneous activity of source and target neuron results in addition of AMPA receptors which increase the synaptic weight. However, there are no dopaminergic synapses located nearby, and the additional AMPA receptors will therefore be removed with a decay time of the order of an hour.

9.5 Episodic Memory

Episodic memory is the ability to recall specific events that occurred at a particular point in time. Such recall can be triggered by words providing an indication of an event (e.g. "what happened at the party yesterday?"), or by other cues such a smell. Once recalled there is an ability to describe what happened during the event, including self actions and feelings.

9.5.1 Description of Episodic Memory at the Level of Major Anatomical Structures

The anatomical architecture for episodic memory recall is illustrated in Fig. 9.12. Visual experience results in receptive field detections in a succession of cortical areas at increasing receptive field complexities. In area TE, receptive fields are effective for discriminating between different types of object. In the posterior parietal areas, receptive fields combine receptive field detections within multiple objects and are able to discriminate between different types of groups of objects [934]. A simple example of different types of groups of objects might be two people who could be discussing, arguing, cooperating in a task, playing a game, fighting, or one chasing the other. Yet more complex receptive fields combining receptive field detections within groups of objects discriminate between different types of groups of groups or complex scenes. Many cortical areas may be required detect receptive fields in ranges of complexity able to make behaviourally useful discriminations between different scenes, and these receptive fields will incorporate information derived from multiple senses. These multiple areas are simplified in Fig. 9.12 into

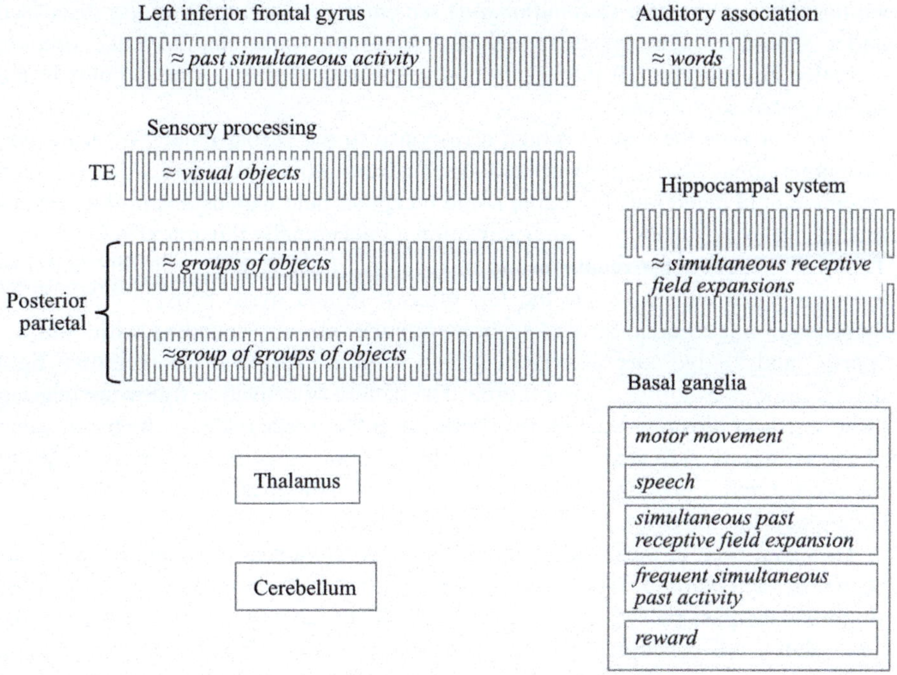

Fig. 9.12 Information model for episodic memory at the level of major anatomical structures. Within visual inputs, receptive fields are defined and detected in different areas that are able to discriminate between (≈) different objects, groups of objects and groups of groups. Within auditory inputs, receptive fields are defined and detected that are able to discriminate between different words. Receptive fields in the left inferior frontal gyrus correlate with groups of columns in other areas that have often been active at the same time in the past. Receptive fields in different parts of the hippocampal system correlate with groups of cortical columns that have tended to expand their receptive fields at the same time in the past. The basal ganglia receive inputs of receptive field detections from cortical areas and interpret each detection as a recommendation in favour of many different behaviours, determining and implementing the most strongly recommended behaviour. The most appropriate detailed cortical information release to implement the behaviour is determined and carried out by the thalamus. Frequently used sequences of attention, indirect activation, speech and reward behaviours are captured by the cerebellum and after initiation by the basal ganglia require no further intervention by that structure

three areas. The symbol ≈ is used to indicate *effective discrimination between*, and the functional roles of the three areas are ≈*objects*, ≈*groups of objects* and ≈*groups of groups*.

Receptive field expansions occur whenever there is a degree of novelty in current experience. In the brain, novelty is defined such that there is some degree of receptive field expansion in response to almost every experience, although the degree will be much greater for some experiences. The process for receptive field expansion generates receptive fields in the hippocampal system able to discriminate between different groups of columns across the cortex that expanded their receptive fields at similar times in the past (≈*simultaneous receptive field expansions*). As discussed under semantic memory, there are also ≈*frequent past simultaneous activity* receptive fields in the left inferior frontal gyrus and ≈*words* receptive fields in the auditory association cortex.

Consider now the way in which a response to the verbal input "What do you remember about Mark's party at the lodge?". During the party there will have been receptive field detections including some receptive field expansions in the sensory processing areas. A pattern of such activation is conceptually illustrated in Fig. 9.13a. The verbal request for recall results in receptive field detections in the auditory association cortex, which via the left inferior frontal gyrus activate columns in the ≈*visual objects* area. Several separate populations corresponding with "Mark", "party" and "lodge" are activated, which in combination drive receptive field detections in the ≈*group of groups* area. The pattern of activation following hearing these words in illustrated in Fig. 9.13b. Because the words relate to the party, some of the indirectly activated columns will be columns also active during the party experience and a proportion will have expanded their receptive fields during the experience, but the overlap will be relatively small.

However, each of these columns will have recommendation strengths in favour of indirect activation of other columns in the same and other areas on the basis of simultaneous past receptive field expansion. Such recommendation strengths for one column are conceptually illustrated in Fig. 9.13c. If such recommendation strengths are accepted, a secondary indirectly activated column population will develop. This secondary population will contain a higher proportion of columns that expanded their receptive fields at the same time in the past. Because of the selection of the words in the verbal input, the overlap with the population active during the original experience will be larger. Columns in the secondary population will also possess recommendation strengths in favour of indirect activation on the basis of past simultaneous receptive field expansion, and use of such recommendation strengths can lead to a tertiary population with even greater overlap with the population active during the original experience, and so on. Activation of the columns detected during the original experience will be experienced as a re-experience, but without receptive field detections close to sensory inputs (i.e. not a visual hallucination). The recommendation strengths of the indirectly activated columns include speech recommendations, making it possible to describe the original experience in a manner similar to how it would have been described at the time.

Accessing an episodic memory therefore requires a sequence of behaviours. Firstly, the words must drive populations of indirect column activations at the ≈*objects level*. Secondly, the populations generated by different words must be combined to activate a population at the ≈*groups of objects* level. Thirdly, the populations at the ≈*objects* and ≈*group of objects* levels must drive activation of a secondary population on the basis of simultaneous past receptive field expansion. This secondary population may be located in all three levels. Fourthly, perhaps the secondary population must drive a tertiary population on the same basis, and further populations. Fifthly, the recommendation strengths of the final population in favour of speech behaviours must be accepted.

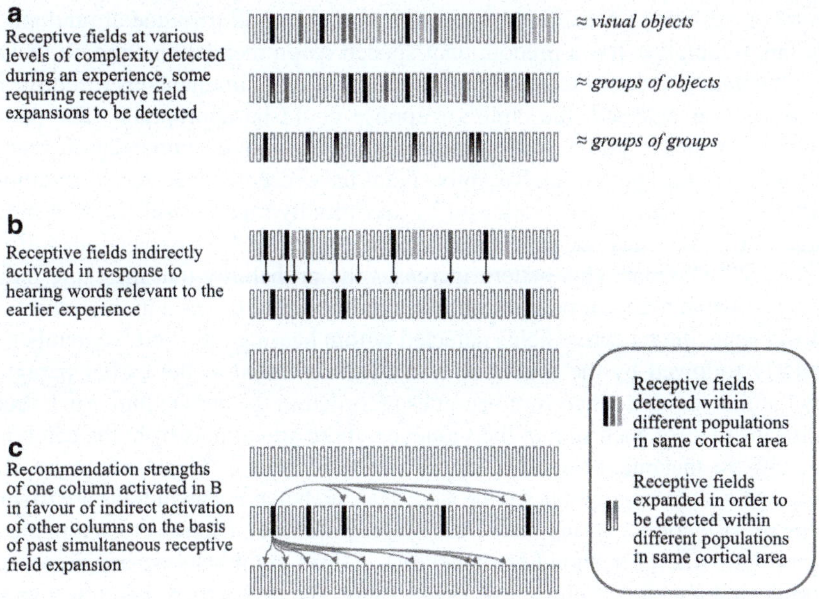

Fig. 9.13 Cortical area and column level information model for recording and accessing an episodic memory. (**a**) During the original experience of an event there will generally be some degree of novelty. As a result, there will be a degree of receptive field expansion, typically greatest at the ≈*groups of groups of objects* level, less at the ≈*groups of objects level*, and least at the ≈visual objects level. (**b**) When words are heard later, they tend to indirectly activate columns in those same visual areas on the basis of frequent simultaneous past activity. If the words have any relationships with the original experience, a few of the columns activated during that experience will be included. (**c**) The columns indirectly activated in (**b**) will have recommendation strengths in favour of indirect activation on the basis of past simultaneous receptive field expansion. If a bias is placed on this type of behaviour, a secondary indirectly activated column population will develop which will tend to be more consistent on the basis of past simultaneous receptive field expansion. Another stage of indirect activation on the same basis will result in a tertiary population highly consistent on that basis. With appropriate choice of the initial words, there is a good probability of significant overlap between the tertiary population and the population activated in the original experience. Such a tertiary population will have recommendation strengths in favour of describing that experience

For a successful episodic memory retrieval, at each stage in the sequence the predominant recommendation strength must be in favour of the appropriate next behaviour. As illustrated in Fig. 9.12, different components in the basal ganglia correspond with different types of behaviour including motor, speech, and various indirect activation behaviours. Outputs from the different cortical areas may have different recommendation strengths in favour of any or all of these types of behaviour. The selected behaviour is implemented by the thalamus releasing the outputs of selected cortical areas to selected targets.

Consider how such a behaviour sequence could be learned. A small child is asked by a parent "Do you remember what you did this morning?". The child's brain has a number of indirect activation behaviours available, and tries one at random. The resultant population has a predominant speech recommendation strength which is implemented. If indirect activation on the basis of past simultaneous receptive field expansion was selected, the child's response could be appropriate. The parent's reaction activates receptive fields in the child's brain with recommendation strengths in favour of reward behaviour. Receptive fields have recently been detected within the auditory input "remember", followed by an (initially random) selection of indirect activation on the basis of past simultaneous receptive field expansion, followed by reward behaviour. This pattern increases the probability that the same indirect activation behaviour will be selected when "remember" is heard in the future.

Over time, the receptive fields detected within hearing the word "remember" are regularly followed by the sequence of behaviours listed in the earlier paragraph. This regular sequence therefore becomes recorded in the cerebellum. Such recording means that for each step in the sequence, there are outputs from the cerebellum targetting the thalamic nuclei that release currently cortical receptive field detections between the areas appropriate for the step. The cerebellar nuclei outputs are triggered by the precise combination of cortical receptive fields that were detected in the past just prior to the corresponding behaviour. These combinations are recorded in Purkinje cell receptive fields in the cerebellum. The cerebellum therefore bypasses the selection of each individual behaviour by the basal ganglia.

The higher the degree of novelty in an experience, the greater the degree of receptive field expansion. Hence experiences with a high degree of novelty will in general be easier to recall. Highly emotional situations also trigger a higher level of receptive field expansion, and such situations are also therefore easier to recall. The combination of high emotion and high novelty can lead to flashbulb memories. An extreme degree of receptive field expansions is triggered in many areas, resulting in a subsequent ability to recall a wide range of details about the event and the circumstances in which the person heard about the event.

Episodic memory allows description of the sequence of events in a past experience. Such a sequence is generated by an initial process of indirect activation that tends to recover the most novel event in the sequence. The earliest event can then be recovered by several stages of indirect activation of columns that expanded their receptive fields just before the currently active column population. The event can then be re-experienced by a sequence of indirect activations on the basis of receptive field expansion just after the currently active column population.

If an episodic memory is frequently recalled, then hearing the trigger words will activate a population of receptive fields with predominant recommendation strength in favour of activation the population of columns corresponding with the first event in the memory. This recommendation strength is on the basis of frequent past column activity shortly after current column activity. The sequence of events in the memory can then be recovered by the same type of indirect activation. Hence for a frequently recalled memory, the indirect activation mechanism can shift from past simultaneous receptive field expansion to frequent past simultaneous activity. The region with the supporting receptive fields will therefore shift from the hippocampal system to the left inferior frontal gyrus. In other words, in information terms memory access has shifted from episodic to semantic. The sequence of indirect activations on the basis of frequent past simultaneous (or shortly later) activity can then be recorded in the cerebellum, and be accessed much more smoothly and reproducibly.

In fact, even regular semantic memories will generally be initiated as episodic memories. For example, suppose that a word is heard for the first time, and a novel image experienced simultaneously. The novelty of verbal and visual experiences results in significant receptive field expansions in auditory and visual cortical association areas. As a result, hearing the word will activate a pseudovisual image of the object on the basis of past simultaneous receptive field expansion. The population of columns will include receptive fields containing information about the circumstances of the first experience (where, who spoke the word etc.). With usage of the word, access will shift to frequent past simultaneous activity, and the information about circumstances will no longer be accessible.

9.5.2 Description of Episodic Memory at Column and Neuron Level

Episodic memory as described in the previous section includes processes performed by the cortex, hippocampal system, basal ganglia, thalamus and cerebellum, and touched indirectly on processes performed by the amygdala. The approach in this section will be to provide a description of selected more detailed processes at the level of neurons and columns in the cortex and hippocampal system. At more detailed levels the descriptions are similar to those for receptive field definition and neurochemistry provided earlier.

There are a number of different ways in which the information processes supporting episodic memory could be implemented within the cortex at the neuron level. Not enough is known about detailed connectivity to pin down the exact mechanisms. The aim in this section is to describe one possible mechanism that is consistent with known anatomy and physiology. A key objective is to demonstrate that the available neuron level mechanisms, using only information that is available to them at this level of detail, are capable of supporting the higher level phenomena.

A possible model of this type at the neuron receptive field level is illustrated in Fig. 9.14. Two columns are illustrated, one in the cortex and the other in the perirhinal cortex of the hippocampal system. In the cortical column, an output signal indicating activity in layer II/III and no activity in layer V/VI is required to indicate that receptive field expansion in the column is appropriate. This signal is generated by a subpopulation (P_2) of pyramidal neurons in layer II/III that are inhibited by layer V/VI pyramidal neurons via interneurons, and communicated to the column in the perirhinal cortex.

P_2 outputs from a range of cortical columns, and from multiple neurons in each such column, are the inputs to layer II/III neurons in the perirhinal cortex. These layer II/III perirhinal neurons therefore have receptive fields corresponding with groups of columns that have often had internal activity and no outputs at same time

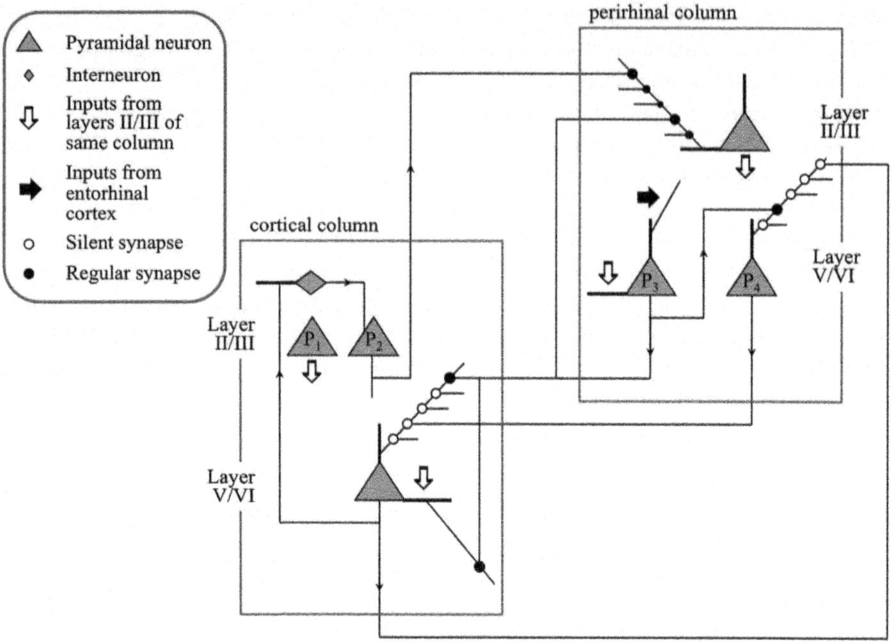

Fig. 9.14 A possible model to manage receptive fields for indirect activation on the basis of past simultaneous receptive field expansion. In one cortical column there are two populations of pyramidal neurons in internal layers II/III with similar receptive fields. Population P_1 is the source of inputs to the basal dendrites of layer V/VI neurons in the same column. Population P_2 is inhibited by layer V/VI neurons in the same column via interneurons. This second population therefore indicates strong internal activity but no output activity by the column. Outputs from P_2 contribute to the receptive fields of layer II/III neurons in perirhinal cortex columns, and the outputs from these perirhinal cortex neurons go on to the hippocampus proper via the entorhinal cortex to determine the appropriateness of receptive field expansions. Layer V/VI pyramidal neurons in the column have direct basal dendritic receptive fields that are combinations of P_1 neurons, and indirect apical receptive fields that are combinations of inputs from population P_4 of layer V/VI pyramidal neurons in the entorhinal cortex. There are two populations of neurons in layers V/VI of the perirhinal cortex column, P_3 and P_4. Neurons in both of these populations only target neurons in columns from which layer II/III neurons receive inputs, including provisional connectivity. P_3 neurons receive inputs from layer II/III neurons in the same perirhinal column, and fire if there is strong activity by their layer II/III inputs plus inputs from higher in the hippocampal system indicating selection for receptive field expansions. P_3 neurons in one perirhinal column target the cortical columns from which the layer II/III neurons in the same perirhinal column receive inputs, and drive receptive field expansions of both the direct and indirect receptive fields of the layer V/VI neurons in those columns. P_3 neurons also drive receptive field expansions in the P_4 population in the same entorhinal column. The P_4 population has receptive fields that are combinations of layer V/VI neurons in the same cortical columns that define the receptive fields of the layer II/III neurons. The perirhinal column drives receptive field expansions in a group of cortical columns. If such receptive field expansions are occurring, the receptive field of P_4 layer V/VI neurons will also tend to expand, adding combinations of inputs from cortical columns in its group that were active at the time of the expansion. The indirect receptive fields of the layer V/VI neurons in active columns will also expand. In the future, if outputs from layer V/VI neurons in a currently active cortical column are released to the perirhinal cortex, P_4 neurons will be activated and drive indirect activations in cortical columns that tended to be active at the time of a past receptive field expansion that included the active cortical column

in the past. One population (P_3) of layer V/VI neurons in the perirhinal column have receptive fields defined on their basal dendrites that are combinations of layer II/III neurons in the same perirhinal column. The two populations of perirhinal column layer V/VI neurons (P_3 and P_4) both provide outputs back to cortical columns. Only cortical columns that send outputs to layer II/III neurons in the perirhinal column can receive inputs from the layer V/VI neurons.

P_3 outputs drive receptive field expansions in its target cortical columns. These P_3 outputs also drive receptive field expansions in layer II/III neurons and in P_4 layer V/VI neurons of the same perirhinal column. A neuron in the P_3 population can only be activated if it is both receiving significant input from layers II/III on basal dendrites and inputs from the entorhinal cortex on apical dendrites indicating that receptive field expansion is appropriate for the group of cortical columns targetted by the perirhinal column.

Layer V/VI neurons in P_4 receive inputs from layer V/VI neurons in the same cortical columns that provide inputs to layer II/III. These P_4 neurons target the apical dendrites of layer V/VI neurons in those columns.

In the cortical column, layer V/VI neurons have basal dendrite receptive fields defined by groups of layer II/III neurons within the same column. They also have apical dendrite receptive fields defined by groups of population P_4 neurons across a range of perirhinal cortex columns. The terminal branches making up the apical receptive field include branches with silent synapses from P_4 neurons in a range of perirhinal columns plus inputs with substantial synaptic weights from P_3 neurons.

The two episodic memory information processes are recording of a memory and access to the memory. A memory is recorded because there is less than the minimum number of receptive field detections in a number of areas. The internal activity of cortical columns that are not producing outputs is communicated to columns in the perirhinal cortex. This triggers activity in the layer II/III pyramidal neurons. Activity of these neurons is communicated to the entorhinal cortex and beyond, but alone is not sufficient to trigger layer V/VI activity in those perirhinal columns. If enough signals come back from higher in the hippocampal system, activity of P_3 will be generated. This activity has a number of effects. Firstly, P_3 inputs trigger expansions in the basal dendritic receptive fields of layer V/VI pyramidals in its target cortical columns. These basal dendritic receptive fields are defined by combinations of P_1 inputs to layer V/VI pyramidals, and receptive field expansions by these layer V/VI pyramidals therefore expand the receptive field of the column. Secondly, the presence of P_3 activity means that P_4 population neurons will be activated by having their apical receptive fields expanded by additions of groups of currently active columns within the set of columns targetted by the perirhinal column. Thirdly, the P_4 activity plus P_3 activity will drive expansion of the apical receptive fields of active V/VI neurons in the active columns. The second and third processes mean that later, if a group of cortical column layer V/VI neurons are active and if their outputs are released to P_4 neurons, the P_4 activity will encourage indirect activations in other columns that expanded their receptive fields at the same time in the past as the currently active column. Note that at this more detailed level of description, the mechanisms support later indirect activation of all the columns in a group that were active during a novel experience, not just the columns in the group that expanded their receptive fields.

This model addresses how the perirhinal cortex could support episodic memory. The parahippocampal cortex will play a similar role for other cortical areas. The entorhinal cortex and hippocampus proper will play similar roles in indirect activations on the basis of simultaneous receptive field expansions across broader ranges of cortical areas.

In summary, if a specific cortical column has strong internal activity and no output, and if it is part of a number of groups defined in the perirhinal cortex in which many of the columns have strong internal activity and no output, the specific column is likely to expand its receptive field. If later that cortical column is detecting its direct receptive field and if its outputs are released to all the perirhinal cortex columns with which it has reciprocal connectivity, those columns will produce outputs encouraging indirect activation of any columns that were active when specific column was part of a receptive field expansion event in the past.

9.5.3 Episodic Memory and Navigation

When a brain is viewing a novel location, the relative orientation of the objects in the vicinity and perhaps some of the objects will be novel. As a result, there will be a number of columns that will all expand their receptive fields at the same time. Some columns and/or neurons in the hippocampal system will therefore carry information about the location. There are therefore neurons in the hippocampal system that have receptive fields (called place fields) corresponding with different locations. Furthermore, when in an adjacent location there will also be receptive field expansions, and these expansions will occur slightly before or after the expansions in the first location. Hence indirect activation on the basis of past temporally correlated receptive field expansion can access information needed for navigation.

9.6 Procedural Memory

Procedural memory is the memory for skills. Many skills involve generating complex muscle movements in response to complex combinations of visual and somatosensory inputs. An example of a relatively simple skill often investigated in the laboratory is mirror tracing, in which a subject attempts to trace an outline image when the outline, tracing pencil and hand holding the pencil are only visible reflected in a mirror. A much more complex skill is finding the way to move around a novel house. Another skill is the ability to move mouth and throat muscles to generate speech. Unlike the ability to remember different occasions on which a skill was learned, it is not generally possible to revert to the skill level used at different times in the past, there is no memory for the past state of skills. Furthermore, it is generally not possible to describe exactly how the skill is performed.

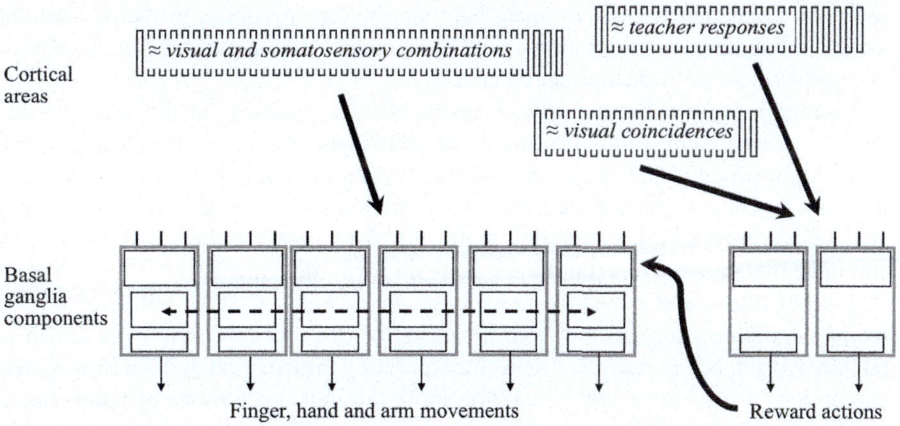

Fig. 9.15 Architecture for understanding the learning of mirror drawing. Receptive fields are defined within combinations of propioceptic information about arm, hand and finger positions plus visual information derived from attention to a visual image. These receptive fields discriminate between situations in which different combinations of muscle movements are appropriate, and each such receptive field recommends a range of such movements. Receptive fields discriminating between complex circumstances corresponding with different behaviours of another person recommend reward behaviours. Receptive fields detected within the degree of coincidence between image and results of drawing movements also recommend reward behaviours. The basal ganglia maintains the recommendation strengths and drives behaviours, with a final more detailed selection of exact muscle movements by the thalamus (not shown)

Measurement of the capabilities of patients who have lost the ability to create new semantic or episodic memories demonstrate that new skills can be acquired in the absence of new declarative memory capability. However, the learning time needed to acquire a skill and the eventual skill level are inferior to normal subjects. Only simple skills such as mirror tracing can be acquired in reasonable time. Thus for the patient HM with no ability to create new declarative memories, a simple skill like mirror tracing could be learned, although more slowly than in normal subjects. Learning to get around a new house took months and performance remained inferior to normal subjects.

9.6.1 Description of Procedural Memory at the Level of Anatomical Structures

A skill such as mirror tracing requires cortical column receptive fields combining information from current visual and somatosensory (finger and hand position) inputs. These receptive fields in the motor cortex must acquire recommendation strengths in favour of finger and hand movements to achieve the next tracing step. The relevant anatomical architecture involved is illustrated in Fig. 9.15.

To perform the mirror tracing skill, appropriate components in the dorsal basal ganglia corresponding with different muscle movements relevant to the skill must be selected at the correct time. Appropriate selections require cortical column receptive fields that can discriminate between the circumstances in which different muscle movements are appropriate, and these fields must acquire recommendation strengths in favour of the correct behaviors.

Furthermore, to achieve learning, appropriate components in the ventral basal ganglia corresponding with different reward behaviours must be selected at the correct time. Appropriate selections require cortical receptive fields that can discriminate between recent correct and incorrect pencil marks, and these receptive fields must acquire recommendation strengths in favour of changes to the recommendation strengths that generated recent finger, hand and arm movements.

Initially, the visual and somatosensory receptive fields detected will tend to have recommendation strengths in favour of tracing behaviours appropriate if a mirror is not being used. Novel receptive fields incorporating information derived from visual perception of the mirror combined with more familiar perceptions of figure shape and hand position will develop. These novel receptive fields will acquire provisional recommendations strengths on to the movement components.

When a selected movement results in an incorrect pencil mark, the teacher will indicate that it is wrong. Correct pencil marks will be followed by positive feedback from the teacher. Existing receptive fields (labelled ≈ *teacher circumstances* in Fig. 9.16) that detect and discriminate between positive and negative teacher feedback will already have recommendation strengths in favour of appropriate changes to recommendation strengths in favour of recent movements. Such receptive fields are located in the prefrontal cortex, such as area 24. Visual receptive fields discriminating between correct and incorrect recently made pencil marks (labelled ≈ *visual coincidences* in Fig. 9.16) will be detected at the same time as the reward behaviour selections. This temporal coincidence means that they will acquire recommendation strengths in favour of those reward behaviours. Learning can then proceed without constant teacher feedback.

The selection of a motor behaviour in the basal ganglia occurs in three stages as illustrated in Fig. 9.15. The first stage is determination of the total recommendation weights for each behaviour across all currently detected receptive fields. The second stage is a competition to select one and only one behaviour. The third stage is implementation of that behaviour. Reward behaviors do not compete with other types, and are therefore determined by a two stage process: determination of the total recommendation weights; then implementation if the total weight is sufficient.

In a subject who has lost the ability to expand receptive fields, previously existing fields will remain as they were prior to the loss. The recommendation strengths associated with those fields will also remain, but can still be changed. Hence a simple new skill could potentially be learned by changes to the recommendation weights of existing receptive fields. Learning is slower because the discrimination of existing receptive fields between circumstances in which different skilled behaviours are appropriate cannot be evolved and will therefore generally be less effective that for normal subjects. Hence the number of practice sessions needed to converge on a set of recommendation weights will be longer, and the end point may be less effective. For a complex new skill, the previously existing receptive fields may be inadequate for the required discriminations.

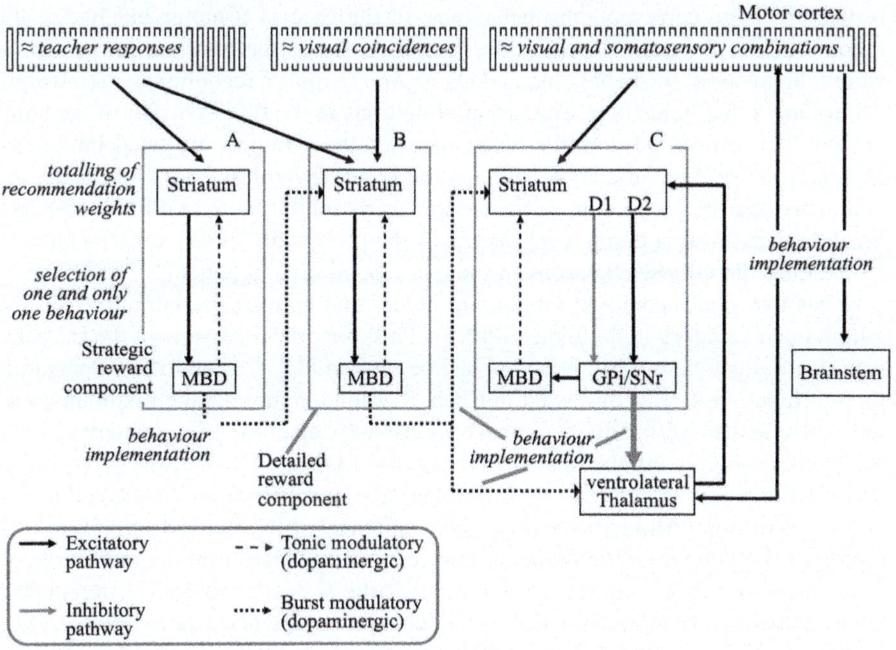

Fig. 9.16 Architecture for understanding management of rewards in the learning of mirror drawing. Initially, appropriate finger movements are rewarded following detection of receptive fields discriminating between different teacher responses. These rewards are selected when certain receptive fields are detected within visual coincidences, and these receptive fields acquire provisional connectivity on to the detailed reward component. With repetition, these visual coincidence receptive fields acquire sufficient recommendation strength to reward finger movement behaviours in the absence of the teacher. Abbreviations: *GPi* globus pallidus internal segment, *MBD* midbrain dopamine neurons (includes substantia nigra pars compacta and ventral tegmental area), *SNr* substantia nigra pars reticulata

9.6.2 Description of Procedural Memory at the Level of More Detailed Anatomical Structures

Definition of receptive fields occurs as described previously. The three stage behaviour selection in the basal ganglia is illustrated in more detail for two reward components and one motor component in Fig. 9.15. Within the dorsal striatum, motor components are located in the matrisomes while reward components are located in the striosomes. Components that reward general circumstances (i.e. strategic rewards) are located in the ventral striatum, components that reward specific movements (i.e. tactical and detailed rewards) in the dorsal striatum. Strategic reward behaviours reward more tactical reward behaviours, giving rise to the spiral of connectivity between the striatum and the midbrain dopamine neurons (see Fig. 8.16).

The group of projection neurons in the striatal part of one component receive a range of inputs from the cortex. Each such input corresponds with a recommendation in favour of the corresponding behaviour. In the case of a motor behaviour, the outputs of the D1 striatal projection neurons indicate the total recommendation weight in favour of the behaviour, and D2 neuron outputs a recommendation weight against any other behaviour. The group of neurons in the GPi/SNr part of the component also correspond with the behaviour, and their outputs are tonal inhibitory. Hence in the resting state all behaviours are blocked. Within a component the D1 path increases the probability of the behaviour by inhibiting the GPi/SNr neurons. By the D2 pathway a component decreases the probability of any other behaviour by exciting the GPi/SNr neurons in the corresponding components. The D2 pathway has two intermediate basal ganglia nuclei that convert the inhibitory striatal output into excitatory input to the GPi/SNr. This more detailed view of the D2 pathway is not illustrated in Fig. 9.16 but can be seen in Fig. 8.15 and how it performs its information role was discussed in Chap. 8. The loop through the midbrain dopamine neurons (MBD) influences the background dopaminergic concentration in the striatum, regulating the relative activity of the D1 and D2 pathways. This regulation ensures that in general one and only one motor behaviour is selected at any one point in time. Motor behaviours are implemented by reduced inhibition of a region of the ventral lateral thalamic nucleus. Motor cortical inputs to that region recommend release of motor cortical outputs to the brainstem to drive different possible detailed muscle movements which could all implement the behaviour selected by the basal ganglia. The reduced inhibition means that the group of motor cortical outputs corresponding with the muscle movements with the strongest total recommendation strength are released to the brainstem to drive the behaviour.

Reward behaviours are implemented by striatal activity triggering burst dopamine activity targetting the striatal part of a wide range of components. Reward behaviours are implemented independent of the selection of other types of behaviour, and their selection depends only on the total strength of corticostriatal inputs to the component. There are therefore no D1/D2 pathways to resolve a complex competition process. Reward components make long term increases to recently active recommendation weights in the component with the highest activity, corresponding with the selected behaviour. These long term increases are implemented by burst dopaminergic activity. Feedback from the thalamus to the striatum ensures that the selected component in the striatum has the highest activity. Burst dopamine reward signals also target the thalamus to increase the recommendation weights in favour of the recently implemented releases of motor cortex outputs to the thalamus in similar circumstances in the future.

Detection of receptive fields correlating with different types of teacher behaviour result in outputs from reward components A and B in Fig. 9.16. The output from B increases the weights of recently active corticostriatal synapses in the selected motor component, and also the weights of recently active corticothalamic synapses on to recently active thalamocortical projection neurons in the ventrolateral thalamus. The output from component A increases the weights of recently active corticostriatal synapses in component B. Because some ≈ *visual coincidences* receptive fields have recently been detected, these receptive fields gain recommendation weights into reward component B.

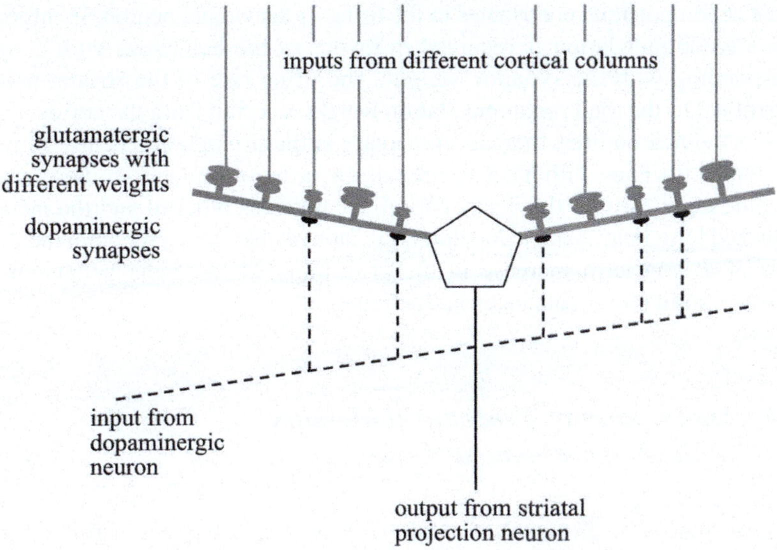

Fig. 9.17 Management of recommendation weights by synapses on a striatal projection neuron. Glutamatergic inputs from the cortex have synaptic weights corresponding with the recommendation weight of the cortical column in favour of the behaviour corresponding with the striatal neuron. If the striatal neuron fires, an LTD-1 mechanism increases the weights of all recently active synapses, but the increases decay over a period of less than an hour. If a burst dopamine signal is received within that decay period, the duration of LTD is extended longer term

Inputs from appropriate cortical areas to appropriate striatal components are required to achieve skill learning. Genetic bias can provide some connectivity, but more specific bias is provided during sleep. Sleep includes partial reruns of past experience, with a bias in favour of the most recent past. Cortical outputs that have often been active in the past at the same time as striatal neurons tend to establish synapses on to those neurons. These synapses are initially low weight, but can be increased by future rewards. Hence if in an early learning period the ≈*visual coincidences* cortical outputs were often active at the same time as striatal neurons in component B, provisional connectivity from the cortical neurons to the striatal neurons will be more extensively available after the next sleep period.

9.6.3 Description of Procedural Memory at the Neuron Level

As an illustration of this level of detail, a striatal projection neuron corresponding with one motor behaviour is illustrated in Fig. 9.17. The neuron receives glutamatergic inputs from a very large number of pyramidal neurons in different cortical columns. Each such input has a synaptic weight corresponding with the recommendation weight of the receptive field detection in favour of the behaviour. When a cortical action potential contributes to the firing of the striatal neuron, in information terms a recommendation is received in favour of the behaviour with a strength corresponding with the synaptic weight. The firing rate of the striatal neuron is proportional to the total recommendation weight received from the cortex.

When the neuron fires, there is an increase in the strength of recently active synapses which declines with a decay time of the order of minutes. If there is a burst dopamine input early in the decay period, the decay is blocked and the increase is established long term. Such a dopaminergic input is therefore a recommendations in favour of a long term increase in the probability of the same behaviour being selected in similar circumstances in the future.

9.6.4 Description of Procedural Memory at the Neurochemical Level

As an illustration of this level of detail, LTP at glutamatergic synapses on striatal neurons following coincident synaptic and neuron activity results from increases in the number of AMPA receptors. This increase decays unless there is a dopaminergic input triggering transcription of genetic information leading to expansion in the spine size.

9.6.5 Capturing Behavioural Sequences

Once learning has been established as cortical receptive fields and striatal recommendation weights, control of the muscle movements in frequently used sequences of such movements can be captured by the cerebellum as described in Chap. 8.

9.7 Hierarchies of Description

At a detailed level, there is an immense complexity to neurochemical processes. There are very large number of neurotransmitters and receptors, voltage gated channels in some cases modulated by neurotransmitters, kinases and other chemicals. The information algorithms implemented are very complex. However, the constraints imposed on brain architecture by natural selection mean that the information processes are constrained into the condition definition/detection and behavioural recommendation forms. High level descriptions of cognitive processes can be constructed using these forms. The validity of the high level model can be tested by mapping selected sections to more detailed levels, sections of the more detailed descriptions to even more detailed levels and so on. If such mapping can be achieved within the constraints of knowledge of anatomy, physiology and neurochemistry, and if mapping is possible using only information available at the more detailed level, there can be confidence in the validity of the high level model.

In Chap. 10, a similar approach will be taken to understanding attention and working memory. Then in Chap. 11 a range of cognitive phenomena will be discussed, using high level descriptions of the type developed in this chapter and Chap. 10. The mapping to more detailed levels will largely be omitted in Chap. 11, because the ways in which such mapping can occur have been established in earlier chapters.

Chapter 10
Attention and Working Memory

The environment in which a brain operates contains many different objects. Some objects urgently require a behavioural response, for others a response is less urgent or not required. Developing a response to an object requires detection of receptive fields that can discriminate between different types of object (≈ *objects*) within the sensory information derived from the object, and determination of the predominant recommendation strength across these receptive fields. This response development occupies cortical and other resources for a period of time, and a process must be established to select the highest priority object and to detect receptive fields within the information derived just from that object. Furthermore, in some situations the receptive fields detected directly within the sensory information derived from the object do not have the type of recommendation strengths needed to generate an appropriate behaviour. In this situation, other receptive fields must be indirectly activated on the basis of past temporally correlated activity with the current sensory receptive fields.

In many situations, the most appropriate behaviour depends not just on one object but on a number of different objects present at the same time. To guide behaviour in these situations, receptive fields within sensory information derived from multiple objects must be detected. These receptive fields must discriminate between different groups of objects (≈ *groups of objects*). Such fields will generally be higher complexity than the≈*objects* fields and be defined largely by combinations of≈*objects* receptive fields. In order to detect such≈*groups of objects* fields, detections of≈*objects* fields within different objects must be active at the same time. However, the inputs and activity generating detections must be kept separate in order to avoid detection of misleading≈*objects* receptive fields corresponding with mixtures of information from two or more separate objects. Hence it must be possible to maintain independent populations of receptive field detections in the same≈*objects* cortical area, and combine the outputs of the separate populations to drive more complex receptive field detections in a different≈*groups of objects* area.

Attention is the ability to select a subset of current sensory inputs to be processed to recommend alternative behaviours. Working memory is the ability to maintain a

L.A. Coward, *Towards a Theoretical Neuroscience: from Cell Chemistry to Cognition*, Springer Series in Cognitive and Neural Systems 8, DOI 10.1007/978-94-007-7107-9_10,

number of simultaneously active populations of receptive field detections independently in the same cortical area. In the following sections we will develop high level descriptions of attention and working memory in the condition/recommendation paradigm, and give examples of how this paradigm makes it possible to create descriptions at more detailed levels for different parts of the high level description.

10.1 Attention

Visual attention has been extensively studied. Suppose that a subject views a scene. Light from the room enters their eyes, derived from many different objects. This input drives a sequence of rapid eye movements, with pauses when the gaze is stationary for a period of time. During these pauses in gaze direction, the sensitive central part of the eye (the fovea) is centred on the object. When the gaze is fixated on an object for a longer period, responses to the object such as naming it are possible. For example, if the visual field contained the collection of objects illustrated in Fig. 10.1a, and the subject was asked "What is the darker object?", the

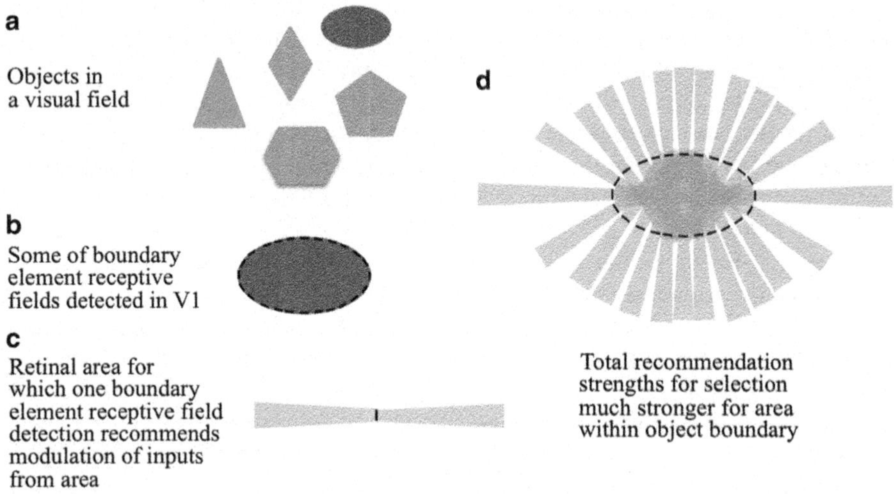

Fig. 10.1 The process for selecting only visual information derived from within one object for detailed receptive field detections. (**a**) A subject sees a visual field with a number of different objects. An image of the group of objects is projected on the retina. Receptive fields corresponding with all the boundary elements in the group of objects ate detected. (**b**) The boundary element receptive fields detected around the oval are illustrated. Such boundary elements are detected at the same time around all the other objects. (**c**) Each boundary element has recommendation strengths in favour of more detailed processing of visual information derived from a band on the retina extending perpendicularly on each side of the element. (**d**) There will therefore be more recommendation strengths in favour of the area inside a closed boundary than the area outside the boundary

response would be to shift attention rapidly from object to object, pausing at each object, perhaps counting the number of sides, then generating the response "oval" from looking at the correct diagram. If further asked "what do you think about when you see the oval?" the response could be to generate and verbalize associations such as rugby balls or the orbits of planets.

There can therefore be three steps to the attention process for a single object. The first is selection of a target, with eye and head movements to centre the eye on that target. The second is deeper processing of information derived from that target to generate a behaviour. A possible third step is generation of information associated with the target object. In this section, a hierarchy of description covering the management of eye movements (called saccades) and head movements, and the mechanisms of information releases for deeper processing will be developed. Finally, the activation of associated information, sometimes called "conscious attention", will be described at relatively high level.

10.1.1 Description of Attention at Anatomical High Level

The high-level architecture for the management of attention is illustrated in Fig. 10.2. The architecture covers the selection and implementation of a saccade driven by appropriate eye muscle movements to register one object with the centre of the fovea, and the release of information derived from just that object for detailed processing.

This architecture shows a range of cortical areas that detect receptive fields relevant for recommending eye movement behaviours. The frontal eye fields (FEF) are particularly effective for discriminating between situations in which different eye movements are appropriate, and therefore have well developed recommendation strengths in favour of different eye movements or saccades [935]. The primary mechanism for implementation of the currently most appropriate behaviour is release of the FEF receptive fields with high recommendation strengths in favour of that behaviour to the brainstem where the predominant eye muscle movements are determined and released to drive movements.

An eye movement shifts the gaze from one visual object to another object elsewhere in the visual field. The other object must be selected from among a range of possibilities. The relevant information needed includes the current position and state of the muscles in the eye, and the positions and types of other objects in the visual field. The position and type information must be available very rapidly. The somatosensory cortex detects receptive fields that discriminate between different eye positions, and communicates current receptive field detections to the FEF [936]. The FEF also receives receptive field detections from the lateral intraparietal cortex (LIP) [937]. The LIP receives information from the dorsal stream in the visual cortex, which provides "where" information and therefore discriminates between objects in different positions. LIP receptive fields are also able to discriminate between the shapes of different objects [938]. Area TE in the ventral ("what")

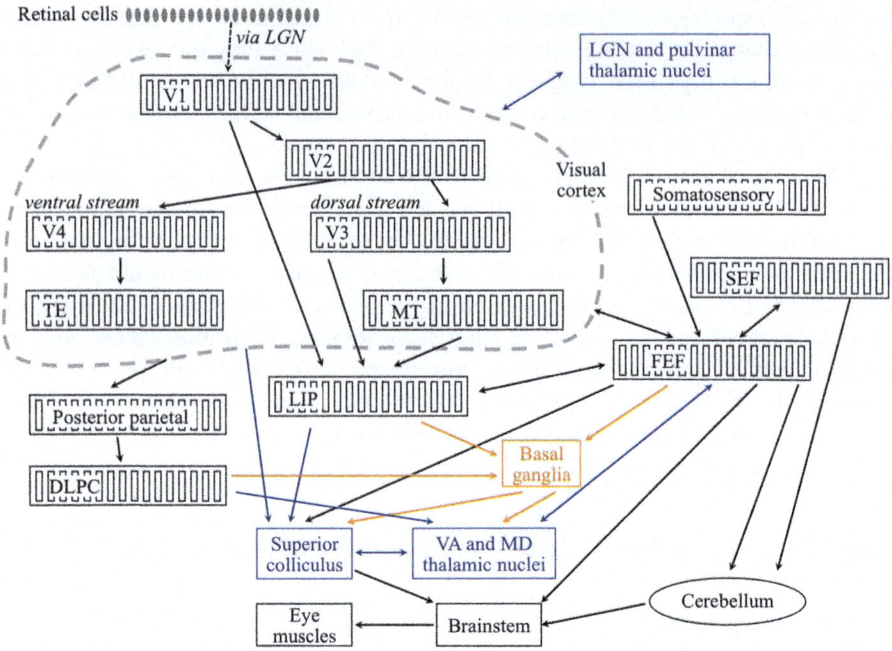

Fig. 10.2 Anatomical architecture for attention. Brain regions participating in the selections of visual objects to be processed for detailed receptive field detections are illustrated. Behaviours implementing such selections include eye movements, head movements and frequency modulations of inputs from limited areas of the retina. Multiple cortical areas detect receptive fields on levels of complexity useful for discriminating between circumstances in which different such behaviours are appropriate. The behaviour most strongly recommended by the cortex is determined by the basal ganglia, and the detailed implementation of the behaviour determined and implemented by the thalamus and superior colliculus. Frequently utilized sequences of attention behaviours can be recorded in and implemented by the cerebellum

stream of visual processing also discriminates between types of objects. However, the LIP receives information directly from V1 and can make less precise but faster type discriminations than TE. A number of areas in the visual cortex also provide direct inputs to the FEF to assist in discrimination effectiveness by providing information on recently viewed objects.

Current FEF receptive field detections are communicated to a specific striatal region of the basal ganglia where they are interpreted as recommendations in favour of different eye movement behaviours. Two other cortical areas also provide strong recommendations to the same striatal region. One of these is the LIP, which provides faster information on current shapes and locations than the information available indirectly through the FEF and can recommend the relative priorities to be placed on different objects [939]. The second is the dorsolateral prefrontal cortex (DPFC). The DPFC detects receptive fields that discriminate between situations in which different attention priorities are appropriate. These receptive fields therefore also recommend gaze shifts to different types of object. The predominant recommendation strength in the basal ganglia will therefore be in favour of the most appropriate object on the basis of current eye position, other recently viewed objects and groups of objects, shapes of objects in the visual field and current attention priorities. Information on recently viewed objects and groups of objects is provided to the DPFC by the posterior parietal cortex.

This predominant recommendation strength is communicated to the thalamus, resulting in release of the subset of FEF receptive field detections that have strong recommendation strengths in favour of the selected behaviour to the brainstem. The selection of the precise subset is also directly influenced by DLPC detections recommending priorities based on the current general situation, but these detections are not relevant to the exact definition of the appropriate eye movements and are not released to the brainstem. In the brainstem different neurons drive different eye movements. The selection of an FEF subset by the basal ganglia means that the brainstem neurons corresponding with the appropriate eye movements have much larger total input that any others, resulting in unambiguous implementation of those eye movements. For eye movement behaviours, visual information more current than that received via the FEF may be relevant. The superior colliculus receives LIP inputs corresponding with recommendations in favour of different types of eye movements based on the different shapes present in the visual field. The superior colliculus also receives more direct inputs from the visual cortex. Selection of action on these recommendations is managed by the basal ganglia in the basis of LIP inputs to the striatum. The basal ganglia therefore ensures that thalamus and superior colliculus receive inputs based on consistent recommendations. The superior colliculus and thalamus also communicate to ensure consistency on a more detailed level. The brainstem integrates the superior colliculus and FEF inputs to an appropriate eye movement.

The supplementary eye field (SEF) discriminates between circumstances in which the end point of a saccade results in exact correspondence between a visual object and the centre of the fovea, and circumstances in which there are small offsets. These receptive field detections can therefore effectively recommend small eye movements to correct for saccade errors [940]. These detections are communicated to the FEF, where detections incorporating this information recommend the appropriate small movements. In addition, the SEF detects receptive fields correlating with the difference between eye direction towards the currently selected visual object and head direction, and each such receptive field has recommendation strengths in favour of a range of head movements to centre the eyes in the orbits [941].

Frequently utilized sequences of eye movements can be recorded in the cerebellum. Such sequences could include patterns of eye movements to scan the features of an often viewed object type (such as a face). Another set of important patterns could be movements to correct registration errors, and the SEM therefore projects to the visuomotor region of the cerebellum [942].

Once an object has been centred on the fovea, the information derived just from the area corresponding with the object is released to detect a more extensive range of receptive fields in many visual cortical areas. Since the selection of the object has already occurred, there is no need for another behaviour selection by the basal ganglia, and this step is therefore managed by the LGN and pulvinar nuclei of the thalamus.

It is important to recognize that this type of architecture, while a good high level description, has a number of inherent approximations even at this level of description. Receptive fields are defined heuristically. Receptive fields at some levels of complexity may be particularly effective for discrimination between circumstances in which different attention behaviours are appropriate. However, some individual receptive fields with other complexities could also be very effective. Genetic information will tend to result in an initial connectivity between areas to define effective receptive field complexities, and from areas effective for a behaviour to the striatal area corresponding with than behaviour. However, with experience, provisional connectivity outside this genetically defined starting point will be established based on frequent simultaneous activity. For example, initially the main attention related connectivity to the striatum comes from the LIP and FEF. However, if a receptive field in some other area is often active just before an attention behaviour is selected and rewarded, that field may establish connectivity on to the striatal region and develop recommendation strengths. Extra connectivity may also be established between areas. The problem with the acquisition of additional connectivity is that receptive field definitions can evolve with experience. Additional connectivity therefore raises the risk of undesirable side effects of later receptive field change. Hence constraints on the degree of such connectivity probably exist.

In the following sections, parts of the attention information process will be described in more detail, focussing on processes different from those discussed in Chap. 9 for memory phenomena. In particular we will focus on the mechanisms by which information releases are carried out by the thalamus.

10.1.2 Description of Attention at the Level of Receptive Fields

At a relatively high level, the concept for using receptive fields to select the area of the retinal from which inputs will be processed in more detail is illustrated in Fig. 10.1. Receptive fields are detected by area V1 that correspond with boundary elements. Receptive fields exist corresponding with boundary elements in all orientations at every point on the retina. Some receptive fields detected in response to the dark oval are illustrated in Fig. 10.1b. Each detected receptive field has recommendation strengths in favour of releasing information derived from a band on the retina extending perpendicularly on both sides (Fig. 10.1c). As a result, there are relatively large recommendation strengths in favour of releasing information derived from the area within the visual object, and relatively small recommendation strengths outside.

Inputs from the retina to the cortex have centre-surround receptive fields. Such a field receives inputs from an area of the retina, and is detected if the average

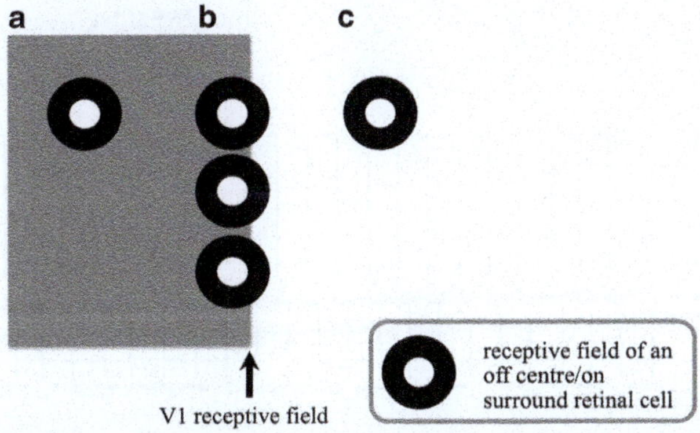

Fig. 10.3 Generation of boundary element receptive fields in cortical area V1 from centre-surround receptive fields of ganglion cells located behind the retina. Ganglion cells receive inputs from a circular region of the retina, and detect their receptive field if the illumination of the centre of the region is different from the illumination of the surround. On-center/off-surround and off-center/on-surround fields exist. If the region for a ganglion cell is located within the boundary of an evenly illuminated object (**a**), there is no contrast and the cell produces no output. If the region is located well outside the object (**c**), again there is even illumination and no receptive field detection. If the region is located on the boundary (**b**), the cell will produce output. Some V1 neurons receive inputs from rows of ganglion cells with different orientations. If the boundary coincides with a row, the corresponding V1 neuron will detect its receptive field

illumination in the centre of the area is different from the illumination around the periphery. These fields can be centre-on surround-off or centre-off-surround-on. Combinations of such fields can detect the presence of a boundary on the retina between darker and lighter areas as shown in Fig. 10.3. Simple connectivity between the retina, LGN and V1 that could implement such receptive field definitions and recommendation strengths is illustrated in Fig. 10.4. In this simple model, one ganglion cell targets one thalamocortical projection neuron in the LGN, and one LGN neuron targets one layer IV pyramidal (spiny stellate) neuron in cortical area V1. The layer IV neurons corresponding with a boundary element on the retina target one layer VI pyramidal. Hence the receptive field of that layer VI pyramidal is the presence of a boundary in the retinal image at the corresponding place on the retina.

This very simple model illustrates the information concept. If there is a boundary in the visual image as illustrated by the darker centre-surround cells in Fig. 10.4, the corresponding LGN neurons will be activated. The layer VI neuron with a receptive field corresponding with the presence of a boundary at the location will then be activated. This layer VI neuron targets the LGN neurons corresponding with a band on the retina extending on either side of the boundary. The strength of this feedback connection will be greatest at the boundary, and diminish with distance from the boundary. Hence the strongest total input will be to the LGN neurons within the boundary.

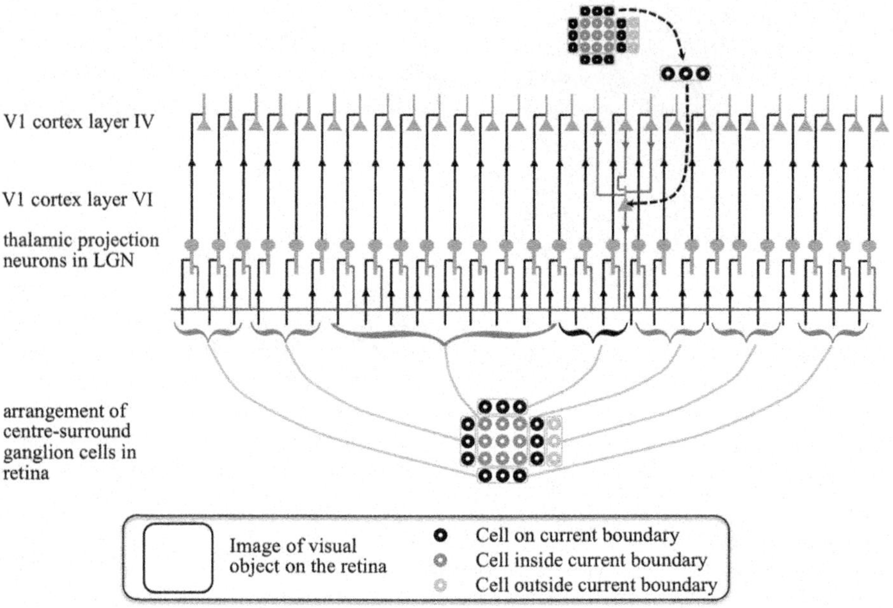

Fig. 10.4 Simple connectivity to support selection of the inputs derived from a limited retinal area corresponding with a visual object for more complex receptive field detections. Each ganglion cell has a receptive field corresponding with contrast on a specific retinal area. Each such ganglion cell targets one thalamic projection neuron. Each thalamic projection neuron targets one V1 cortical neuron in layer IV. The three layer IV neurons corresponding with one row on the retina target one layer VI cortical neuron. Hence a layer VI neuron has a receptive field corresponding with the presence of a boundary along its corresponding row. Each such layer VI neuron targets all the thalamic projection neurons corresponding with regions on the retinal extending on each side of its corresponding boundary. The layer VI activity plus the increased thalamic projection neuron activity drives TRN interneuron (not illustrated) activity which places a gamma band frequency modulation on the appropriate layer IV neurons, which is propagated through to the layer VI neurons

Within a major boundary surrounding an object, there will be weaker centre-surround field detection corresponding with the detailed appearance of the object. The activity in the LGN corresponding with this information will be increased. The increased activity results in increased activity downstream in layers IV and VI. In addition (see Fig. 8.16), axons in both directions between the LGN and V1 have side branches on to interneurons in the TRN, and these TRN neurons in turn target the corresponding LGN neurons. Activation of these interneurons places a gamma band modulation on the LGN and downstream V1 neurons, hence the layer VI outputs to V2 etc. are modulated at the gamma frequency. As discussed in the next section, the increased activity and modulation implement the release of information derived from within the boundary deeper into the cortex.

In practice the receptive field complexities will be more complex than in this simple model. For example, a layer IV receptive field could be a short boundary element, and a layer VI receptive field could be a longer element made up of a combination of layer IV receptive fields. One advantage of this more complex arrangement would be the ability to bridge gaps in a visual boundary.

10.1.3 Description of Attention at the Level of Frequency Modulation of Cortical Action Potentials

As illustrated in Fig. 10.5, frequency modulation of the action potentials generated by layer VI pyramidals in V1 means that the action potentials are bunched within the decay time of injected postsynaptic potentials. There is therefore a higher probability that the postsynaptic potentials will overlap and reinforce each other, reaching threshold for target neuron firing. With appropriate thresholds on the target neurons, modulated inputs will generate a much higher level of target neuron firing than unmodulated inputs. Hence modulation is effectively a release mechanism from V1 to V2.

Modulation of inputs will tend to result in the same modulation of outputs, with a time delay caused by the neuron integration time. Hence modulation of V1 outputs could allow receptive field detection to propagate through V2 and beyond, for example through V4 to TE. Some reinforcement of the modulation may be required by TRN interneuron targetting of later area outputs.

10.1.4 Description of Attention at the Level of Synaptic Activity

TRN modulation is imposed by burst firing of the GABAergic interneurons at the gamma frequency. This firing inhibits the target thalamocortical projection neurons from firing coincident with the TRN action potentials. The activity of the thalamocortical neurons is increased by the feedback from layer VI as described in the previous section, but in addition the GABAergic signals are not purely inhibitive.

An action potential arriving at a $GABA_A$ receptor can be either inhibitive or excitatory depending on the membrane potential of the target neuron in the vicinity. If the membrane is already depolarised, for example by recent arrival of an action potential at a neighbouring glutamatergic synapse, the effect is hyperpolarising and therefore inhibitory. However, if the membrane is at resting potential, the membrane is depolarised and the effect is excitatory. If a glutamatergic input arrives after a nearby GABAergic input, the earlier input may reinforce the depolarising effect of the glutamatergic input. Similarly, if a calcium action potential propagating into the soma arrives shortly after a GABAergic to the soma, it may be reinforced. Hence the GABAergic TRN interneuron outputs at the gamma frequency do not just inhibit action potential generation by their targets in phase with their outputs, but can excite their targets out of phase with their outputs.

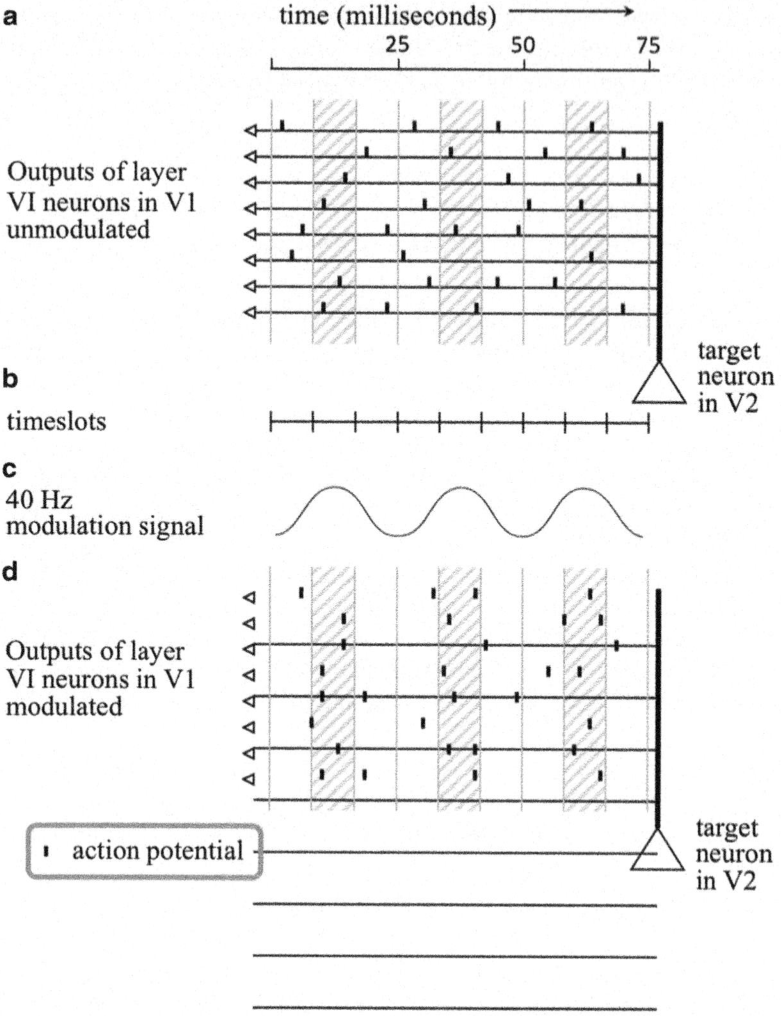

Fig. 10.5 Conceptual illustration of the effect of frequency modulation of V1 outputs to neuron targets in V2. A neuron is targetted with the excitatory outputs of eight source neurons. Because of the decay time of EPSPs, unless enough input action potentials arrive within a period of less than about 8 ms there will be no output by the target neuron. Time is divided into a sequence of about 8 ms timeslots (**b**). If there is no frequency modulation (**a**) the action potentials arriving at the target neuron are roughly even across any timeslot, regardless of their arrangement in time. The application of a 40 Hz modulation (**c**) to all of the input neurons shifts action potentials towards the nearest modulation peak, as shown in (**d**). The effect is that in the timeslot centred around the modulation peak, there are more action potentials than in any other timeslot, and more than in any similar time period without modulation. Hence when modulation is imposed, the variability of integrated signal arriving at a neuron is affected. In the illustration, the number of peaks within one integration window could vary from three to four without modulation, but from one to seven with modulation. Thus if the threshold for a neuron receiving the inputs was at five, only the modulated combination would generate any output. If in a set of sensory inputs or outputs from a cortical area were modulated with the same frequency and phase and others unmodulated, the result would be that conditions would be detected within the modulated subset but not within the unmodulated subset

10.1.5 Description of Attention at the Level of Neurochemistry

As discussed in the previous section, $GABA_A$ receptors can be either depolarising or hyperpolarising depending on local membrane potential. $GABA_A$ receptors allow both Cl^- and HCO_3^- currents, but the relative conductance depends on the membrane potential. Resting potential favours HCO_3^- current, depolarisation favours HCO_3^- current. Cl^- ion concentration is higher outside a neuron, and a Cl^- current is hyperpolarising. However, HCO_3^- ion concentration is generally lower outside a neuron, and an HCO_3^- current is therefore depolarising.

10.1.6 Conscious Attention

The ability to talk about the associations of a visual object requires access to speech recommendation strengths that are not present in the column receptive fields detected within the sensory inputs. Hence indirect activation of other columns on the basis of past temporally correlated activity is required, using mechanisms similar to semantic and episodic memory retrievals discussed in Chap. 9. Such indirect activations are recommended by the sensory receptive fields, but acceptance of these behaviours requires supporting recommendations from the prefrontal cortex. This topic will be considered further when human consciousness is discussed in Chap. 11.

10.2 Working Memory

One important category of experiments measures the ability to recall lists of words for a brief interval after being presented with the list. Early experiments estimated that there was an ability to recall about seven words for a short while after viewing. Later experiments have shown that the number is very dependent on the existence of associations between the objects. If there is a semantic relationship between some of the words on the list, the number that can be recalled is increased. In the extreme, all the different words in a meaningful sentence can generally be recalled. In order to measure the underlying limit when there is no such semantic chunking, subjects are given many lists of different lengths, and the maximum length for which no errors ever occur is determined. This maximum length is found to be between three and four words.

Another category of experiments measures the ability to recall visual images. For example, an arrangement of rectangles may be shown, and the subject asked to draw the arrangement after a short delay. Yet another category of experiments measures the ability to recall sequences of finger movements. In both cases there are analogous limits on recall capacity. It is also found that there is a degree of independence between verbal, visual and motor working memories. For example, a distracting visual task does not reduce verbal working memory capacity and vice versa.

10.2.1 Description of Working Memory at Anatomical High Level

When an object is perceived, cortical column receptive fields detected within the sensory inputs have strong recommendation weights in favour of correctly naming the object. Each object on a list will activate such a population of receptive fields. For an object to be named a little later, two conditions must be met. Firstly, some of the receptive fields activated by perceiving the objects must have remained active, and these receptive fields must together have sufficient total appropriate speech recommendation strength. Secondly, the receptive fields activated by each object on the list must be separated, so that separate total recommendation strengths can be determined. Hence the fundamental limit for working memory is the number of independent column populations that can be maintained active at the same time.

However, there is a way to exceed this limit under some circumstances using indirect activation mechanisms. As discussed in detail under episodic memory in Chap. 9, it is often necessary to develop a behaviour that is appropriate to a group of objects or even a group of groups. Furthermore, the appropriate behaviour towards one object may be influenced by the presence of certain groups and groups of groups. Parietal cortical areas with receptive fields able to discriminate between different types of groups (≈*groups*) and between groups of groups (≈*groups of groups*) are necessary. Detection of≈*groups* receptive fields requires inputs from≈*objects* receptive fields derived from multiple objects, but the processes for detection of these different≈*objects* detections must not interfere with each other. Detection of≈*groups of groups* receptive fields requires inputs from≈*groups* receptive fields derived from multiple groups. When≈*groups* receptive fields have been activated, they are active at the same time as the≈*objects* receptive fields that provided their input. Hence a little later if the≈*objects* receptive fields are inactive, but the≈*groups* fields are still active, the≈*objects* receptive fields can be reactivated on the basis of recent simultaneous activity with the currently active≈*groups* fields. Similarly, recently active≈*groups* receptive fields can be indirectly activated by currently active≈*groups of groups* fields. In addition, if there is novelty in the groups or groups of groups, some receptive field expansion will occur. Indirect activation on the basis of simultaneous past receptive field expansion can then recover, for example, inactive groups from currently active groups of groups.

A primary behavioural value of the information mechanisms that result in the phenomena of "pure" working memory is therefore the ability support multiple but independent activations in the same cortical area, making it possible to define and detect higher complexity receptive fields made up of combinations across such multiple populations. Because there are limits to the number of such populations, maintaining a population active must be a behaviour that is recommended by column receptive field detections and has sufficient total recommendation strengths across those detections to be accepted.

Some key aspects of the high level anatomical architecture supporting working memory mechanisms is illustrated in Fig. 10.6. Independent populations of column

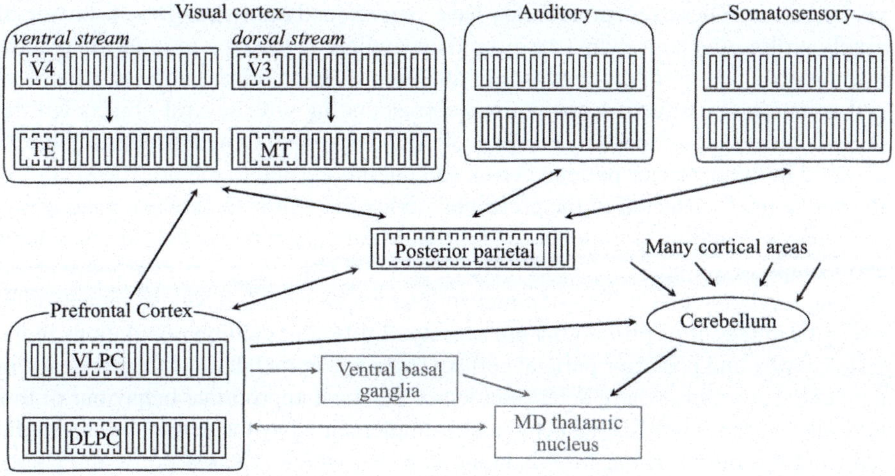

Fig. 10.6 Working memory information architecture. Recommendations in favour of prolonging the activity of a population of active columns in a visual area are generated by the prefrontal cortex. The ventral basal ganglia determines the total recommendation strengths in favour of such behaviours and selects the behaviour with sufficient predominant total strength. The thalamus performs detailed implementation of the selected behaviour. The cerebellum can acquire and implement frequently utilized sequences of behaviours that include working memory behaviours

activations in response to different visual objects can be supported in the higher areas of the visual cortex, up to a limit of about three. An active column in the visual cortex has recommendation strength in favour of continuing its own activation, but this is only sufficient to prolong its activity very briefly, just long enough to continue while gaze is focussed on the next few objects. The posterior parietal cortex [943] (PPC) detects receptive fields discriminating between different groups of objects and groups of groups, and can include auditory and somatosensory information in such discriminations. Various areas in the prefrontal cortex [944] discriminate between very general circumstances, including the current and recent presence of different types of groups of objects and groups of groups derived from the posterior parietal area. These receptive field detections have recommendation strengths (among others) in favour of prolongation of the current activity of a column population in the visual association areas. These recommendation strengths are instantiated in the striatum, in this case the ventral striatum [945] because of the more strategic nature of the recommended behaviours. If there is sufficient recommendation strength in favour of a prolongation behaviour, the ventral basal ganglia triggers the mediodorsal thalamic nucleus [946] to release prefrontal cortex detections with the strongest recommendation strengths in favour of the behaviour to the visual association areas [947].

Consider the process for handling a visual scene within this architecture. An object is selected by the attention process and receptive fields detected within the visual cortex. These receptive fields have recommendation strengths in favour of briefly prolonging their own activity, and depending on the type of object may have recommendation strengths in favour of attention shifts to related objects. After several attention shifts, active populations corresponding with several objects will be present in the visual areas and receptive fields corresponding with groups of objects detected in the posterior parietal cortex and communicated to the prefrontal cortex. Receptive fields detected in the prefrontal cortex discriminate between general circumstances including combinations of different recently perceived groups of objects and groups of groups, and other behavioural priorities (e.g. on aggressive, food seeking etc. behaviours). These frontal cortex receptive fields have recommendation strengths in favour of prolonging the activity of different column populations in the visual cortex and posterior parietal cortex. These recommendation strengths exist in the ventral striatum, where determination of the most appropriate behaviour of this type takes place. The selected behaviour is implemented by the ventral basal ganglia reducing inhibition of the MD thalamic nucleus, which releases prefrontal cortex outputs with the strongest recommendation weights in favour of the behaviour to the visual cortex and posterior parietal cortex to prolong the activity of the selected populations. These prolongations define which≈*objects*, ≈*groups* and≈*groups of groups* receptive fields will continue to be active and can therefore be combined with the receptive fields detected within the next attention object. At some point, the latest combination of≈*objects*, ≈*groups* and≈*groups of groups* receptive fields may have a predominant recommendation strength in favour of, for example, a motor behaviour relative to the overall perceived environment. This motor behaviour will then be implemented.

In the case of "pure" working memory, the number of visual objects that can be recalled is firstly dependent on the number of separate populations of column receptive field detections that can be maintained. This number is between three and four. For visual working memory these populations are maintained in the visual cortex, for verbal memory in the auditory cortex, and for motor memory in the somatosensory cortex. Hence these different types of working memory can operate relatively independently. Recall of larger groups depends upon prolongation of the activity of≈*groups* receptive fields in the posterior parietal cortex, and the number of such≈*groups* populations is subject to the same limit as in the sensory cortices. However, such enhanced recall depends on the appropriate≈groups receptive fields, which will generally exist only if some prior association has been established between the types of objects. Even more enhanced recall could be achieved via multiple≈groups of groups populations, but at this level the reconstruction of the original objects (on the basis of frequent past simultaneous activity) might be less precise.

The cerebellum also plays a role in working memory [765]. There are sequences of information processes such as prolongation of activity, indirect activation on the basis of recent simultaneous activity, release of outputs of multiple populations from one cortical area to another and so on. Each of these information processes is a behaviour which must be recommended by cortical receptive field detections and accepted and implemented by the basal ganglia. However, as for motor movements, frequently implemented sequences of behaviours can be captured by the cerebellum, which can then implement such behaviours following an initial selection of the sequence by the basal ganglia. Each behaviour in the sequence is initiated by the detection of a very complex Purkinje cell receptive field, and implemented by cerebellar nuclei outputs targetting the appropriate thalamic nucleus.

Some examples of more detailed levels of description will now be discussed. Much of the more detailed description of attention also applies to working memory.

10.2.2 Description of Working Memory at the Level of Frequency Modulation of Cortical Action Potentials

One key mechanism at a more detailed level is the support for multiple populations of receptive field detections within the same cortical area. This mechanism is illustrated in Fig. 10.7. As discussed under attention, the postsynaptic potential injected by an incoming action potential decays with a half life of ~8 ms. Because a frequency modulation imposed on the inputs to a cortical area results in bunching of

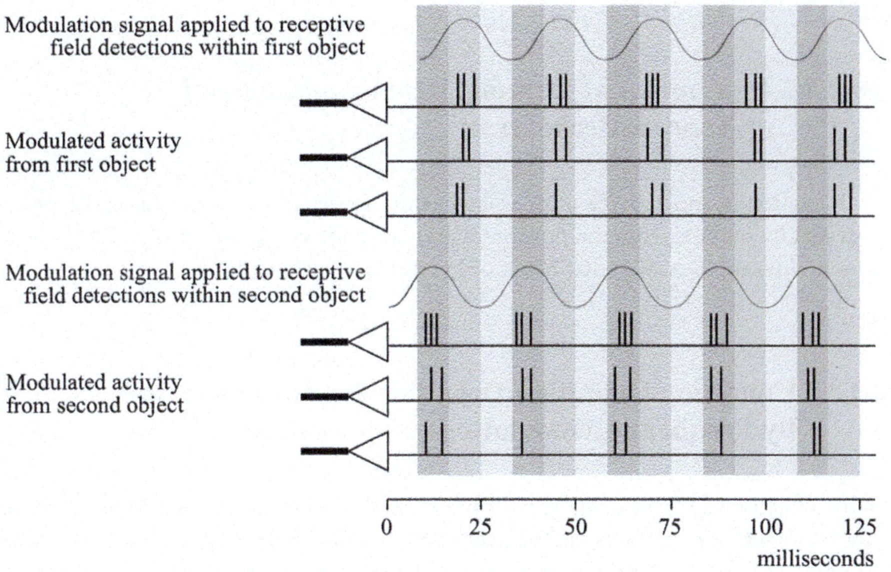

Fig. 10.7 Frequency modulation mechanism for working memory. Receptive field detections within one visual object result in output action potential spikes that are frequency modulated to mainly occur within one timeslot in each modulation period. Receptive field detections within a second visual object result in output action potential spikes that are frequency modulated to mainly occur within a second timeslot in each modulation period. The separation of the timeslots is such that postsynaptic potentials resulting from one object have substantially decayed by the time postsynaptic potentials resulting from the second object occur. Hence neurons separately process information derived from the two objects

action potentials, a higher level of outputs results because the input action potentials all tend to occur within one decay time.

Suppose now that the peak to peak interval of the modulation frequency is several times longer than the postsynaptic potential decay time. It is then possible to notionally divide each interval into several "timeslots" that are each one decay time long. If a frequency modulation signal is applied to one set of inputs to a cortical area, the action potentials will be bunched within one timeslot. If a second modulation signal with the same frequency but different phase is applied to a different set of inputs, their action potentials will be bunched within a different timeslot. The postsynaptic decay time means that if the timeslots do not overlap, and the bunching is fairly tight, the postsynaptic potentials resulting from one set of inputs will have relatively small overlap with the postsynaptic potentials resulting from the other set, even within the same neuron. There will be some overlap, or crosstalk, but because receptive field detections are recommendations and only the predominant recommendation is accepted, a degree of error of this type is tolerable.

If the decay time is ~8 ms, and the modulation frequency is in the gamma band at 40 Hz, a little over three relatively separate timeslots can exist within the 25 ms peak to peak interval. The ratio of decay time to modulation interval is the observed working memory capacity with no chunking.

10.2.3 Description of Working Memory at the Level of Neuron Mechanisms

Prolongation of the activity of neurons in the sensory or posterior parietal cortices is driven by outputs from the prefrontal cortex. In the prefrontal cortex, local recurrent circuits support extended firing of pyramidal neurons [948].

10.3 Condition Definition/Detection Resource Sharing by Frequency Modulation

In this chapter we have shown how simple attention and working memory phenomena can be described in terms of the condition definition/detection and behavioural recommendation information models, using a frequency modulation mechanism.

Attention uses frequency modulation to identify the subset of available information to be processed by the cortex at each point in time. Working memory reflects the use of frequency modulation to allow the sharing of one cortical resource across multiple simultaneous processes without interference between the processes.

Use of frequency modulation based mechanisms is a general feature of brain processing, with attention and working memory being phenomena in which the mechanism can to some degree be investigated in isolation. However, an actual cognitive process typically employs frequency modulation mechanisms along with all the information recording and accessing mechanisms described in the previous chapter.

In the next chapter we will discuss a range of more realistic cognitive phenomena, and demonstrate how these phenomena can be understood at higher levels using the condition/recommendation paradigm. Occasional examples of mapping to more detailed levels will be given. However, in most cases the information processes for these cognitive phenomena are combinations of the simple attention and memory processes described in this chapter and the previous chapter, and detailed mapping would occur in a manner analogous with that for the simple processes.

Chapter 11
Understanding Complex Cognitive Phenomena

We have established in Chaps. 9 and 10 that causal descriptions of relatively simple cognitive phenomena at high-level using the condition definition/detection and behavioural recommendation information models can be mapped into more detailed causal descriptions at anatomical, physiological and neurochemical levels. Hence we can have confidence that higher level descriptions of more complex cognitive phenomena made up of combinations of the information processes used by the simpler phenomena could also be mapped to more detailed levels as required. In this chapter we will develop such higher level descriptions of more complex phenomena, only rarely going down to the more detailed levels. These higher level descriptions will be in terms of cortical column receptive field detections at different levels of complexity and behavioural interpretation of these receptive field detections in various subcortical structures.

The key requirement is to develop descriptions of cognitive phenomena using only the information processes available to a complex learning system. Such a system heuristically defines receptive fields that can discriminate between but not correlate precisely with circumstances in which different behaviours are appropriate, associates each receptive field detection with a range of behavioural recommendations, and determines and implements the behaviour (or combination of consistent behaviours) most strongly recommended across currently active receptive fields. Implementation generally occurs in two stages. First, the most strongly recommended general type of behaviour is determined by the basal ganglia and implemented in the thalamus. Next, the thalamus implements the most strongly recommended behaviour of the selected general type. For frequently used sequences of behaviours, receptive fields detections recommend initiation of the behaviour sequence into the basal ganglia and cerebellum. If there is a predominant recommendation strength in favour of the sequence, the basal ganglia surrenders control to the cerebellum. Then on the completion of each type of process in the sequence, the cerebellum detects a receptive field that immediately implements the next type into the thalamus, which in turn implements the most strongly currently recommended behaviour of the type.

L.A. Coward, *Towards a Theoretical Neuroscience: from Cell Chemistry to Cognition*, Springer Series in Cognitive and Neural Systems 8, DOI 10.1007/978-94-007-7107-9_11,

Table 11.1 The possible types of information processes in a complex learning system

1.	Changes to receptive fields
2.	Changes to recommendation weights
3.	Changes to the thresholds for detection of receptive fields in current inputs
4.	Prolongation of the activity of a receptive field detection
5.	Indirect activation of receptive fields on the basis of recent activity: at the same time; earlier; later
6.	Indirect activation of receptive fields on the basis of frequent past activity; at the same time; earlier; later
7.	Indirect activation of receptive fields on the basis of receptive field expansion: at the same time; earlier; later
8.	Releases of receptive field detections from one area to another, or to muscle control
9.	Recording of frequently used sequences of behaviour types
10.	Activation of a recorded sequence

All cognitive processes must be supported by sequences and combinations of these processes. To be implemented, a process must be adequately recommended by currently active receptive field detections

There are a limited number of different types of information process available to a complex learning system, and in all cases an information process is itself a behaviour which must have adequate recommendation strength across currently active receptive fields in order to take place. There are a number of different types of information process summarized in Table 11.1. *Firstly*, changes to receptive fields. Such changes are generally expansions and are behaviours which must be recommended by combinations of other receptive field detections. *Secondly*, changes to recommendation weights. Again, these are behaviours which must be recommended by combinations of receptive field detections. *Thirdly*, direct activation of receptive fields by the presence of the field in current sensory inputs. Any changes to the threshold for detection are behaviours. *Fourthly*, prolongation of the activity of a receptive field detection, which is a behaviour that must be recommended by combinations of receptive field detections. A limited number of independent populations of activated receptive fields can be supported in the same receptive field detection resources. *Fifthly*, indirect activation of receptive fields on the basis of past activity temporally correlated with the past activity of currently active receptive fields. Temporal correlations can include recent activity at the same time, earlier or later; frequent past activity at the same time, earlier or later; and past receptive field expansion at the same time, earlier or later. Each type of indirect activation is a different type of behaviour which must be sufficiently recommended to be implemented. *Sixthly*, releases of receptive field detections from one area to another, or to muscle control. Such releases may include combining the detections by independent populations. *Seventhly*, frequently used sequences of behaviours can be recorded to make them more reproducible and rapid. Sequences can include reward and indirect activation behaviours.

In earlier chapters we have shown how these information processes are implemented at more detailed levels in the brain. If a cognitive process can be described using only these seven types of information process, there can be confidence that the description can be mapped through anatomy and physiology to neurochemistry as required. In this chapter we will develop descriptions in terms of these information processes for a number of cognitive phenomena, only rarely shifting to more detailed levels.

In many cases a cognitive process could be supported by different sequences and combinations of these information processes. The actual process used in different brains may be different, or different for the same brain in slightly different circumstances. The exact process will be very sensitive to the detailed learning experience of the specific brain.

As discussed in earlier chapters, a receptive field is instantiated in a cortical column and is defined by some similarity circumstance in sensory inputs or other input information that has occurred many times in the past. Such a receptive field is detected if current experience contains this circumstance. Each cortical area has a specified source of inputs, and receptive fields are initiated by assigning to each column within the area a different, narrowly defined similarity circumstance. Then in every experience the detection of at least a minimum number of receptive fields within the area is forced. If an experience does not result in the minimum directly, the receptive fields most similar to circumstances present in the experience are expanded until they are detected. This definition process thus divides up experience into relatively orthogonal similarity "chunks" in such a way that at least a minimum number of chunks are present in every experience. The process does not lead to similarity chunks that correspond with "meaningful" cognitive features. However, by appropriate specification of the complexity of the receptive fields, similarity chunks can be defined that are effective for discriminating between situations in which different cognitive behaviors are appropriate. Receptive field complexity for an area is specified by definition of a group of major sources of inputs to be the elements making up the fields. Major sources could be raw sensory inputs or inputs from a specific layer of a specific other area. Receptive field complexity is further specified by the typical number of elements that must be present for detection to occur. Any one area has a genetic bias in favour of receiving most of its inputs from a specific group of other areas, and this genetic bias reflects natural selection pressures in favour of a set of areas that together can discriminate between a very high proportion of situations in which different behaviours are appropriate.

An appropriate behaviour might be speaking a category name when an instance of the category is perceived. Because of the heuristic definition process, no one receptive field will be detected within all instances of the category and never detected within instances of any other category. Effective discrimination in this example means that the groups of receptive fields detected within instances of the category are sufficiently similar to each other, and sufficiently different from the groups detected within instances of any other category. "Sufficiently" similar and "sufficiently" different means that recommendation weights can be assigned in such a way that high integrity behaviour selection is possible.

The simplified highest level information architecture which is the starting point for descriptions is as illustrated in Fig. 11.1, a simplified version of Fig. 8.1. In this

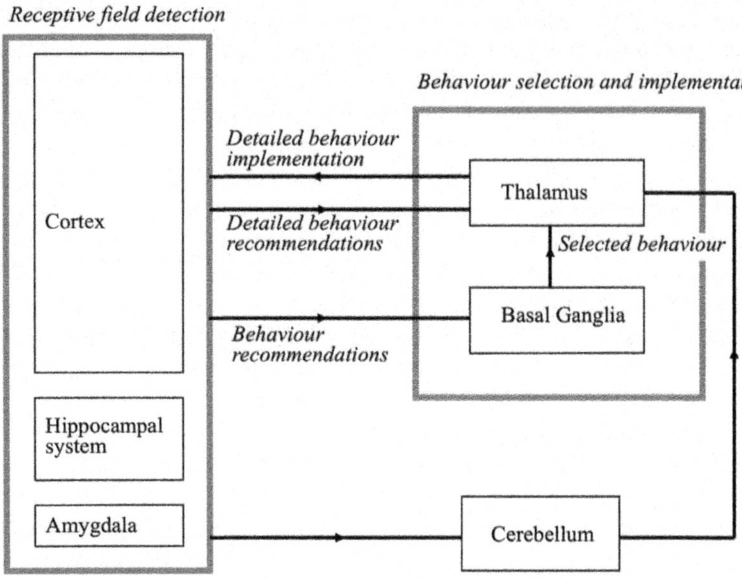

Fig. 11.1 Simplified highest level architecture adequate for understanding many complex cognitive processes

figure the cortex, the hippocampal system, and the amygdala define and detect receptive fields. Receptive field detections are released to the thalamus, basal ganglia and cerebellum. The basal ganglia interprets receptive field detections as behavioural recommendations in favour of different types of behaviour and selects and implements the type with the strongest total recommendation strength. The selected type is communicated to the thalamus. The thalamus receives receptive field detections and interprets these detections as recommendations in favour of very specific behaviours. The basal ganglia indicates to the thalamus the type of behaviour that has been selected, and the thalamus implements the most strongly recommended specific behaviour that is within the selected type. This selected specific behaviour is implemented by release of receptive field detections between different receptive field detection regions, or externally to drive a motor implementation such as attention or speech. Frequently used sequences of behaviour types are recorded in the cerebellum. Such frequent use means that for each behaviour type in the sequence a very specific receptive field can be defined in the cerebellum with very strong recommendation strength in favour of initiation of that type. These receptive field detections can then bypass the general behaviour selection in the basal ganglia and act directly on the thalamus or muscle control.

For simplicity, the amygdala in this information architecture will be taken to include some hypothalamic receptive field detection functions, and the striatal like nuclei in the amygdala with a behaviour selection role will be taken to be included

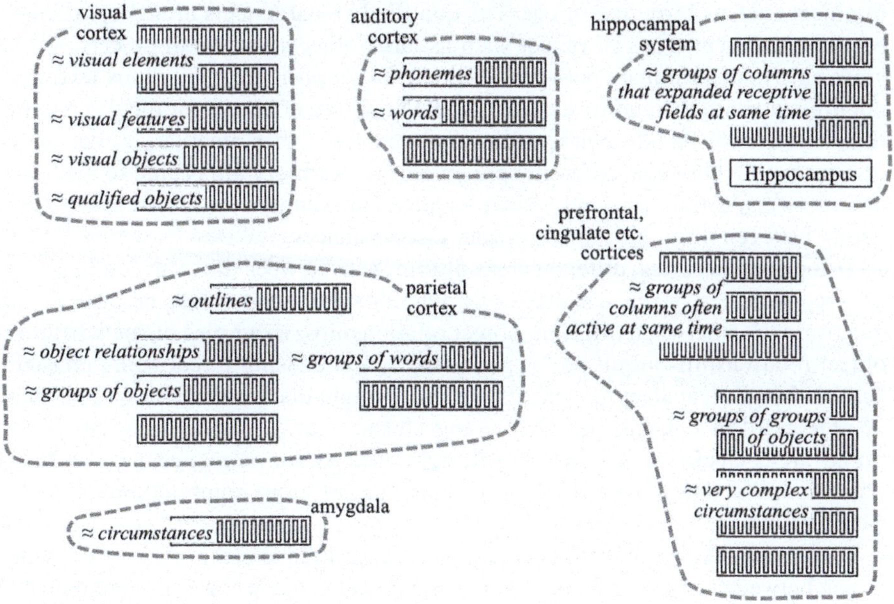

Fig. 11.2 Cortical architecture for understanding higher cognitive processes. The illustrated cortical areas each define and detect receptive fields in a different range of complexity. Each range of complexity is effective for discriminating between different types of circumstances in which different behaviours are appropriate. The circumstances discriminated by some areas are indicated, although these indications are approximate and in practice multiple areas will contribute to discriminations of each type. The circumstance discriminations mean that receptive field detections in one area can be assigned recommendation strengths which allow high integrity selections between certain types of behaviors. Again, in practice multiple areas will contribute recommendation strengths to any one type of behaviour

in the basal ganglia. The general neurotransmitter distribution systems that modulate behaviour will be largely omitted.

Some receptive field types may be particularly effective for discriminating between certain types of behavioural situations. Such types will have a bias in favour of output connectivity on to striatal regions corresponding with those types. Again, such a bias will be the result of natural selection. However, although one receptive field type may be particularly effective, it may not be sufficient without receptive fields detected by another area.

Some of the key receptive field detection areas and complexities discussed in previous chapters and relevant to complex cognitive processes are illustrated in Fig. 11.2. For a number of areas, possible discrimination types have been indicated. In the visual cortex different areas discriminate between visual elements, visual features, visual objects, and qualified visual objects. The ≈ *visual elements* area discriminates between different boundaries on the retina. The ≈ *visual features* area discriminates between different features of visual objects such as wings, feathers or beaks. The ≈ *visual objects* area discriminate between different categories of object such as bird, dog or cat. The qualified visual objects area discriminates between different types of object such as large objects and small objects. Visual areas able to discriminate between different object positions on various levels are not illustrated. In the auditory cortex, some areas are effective for the discrimination between different sounds such as phonemes. Other areas are effective for discriminating between different spoken words. Auditory areas able to discriminate between different sound source locations are omitted, as are areas discriminating between different somatosensory circumstances.

In the parietal cortex, different areas discriminate between different combinations of circumstances in the sensory areas. One or more areas might be effective in discrimination between different object relationships. Examples of such different object relationships might be "dog chasing", "cat chasing", "cat being chased", "cat stalking", "cat sleeping" etc. Other areas might discriminate between groups of objects ("cat chasing squirrel", "dog chasing cat", "dog fighting cat" etc.). Some areas distinguish between different groups (or sequences) of auditory objects, while other areas distinguish between different combinations of poly-sensory objects.

In reality a number of different areas may be required to make effective discrimination between one type of situation and one area may contribute to more than one type of discrimination. On the other hand, the more different types of behaviour a area supports, the greater the risk that receptive field changes in an area will result in undesirable side effects. Furthermore, there are some constraints that result from the ways in which information must be combined. For example, because receptive fields able to discriminate between groups of objects must include receptive fields detected within the individual objects, ≈ *visual objects* areas will tend to be separate from ≈ *groups of objects* areas. Hence the cognitive discrimination type labels given to the information processes performed by some areas in Fig. 11.2 are a useful approximation for describing cognitive processes, but the descriptions become much more complex if greater precision is required.

In the hippocampal system it is less possible to assign cognitive discrimination labels. Receptive fields are groups of columns in other areas that have expanded their receptive fields at related times in the past. These receptive fields are useful for a number of cognitive purposes. As mentioned in Chap. 10, when a specific physical location is visited for the first time, the combination of visible objects and relative positions in the visual field will be novel and will drive receptive field expansion in a group of visual and parietal cortical columns. These receptive field expansions will all occur at the same time, and hence some hippocampal receptive fields are place fields discriminating between different locations (≈*locations*).

In the prefrontal and cingulate cortices, the receptive fields correspond with complex combinations of other receptive fields that have been detected at correlated times in the past. In some areas these receptive fields link groups of sensory columns with other groups of sensory columns. These receptive fields support semantic memory as discussed in Chap. 9. Yet other areas might distinguish between different groups of groups of objects. Groups of groups could combine visual, auditory and other sensory inputs. Some such ≈*groups of groups* receptive fields might alternatively be located in the parietal cortex. Other prefrontal and cingulate receptive fields make it possible to discriminate between different very complex, often polysensory circumstances. One example is receptive fields in the cingulate cortex able to discriminate between very complex social circumstances in which a positive or negative reward is appropriate. Another is the receptive fields in the dorsolateral prefrontal cortex which are able to discriminate between circumstances in which different types of behaviour should have priority (i.e. selection of current emotional state).

In the amygdala, receptive fields are also able to discriminate between circumstances in which different rewards are appropriate, and circumstances in which different behavioural priorities are appropriate. However, in the amygdala there is a higher degree of genetic predefinition of the receptive fields than in the cortical areas.

Note, however, that any discrimination labels placed on cortical and other areas are only approximate. Receptive fields in one area are established by genetically controlled bias on input sources, followed by an heuristic definition process which could add a small proportion of inputs from other sources. The result is a set of receptive fields that are able to discriminate between different types of activity patterns in the input sources. The combination of genetic bias and experience results in these fields being able to discriminate between circumstances in which different behavioural responses are appropriate, for a range of different types of behaviour. Genetic bias results in outputs from any one area tending to target the regions in the striatum corresponding with the types of behaviour that are effectively discriminated, but again other areas may heuristically acquire a small degree of connectivity to those regions. However, the approximations make it possible to describe cognitive processes, and because the approximations are understood a description could be made more precise as required.

11.1 Imagined Situations and False Memories

Human beings can imagine situations that have never occurred, and even situations which could not occur in reality. To understand this phenomenon, consider a specific example. Suppose a subject is asked to imagine being at a barbeque on a beach with Winston Churchill. Hearing the word "barbeque" directly activates receptive fields in the auditory cortex, initially in an area discriminating between individual sounds like phonemes, then independent populations of such receptive field detections corresponding with different phonemes are combined to drive a population of receptive field detections at the ≈*words* level. These ≈*words* columns drive activation of a population of columns in the ≈*visual objects* area on the basis of frequent simultaneous past activity, acting via the appropriate areas in the prefrontal cortex as described in Chap. 9. This population will result in a subjective experience similar to looking at a barbeque. The subjective experience is a weighted average of barbeques seen in the past, but without activity in the visual cortical areas close to sensory inputs (i.e. not an hallucination). The activity of the population is prolonged at a specific phase of frequency modulation. Hearing the word "beach" by a similar process results in another population in the ≈*visual objects* area at a different phase of frequency modulation. Then hearing the words "Winston Churchill" by the same process results in a third population at a third phase of modulation. Next, the outputs from these populations that target the ≈*groups of objects* area are brought into the same modulation phase. A population of receptive field detections is driven in this target area. Although the individual objects are familiar, the combination has never been experienced. The number of columns detecting their receptive fields in the ≈*groups of objects* area will therefore be below the minimum and receptive field expansion will occur. The populations across the ≈visual objects and ≈groups of objects areas will have some similarity with the populations that would be activated in a real experience. However, there will be much less activity in the areas close to sensory inputs than in a real experience. The receptive field expansions make later recall of the experience possible, in the same way as an episodic memory described in Chap. 9.

There are a number of complicating factors that modify this simple description. Firstly, rather more areas will be involved. For example, receptive fields could be detected on levels of complexity intermediate between the levels we have labelled ≈*visual objects* and ≈*groups of objects* or below the ≈*visual objects* level, and more than one level of complexity may contribute to these types of discrimination. Secondly, receptive field detections could be activated closer to sensory inputs on the basis of frequent past activity simultaneous with ≈*visual objects* activity, providing a richer sensory experience. Thirdly, each word could also result in indirect activations at the higher complexity levels such as ≈*groups of groups*, leading to activity and receptive field expansions at higher levels. This higher level activity could lead to further indirect activations and a richer subjective experience of the overall pseudoevent.

Although receptive field expansions could lead to recall of the pseudoevent, a couple of factors usually make it possible to distinguish such a recall from recall of a real event. Firstly, the degree of activity at the levels of complexity close to sensory inputs will generally be lower for an imagined event. Secondly, for a real event, indirect activation on the basis of past receptive field expansion just before or just after will generate recall of circumstances that lead to or follow the event in a plausible fashion. Receptive fields at a very high complexity able to discriminate between differences in these factors will acquire recommendation strengths in favour of behaviours appropriate to the recall being real or imaginary. However, if the imagined event has some aspects very similar to a number of real events, it can be possible to create pseudoevents that are perceived as real.

11.2 Familiarity and Déjà Vu

Human beings have a very high capability to distinguish between novel and familiar objects and situations. The basis for this capability is that in any novel situation there will be receptive field expansion. Receptive fields able to discriminate between the presence and absence of receptive field expansions in different areas will be able to recommend different types of appropriate behaviour. Such behaviours include saying the experience is novel or familiar.

However, suppose that in some situation there is receptive field expansion in some areas but not in others. For example, suppose that there is receptive field expansion in the ≈*groups of objects* area, but once that expansion has occurred, the outputs from the area drive adequate receptive field detection at the ≈*group of groups* level without the need for receptive field expansion. Such an unusual situation could give rise to the déjà vu experience, with some areas indicating novelty and other areas familiarity.

Recognizing the familiarity of an object is a simple receptive field detection process. Describing a familiar object when given its name is a much more complex process involving sequences of indirect activations. Recall is therefore much more difficult than recognition.

11.3 Prospective Memory

We will now consider the more complex processes of prospective memory. As discussed in Chap. 2, prospective memory is a simple example of the ability establish a plan and later to carry out that plan. As an example we will consider how receiving a telephone call in which an appointment was scheduled with a doctor at 11 a.m. on Thursday can lead to being at the appointment.

First we will give an overview of the information processes, then go through in more detail. The overall process is that when the appointment is made, words are heard and spoken about the doctor, the symptoms, and the date and time of the appointment. These words indirectly activate receptive fields at many levels of complexity, including initially ≈*visual words*, then through ≈*groups*, ≈*groups of groups* to ≈*complex circumstances*. The subjective experience of this indirectly activated secondary population has some similarities with an actual experience of the appointment. The processes are much the same as for an imaginary event as discussed in an earlier section, and because the future appointment is not exactly the same as any past appointment, there are receptive field expansions to establish the secondary population. The secondary population then activates a tertiary population at roughly the same levels of complexity on the basis of frequent past activity somewhat before the activity of its own columns. This tertiary population will be similar to the population activated when setting off for an appointment in the past, and has recommendation strengths in favour of setting off. However, these recommendation strengths are not sufficient to initiate the behaviour. Later, hearing one or two of the words activates some of the columns in the secondary population, which via recommendation strengths in favour of indirect activation on the basis of past simultaneous receptive field expansion reconstructs a population similar to the original secondary population, as in episodic memory described in Chap. 9. This secondary population again activates the tertiary population with recommendation strengths in favour of starting off for the appointment. Again, this population does not have sufficient recommendation strength to initiate the behaviour unless some additional strength is present. If inputs providing information on current time and date are present, these inputs result in activation of a complex population at a level that can discriminate between different times. If there is an overlap between this population and columns in the secondary and tertiary populations, the receptive fields correlating with that overlap have recommendation strengths in favour of initiation of a motor behaviour. These recommendation strengths add to those of the tertiary population to initiate setting off.

A more detailed view of the process for acquisition of a prospective memory is summarized in Table 11.2, and for access in Table 11.3. If the process for creating procedural memories has been learned in past experience, an important first step is detection of receptive fields at very complex levels that are able to discriminate between situations in which the sequence of information processes that result in a procedural memory should be initiated and other situations. These receptive fields have recommendation strengths in favour of initiation of different sequences of information processes. If the predominant recommendation strength is in favour of one particular sequence, the recording of that sequence in the cerebellum is activated. Learning to acquire and access procedural memories includes acquisition of a receptive field in the cerebellum that is detected at the completion of each type of information process in the sequence. This field implements the next type directly into the thalamus.

During the telephone call, the words "doctor", "Thursday", "11", "morning" and "a.m." are heard and spoken, plus words describing the symptoms or medical condition. Each word results in direct detection of receptive fields through the auditory cortex to levels of complexity ≈*words*. Each word has often been heard a number of times in the past, and for each word the visual circumstances had some similarity. Hence each word has recommendation strengths (via the ≈*groups of columns often active at the same time* areas) in favour of indirectly activating columns in the ≈ *visual objects* area, and sometimes also in the ≈ *group of visual objects* area and perhaps at even higher complexities. For example, the word "doctor" may have been heard in the past a number of times when a medical doctor has been seen. Hearing the word will indirectly activate visual columns in the ≈*visual objects* area often directly detected in the past when looking at a doctor. Hence the word will activate a pseudoimage that is a weighted average of the visual images of doctors seen in the past. In addition, the word "doctor" may have been heard in the past while in the reception rooms of different doctors. A doctor's reception room is a group of visual objects including a receptionist, a counter, a seating area, magazines etc. Hence the word could activate a pseudoimage in a ≈*groups of visual objects* area that is a weighted average of the columns directly activated during visits to different doctors' reception rooms in the past. During past appointments both the reception room and the examination room will have been visited. These are different groups of objects, and the word could therefore activate a pseudoimage in a ≈*groups of visual groups* area that is a weighted average of the doctors' suites. Hearing the word "Thursday" will similarly activate a range of visual columns at different levels of

Table 11.2 Sequence of information process steps supporting creation of a prospective memory

Information process step	External inputs	Receptive field detections	Information events	Subjective experience	Some recommendation strengths	Accepted recommendation
1	Circumstances indicating that creation of a prospective memory is needed	≈complex circumstances		Weighted average of past experiences in which a prospective memory has been acquired		Initiate a sequence of information processes recorded in the cerebellum
2	Words including "doctor", "11", "am", "Thursday", symptom description	≈auditory words		Hearing words	Indirect activations on the basis of frequent past simultaneous activity	Indirect activation of visual columns on the basis of frequent past simultaneous activity
3		Several populations in ≈visual objects		Pseudovisual images	Release recommendations	*Multiple* ≈visual objects populations released to ≈groups
4		Several populations in ≈groups of objects	Receptive field expansions	Pseudovisual images of groups such as doctor's reception area, consultation room		Multiple ≈groups populations released to ≈groups of groups

(continued)

Table 11.2 (continued)

Information process step	External inputs	Receptive field detections	Information events	Subjective experience	Some recommendation strengths	Accepted recommendation
5		Several populations in ≈groups of groups	Receptive field expansions	Pseudovisual images of groups of groups such as doctor's suite		Detections within multiple groups of groups synchronized and released to ≈ very complex circumstances
6		"***Future appointment population***" including receptive field detections at many levels of complexity	Receptive field expansions	Being at doctor at date and time and describing current symptoms		Indirect activations in same areas on the basis of frequent past activity somewhat earlier
7		"***Future departure population***" including receptive field detections at many levels of complexity	Receptive field expansions	Setting off for doctor appointment		Outputs of ***future appointment population*** and ***future departure population*** synchronized and released to ≈rewards area
8		≈rewards				Increase recommendation strengths of recently active population A columns in favour of indirect activation of recently active population B columns

Table 11.3 Sequence of information process steps supporting access to a prospective memory

Information process step	External inputs	Receptive field detections	Information events	Subjective experience	Some recommendation strengths	Accepted recommendation
1	Stimuli such as awareness of symptoms, hearing name of day, thinking about doctor etc.	≈visual objects		Separate visual and pseudovisual images	Various indirect activations; Speaking words; Releases to other areas	Releases to more complex other areas
2		≈visual objects and more complex areas		Pseudovisual images of vague combinations of input stimuli	Various indirect activations	Indirect activation of columns in same and other areas on the basis of past simultaneous receptive field expansion
3		***Future appoint-ment population***		Being at doctor at date and time and describing current symptoms	Prolong activity; Indirect activations on the basis of frequent past activity just before; Attention to current date and time	Prolong activity; Attention to current date and time
5	Current date and time	Population A ≈very complex circumstances	Receptive field expansions	Pseudovisual images	Determine overlap between parts of ***future appointment population*** and population A	Release of activity in specific areas of ***future appointment population*** and population A

(continued)

Table 11.3 (continued)

Information process step	External inputs	Receptive field detections	Information events	Subjective experience	Some recommendation strengths	Accepted recommendation
6	[Current and appointment date and time correspond]	Overlap indicated in ≈overlap ***future appointment population*** and population A			Reinforce indirect activations on the basis of previous activity somewhat earlier; reinforce motor recommendations	
7		***Future departure population***		Setting off for doctor appointment	Motor movements to set off for appointment	Motor movements to set off for appointment
6a	[Current and appointment date and time do not correspond]	Overlap not indicated in ≈overlap between A and C			Releases leading to rewards	Outputs of A and B synchronized and released to ≈rewards area
7a		≈rewards			Various changes to recommendation strengths in favour of recently accepted behaviours	Increase recently utilized recommendation strengths in favour of indirect activations on the basis of past simultaneous receptive field expansion, and recent simultaneous activity

complexity that are the columns most frequently activated in the past when the word "Thursday" was also heard. In addition, if columns in the ≈ *words* area corresponding with "Thursday" have been indirectly activated in the past, perhaps on some other day, then columns often active at such times (themselves perhaps indirectly activated) could be activated later by hearing the word. Hence even thinking about Thursday might create some indirect activation strengths. The exact combination of indirectly activated columns will depend not just on the frequency of past simultaneous activity but also on the reward behaviours that have followed the use of the recommendation strengths in the past.

The indirectly activated populations can then be combined in various ways to generate receptive field detections at more complex levels. For example, the ≈*visual objects* column populations indirectly activated in response to hearing words describing symptoms and the word "doctor" could be held active at different phases of frequency modulation, then the outputs brought into the same modulation phase and released to the ≈*groups of visual objects* area. The activation in that area might correspond approximately with a visual experience of describing the symptoms to a doctor. If the specific symptoms have not been described to a doctor in the past, the number of columns activated may be below the minimum for a population in this area, resulting in receptive field expansions driven through the hippocampal system. Similarly, the outputs from separate ≈*visual objects* populations indirectly activated by hearing "11", "a.m." and "Thursday" may be synchronized and released to the ≈*groups of visual objects* area where they drive activation of a population at a different phase of modulation from the doctor-plus-symptoms population. A separate population indirectly activated by the individual words as described earlier may also be present. The outputs from these ≈*groups of visual objects* populations are then brought into the same modulation phase and released to ≈*groups of visual groups* areas.

In the ≈*groups of visual groups* and ≈*very complex circumstances* areas, a column population is activated which will have some similarities to a population that would be activated if experiencing an 11 a.m. Thursday appointment describing current symptoms to a doctor. We will call this secondary population the ***future appointment population***. Because such an event has not happened in the part, it is probable that the number of activated columns is less that the minimum number for the area, and receptive field expansion will take place.

Each column in this ***future appointment population*** will have a range of different recommendation strengths into the striatum. One such strength is in favour of activating other columns often active in the past somewhat before currently active columns. In the past, leaving to go to an appointment occurred prior to being at an appointment. Hence acceptance of this type of recommendation strengths would lead to a column population including columns that were active when setting off for appointments in the past, but incorporating information about the future appointment. We will call this tertiary population the ***future departure population***. Receptive field expansion would also be required to generate this population. The active columns in this ***future departure population*** will have recommendation strengths in favour of setting off, because such behaviour in the past was followed by arriving on time at an appointment, and receptive fields detected within that circumstance recommended a reward behaviour which was accepted. However, these recommendation strengths are not sufficient alone to initiate departure, additional recommendation strengths are required.

There are other ≈*complex circumstances* receptive fields that discriminate between situations in which a prospective memory process is under way and has reached the stage at which the ***future departure population*** has been activated, and other situations of comparable complexity. The receptive fields in this area correlating with this stage in the prospective memory process have recommendation strengths in favour of a reward that temporarily (for example, days rather than weeks or months) increases recently utilized recommendation strengths in favour of activating other columns often active in the past somewhat before currently active columns. The recommendation strengths of ***future appointment population*** columns that activated ***future departure population*** are therefore increased, increasing the probability of a similar indirect activation in the near future.

Some ≈*very complex circumstances* columns discriminate between different dates and times. Columns in this area will be activated as part of the ***future appointment population***, and also as part of the ***future departure population***. Columns in the same area can also be activated in response to inputs carrying information about current date and time. Even more complex ≈*very complex circumstances* columns discriminate between circumstances when there is overlap between columns activated in secondary or tertiary populations of the type described and columns activated in response to current sensory inputs. Columns often activated when there is such overlap have acquired recommendation strengths that reinforce motor behaviours recommended by the currently active secondary or tertiary population. Overlap correlates with current time being the time of the appointment, and supports the recommendation strengths of the ***future departure population*** columns in favour of immediate departure. Lack of overlap leaves the total recommendation strengths too low for behaviour initiation.

Because receptive field expansion was required to establish the ***future appointment population***, the columns have relatively strong recommendation strengths in favour of indirect activation of each other on that basis. Hence after learning about the appointment, if there is a current sensory input which activates some of the columns in the ***future appointment population***, a significant proportion of the full population will tend to be activated by the episodic memory process described in Chap. 9. For example, noticing the symptoms or hearing "Thursday" could lead to such activation. The more aspects of the appointment that were considered in the activation of ***future appointment population***, the more likely that some combination of sensory inputs will tend to activate it. Each time it is activated, it will in turn tend to activate a significant proportion of the ***future departure population***. The relevant recommendation strengths for this indirect activation will have been somewhat increased by the activity when the appointment was made. In addition, receptive field expansions were required to establish the ***future departure population***. Recommendation strengths on the basis of receptive field expansion of one column just before expansion of another column will reinforce the indirect activation of

future departure population from ***future appointment population***. Furthermore, expansions within the original ***future departure population*** will result in indirect activation strengths within the population tending to converge later reactivations more closely with the original. Hence by the combination of the various indirect activation mechanisms a regenerated ***future departure population*** will be activated that approximates to the original.

Each regeneration results in the presence of the recommendation strengths in favour of departure. Unless there is overlap between the regenerated ***future departure population*** and a population activated at the same level of complexity in response to inputs correlating with current time and date, these recommendation strengths will be inadequate. However, when there is overlap, the total recommendation strengths are adequate and departure is initiated. The sequence of information processes can thus lead to prospective memory.

11.3.1 Learning to Create Prospective Memories

Cortical columns do not "know" about events, plans etc. All they have is a receptive field, and if activated directly or indirectly receptive fields recommend various behaviours in the basal ganglia, thalamus and cerebellum. Behaviours include receptive field expansions lasting for various periods of time, activity prolongations, indirect activations of other columns, changes to recommendation weights, or generation of muscle movements. Each behaviour is implemented by release of a subset of current receptive field activations from one specific cortical area or group of areas to another, or in the case of muscle movements to the brainstem, or in the case of rewards by activation of dopaminergic neurons. Learning to create prospective memories therefore involves establishing a long sequence of such behaviours, in the appropriate order. At each point in the sequence the predominant recommendation weight must be in favour of the next behaviour.

The sequence must first be learned by associating receptive field activations with appropriate recommendation weights in the basal ganglia and thalamus. Weights in the basal ganglia recommend a behaviour type, such as release of receptive field detections from area A to area B. Weights in the thalamus recommend release of specific individual receptive fields from area A. In this example, successful learning results in receptive fields detected towards the end of the previous behaviour recommending the release from area A to B type of behaviour into the basal ganglia, and receptive field detections across a number of cortical areas including A recommending release of specific individual fields in area A. The overall effect is release of a subset of the currently active receptive fields in area A.

A learned sequence can be transferred to the cerebellum, where different Purkinje fields acquire correspondence with the exact circumstances in which each behaviour type must be initiated, and cerebellar nuclei neurons drive the appropriate thalamic nucleus to implement the specific release.

The brain is initially configured so that columns in any one area tend to have certain types of recommendation strengths, such as strengths in favour of indirect activation of columns in a specific group of areas (perhaps including the same area). This genetic bias makes learning of certain types of behaviour easier and more likely.

Individual segments of the sequence will first be learned separately as discussed for learning episodic memory in Chap. 9. For example, learning to activate ***future appointment population*** types could proceed by a child being asked "What are we doing tomorrow?". The child's brain experiments with various types of indirect activations, and following a successful response, the adult behaviour results in receptive field detections with reward recommendation strengths, increasing the probability of the appropriate indirect activation behaviours in the future. Learning to activate ***future departure population*** types could proceed by a child being asked "What do we have to do before we do that?". The segments can then be used in different combinations to attempt an appropriate response to more complex cognitive problems. If there is often a reward at the end of a particular combination, the receptive fields active at the end of one segment will tend to recommend the next segment.

Learning is also required for reward behaviours. This learning is also made more effective by initial genetic programming. For example, as illustrated in Fig. 11.3, there is a hierarchy of reward behaviours with genetically predefined connectivity. Receptive fields corresponding with the presence of a smiling face are defined genetically in the amygdala. These fields have strong genetically defined connectivity on to a strategic reward behaviour component. In addition, receptive fields in specific cortical areas such as D in Fig. 11.3 (which could be Brodmann area 24) have genetically defined weak connectivity on to the same component. This cortical area has a genetically defined receptive field complexity that is effective for discriminating between different complex social situations. Hence among other capabilities it can discriminate between favourable and unfavourable situations following a behaviour. Initially, the strategic reward behaviour is triggered only by relatively specific receptive fields in the amygdala corresponding with circumstances like a smiling face. However, any receptive fields in areas like D often active when the reward behaviour occurs gain recommendation strength in favour of that behaviour. This allows bootstrapping of very complex reward circumstances. For example, initially the smile of a parent triggers the reward behaviour, but the body language corresponding with approval in the culture in which the child is living will also often be present. Hence receptive fields correlating with that body language will acquire recommendation strengths, eventually sufficient to generate the strategic reward behaviour without needing the perception of smile. Other receptive fields correlating with the perception of a group of people applauding could gain recommendation strengths on to the strategic reward behaviour by simultaneous detection with the occurrence of the reward behaviour driven by other receptive fields.

In a long sequence of behaviours, a strategic reward at the end may not be sufficiently discriminating to reward each behaviour in the sequence. As illustrated in Fig. 11.3, the behaviour of releasing receptive field detections from area A to area B

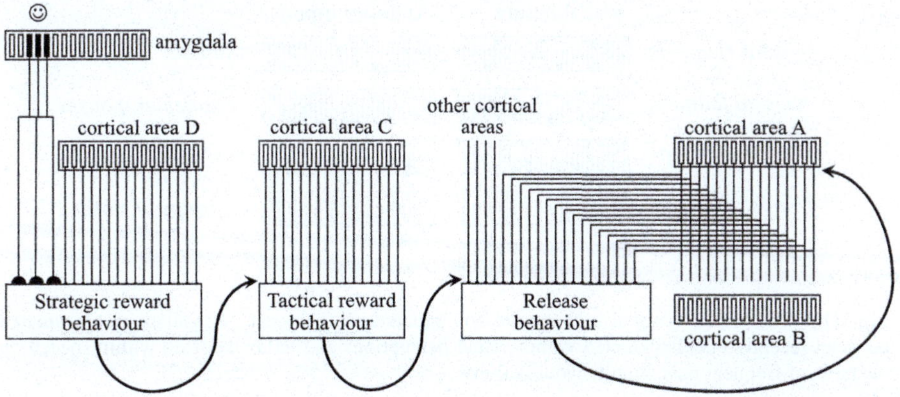

Fig. 11.3 Bootstrapping of reward behaviours. Certain genetically programmed receptive fields in the amygdala and perhaps elsewhere have genetically programmed recommendation strengths in favour of strategic reward behaviours. Receptive fields in other cortical areas have genetically programmed connectivity to the strategic reward component, and can acquire recommendation strengths if active when the amygdala receptive fields result in the reward behaviour. Yet other cortical areas have connectivity to more tactical reward components, and the recommendation strengths are modulated by the strategic reward component

will be recommended by area A detections and detections by a limited number of other areas. The identities of these other areas are limited by genetic information. All the inputs to the release behaviour component initially have relatively weak recommendation strengths. A tactical reward component exists which targets these inputs to the release behaviour. The inputs to this reward component come from a small, genetically specified number of cortical areas like C in the diagram. These inputs initially have enough weight to cause the tactical reward component to be activated whenever there is sufficient activity in area C. However, if later a strategic reward occurs, the weights of these inputs to the tactical reward component are modulated. Hence over a number of experiences the tactical reward becomes appropriate to reward the release behaviour, independent of whether the strategic reward occurs. Tactical rewards for the individual steps in the prospective memory will generally be established when the steps are learned, but may be refined as the overall process is learned.

11.4 Speech Processing

Speech capabilities are fundamental to cognition, and these capabilities can be understood in terms of sequences of the same attention, indirect activation, information release, etc. behaviours. Understanding speech requires a way to associate receptive fields activated in response to hearing words with receptive fields activated in response to seeing the corresponding visual situations. The mechanism

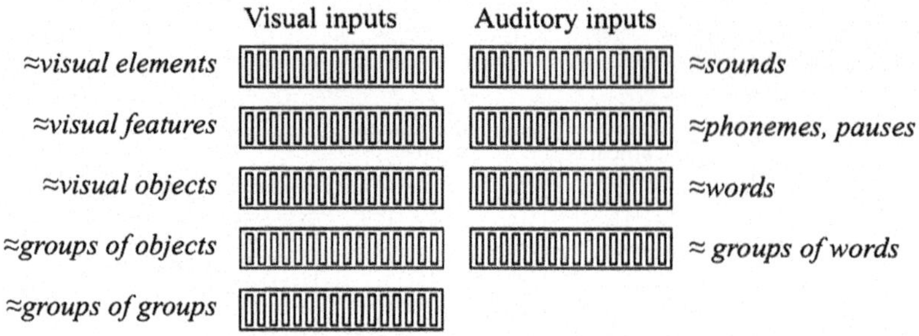

Fig. 11.4 Simplified cortical architecture for understanding speech. Associations are created between receptive fields detected within visual perceptions and fields detected within speech on the basis of frequent past simultaneous activity

for creating such associations is indirect activation on the basis of frequent past simultaneous activity as discussed under semantic memory.

A simplified cortical model to understand how speech understanding is supported is illustrated in Fig. 11.4. In this model visual inputs directly drive receptive field detections at different levels of complexity which can discriminate between different types of visual situations. The primary inputs to one area are receptive field detections in the previous area, but inputs from other areas are possible. Auditory inputs also drive receptive field detections at different levels of complexity. The early levels process information derived from just one sense, but the higher levels of complexity will be polysensory to varying degrees.

As described in Chap. 9, understanding of nouns is achieved by frequent simultaneous presence of the word and the corresponding visual object, leading to the ability of columns in a ≈*words* cortical area being able to indirectly activate appropriate columns in a ≈*visual objects* area. The indirectly activated visual columns have recommendation strengths in favour of a range of behaviours appropriate to the presence of the object. In the case of a verb like "chase", many scenes in which one object chased another are viewed when the word is heard. The chasing and chased object will be different in each scene, and there will be little temporal consistency between columns activated in the ≈*words* and ≈*visual* objects areas. However, there will be consistency between columns activated in the ≈*words* area and some columns activated in the ≈*groups of objects* area. The receptive fields activated in response to the word therefore acquire recommendation strengths to activate that group of columns in the ≈*groups of objects* area, resulting in a subjective experience as if perceiving a vague object chasing another vague object.

Consider how the spoken sentence "white dog chased black and grey cat" could be processed in this model. In simplified terms, each spoken word directly activates a population of auditory columns, and the auditory columns indirectly activate a population of visual columns at one or more specific levels of complexity, on the basis of frequent past simultaneous activity. The visual populations corresponding with each word in the current phrase are maintained active at different phases of frequency modulation at appropriate levels of receptive field complexity. When the phrase is complete, the outputs from all the populations are synchronized and released to a higher level of complexity. The resultant visual population at those higher levels is a pseudoexperience as if visually perceiving the situation corresponding with the phrase. The same process is repeated until the three phrases have indirectly activated populations on the basis of frequent past simultaneous activity, in the same or different target areas. When the sentence is complete, outputs from the three populations are synchronized and released to an area detecting receptive fields at a yet higher level of complexity. The resultant visual population at this point is a pseudoexperience as if visually perceiving the situation corresponding with the sentence. Populations corresponding with different sentences can be combined to generate receptive field detections at even more complex levels, corresponding with pseudoexperiences of being in the circumstances described verbally.

These steps in one possible sequence of information mechanisms are described in somewhat more detail in Table 11.4. This sequence illustrates a number of points. Firstly, a population which will give a pseudovisual experience as if the scene described by the sentence was being viewed in reality can be generated by a sequence of the types of information processes which can be supported in the brain. This pseudovisual experience does not include receptive field detections close to raw visual input, but makes available a similar range of behavioural recommendations to that which would result from viewing the scene. Secondly, the order in which information synchronisations and releases to other areas are performed is critical to achieving a meaningful end activation. The timing of such information releases is determined by detection of other receptive fields within circumstances such as pauses, existence of specific combinations of active populations, and the presence or absence of certain words such as "and". For example, in the phrase "black and grey cat", hearing the "and" generates receptive fields that recommend not synchronizing and releasing receptive field detections to higher levels of complexity until one extra population has been indirectly activated.

The simplified Fig. 11.4 architecture is a useful approximation for understanding, but there are some issues which require a more detailed view. For example, it is not obvious how the ≈groups of objects population activated in response to hearing "white dog chased black and grey cat" would be different from hearing "black and grey cat chased white dog". Resolving this problem requires receptive fields to be defined on rather more levels of complexity, able to discriminate between various intermediate situations. The visual system has a large number of different areas providing such discrimination possibilities. Some such areas are illustrated in Fig. 11.5. This figure illustrates some cortical areas in the ventral and dorsal stream of the visual system and in the parietal and prefrontal cortices. Possible types of discrimination that could be performed by different areas are shown. Receptive fields in the ventral stream are involved in identification of objects. Such receptive fields have recommendation strengths including in favour of appropriate responses to a type of object, such as category naming. Receptive fields in the dorsal stream are involved in identification of the positions and motions of objects. Such receptive

Table 11.4 Possible sequence of information processes in simple Fig. 11.4 cortical model to support understanding of speech.

	Step	Comment	Example "white dog chased black and grey cat"
1	Detection of ≈*phonemes* column population in hearing a phoneme	Population activated at a specific modulation phase	"wh"
2	Detection of ≈*words* column population in hearing a word	Outputs from multiple populations in ≈*phonemes* synchronized and released to ≈*words* where they activate a column population at a specific modulation phase corresponding with the word; ≈*phonemes* populations then extinguished	≈phonemes "wh" + "i" + "te" → ≈features white
3	Indirect activation of visual columns by ≈*words* columns on the basis of frequent past simultaneous activity	Visual column population spread across multiple levels of complexity. Depending on the word, could be strongest in one level	≈*words* "black" → ≈*visual features*
4	Prolong activity of visual column population; extinguish population in ≈*words*	Activity prolonged at a specific modulation phase	
5	Repeat steps 1–4 for each word in phrase	Separate populations at different phases of frequency modulation	≈*words* "white" → ≈*visual features* ≈*words* "dog" → ≈*visual objects*
6	Detection of receptive fields correlating with the end of the phrase (such as within ≈*phonemes, pauses* and/or a specific combination of active populations)	Outputs from visual column populations at each level of complexity bought into same modulation phase and released to next level. Result is generation of a population corresponding with phrase, generally strongest at ≈*visual objects* or ≈*groups of objects* level	
7	Receptive field expansions if required to achieve minimum number of column activations in phrase population in ≈*visual objects* or ≈*groups of objects* area	Subjective experience as if visually perceiving an experience corresponding with phrase, without receptive fields close to sensory inputs.	"white dog" → ≈***visual objects***

8	Prolong visual population corresponding with phrase at a specific modulation phase; extinguish visual populations corresponding with words		
9	Repeat steps 1–8 for all phrases in sentence	Separate populations corresponding with each phrase, strongest in ≈*visual objects* or ≈*groups of objects*, at different phases of modulation.	"white dog" → ≈*visual objects* "chased" → ≈*groups of objects* "black and grey cat" → ≈*visual objects*
10	Detection of receptive fields correlating with the end of the sentence (such as within ≈*phonemes, pauses* and/or a specific combination of active populations)	Outputs from visual column populations at each level of complexity bought into same modulation phase and released to next level. Result is generation of a population strongest at ≈*group of objects* or ≈*group of groups* level corresponding with sentence	
11	Receptive field expansions if required to achieve minimum number of column activations in sentence population in ≈*group of objects* or ≈*groups of groups* area	Subjective experience as if visually perceiving an experience corresponding with sentence, without receptive fields close to sensory inputs.	"white dog chased black and grey cat" → ≈*group of objects*
12	Extinguish indirectly activated visual populations corresponding with phrases and prolong population corresponding with sentence at a specific modulation phase		"owner of cat yelled at dog" → second ≈*group of objects*
13	Repeat steps 1–12 for next sentence	Populations of columns at ≈*groups of groups* or ≈*complex circumstances* activated, similar to column activations if visually experiencing the speech topic, but without activations closer to sensory inputs	"white dog chased black and grey cat" + "owner of cat yelled at dog" → ≈*groups of groups*
14	Repeat for multiple sentences, receptive field expansions if required	Populations of columns at ≈*complex circumstances* and ≈*very complex circumstances* levels activated; Subjective experience as if visually perceiving an experience.	White dog chased black and grey cat. Owner of cat yelled at dog. Owner of dog called dog. Owner of cat complained to owner of cat. I was watching from my garden.

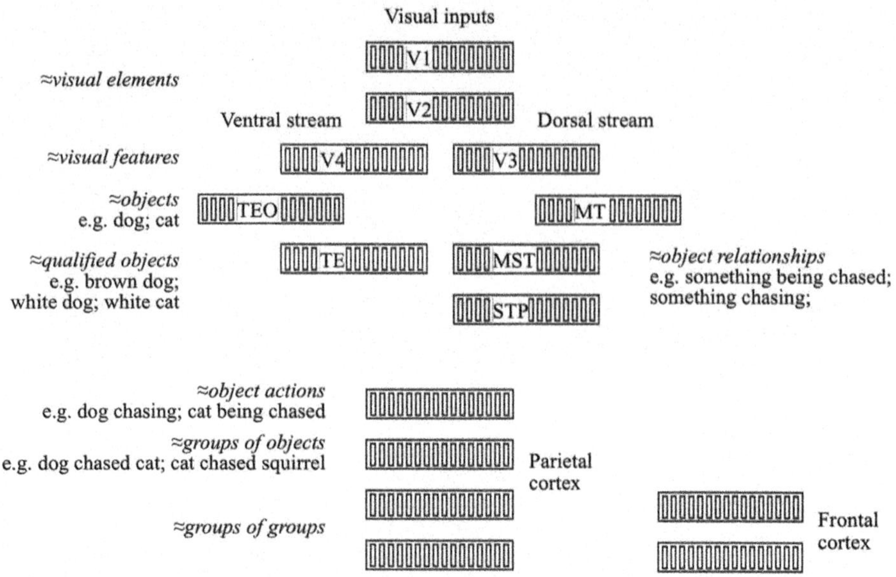

Fig. 11.5 Intermediate complexity receptive fields able to discriminate between circumstances intermediate between, for example, objects and groups of objects. These type of receptive fields are needed to be able to discriminate between, for example, a dog chasing a cat and a cat chasing a dog

fields have recommendation strengths in favour of appropriate responses to the motion and position of an object, including eye tracking. There is significant information exchange between the areas.

In the architecture illustrated in Fig. 11.5, two populations corresponding with "white dog" and "black and grey cat" are created in the ≈*qualified objects* area. For example, outputs from ≈*visual features* activated indirectly by hearing "brown" and outputs from ≈*objects* activated indirectly by hearing "dog" are synchronized and released to ≈*qualified objects*. The activity of the resultant "white dog" population is prolonged at a specific modulation phase, and a second population "black and grey cat" generated at a different modulation phase. Hearing the word "chased" activates two populations of columns in ≈*object relationships* at different phases of frequency modulation. The first is made up of columns often active in the past when observing something chasing, the second of columns often active in the past when observing something being chased. Outputs from the first population in ≈*object relationships* are synchronized from the population in ≈*qualified objects* corresponding with "brown dog" and released to the ≈*object actions* area. This release drives activation of a population of columns that have tended to be active in the past when a white dog chasing has been viewed. Receptive field expansion will occur if there are novel aspects to the scene. Outputs from the second population in ≈*object relationships* are synchronized from the population in ≈*qualified objects* corresponding with "black and grey cat" and also released to the ≈*object actions* area. This release drives activation of a population of columns at a different modulation phase from the first, a population that has tended to be active in the past when a black and grey cat being chased has been viewed. Outputs from the two populations in ≈*object actions* are then synchronized and released to the ≈*groups of objects* area. The resultant population will be similar to the one that would be activated if the dog chasing cat scene actually took place. Combinations of multiple populations in the ≈*groups of objects* could be combined at higher levels of complexity to discriminate between different groups of groups.

The word order in the sentence follows the English grammar rule of subject-verb-object. Extracting the appropriate meaning depends upon generating an activation in ≈qualified objects in response to the subject, two populations in ≈object relationships in response to the verb, another population in ≈qualified objects in response to the object, then generating the correct combinations in ≈object actions. The correct sequence of behaviours must be learned and recorded in the cerebellum. Different rules (subject-object-verb, verb-object-subject or object-verb-subject, all of which occur in other languages [949]) simply mean learning different sequences of information releases between cortical areas. An information release is a behaviour which must be the predominant recommendation of currently active columns. Hence an English speaker will learn one set of sequences of indirect activations, activity prolongations, and information releases. For a different language, different sequences will be learned.

11.4.1 Bootstrapping Speech Generation

Acquisition of an appropriate combination of behaviours is made more efficient by genetic information. However, given the complexity of the experienced environment there are limits to the degree to which behaviours can be preprogrammed. For receptive fields, genetic information can place biases on initial connectivity, but can rarely specify receptive fields in detail. Behavioural components can be specified to a higher degree, and biases places on their receptive field inputs. These concepts are illustrated for speech production in Fig. 11.6. In the figure there are a number of cortical areas with receptive fields relevant to speech production, and a number of basal ganglia and cerebellar components corresponding with specific behaviours and behaviour sequences. Genetically defined connectivity results in receptive fields with different complexities being defined through experience in different areas. Some of the key genetically defined biases on connectivity between areas are illustrated, and the type of discrimination which is supported as a result. There is also genetically defined bias in favour of connectivity between certain areas and the components corresponding with certain types of behaviour. Key biases of this type are also illustrated in Fig. 11.6.

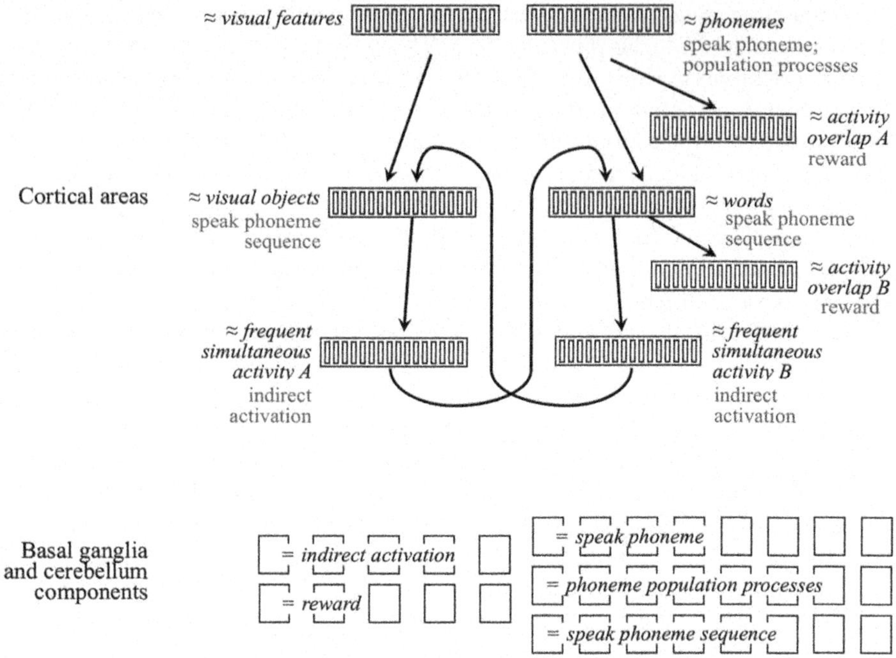

Fig. 11.6 Genetically defined initial connectivity to make bootstrapping of speech capabilities more effective. Genetically defined connectivity between cortical areas is indicated by *arrows*. Genetically defined behavioural component targets of cortical areas indicated by *grey text*

11.4.1.1 Learning to Imitate Phonemes

Inputs to receptive fields in the area ≈*phonemes* have a genetic bias in favour of combinations of inputs from the primary auditory cortex effective to discriminate between different phonemes. Early auditory experience will therefore generate a set of receptive fields able to discriminate between the phonemes heard in the infant's environment, in other words between phonemes used in the local language. The basal ganglia components =*speak phonemes* are hard wired to correspond with a specific set of phonemes, and any human language can therefore only use some subset. Each component drives the combination of muscle movements required to generate the corresponding phoneme. Any one column in the ≈*phonemes* area has initial connectivity (and therefore recommendation strengths) on to all the =*speak phonemes* components. A column also has recommendation strengths in favour of various behaviours (=*phoneme population processes* components), including prolonging activity. Pauses also activate receptive fields and have recommendation strengths in favour of releasing outputs from several populations active in ≈*phonemes* to ≈*words*. The area ≈*activity overlap A* has input connectivity that tends to discriminate between situations in which there is strong overlap between two successive populations in ≈*phonemes*. These columns have genetically defined recommendation strengths on to a reward component that rewards recently accepted recommendation strengths of ≈*phonemes* columns on to =*speak phoneme* components. The reward component also has inputs from columns that discriminate between the presence and absence of recent speech generation muscle movements. A baby's auditory environment includes many spoken phonemes. Such a phoneme will activate columns in the ≈phonemes area. Each such column has recommendation strengths in favour of speaking phonemes, and also in favour of prolonging its own activity. In response to hearing a phoneme there will therefore be recommendation strengths in favour of speaking a range of phonemes. The phoneme component that happens to have a slightly higher total weight will be selected and the phoneme spoken. This self-spoken phoneme will also be heard, and will activate a second column population in ≈*phonemes*. If there is a strong overlap between the two populations in ≈*phonemes* (i.e. the phoneme heard first is similar to the spoken phoneme), columns in ≈*activity overlap A* with recommendation strengths in favour of a reward behaviour are detected. If columns correlating with a recent speech generation muscle movement are also detected, the total recommendation strength is sufficient to trigger the reward. This reward increases the recommendation strengths of recently active ≈*phonemes* columns into the recently accepted =*speak phoneme* behaviour. Hence the overall effect is to cause imitation of phonemes.

11.4.1.2 Learning to Imitate Words

Columns in the ≈*words* auditory area have genetically defined inputs from the ≈*phonemes* area, and thresholds such that multiple phonemes are required for receptive field definition. Receptive fields correlating with pauses have recommendation strengths in favour of synchronisation of all outputs from ≈*phonemes* and then extinguishing the ≈phoneme activity. Hence columns in ≈*words* area will develop receptive fields that can discriminate between different frequently experienced sequences of phonemes, i.e. words. Furthermore, the frequently used sequence of muscle movements associated with speaking a word develops a =*speak phoneme sequence* component in the cerebellum. Columns in the ≈*words* area have genetically established connectivity on to the =*speak phoneme sequence* components. By a similar process to that described for phonemes in the previous section, ≈*words* columns acquire recommendation strengths in favour of imitation of a word just spoken.

11.4.1.3 Learning Associations Between Spoken Words and Visual Experiences

As described under semantic memory, inputs to the ≈*visual objects* area are genetically defined in such a way that receptive fields develop that are able to discriminate between different categories of visual object. Genetically defined bias on connectivity from columns in area ≈*words* to area ≈*frequent simultaneous activity B* results in heuristically defined receptive fields corresponding with groups of columns in ≈*visual objects* that are often active at the same time. Genetic bias results in these columns targetting columns in ≈*visual objects* that are often active at the same time. As a result, simultaneous experiencing of visual objects and corresponding words results in hearing the word tending to activate a pseudovisual image as discussed under semantic memory. Genetically defined bias on connectivity from area ≈*visual objects* to area ≈*frequent simultaneous activity A* results in heuristically defined receptive fields corresponding with groups of columns in ≈*visual objects* that are often active at the same time. Genetic bias results in these columns targetting columns in ≈*words* that are often active at the same time. As a result, seeing the word tends to result in a pseudoauditory experience as though the word was being heard. As described earlier, the ≈words columns have acquired recommendation strengths in favour of speaking words. If these speech behaviours are accepted and rewarded, the ≈visual objects columns will acquire recommendation strengths in favour of speaking the corresponding word. Hence seeing an object can result in speaking its name.

11.4.1.4 Simplifications in the Description of Bootstrapping Speech

This architecture thus demonstrates that by placing specific biases on the connectivity between cortical areas and between areas and types of behavioural component, the process of learning speech can be made much more efficient. However, the architecture is a simplification which makes understanding easier. For example, rather than the two areas illustrated for processing of auditory inputs, it is likely that there are several areas at various receptive field complexities needed to provide adequate discrimination between phonemes and between words.

The key point is that bootstrapping of speech behaviour can be understood in terms of the information mechanisms available to a complex learning system, and the ways in which the description could be made more precise and more detailed are clear.

11.5 Cognitive Deficits Following Brain Damage

Specific cognitive deficits often occur following damage to specific brain areas. Such associations have long been viewed as a way to understand the functions of different brain regions. When damage to a brain region is seen as affecting the information processes supported by that region, the reasons for the resultant combinations of deficits can be understood.

11.5.1 Hippocampal System Damage

One of the most dramatic associations is between damage to the hippocampal system and a specific group of cognitive deficits. This group of deficits was extensively investigated in the patient HM. This patient had experimental surgery bilaterally resecting the hippocampus proper and a number of associated areas in order to treat intractable epilepsy. Several deficits followed this surgery, but also a number of cognitive capabilities were unaffected. Firstly, the capability to create new episodic and semantic memories was lost, a deficit known as anterograde amnesia. All semantic memories acquired prior to surgery were retained. Episodic memories covering a period of 11 years prior to surgery were lost, a deficit known as retrograde amnesia. However, older episodic memories were retained. All skills acquired prior to surgery were retained, and new very simple skills could still be acquired. Attention, working memory and priming were unaffected.

The hippocampal system supports two types of information process. Firstly, driving receptive field expansions in appropriate cortical columns. Secondly, selecting columns to be indirectly activated on the basis of past temporally correlated receptive field expansion. Loss of the ability to expand receptive fields means loss of all new semantic or episodic memories. Loss of the ability to indirectly activate on the basis of past temporally correlated receptive field expansions means loss of access to episodic memories. Past semantic memories will still be accessible based on frequent past simultaneous activity. If an episodic memory is often accessed, retrieval could shift to being based on frequent past simultaneous activity (i.e. semantic memory). Hence more distant episodic memories can be retained on this basis.

All existing cortical receptive fields and their associated recommendation strengths will be retained. Hence skills acquired in the past will be unaffected. Learning of a complex skill will generally require receptive field expansions and will not be possible. Learning of a simple skill could be supported by changes to the recommendation weights of existing receptive fields. However, even simple skill learning will be affected by the inability to adjust receptive fields, and will be less efficient. The ability to indirectly activate columns on the basis of recent simultaneous activity will be unaffected. Hence priming memory will remain. Working memory based on the ability to support multiple independent column populations in the same cortical area will also be unaffected.

Individual structures within the hippocampal system support more detailed information processes in support of the general system processes, and the cognitive deficits limited to the individual structures can be understood in these terms. For example, neurons in CA1 produces the outputs that drive receptive field expansions. If such outputs were also used to drive indirect activations, unmanaged receptive field expansions would also occur. Hence as expected, damage limited to that structure results in anterograde amnesia but no retrograde amnesia [950, 951]. The hypothalamus influences the location of receptive field expansions in favour of regions supporting current behavioural priorities, and damage strictly limited to the mammillary bodies results in anterograde amnesia but minimal retrograde amnesia [526]. The anterior thalamic nucleus manages release of hippocampal system outputs, and damage therefore results in both anterograde amnesia and retrograde amnesia [952].

Hence understanding the information processes supported by the hippocampal system makes it possible to understand the exact combination of cognitive deficits that follow hippocampal damage.

11.5.2 Basal Ganglia Damage

The key information role of the basal ganglia is to determine the total recommendation strengths in favour of different behaviours, and to implement the most strongly recommended behaviour. Part of this role is to ensure that one but not more than one behaviour of a particular type is selected. In other words, a motor behaviour and an indirect activation behaviour could be selected simultaneously, but not two incompatible motor behaviours. Hence the symptoms of Parkinson's disease, Tourette's syndrome, and Hemiballism can be understood in terms of the specific information processes affected.

In Parkinson's disease, degeneration of dopamine neurons means that the activity of the indirect path is increased relative to that of the direct path. The direct path recommends in favour of different behaviours and the indirect path recommends against behaviours. Hence the imbalance results in no behaviour being selected, corresponding with the inability to initiate voluntary movements.

In Hemiballism, damage to the subthalamic nucleus results in reduced recommendation strengths against behaviours, and selection of more than one behaviour of the same type. It is not possible to actually make muscles move limbs in two directions at once, and the conflict will be resolved closer to muscle control. However, such resolution depends on slight differences in incompatible basal ganglia outputs, and small almost random changes result in sudden large shifts in behaviour, corresponding with the wild, uncontrollable flinging movements of arms and legs. In Tourette's syndrome, excess delivery of dopamine to the striatum also results in selection of more than one behaviour type, resulting in the sudden introduction of "unselected" behaviours.

11.5.3 Sensitivity of Prospective Memory to Dementia

Loss of the ability to create and access prospective memories is one of the most frequent and early signs of dementia. The dependence of prospective memory on many different types of information processes means that whatever type of process is affected, there is likely to be an effect on prospective memory capabilities. Hence the observation that prospective memory loss is one of the earliest signs of dementia.

11.6 Consciousness and Self Awareness

Consciousness is often regarded as a great mystery, sometimes thought of as something that permeates life like some "aether". Debates are ongoing on the degree to which different forms of life are "conscious". Arguments occur about the degree to which human consciousness differs from anything experienced by other animals. In this section we will treat human consciousness in the same way as any other cognitive phenomenon: description at high level, indicating how mapping could occur through anatomy, physiology and neurochemistry.

As discussed in Chap. 2, there are a number of phenomena that are regarded as exhibiting human consciousness. Firstly, the vividness of some experiences of sensory circumstances, and the inability to describe that vividness in any detail. Related to this phenomenon is the question of whether individual experience is unique to each individual. Secondly, the often experienced stream of mental images, only weakly linked to current sensory inputs. This stream is sometimes visual and sometimes verbal, and is experienced as moments when images are sharply focussed separated by periods of image vagueness. Thirdly, the ability to express internal mental states in words, an ability that is not available for all internal states. For example, if an important decision needs to be made, the process of thinking about alternatives can be described, but the instant in which the selection of one alternative is made seems inaccessible. Fourthly, the ability to experience self, as if self were something inside, observing what is going on and being observed. We will discuss each of these phenomena separately, although there are relationships between them.

11.6.1 Vividness of Subjective Experience

Looking at an object can result in a range of possible immediate behaviours, such as naming the object, avoiding the object, picking up the object etc. One additional possibility is to increase the focus of attention on the object, resulting in a more vivid awareness of that object. The example often given in philosophical literature is the colour red. Focussing on a sensory experience of the colour results in a subjective experience that it is difficult to describe other than to say it is more vivid. It is argued that such experiences of the same sensory object are different for each individual.

At the cortical receptive field level of description, viewing an object results in direct detection of receptive fields. Each receptive field has a range of recommendation strengths, and behaviours appropriate to the object tend to have the strongest recommendation strengths. For example, viewing a red object would result in significant recommendation strengths in favour of saying "red". Viewing a tree while out on a forest walk would result in significant recommendation strengths in favour of saying "tree" and motor behaviours to avoid walking into it.

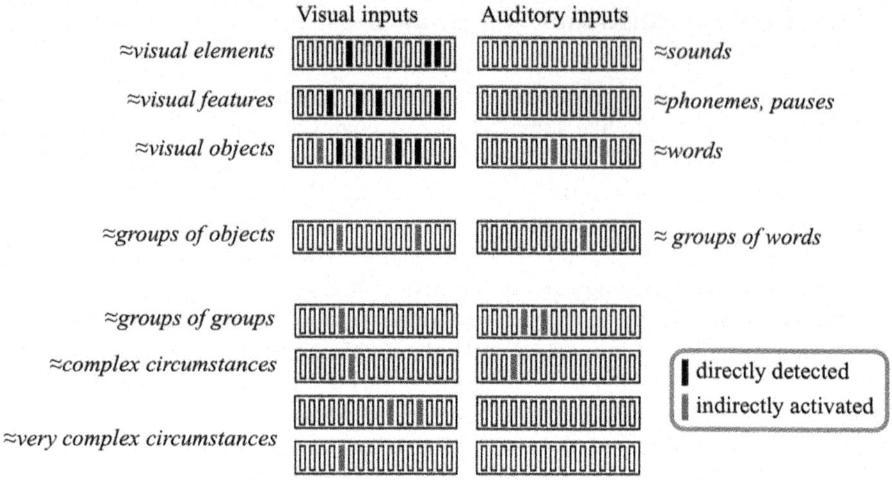

Fig. 11.7 Subjective experience (quale) triggered by a direct visual perception. The directly detected receptive fields have a wide range of indirect activation recommendation strengths. If indirect activation behaviours are generally detected, the result will be a "halo" of indirectly activated columns. These columns will be fragments of column populations activated in the past when the visual columns were also active. Examples could by fragments of the populations activated in the past in response to objects often present at the same time as the current object, or fragments of populations activated in response to situations in which objects similar to the current object were present. The subjective experience will be as if perceiving a weighted average of past experiences in which the current object being perceived was also present

All the directly activated columns also have recommendation strengths in favour of indirect activation of other columns on the basis of past temporally correlated activity. If such recommendations were accepted while viewing the colour red, the result would be activation of a "halo" of other receptive fields as conceptually illustrated in Fig. 11.7. This halo would include columns that expanded their receptive fields in the past close to times when many of the columns activated directly also expanded their receptive fields. Other columns would be indirectly activated on the basis of recent or frequent past activity at similar times. This halo would result in a much richer subjective experience derived from the current sensory object. Its behavioural value would be making available a much wider range of potentially relevant recommendation strengths. However, there will generally be no predominant recommendation strength without further processing. For example, some of the columns will have been indirectly activated on the basis of past temporally correlated receptive field expansion. The recommendation strengths resulting in those activations would be the result of many different past experiences in most of which the colour red was present. The population could therefore be viewed as containing fragments of the populations active during many different past experiences. However, no speech behaviour describing one of the past experiences will have a predominant stronger total recommendation strength. Hence it is not possible to describe any experience in detail verbally. Further stages of indirect activation would be required to generate a predominant recommendation, such as to describe a past experience as discussed under episodic memory in Chap. 9. Such a description of past experience could be an appropriate response depending on the social situation, and such a behaviour could not have been generated just from the recommendation strengths available in the directly activated receptive fields.

Two factors make the halo population unique to the individual. One is that receptive fields are defined heuristically, and there will generally be little correspondence between the column receptive fields in one brain and those in another. Secondly, the exact population indirectly activated will depend on the past experience of the brain, which is again unique to the individual. A subjective experience in response to a specific visual input has been called a quale (plural qualia), and ascribed the properties of being impossible to communicate, independent of other current experiences, and probably unique to each individual. Indirect activations of cortical columns on the basis of temporal correlations in past activity can account for these properties of qualia.

11.6.2 Internal Stream of Mental Images

The experience of a constant stream of internal mental images can be understood by an extension of the descriptions in the previous section. At any point in time a population of columns will be active. That population will have recommendation strengths in favour of indirect activation of a secondary population. The secondary population will have recommendation strengths in favour of indirect activation of a tertiary population and so on. The basis for the indirect activation could vary. Indirect activation on the basis of past temporally correlated receptive field expansion could lead to an episodic memory of a real event. Alternatively, depending on the content of the starting population, it could lead to imagining an event. Further indirect activations on the basis of past frequent activity or receptive field expansion shortly after columns in the active population could then lead to imagining events following the initial event, in other words an imaginary narrative. Indirect activations on the basis of frequent past simultaneous activity could lead to addition of receptive fields closer to sensory inputs, enriching a more conceptually imagined event. Any population could require receptive field expansion to achieve the required minimum size, providing the basis for later recall of the imagining.

One problem could be that after a long sequence of indirect activations, the population would become chaotic and contain little behavioural meaning. A way to avoid such chaotic populations would be to exploit the recommendation strengths that exist as a result of speech capabilities. As discussed in an earlier section, speech capabilities depend upon visual columns acquiring recommendation strengths in favour of indirect activation of auditory columns, and vice versa. These recommendation strengths can be used to maintain the focus of the stream of images. Suppose that after a few stages of indirect activation, acceptance of the predominant recommendation of the visual population in favour of indirect activation of an auditory population was forced. The auditory activation could be at levels of complexity corresponding with ≈*words*, ≈*groups of words*, or even ≈*groups of groups*. The result would be a subjective experience of hearing words that were the best fit to the current visual population. In other words the best available meaningful verbal description of that population. Now suppose that the visual population was extinguished and a new visual population indirectly activated by the auditory population. The new visual population will tend to have more meaning because the less related column activities will have been dropped. Focusing by means of speech capabilities will therefore allow much longer streams of useful indirect activations. The subjective experience would be of periods of time during which the mental content was fuzzy separating relatively sharp content. The sharp content could be verbal. Experience of the stream of consciousness appears to correspond with this model.

11.6.3 Verbal Accessibility of Internal Mental States: Dichotic Listening

Mental states which do not generate any verbal output but nevertheless exert an influence on behaviour are often called "unconscious". A classic example is dichotic listening. Suppose a subject is listening to earphones, different meaningful texts are presented to the left and right ears, and the subject is asked to repeat the text heard in one specific ear. This task can be successfully performed, and the subject will have memory for the content of the attended text and not for the unattended. However, if at some point the texts are switched between ears, the subject will continue to repeat the meaningful text, without being aware of having switched ears. The implication is that although there is no "consciousness" of the content presented to one ear, some information is reaching the brain from that ear to allow the switch when it occurs.

The information processes resulting in dichotic memory can be understood from Fig. 11.8. Auditory columns directly activated by hearing words in the left and in the right ears all have recommendation strengths in favour of indirect activation of columns at various levels of complexity. Hearing the instructions to attend to the left ear results in recommendation strengths associated with the left ear being increased. As discussed for speech, each word activates a population of columns in the ≈*visual objects* areas, combinations of such populations drive an activation of receptive fields in ≈*groups of objects* areas corresponding with phrases. Combinations ≈*groups of objects* populations in turn drive activation of populations in ≈*groups of groups* areas. Receptive field expansions in these areas mean that episodic memory mechanisms can retrieve information about the attended text. Indirect activation recommendation strengths of right ear column activations are weakened and do not lead to these types of indirect activations. Hence there is no ability to recall the unattended text.

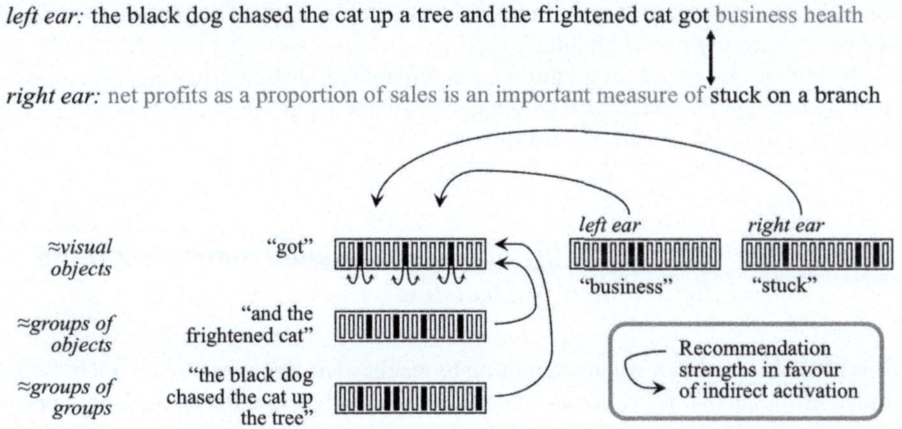

Fig. 11.8 The point at which the echoed text switched from left to right ear in a dichotic listening experiment. Different auditory columns activated in response to hearing the different words all have recommendation strengths in favour of indirect activation of ≈visual objects columns. Because of the original instructions to echo the left ear, auditory columns activated in response to sounds of "business" in the left ear have greater such recommendation strengths than those in favour of "stuck" derived from the right ear. However, already activated columns in the visual areas also have recommendation strengths on the basis of frequent past simultaneous activity. These are predominantly in favour of the columns that would be activated in response to "stuck", and it is therefore these columns that are activated

How can the switch between ears be understood? In Fig. 11.8, a possible pattern of column activation at the moment the texts were switched is conceptually illustrated. Auditory columns activated in response to the sequence of words have resulted in indirectly activated columns in various areas as described in more detail in the speech section. Columns have been activated in ≈*groups of groups* areas corresponding with "the black dog chased the cat up a tree". Columns have been activated in ≈*groups of objects* areas corresponding with the phrase "and the frightened cat". Columns have been activated in ≈*visual objects* areas corresponding with hearing the word "got". At this point the next word in the meaningful continuation is "stuck". However, the word "business" is presented to the left ear while the word "stuck" is presented to the right ear. Auditory columns activated in response to both words have recommendation strengths in favour of indirect activation of visual columns in the same area that the word "got" has created an activation, but because of the bias in favour of the left ear, the recommendation strengths in favour of columns correlating with "business" are stronger. However, this is not the only source of indirect activation recommendation strengths. Columns activated in ≈groups of groups, ≈groups of objects, and ≈visual objects in response to hearing the attended text so far also have recommendation strengths in favour of indirect activation of columns in ≈visual objects. In the past, when words and phrases similar to those that have generated the existing indirect activations were heard, the word "stuck" has been heard much more often than the word "business". Hence these additional recommendation strengths will reinforce those generated via the right ear, resulting in the recommendation strengths in favour of a "stuck" derived activation in ≈*visual objects* becoming predominant.

Note that this source of additional recommendation strengths is also the reason that speech can be understood in noisy environments or when a proportion of the words are mispronounced or even missed.

11.6.4 Verbal Accessibility of Internal Mental States: Planning and the Moment of Decision

It is also observed that when attempting to reach a decision on some complex issue, much of the mental process can be described verbally. However, it does not appear possible to describe the actual moment when the decision is made in any detail.

The process for reaching a decision can be viewed as an example of a stream of consciousness. The difference from the description in the earlier section is that the stream is managed much more explicitly. Consider for example the mental processes that could lead to choosing to go to a restaurant after a concert. The processes could be initiated by a question asked by a companion. Hearing words like "do", "after" and "concert" will directly activate auditory columns, which will in turn indirectly activate visual and other columns. Among other recommendation strengths, the columns activated in response to "do" and "after" have acquired recommendation strengths in favour of activating a sequence of information processes of different types. Acquisition of such sequences is acquisition of the ability to plan. Indirect activation of column populations on the basis of activity somewhat after the population activated in response to "concert" will result in populations similar to those activated in the past when starting out for, travelling to and being at different locations after concerts. Column populations corresponding with the rewards implemented following different locations can be activated. Column receptive fields will also be detected that correlate with blood sugar level. A wide range of recommendation strengths in favour of departing for different locations will therefore be available. All of these populations will also have speech generation recommendation strengths and can therefore be described in words. However, at some point the predominant recommendation strength is determined in the basal ganglia and followed by departure to the selected destination. Receptive fields are not detected that discriminate between different processes occurring in the basal ganglia, and the behaviour selection process therefore cannot be described verbally.

11.6.5 Self Awareness

Self awareness includes a number of phenomena. Firstly, when attention is directed to self, there is a sense of someone being there. Secondly, in planning processes, a representation of self can be included. Situations featuring self can be imagined in

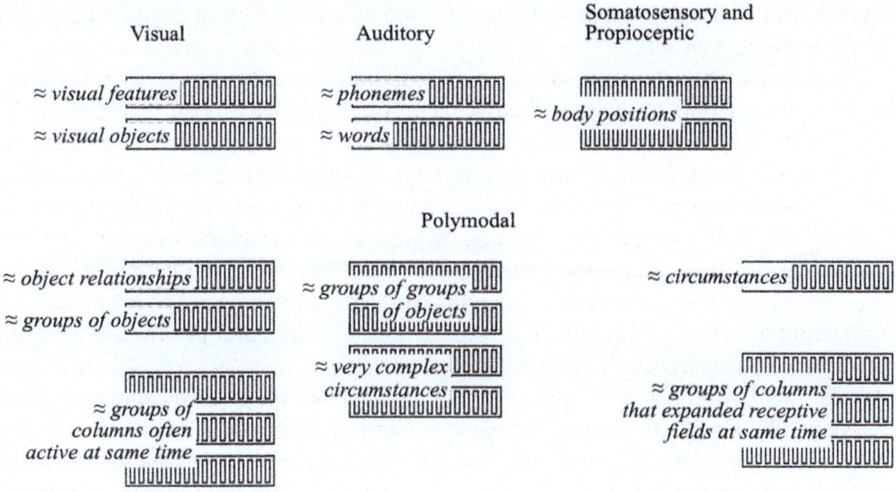

Fig. 11.9 Cortical areas relevant to the learning and experience of self awareness

two ways. For example, if you are asked to imagine the last time you went swimming, often the first mental image will be of watching yourself swim, perhaps in the ocean. This "outside" or "I" perspective is not experienced in reality. The other way is to visualize the feel of water splashing in your face, what we will call the "inside" or "me" perspective. One behavioural value of the I and me perspectives is that they can be utilized in planning to imagine self in different situations and determine which situation has the highest reward potential.

A set of cortical areas relevant to self awareness is illustrated in Fig. 11.9. Within those areas, consider the learning that occurs in a child in response to hearing his name, and suppose that name is Billy. The name is heard frequently, and each time it activates columns in the ≈*words* areas. On many of these occasions, an adult is encouraging the child to focus attention on himself. Each time the name is heard, columns throughout the cortex will also be active. However, columns activated at more complex levels when attention is focussed on somatosensory and propioceptic inputs from Billy's body, and visual inputs from looking at his own body will tend to be active more often. Over time, the columns directly activated by hearing the word "Billy" will acquire recommendation strengths in favour of indirect activation of a population of complex columns that represent a weighted average of the columns active in the past when the word was heard. The population activated in response to the name would be subjectively experienced as if the brain were experiencing an average of past experiences when attention was being paid to sensory inputs from the body. At such times, the brain would be receiving actual current sensory inputs from the body, which could activate another population in the same areas at a different phase of frequency modulation. Both populations would have a wide range of recommendation strengths, including in favour of speech. Hence it would be possible to describe either the current experience of self or the average past self.

In the past when the name has been heard, priority has been placed on different types of behaviour. More specifically, receptive fields have been detected in the amygdala (≈circumstances) and in prefrontal areas (≈very complex circumstances) that recommend different behaviour type such as aggressive, fearful, food-seeking, etc. A recommendation in favour of a behaviour type is subjectively experienced as an emotion such as anger, fear, hunger, disgust, surprise, sadness or pride. If some such receptive fields have often been active in the past when the name was heard, they can be indirectly activated in response to hearing the name. The recommendation strengths of these receptive fields in favour of behavioural priorities will result in subjective experience of emotions.

The exact "pseudoself" population will also depend on whatever other columns happen to be active often when the name is heard. Receptive fields detected within the current circumstances will also have recommendation strengths in favour of indirect activations, which when added to the recommendation strengths of columns activated by hearing the name will result in activation of columns active in the past when attention was focussed on self in similar circumstances.

Speaking phrases like "I am Billy" or "Billy is me" results in columns indirectly activated in response to hearing "Billy" being active at the same time as columns directly activated in response to hearing "I" or "me". Hence the pronouns can acquire recommendation strengths in favour of activation of pseudoself populations. Another spoken phrase with significant consequences is "Billy is a boy". The word "boy" will often be heard when Billy's attention is being directed to other boys. Hence the receptive fields directly activated by seeing a boy, sitting, running, performing a wide range of activities, will be active at the same time as the columns directly activated by hearing the word "boy". As a result, the columns activated in response to hearing "Billy" will acquire recommendation strengths in favour of indirectly activating a column population as if observing a boy. Such recommendation strengths result in a pseudoself population that includes the perspective from outside. Over time, rewards can adjust recommendation strengths of columns directly activated by hearing "I" and "me" such that they tend to activate an externally viewed and internally viewed pseudoself respectively.

Stream of consciousness processes can sometimes result in activation of ≈*words* columns associated with name, I or me, leading to activation of pseudoself populations. Combination of pseudoself populations with, for example, populations corresponding with events can lead to populations corresponding with self participating in events. Simultaneous activity during a stream of consciousness can lead to additional recommendation strengths in favour of indirect activation. Hence the populations activated in response to hearing the name or pronoun, or indirect activation of the corresponding columns in ≈*words* can evolve. Thus the self image can change based upon both external and stream of consciousness experiences.

The behavioural value of this capability is enhancement of planning capability through using a wider range of more detailed information about past responses of self to different circumstances. Alternatives that integrate a self "model" can be explored. Note that analogous processes can result in models for other people, initiated by columns directly activated in response to hearing their name being active at the same time as columns at many levels of complexity activated in response to observing them.

11.6.6 Understanding Human Consciousness

The above discussion has demonstrated that the phenomena of human consciousness can be described in terms of combinations and sequences of information processes. Earlier in the book we have demonstrated how these information processes are implemented by combinations and sequences of more detailed processes in brain anatomy, and so on through physiology to neurochemistry. A scientific understanding of human consciousness has therefore been presented.

11.7 Literature, Music and Art

Every cortical column recommends an extensive range of behaviours, and can be activated directly by sensory inputs or indirectly on the basis of temporal correlations in past activity with other columns. The indirect activation of a column results in outputs that are indistinguishable from outputs that can result from direct activation. Indirect activations are therefore subjectively identical with the results of sensory experiences. The difference from an actual sensory experience is that in general columns close to raw sensory inputs are indirectly activated to a much smaller degree, and indirect activations are therefore not sensory hallucinations. Indirect activations can result in receptive field expansions just as in the case of sensory experiences, and can therefore be recalled later as if they were sensory experiences, again with the exclusion of activity close to raw sensory inputs.

Chains of indirect activations may lead to subjective experiences that would be difficult or impossible to reach by direct activation from sensory inputs. Hence literature can generate pseudoexperiences of non-existent circumstances. The indirectly activated receptive fields may have recommendation strengths in favour of emotional states, and such states are therefore part of the pseudoexperiences.

Music activates auditory receptive fields. Speech also activates auditory receptive fields, and those receptive fields have strong recommendation strengths in favour of indirect activation of visual receptive fields on many different levels of complexity. In addition, some auditory receptive fields may have recommendation strengths in favour of different types of behaviours, such as alarm or cooperative behaviours. Hence music also can activate long chains of indirect activation, with the chains modulated by various behavioural biases. Once again, subjective experiences can result, and these subjective experiences may be distinct from any achieved by direct experience.

Viewing art results in direct activation of visual receptive fields, leading to direct activation of receptive fields on many higher levels of complexity. All these receptive fields have indirect activation recommendation strengths, and both directly and indirectly activated receptive fields may have recommendation strengths in favour of emotions or rewards. Past visual experience will strongly influence many of these recommendation strengths. Once again, chains of indirect activations may lead to complex subjective experiences.

The receptive fields and recommendation strengths utilized to achieve the subjective experiences generated by literature, music and art, are determined to a degree by genetic information and heavily by past experience. These subjective experiences are therefore determined by a complex interplay between these factors.

Chapter 12
Towards a Theoretical Neuroscience

What can we expect from a theoretical neuroscience? Experimental work on brains has been carried out on many levels of detail, including behavioural psychology, major anatomical structures, detailed anatomical structures, neurons, and detailed chemical processes within neurons. However, for human understanding of how cognition results from activity on more detailed levels, experimental work needs to be integrated in a theoretical framework. Such an integration requires insight into the information processes performed by brain structures at the different levels. A theoretical neuroscience must therefore address the nature of these information processes.

One problem is that, perhaps especially in the case of human brains, each brain is unique. We would like to understand the nature of the differences between individual brains, and any general similarities. We would also like to understand the similarities and differences between human brains and the brains of other species. We would like to understand the reasons for the different anatomical structures, and what these different structures contribute to higher cognition. We would like to understand how neurons contribute to higher cognition. We would like to know how higher cognition depends upon different neurochemical processes. We would like to be able to quantitatively model the brain on various levels of detail including neurochemical, neuron, and cognition. Such models need to make simplifications and approximations, and we need to know how to make such simplifications and approximations in a way that does not result in models that are qualitatively misleading.

Theoretical physics has been very successful in quantitatively modelling phenomena at a very detailed level and at a number of intermediate levels. However, the models created by physics tend to be of relatively simple systems. Attempts to apply mathematical physics to very complex systems such as atmospheric weather have had much more limited success. This book has argued that in order to achieve adequate models for human brains, techniques must be imported from the technologies for designing, manufacturing, and maintaining extremely complex electronic real-time systems. These techniques are applicable because the constraints imposed

L.A. Coward, *Towards a Theoretical Neuroscience: from Cell Chemistry to Cognition*, Springer Series in Cognitive and Neural Systems 8, DOI 10.1007/978-94-007-7107-9_12,

by natural selection on brains are analogous with constraints on the architectures of human designed electronic systems. Although the architectures that result from these constraints for electronic systems and brains are qualitatively different, in both cases they result in separations of the system resources into subsystems performing different information processes. Furthermore, just as the information processes performed by an electronic system can be classified as either data read/writes or instructions, so the information processes performed by a brain can be classified as condition definition/detections or recommendations.

A theoretical framework therefore exists as a result of natural selection pressures. These pressures come from the advantages possessed by one species over another if it can learn and perform a given set of behaviors with fewer resources, learn with less interference with prior learning, and requires less information in a simpler development process for brain development.

Any complex control system must detect conditions within its available information and associate conditions with behaviours. A key result of natural selection pressures is that for brains that learn a high proportion of their behaviours, these information processes are constrained. Firstly, condition definition and detection are a single process, with condition definition taking place as part of the condition detection process. Secondly, condition detections are associated with behavioural recommendations, in contrast with electronic computer systems in which condition detections are linked with behavioural commands. Thirdly, there is a major separation between resources that define/detect conditions and resources that maintain recommendation strengths and determine the most strongly recommended behaviours. This separation, called the recommendation architecture, is necessary because feedback of the consequences of behaviours is the primary guide for recommendation strengths but such feedback cannot be applied directly to condition definitions without excessive undesirable side effects. Finally, resources are further subdivided into major components that perform condition definition/detection and behavioural recommendation information processes to support condition change management, behavioural type prioritisation, and efficient performance of behavioural sequences.

These constraints make clear the reasons for the existence of the major anatomical structures in the mammal brain: the cortex defining and detecting conditions; the basal ganglia and thalamus assigning recommendation strengths to condition detections and determining behaviours; the hippocampus managing changes to cortical conditions; the amygdala/hypothalamus managing priorities placed on different types of behaviour; and the cerebellum managing frequently used sequences of behaviour.

These constraints also indicate why mammal brains all exhibit similar general structures: a cortex with areas and columns; thalamus with separate very similar nuclei; the basal ganglia with different nuclei; the hippocampus with a cortex like structure; the amygdala with cortex like and basal ganglia like divisions etc. Even in bird brains, recent studies indicate that there are anatomical structures performing different information processes, and these structures correspond with cortex, basal ganglia, thalamus, hippocampus, amygdala, cerebellum etc. at the levels of connectivity and neuron type [953]. The avian equivalent to the cortex is not layered, but has different regions that correspond with cortical layers IV, II/III, and V/VI. Thus the evidence suggests that the bird and mammal brains have similar information processing architectures despite separate evolutionary paths for about 300 million years.

The condition definition/detection and behavioural recommendation information models result in the existence of a hierarchy linking high level descriptions of cognitive phenomena through various anatomical levels of detail to descriptions in terms of neurochemistry. Just as in the physical sciences, more detailed descriptions down to the level of quantum mechanics could in principle be constructed. The existence of a rigorous route to fundamental levels is necessary for the integrity of the theory. However, again as in the physical sciences, the approximate descriptions at higher level are fully adequate for understanding, and direct use of quantum mechanics in the higher descriptive levels is almost always unnecessary.

The existence of the condition/recommendation based hierarchy of descriptions means that if a description of a cognitive process is constructed within this information paradigm, there can be confidence that detailed physiological processes exist to support the process. However, the number of different combinations of very detailed processes that could implement the same higher level process may be very large. Different individual brains may implement similar cognitive processes by different combinations of detailed processes. Hence the same limits exist for brain science as for physical sciences: high level descriptions that are consistent with detailed understanding can be created, but full modelling at the most detailed levels will in general be impractical. Nevertheless, intuitively satisfying understanding of cognitive phenomena in terms of anatomy etc. can be achieved, along with detailed modelling as required.

What are the sources of the differences between brains? The primary capabilities are the abilities to define and detect sets of receptive fields on different levels of complexity, to associate receptive fields with behavioural recommendations, and to indirectly activate receptive fields on the basis of temporal correlations in past activity.

Tool making is a characteristic human activity, but has also been observed in other animals including chimpanzees and crows. If a tool is made when and where it is to be used, and from materials available at the site, the recommendation strengths associated with receptive fields detected within current sensory inputs could be sufficient to manage the toolmaking behaviour. However, if the tool must be made in advance from materials that are not available where it will be used, then chains of indirect activations of receptive fields will be required. Human beings have much greater abilities to support such long indirect activation chains, largely because of speech derived mechanisms to maintain useful degrees of meaning within such chains as discussed in Chap. 11. Thus examples of toolmaking in other animals generally involve on-site manufacturing from available materials. Only chimpanzees demonstrate a limited ability to collect materials for tool making away from the site of use, but even in this case the actual toolmaking occurs at the site. Human consciousness and self awareness also depend upon long chains of meaningful indirect activations. Hence these phenomena in other species without speech capabilities will also be limited.

Individual differences between human beings has been a controversial topic. Acquiring capability in an intellectual discipline like physics or history requires acquisition of receptive fields with sufficient discrimination to support appropriate recommendation strengths. These receptive fields are defined heuristically through extensive experience. Genetic information specifies the ranges of complexity within which receptive fields are initially defined. Differences in these genetic specifications may result in different intellectual capabilities. For example, research skills in history and in physics will both require the ability to define/detect relatively large numbers of receptive fields that discriminate at the level of extremely complex circumstances. However, the most effective complexity levels may be somewhat different. The exact complexities and the numbers of receptive fields that can be created at each complexity level may differ between brains. Furthermore, there may be differences in the degree to which the receptive fields within one area can vary from initial starting points. One brain may have the capability to allow receptive fields in one area to draw a higher proportion of inputs from a somewhat larger number of areas than another brain. Another brain may have a capability to define more receptive fields in some areas. Increased connectivity or receptive field numbers will have resource costs, and different compromises may be specified by different genes. Skill in physics research may require the ability to define/detect relatively large numbers of novel receptive fields at certain levels of complexity relative to sensory inputs. Skill in history might require the ability to define/detect receptive fields at extremely complex levels. Hence genetic differences could give rise to different capabilities in these areas. Some skills may require receptive fields that can only be defined through more extensive experiences than other skills, and peak human capability in such skills will be reached later in life.

Limits to human capabilities may occur as a result of unavailability of receptive fields at an appropriate level of complexity. One example could be arithmetic. As discussed in Chap. 2, human beings cannot immediately discriminate between groups of objects on the basis of the number of objects, unless there are fewer than six objects in the groups. Above this limit, counting of the objects in each group is necessary. This is in contrast with the ability to discriminate between many thousands of only slightly different human faces. Hence humans do not have receptive fields on a level of complexity that can easily discriminate between such number based groups. Human difficulties with arithmetic reflect this lack of appropriate receptive fields.

Thus very general psychological questions can be addressed by the approach outlined in this book. Understanding must be based on a system perspective, not on associating different brain structures with specific cognitive phenomena. For example, in the brain the most complex information processing is performed by the cortex, and cortical information carries very complex behavioural meanings. However, any element of cortical information is behaviourally ambiguous, and to determine behaviour this cortical information must be interpreted by subcortical structures. Hence assignment of simple behavioural meanings to cortical activity will in general be misleading. At the other extreme of detail, chemical changes to create memory are the result of recommendations from multiple sources, these recommendations typically instantiated by different kinase cascades. As a result, memory will not depend on one molecule, and most memory processes will be able to continue when almost any one contributing molecule is inactivated. The only way to understand cognitive processing by the brain is therefore to maintain a systems perspective, avoiding associations between specific cortical structures on any level and unambiguous cognitive behaviours.

A conceptual framework is a prerequisite for simulations of detailed anatomy and physiology, because it identifies simplifications and approximations that will not result in catastrophic deviations between the simulations and actual biology. For example, in the brain, condition definition and detection are entangled and simulations that attempt to separate detection from definition will be misleading. Simulations that associate conditions with commands will also be misleading, because such unambiguous associations can only exist in computer systems. Simulation approaches that apply large computational resources without a conceptual framework can result in electronic systems with very complex behaviours. However, even if the behaviours appear to resemble cognition, it will be difficult to understand how those behaviours are supported by the detailed information processes within the system, and the contribution of such simulations to human understanding of biological brains is therefore limited.

Key types of information process supporting higher cognition include definition of receptive fields by detection within experience, association of receptive fields with behavioural recommendations by consequence feedback applied following behavioural selections, and indirect activation of receptive fields on the basis of prior temporal correlations in activity. Simulations of these information processes have demonstrated that they can support cognitive-like behaviours [932]. Support for managing access to a condition definition/detection resource using the frequency modulation mechanism has also been simulated [954].

The previous chapters have shown that once a complex behaviour is causally described in terms of the condition definition/detection and behavioural recommendation information models, that description can be mapped through major anatomical structures, detailed anatomical structures, and neurons down to neurochemistry. The condition definition/detection and behavioural recommendation information models within the recommendation architecture framework thus provides the required framework for theoretical neuroscience.

What is the way forward from here? The system architecture approach needs to be embedded throughout experimental and computational neuroscience. Experimental results on all levels of detail need to be interpreted in terms of the condition definition/detection and behavioural recommendation information paradigms, and mapping carried out between levels. Computational modelling needs to be based on the same information paradigms, with those paradigms implemented on varying levels of detail from kinase cascades through neurons and neuron assemblies to major anatomical structures. This approach can lead to far more effective understanding of how physical damage on all levels leads to cognitive deficits and how the effects of such damage can most effectively be corrected.

Design of systems that possess human-like cognitive capabilities appears well within reach, provided that such design is based on the same system architectural approach.

References

1. Coward LA (1990) Pattern thinking. Praeger, New York
2. Coward LA (2005) A system architecture approach to the brain: from neurons to consciousness. Nova Science Publishers, New York
3. Abbott LF (2006) Theoretical neuroscience rising. Neuron 60:489–495
4. Marr D (1982) Vision: a computational investigation into the human representation and processing of visual information. W. H. Freeman and Company, San Francisco
5. Backhaus J, Born J, Hoeckesfeld R, Fokuhl S, Hohagen F, Junghanns K (2007) Midlife decline in declarative memory consolidation is correlated with a decline in slow wave sleep. Learn Mem 14:336–341
6. Sacks O (1970) The man who mistook his wife for a hat. Simon and Schuster, New York
7. Elliot FA (1982) Neurological findings in adult minimal brain dysfunction and the dyscontrol syndrome. J Nerv Ment Dis 170(11):680–687
8. Harlow JM (1868) Recovery from the passage of an iron bar through the head. Publ Mass Med Soc 2:327–347. Republished in Macmillan M (2000) An odd kind of fame: stories of Phineas Gage. MIT Press, Cambridge
9. Hoffman JE, Subramaniam B (1995) The role of visual attention in saccadic eye movements. Percept Psychophys 57:787–795
10. Lamme VAF (2003) Why visual attention and awareness are different. Trends Cogn Sci 7:12–18
11. Schacter JC, Tulving E (1994) What are the memory systems of 1994? In: Schacter JC, Tulving E (eds) Memory systems 1994. MIT Press, Cambridge, MA
12. Rips L, Shoben J, Smith E (1973) Semantic distance and verification of semantic relations. J Verbal Learn Verbal Behav 12:1–20
13. Backman L, Nilsson L-G (1996) Semantic memory functioning across the adult life span. Eur Psychol 1(1):27–33. This study was on people speaking Swedish
14. Spaan PEJ, Raaijmakers JGW, Jonker C (2003) Alzheimer's disease versus normal ageing: a review of the efficiency of clinical and experimental memory measures. J Clin Exp Neuropsychol 25(2):216–233
15. Brown R, McNeill D (1966) The "tip of the tongue" phenomenon. J Verbal Learn Verbal Behav 5:325–337
16. Canli T, Zhao Z, Brewer J, Gabrieli JDE, Cahill L (2000) Event-related activation in the human amygdala associates with later memory for individual emotional experience. J Neurosci 20(RC99):1–5
17. Brown R, Kulik J (1977) Flashbulb memories. Cognition 5:73–99
18. Simons DJ, Chabris CF (1999) Gorillas in our midst: sustained inattentional blindness for dynamic events. Perception 28:1059–1074

L.A. Coward, *Towards a Theoretical Neuroscience: from Cell Chemistry to Cognition*, Springer Series in Cognitive and Neural Systems 8, DOI 10.1007/978-94-007-7107-9,

19. Miller MB, Gazzaniga MS (1998) Creating false memories for visual scenes. Neuropsychologia 36(6):513–520
20. Memoni A, Bartlett J (2002) The effects of verbalization on face recognition in young and older adults. Appl Cogn Psychol 16:635–650
21. Addis DA, Wong AT, Schacter DL (2007) Remembering the past and imagining the future: common and distinct neural substrates during event construction and elaboration. Neuropsychologia 45:1363–1377
22. Loftus EF, Pickrell JE (1995) The formation of false memories. Psychiatr Ann 25:720–725
23. Loftus EF (1993) The reality of repressed memories. Am Psychol 48(5):518–537
24. Standing L, Conexio J, Haber RN (1970) Perception and memory for pictures: single-trial learning of 2500 visual stimuli. Psychon Sci 19(2):73–74
25. Scott A, Stern Y, Tang M, Mayeux R (1999) Selective decline in memory function among the healthy elderly. Neurology 52(7):1392–1396
26. Scoville WB, Milner B (1957) Loss of recent memory after bilateral hippocampal lesions. J Neurol Neurosurg Psychiatry 20:11–21
27. Sagar JH, Cohen NJ, Corkin S, Growden JH (1985) Dissociations among processes in remote memory. Ann N Y Acad Sci 444:533–535
28. Kensinger EA, Ullman MT, Corkin S (2001) Bilateral medial temporal lobe damage does not affect lexical or grammatical processing: evidence from amnesic patient H.M. Hippocampus 11:347–360
29. Corkin S (1968) Acquisition of motor skill after bilateral medial temporal-lobe excision. Neuropsychologia 6:225–264
30. Milner B, Corkin S, Teuber H-L (1968) Further analysis of the hippocampal amnesic syndrome: 14-year follow-up study of H.M. Neuropsychologia 6:215–234
31. Graham KS, Simons JS, Pratt KH, Patterson K, Hodges JR (2000) Insights from semantic dementia on the relationship between episodic and semantic memory. Neuropsychologia 38(3):313–324
32. Miller GA (1956) The magical number seven, plus or minus two: some limits on our capacity for processing information. Psychol Rev 63:81–97
33. Cowan N (2000) The magical number 4 in short-term memory: a reconsideration of mental storage capacity. Behav Brain Sci 24:87–114
34. Della Sala S, Gray C, Baddeley A, Allamano N, Wilson L (1999) Pattern span: a tool for unwelding visuo-spatial memory. Neuropsychologia 37:1189–1199
35. Panegyres PK (2004) The contribution of the study of neurodegenerative disorders to the understanding of human memory. Q J Med 97:555–567
36. Perry RJ, Hodges JR (2000) Differentiating frontal and temporal variant frontotemporal dementia from Alzheimer's disease. Neurology 54:2277–2284
37. Bainbridge L (1977) Verbal reports as evidence of the process operator's knowledge. Int J Man Mach Stud 11:411–436
38. Stadler MA (1989) On learning complex procedural knowledge. J Exp Psychol Learn Mem Cogn 15(6):1061–1069
39. Thomas-Ollivier V, Reymann JM, Le Moal S, Schuck S, Lieury A, Allain H (1999) Procedural memory in recent-onset Parkinson's disease. Dement Geriatr Cogn Disord 10:172–180
40. Kerremans JC, Stroebe W, Claus J (2006) Beyond Vicary's fantasies: the impact of subliminal priming and brand choice. J Exp Soc Psychol 42(6):792–798
41. Bar M, Biederman I (1998) Subliminal visual priming. Psychol Sci 9(6):464–469
42. Gabrieli JDE, Fleischman DA, Keane MM, Reminger SL, Morrell F (1995) Double dissociation between memory systems underlying explicit and implicit memory in the human brain. Psychol Sci 6:76–82
43. Budson AE, Price BH (2001) Memory: clinical disorders. In: Encyclopedia of life sciences. Internet address: http://www3.interscience.wiley.com/aboutus/topics/ELS_landingpage.html
44. Smith G, Della Sala S, Logie RH, Maylor EA (2000) Prospective and retrospective memory in normal ageing and dementia: a questionnaire study. Memory 8:311–321

45. Burgess PW, Quayle A, Frith CD (2001) Brain regions involved in prospective memory as determined by positron emission tomography. Neuropsychologia 39:545–555; Nyberg L, Forkstam C, Petersson KM, Cabeza R, Ingvar (2002) Brain imaging of human memory systems: between-systems similarities and within-system differences. Cogn Brain Res 13:281–292
46. Holowka S, Petitto LA (2002) Left hemisphere cerebral specialization for babies while babbling. Science 297:1515
47. Herschensohn J (2007) Language development and age. Cambridge University Press, Cambridge
48. Marler P (1991) The instinct to learn. In: Carey S, Gelman R (eds) The epigenesis of mind: essays on biology and cognition. Lawrence Erlbaum Associates, Inc., Hillsdale, pp 37–66
49. Chomsky N (1957) Syntactic structures. Mouton, The Hague
50. Kuhl PK (2004) Early language acquisition: cracking the speech code. Nat Rev Neurosci 5:831–843
51. Lechtenberg R, Gilman S (2004) Speech disorders in cerebellar disease. Ann Neurol 3(4):285–290
52. Dronkus NF, Larsen J (2001) Neuroanatomy of the classical symptoms of aphasia. In: Boller F, Grafman F (eds) Handbook of neuropsychology: language and aphasia. Elsevier, New York, pp 19–30
53. Boatman D, Gordon B, Hart J, Selnes O, Miglioretti D, Lenz F (2000) Transcortical sensory aphasia: revisited and revised. Brain 123:1634–1642
54. Dogil G, Ackermann H, Grodd W, Haider H, Kamp H, Mayer J, Riecker A, Wildgruber D (2002) The speaking brain: a tutorial introduction to fMRI experiments in the production of speech, prosody and syntax. J Neurolinguist 15:59–90
55. Hickok G, Poeppel D (2002) Towards a functional neuroanatomy of speech perception. Trends Cogn Neurosci 4(4):131–138
56. Burgess PW, Veitch E, de Lacy Costello A, Shallice T (2000) The cognitive and neuroanatomical correlates of multitasking. Neuropsychologia 38:848–863
57. Cantalupo C, Hopkins WD (2001) Asymmetric Broca's area in great apes. Nature 414:505
58. Warrington EK (1982) The fractionation of arithmetical skills: a single case study. Q J Exp Psychol 34A:31–51
59. Meissner CA, Brigham JC (2001) Thirty years of investigating the own-race bias in memory for faces: a meta-analytic review. Psychol Public Policy Law 7:3–35
60. Pascalis O, Scott LS, Kelly DJ, Shannon RW, Nicholson E, Coleman M, Nelson CA (2005) Plasticity of face processing in infancy. Proc Natl Acad Sci USA 102:5297–5300
61. Kanwisher N, Yovel G (2006) The fusiform face area: a cortical region specialized for the perception of faces. Philos Trans R Soc B 361:2109–2128
62. Bukach CM, Gauthier I, Tarr ML (2006) Beyond faces and modularity: the power of an expertise framework. Trends Cogn Sci 10:159–166
63. Ekman P, Oster H (1979) Facial expressions of emotion. Annu Rev Psychol 30:527–554
64. Adolphs R, Tranel D, Hamann S, Young AW, Calder AJ, Phelps EA, Anderson A, Lee GP, Damasio AR (1999) Recognition of facial emotion in nine individuals with bilateral amygdala damage. Neuropsychologia 37:1111–1117; Heberlein AS, Padon AA, Gillihan SJ, Farah MJ, Fellows LK (2008) Ventromedial frontal lobe plays a critical role in facial emotion recognition. J Cogn Neurosci 20(4):721–733
65. Matsumoto D, Consolacion T, Yamada H, Suzuki R, Franklin B, Paul S, Ray R, Uchida H (2002) American-Japanese cultural differences in judgements of emotional expressions of different intensities. Cogn Emotion 16(6):721–747
66. Parrott WG (2001) Emotions in social psychology: key readings in social psychology. Psychology Press, Philadelphia
67. Plutchik R (2001) The nature of emotions. Am Sci 89(4):344–350
68. Tracy JL, Robins RW (2008) The nonverbal expression of pride: evidence for cross-cultural recognition. J Pers Soc Psychol 94:516–530
69. Haidt J, Keltner D (1999) Culture and facial expression: open-ended methods find more expressions and a gradient of recognition. Cogn Emotion 13:225–266

70. Simonton DK (1991) Career landmarks in science: individual differences and interdisciplinary contrasts. Dev Psychol 27:119–130
71. Sheets DJ, Bradley DB, Hendricks J (2005) Enduring questions in gerontology. Springer, New York
72. Simonton DK (2007) Creative life cycles in literature: poets versus novelists of conceptualists versus experimentalists. Psychol Aesthet Creat Arts 1:133–139
73. Weir AAS, Chappell J, Kacelnik A (2002) Shaping of hooks in New Caledonian crows. Science 297:981
74. Hunt GR, Russell D, Gray RD (2003) Diversification and cumulative evolution in New Caledonian crow tool manufacture. Proc R Soc Lond B 270:867–874
75. Sanz C, Morgan D, Gulick S (2004) New insights into chimpanzees, tools, and termites from the Congo Basin. Am Nat 164(5):567–581
76. Price ER, Lambeth SP, Schapiro SJ, Whiten A (2009) A potent effect of observational learning on chimpanzee tool construction. Proc R Soc B 276:3377–3383
77. James W (1892) Psychology. World, Cleveland/New York
78. Block N (1995) On a confusion about a function of consciousness. Brain Behav Sci 18:227–287
79. Treisman AM (1960) Contextual cues in selective listening. Q J Exp Psychol 12:242–248
80. Tye M (2007) Qualia. In: Edward NZ (ed) The Stanford encyclopedia of philosophy (Fall 2007 Edition). http://plato.stanford.edu/archives/fall2007/entries/qualia/
81. Jaynes J (1976) The origin of consciousness in the breakdown of the bicameral mind. Harvard University Press, Boston
82. Jacobs J (1970) The economy of cities. Vintage Books, London
83. Lehmann J, Korstjens AH, Dunbar RIM (2007) Group size, grooming and social cohesion in primates. Anim Behav 74:1617–1629
84. Lee RB, DeVore I (1968) Man the hunter. Aldine De Gruyter, New York
85. Borgia G (1995) Complex male display and female choice in the spotted bowerbird: specialized functions for different bower decorations. Anim Behav 49:1291–1301
86. Dutton D (2009) The art instinct. Bloomsbury, New York
87. Catchpole CK, Slater PJB (1995) Bird song: biological themes and variations. Cambridge University Press, Cambridge
88. Dale AM, Sereno MI (1993) Improved localization of cortical activity by combining EEG and MEG with MRI cortical surface reconstruction: a linear approach. J Cogn Neurosci 5:162–176
89. London M, Hausser M (2005) Dendritic computation. Annu Rev Neurosci 28:503–532
90. DiStefano PS, Friedman B, Radziejewski C, Alexander C, Boland P, Schick CM, Lindsay RM, Wiegand SJ (1992) The neurotrophins BDNF, NT-3, and NGF display distinct patterns of retrograde axonal transport in peripheral and central neurons. Neuron 8:983–993
91. Whittingstall K, Nikos K, Logothetis NK (2009) Frequency-band coupling in surface EEG reflects spiking activity in monkey visual cortex. Neuron 64:281–289
92. Giuditta A, Kaplan BB, van Minnen J, Alvarez J, Koenig E (2002) Axonal and presynaptic protein synthesis: new insights into the biology of the neuron. Trends Neurosci 25:400–404
93. Connors BW, Gutnick MJ (1990) Intrinsic firing patterns of diverse neocortical neurons. Trends Neurosci 13:99–104
94. Calvin WH, Sypert GW (1976) Fast and slow pyramidal tract neurons: an intracellular analysis of their contrasting repetitive firing properties in the cat. J Neurophysiol 39:420–434
95. Llinas RR (1988) The intrinsic electrophysiological properties of mammalian neurons: insights into central nervous system function. Science 242:1654–1664
96. Bannister NJ, Larkman AU (1995) Dendritic morphology of CA1 pyramidal neurones from the rat hippocampus: I. Branching patterns. J Comp Neurol 360:150–160
97. Elston GN, Benavides-Piccione R, DeFelipe J (2001) The pyramidal cell in cognition: a comparative study in human and monkey. J Neurosci 21(RC163):1–5
98. Kim HG, Connors BW (1993) Apical dendrites of the neocortex: correlation between sodium- and calcium-dependent spiking and pyramidal cell morphology. J Neurosci 13(12):5301–5311
99. Sabatini BL, Regehr WG (1996) Timing of neurotransmission at fast synapses in the mammalian brain. Nature 384:170–172

100. Connors BW, Long MA (2004) Electrical synapses in the mammalian brain. Annu Rev Neurosci 27:393–418
101. Bannister NJ, Larkman AU (1995) Dendritic morphology of CA1 pyramidal neurones from the rat hippocampus: II. Spine distributions. J Comp Neurol 360:161–171
102. Koch C, Zador A (1993) The function of dendritic spines: devices subserving biochemical rather than electrical compartmentalization. J Neurosci 13:413–422
103. Bhatt DH, Zhang S, Gan W-B (2009) Dendritic spine dynamics. Annu Rev Physiol 71:261–282
104. Burnstock G (2004) Cotransmission. Curr Opin Pharmacol 4:47–52
105. Krebs EG, Beavo JA (1979) Phosphorylation-dephosphorylation of enzymes. Annu Rev Biochem 48:923–959
106. Wong W, John D, Scott JD (2004) AKAP signalling complexes: focal points in space and time. Nat Rev Mol Cell Biol 5:959–969
107. Kempermann G (2012) New neurons for 'survival of the fittest'. Nat Rev Neurosci 13:727–736
108. Carafoli E (1991) Calcium pump of the plasma membrane. Physiol Rev 71:129–153
109. Raley-Susman KM, Sapolsky RM, Kopito RR (1993) Cl^-/HCO_3^- exchange function differs in adult and fetal rat hippocampal neurons. Brain Res 614:308–314
110. Staley K, Smith R, Schaack J, Wilcox C, Jentsch J (1996) Alteration of $GABA_A$ receptor function following gene transfer of the CLC-2 chloride channel. Neuron 17:543–551
111. Staley KJ, Mody I (1992) Shunting of excitatory input to dentate gyrus granule cells by a depolarising $GABA_A$ receptor–mediated postsynaptic conductance. J Neurophysiol 68:197–212
112. Nilius B, Droogmans G (2003) Amazing chloride channels: an overview. Acta Physiol Scand 177(2):119–147
113. Magee JC (1999) Voltage-gated ion channels in dendrites. In: Stuart G, Spruston N, Hausser M (eds) Dendrites. Oxford University Press, Oxford, pp 139–160
114. Hausser M, Spruston N, Stuart GJ (2000) Diversity and dynamics of dendritic signaling. Science 290:739–744
115. Williams SR, Stuart GJ (2000) Action potential backpropagation and somato-dendritic distribution of ion channels in thalamocortical neurons. J Neurosci 20:1307–1317
116. Tank DW, Sugimori M, Connor JA, Llinas RR (1988) Spatially resolved calcium dynamics of mammalian Purkinje cells in cerebellar slice. Science 242:773–777
117. Reuveni I, Friedman A, Amitai Y, Gutnick MJ (1993) Stepwise repolarisation from Ca^{2+} plateaus in neocortical pyramidal cells: evidence for nonhomogeneous distribution of HVA Ca^{2+} channels in dendrites. J Neurosci 13:4609–4621
118. Yamada M, Inanobe A, Kurachi Y (1998) G protein regulation of potassium ion channels. Pharmacol Rev 50:723–757
119. Bergson C, Mrzljak L, Smiley JF, Pappy M, Levenson R, Goldman-Rakic PS (1995) Regional, cellular, and subcellular variations in the distribution of D1 and D5 dopamine receptors in primate brain. J Neurosci 15:7821–7836
120. Purcell AL, Sharma SK, Bagnall MW, Sutton MA, Carew TJ (2003) Activation of a tyrosine kinase-MAPK cascade enhances the induction of long-term synaptic facilitation and long-term memory in aplysia. Neuron 37:473–484
121. Huang EJ, Reichard LF (2003) Trk receptors: roles in neuronal signal transduction. Annu Rev Biochem 72:609–642
122. Kole MHP, Ilschner SU, Kampa BM, Williams SR, Ruben PC, Stuart GJ (2008) Action potential generation requires a high sodium channel density in the axon initial segment. Nat Neurosci 11:178–186
123. Hursh JB (1939) Conduction velocity and diameter of nerve fibers. Am J Physiol Cell Physiol 127:131–139
124. Hartline DK, Colman DR (2007) Rapid conduction and the evolution of giant axons and myelinated fibers. Curr Biol 17(1):R29–R35
125. Sah P (1996) Ca^{2+}-activated K^+ currents in neurones: types, physiological roles and modulation. Trends Neurosci 19:150–154

126. Faber ESL, Sah P (2003) Calcium-activated potassium channels: multiple contributions to neuronal function. Neuroscientist 9:181–194
127. Mel BW (1994) Information processing in dendritic trees. Neural Comput 6:1031–1085
128. Oakley JC, Schwindt PC, Crill WE (2001) Initiation and propagation of regenerative Ca^{++}-dependent potentials in dendrites of layer 5 pyramidal neurons. J Neurophysiol 86:503–513
129. Waters J, Schaefer A, Sakmann B (2005) Backpropagating action potentials in neurones: measurement, mechanisms and potential functions. Prog Biophys Mol Biol 87(1):145–170
130. Larkum ME, Rioult MG, Luscher HR (1996) Propagation of action potentials in the dendrites of neurons from rat spinal cord slice cultures. J Neurophysiol 75:154–170. Mol Biol 87(1):145–170
131. Abraham WC, Huggett A (1997) Induction and reversal of long-term potentiation by repeated high-frequency stimulation in rat hippocampal slices. Hippocampus 7:137–145
132. Bi G-q, Poo M-m (1998) Synaptic modifications in cultured hippocampal neurons: dependence on spike timing, synaptic strength, and postsynaptic cell type. J Neurosci 18:10464–10472
133. Zucker RS, Regehr WG (2002) Short-term synaptic plasticity. Annu Rev Physiol 64:355–405
134. Atwood HL, Karunanithi S (2002) Diversification of synaptic strength: presynaptic elements. Nat Rev Neurosci 3:497–516
135. Miller RJ (1998) Presynaptic receptors. Annu Rev Pharmacol Toxicol 38:201–227
136. Malenka RC, Bear MF (2004) LTP and LTD: an embarrassment of riches. Neuron 44:5–21
137. Lynch MA (2004) Long-term potentiation and memory. Physiol Rev 84:87–136
138. Manahan-Vaughan D (2000) Long-term depression in freely moving rats is dependent upon strain variation, induction protocol and behavioral state. Cereb Cortex 10:482–487
139. Berretta N, Nistico R, Bernardi G, Mercuri NB (2008) Synaptic plasticity in the basal ganglia: a similar code for physiological and pathological conditions. Prog Neurobiol 84:343–362
140. Ito M (1989) Long-term depression. Annu Rev Neurosci 12:85–102
141. Matsuzaki M, Honkura N, Ellis-Davies GC, Kasai H (2004) Structural basis of long-term potentiation in single dendritic spines. Nature 429:761–766
142. Patterson MA, Szatmari EM, Yasuda R (2010) AMPA receptors are exocytosed in stimulated spines and adjacent dendrites in a Ras-ERK–dependent manner during long-term potentiation. Pro Natl Acad Sci USA 107:15951–15956
143. Snyder EM, Philpot BD, Huber KM, Dong X, Fallon JR, Mark F, Bear MF (2001) Internalization of ionotropic glutamate receptors in response to mGluR activation. Nat Neurosci 4:1079–1085
144. Raymond CR (2007) LTP forms 1, 2 and 3: different mechanisms for the 'long' in long-term potentiation. Trends Neurosci 30:167–175
145. Reymann KG, Frey JU (2007) Late maintenance of hippocampal LTP: requirements, phases, 'synaptic tagging', 'late-associativity' and implications. Neuropharmacology 52:24–40
146. Calabresi P, Picconi B, Tozzi A, Di Filippo M (2007) Dopamine-mediated regulation of corticostriatal synaptic plasticity. Trends Neurosci 30:211–219
147. Stork O, Welzl H (1999) Memory formation and the regulation of gene expression. Cell Mol Life Sci 55:575–592
148. Pfeiffer BE, Huber KM (2006) Current advances in local protein synthesis and synaptic plasticity. J Neurosci 26:7147–7150
149. Berke JD, Steven E, Hyman SE (2000) Addiction, dopamine, and the molecular mechanisms of memory. Neuron 25:515
150. Lemon N, Manahan-Vaughan D (2006) Dopamine D1/D5 receptors gate the acquisition of novel information through hippocampal long-term potentiation and long-term depression. J Neurosci 26:7723–7729
151. Pastalkova E, Serrano P, Pinkhasova D, Wallace E, Fenton AA, Sacktor TC (2006) Storage of spatial information by the maintenance mechanism of LTP. Science 313:1141–1144
152. Whitlock JR, Heynen AJ, Shuler MG, Bear MF (2006) Learning induces long-term potentiation in the hippocampus. Science 313:1093–1097
153. Abraham WC, Williams JM (2003) Properties and mechanisms of LTP maintenance. Neuroscientist 2003(9):463–474

154. Turrigiano GG (2008) The self-tuning neuron: synaptic scaling of excitatory synapses. Cell 135:422–435
155. Turrigiano GG, Leslie KR, Desai NS, Rutherford LC, Sacha B, Nelson SB (1998) Activity-dependent scaling of quantal amplitude in neocortical neurons. Nature 391:892–896
156. Stephenson FA, Hawkins LM (2001) Neurotransmitter receptors in the postsynaptic neuron. In Encyclopedia of life sciences. http://dx.doi.org/10.1038/npg.els.0000284
157. Salio C, Lossi L, Ferrini F, Merighi A (2006) Neuropeptides as synaptic transmitters. Cell Tissue Res 326:583–598
158. Gudermann T, Schoneberg T, Schultz G (1997) Functional and structural complexity of signal transduction via G-protein-coupled receptors. Annu Rev Neurosci 20:399–427
159. Hamm HE (1998) The many faces of G protein signaling. J Biol Chem 273:669–672
160. Cossart R, Epsztein J, Tyzio R, Becq H, Hirsch J, Ben-Ari Y, Crepel V (2002) Quantal release of glutamate generates pure kainate and mixed AMP/kainate EPSCs in hippocampal neurons. Neuron 35:147–159
161. Traynelis SF, Wollmuth LP, McBain CJ, Menniti FS, Vance KM, Ogden KK, Hansen KB, Yuan H, Myers SJ, Ray Dingledine R (2010) Glutamate receptor ion channels: structure, regulation, and function. Pharmacol Rev 62:405–496
162. Rosenmund C, Stern-Bach Y, Stevens CF (1998) The tetrameric structure of a glutamate receptor channel. Science 280:1596–1599
163. Chung HJ, Steinberg JP, Huganir RL, Linden DJ (2003) Requirement of AMPA receptor GluR2 phosphorylation for cerebellar long-term depression. Science 300:1751–1755
164. Nusser Z, Mulvihill E, Streit P, Somogyi P (1994) Subsynaptic segregation of metabotropic and ionotropic glutamate receptors as revealed by immunogold localization. Neuroscience 61:421–427
165. Jaskolski F, Coussen F, Mulle C (2005) Subcellular localization and trafficking of kainate receptors. Trends Pharmacol Sci 26:20–26
166. Niswender CM, Conn PJ (2010) Metabotropic glutamate receptors: physiology, pharmacology, and disease. Annu Rev Pharmacol Toxicol 50:295–322
167. Cavallero A, Marte A, Fedele E (2009) l-Aspartate as an amino acid neurotransmitter: mechanisms of the depolarisation-induced release from cerebrocortical synaptosomes. J Neurochem 110:924–934
168. Fleck MW, Henze DA, Barrioneuvo G, Palmer AM (1993) Aspartate and glutamate mediate excitatory synaptic transmission in area CA1 of the hippocampus. J Neurosci 13:3944–3955
169. Chen PE, Geballe MT, Stansfeld PJ, Johnston AR, Yuan H, Jacob AL, Snyder JP, Traynelis SF, Wyllie DJA (2005) Structural features of the glutamate binding site in recombinant NR1/NR2A N-methyl-d-aspartate receptors determined by site-directed mutagenesis and molecular modeling. Mol Pharmacol 67:1470–1484
170. Kaila K (1994) Ionic basis of $GABA_A$ receptor channel function in the nervous system. Prog Neurobiol 42:489–537
171. Kaupmann K, Schuler V, Mosbacher J, Bischoff H, Bittiger H, Heid J, Froestl W, Leonhard S, Pfaff T, Karschin A, Bettler B (1998) Human γ-aminobutyric acid type B receptors are differentially expressed and regulate inwardly rectifying K+ channels. Proc Natl Acad Sci USA 9:14991–14996
172. Gulledge AT, Stuart GJ (2003) Excitatory actions of GABA in the cortex. Neuron 37:299–309
173. Staley KJ, Proctor WR (1999) Modulation of mammalian dendritic $GABA_A$ receptor function by the kinetics of Cl and HCO_3 transport. J Physiol 519:693–712
174. Koch C, Poggio T, Torre V (1983) Nonlinear interactions in a dendritic tree: localization, timing, and role in information processing. Proc Natl Acad Sci USA 80:2799–2802
175. Lynch JW (2004) Molecular structure and function of the glycine receptor chloride channel. Physiol Rev 84:1051–1095
176. Danysz W, Parsons CG (1998) Glycine and N-methyl-D-aspartate receptors: physiological significance and possible therapeutic applications. Pharmacol Rev 50:597–664

177. Martina M, Gorfinkel Y, Halman S, Lowe JA, Periyalwar P, Schmidt CJ, Bergeron R (2004) Glycine transporter type 1 blockade changes NMDA receptor-mediated responses and LTP in hippocampal CA1 pyramidal cells by altering extracellular glycine levels. J Physiol 557:489–500
178. Smith KE, Borden LA, Hartig PR, Branchek T, Weinshank RL (1992) Cloning and expression of a glycine transporter reveal colocalization with NMDA receptors. Neuron 8:927–935
179. Zeilhofer HU, Studler B, Arabadzisz D, Schweizer C, Ahmadi S, Layh B, Bösl MR, Fritschy J-M (2005) Glycinergic neurons expressing enhanced green fluorescent protein in bacterial artificial chromosome transgenic mice. J Comp Neurol 482:123–216
180. Beaulieu J-M, Gainetdinov RR (2011) The physiology, signaling, and pharmacology of dopamine receptors. Pharmacol Rev 63:182–217
181. Surmeier DJ, Ding J, Day M, Wang Z, Shen W (2007) D1 and D2 dopamine-receptor modulation of striatal glutamatergic signaling in striatal medium spiny neurons. Trends Neurosci 30:228–235
182. Surmeier DJ, Song W-J, Zhen Y (1996) Coordinated expression of dopamine receptors in neostriatal medium spiny neurons. J Neurosci 16:6579–6591
183. Jay TM (2003) Dopamine: a potential substrate for synaptic plasticity and memory mechanisms. Prog Neurobiol 69:375–390
184. Swant J, Wagner JJ (2006) Dopamine transporter blockade increases LTP in the CA1 region of the rat hippocampus via activation of the D3 dopamine receptor. Learn Mem 13:161–167
185. Centonze D, Picconi B, Paolo Gubellini P, Giorgio Bernardi G, Calabresi P (2001) Dopaminergic control of synaptic plasticity in the dorsal striatum. Eur J Neurosci 13:1071–1077
186. Frey U, Huang YY, Kandel ER (1993) Effects of cAMP simulate a late stage of LTP in hippocampal CA1 neurons. Science 260:1661–1664
187. Caille I, Dumartin B, Bloch B (1996) Ultrastructural localization of D1 dopamine receptor immunoreactivity in rat striatonigral neurons and its relation with dopaminergic innervation. Brain Res 730:17–31
188. Floresco SB, West AR, Ash B, Moore H, Grace AA (2003) Afferent modulation of dopamine neuron firing differentially regulates tonic and phasic dopamine transmission. Nat Neurosci 6:968–973
189. Meiergerd SM, Patterson TA, Schenk JO (1993) D_2 receptors may modulate the function of the striatal transporter for dopamine: kinetic evidence from studies in vitro and in vivo. J Neurochem 61:764–767
190. Grace AA (1991) Phasic versus tonic dopamine release and the modulation of dopamine system responsivity: a hypothesis for the etiology of schizophrenia. Neuroscience 41:1–24
191. Frazer A, Hensler JG (1999) Serotonin receptors. In: Siegel GJ, Agranoff BW, Albers RW, Fisher SK, Uhler MD (eds) Basic neurochemistry: molecular, cellular, and medical aspects, 6th edn. Lippincott-Raven, Philadelphia, pp 263–292
192. Lukyanetz EA, Sotkis AV, Kostyuk PG (2002) Mechanisms of up-regulation of single calcium channels by serotonin in Helix pomatia neurons. Biochem Biophys Res Commun 293:132–138
193. Karlin A (2002) Emerging structure of the nicotinic acetylcholine receptors. Nat Rev Neurosci 3:102–114
194. Dani JA, Mayer ML (1995) Structure and function of glutamate and nicotinic acetylcholine receptors. Curr Opin Neurobiol 5:310–317
195. Felder CC (1995) Muscarinic acetylcholine receptors: signal transduction through multiple effectors. FASEB J 9:619–625
196. Marrion NV (1997) Control of M-current. Annu Rev Physiol 59:483–504
197. Hein L (2006) Adrenoceptors and signal transduction in neurons (2006). Cell Tissue Res 326:541–551
198. Samuels ER, Szabadi E (2008) Functional neuroanatomy of the noradrenergic locus coeruleus: its roles in the regulation of arousal and autonomic function part I: principles of functional organization. Curr Neuropharmacol 6:235–253
199. Sah P, Isaacson JS (1995) Channels underlying the slow afterhyperpolarisation in hippocampal pyramidal neurons: neurotransmitters modulate the open probability. Neuron 15:435–441

200. Nicoll RA (1988) The coupling of neurotransmitter receptors to ion channels in the brain. Science 241:545–551
201. Pankratov Y, Lalo U, Verkhratsky A, North RA (2006) Vesicular release of ATP at central synapses. Eur J Physiol 452:589–597
202. Zimmermann H (2008) ATP and acetylcholine, equal brethren. Neurochem Int 52:634–648
203. Williams M (2002) Purinergic neurotransmission. In: Davis KL, Charney D, Coyle JT, Nemeroff C (eds) Neuropschopharmacology: the fifth generation of progress. Lippincott Williams and Wilkins, Philadelphia
204. Jo Y-H, Role LW (2002) Coordinate release of ATP and GABA at in vitro synapses of lateral hypothalamic neurons. J Neurosci 22:4794–4804
205. Pankratov YV, Lalo UV, Krishtal OA (2002) Role for P2X receptors in long-term potentiation. J Neurosci 22:8363–8369
206. Schiffmann SN, Fisone G, Moresco R, Cunha RA, Ferre S (2007) Adenosine A2A receptors and basal ganglia physiology. Prog Neurobiol 83:277–292
207. Cunha RA (2001) Adenosine as a neuromodulator and as a homeostatic regulator in the nervous system: different roles, different sources and different receptors. Neurochem Int 38:107–125
208. Fredholm BB, Abbracchio MP, Burnstock G, Daly JW, Harden TK, Jacobson KA, Leff P, Williams M (1994) Nomenclature and classification of purinoceptors. Pharmacol Rev 46:143–156
209. Fredholm BB, Dunwiddie TV (1988) How does adenosine inhibit transmitter release? Trends Pharmacol Sci 9:130–134
210. Bongsebandhu-phubhakdi S, Manabe T (2007) The neuropeptide nociceptin is a synaptically released endogenous inhibitor of hippocampal long-term potentiation. J Neurosci 27:4850–4858
211. Balschun D, Reymann KG (1994) Cholecystokinin (CCK-8S) prolongs 'unsaturated' θ-pulse induced long-term potentiation in rat hippocampal CA1 in vitro. Neuropeptides 26:421–427
212. Wilson RI, Nicoll RA (2002) Endocannabinoid signaling in the brain. Science 296:678–682
213. Hillard C, Jarrahian A (2003) Cellular accumulation of anandamide: consensus and controversy. Br J Pharmacol 140:802–808
214. Ronesi J, Gerdeman GL, Lovinger DM (2004) Disruption of endocannabinoid release and striatal long-term depression by postsynaptic blockade of endocannabinoid membrane transport. J Neurosci 24:1673–1679
215. Gerdeman GL, Lovinger DM (2003) Emerging roles for endocannabinoids in long-term synaptic plasticity. Br J Pharmacol 140:781–789
216. Di Tomaso E, Beltramo M, Piomelli D (1996) Brain cannabinoids in chocolate. Nature 382:677–678
217. Baranano DE, Ferris CD, Snyder SH (2001) Atypical neural messengers. Trends Neurosci 24:99–106
218. Toth K (2011) Zinc in neurotransmission. Annu Rev Nutr 31:139–153
219. Sindreu CB, Varoqui H, Erickson JD, Perez-Clausell J (2003) Boutons containing vesicular zinc define a subpopulation of synapses with low AMPAR content in rat hippocampus. Cereb Cortex 13:823–829
220. Perez-Clausell J, Danscher G (1985) Intravesicular localization of zinc in rat telencephalic boutons. A histochemical study. Brain Res 337:91–98
221. Peters S, Koh J, Choi DW (1987) Zinc selectively blocks the action of N-methyl-d-aspartate on cortical neurons. Science 236:589–593
222. Westbrook GL, Mayer ML (1987) Micromolar concentrations of Zn^{2+} antagonize NMDA and GABA responses of hippocampal neurons. Nature 328:640–643
223. Sensi SL, Yin HZ, Carriedo SG, Rao SS, Weiss JH (1999) Preferential Zn^{2+} influx through Ca^{2+}-permeable AMPA/kainate channels triggers prolonged mitochondrial superoxide production. Proc Natl Acad Sci USA 96:2414–2419
224. Li Y, Hough CJ, Frederickson CJ, Sarvey JM (2001) Induction of mossy fiber → CA3 long-term potentiation requires translocation of synaptically released Zn^{2+}. J Neurosci 21:8015–8025

225. Hölscher C (1997) Nitric oxide, the enigmatic neuronal messenger: its role in synaptic plasticity. Trends Neurosci 20:298–303
226. Son H, Hawkins RD, Martin K, Kiebler M, Huang PL, Fishman MC, Kandel ER (1996) Long-term potentiation is reduced in mice that are doubly mutant in endothelial and neuronal nitric oxide synthase. Cell 87:1015–1023
227. Steinert JR, Kopp-Scheinpflug C, Baker C, Challiss RAJ, Mistry R, Haustein MD, Griffin SJ, Tong H, Graham BP, Forsythe ID (2008) Nitric oxide is a volume transmitter regulating postsynaptic excitability at a glutamatergic synapse. Neuron 60:642–656
228. Dieni S, Matsumoto T, Dekkers M, Rauskolb S, Ionescu MS, Deogracias R, Gundelfinger ED, Kojima M, Nestel S, Frotscher M, Barde Y-A (2012) BDNF and its pro-peptide are stored in presynaptic dense core vesicles in brain neurons. J Cell Biol. doi:10.1083/jcb.201201038
229. Arévalo JC, Wu SH (2006) Neurotrophin signaling: many exciting surprises! Cell Mol Life Sci 63:1523–1537
230. Pape HC, McCormick DA (1989) Noradrenaline and serotonin selectively modulate thalamic burst firing by enhancing a hyperpolarisation-activated cation current. Nature 340:715–718
231. McCormick DA, Williamson A (1989) Convergence and divergence of neurotransmitter action in human cerebral cortex. Proc Natl Acad Sci USA 86:8098–8102
232. Angelino E, Brenner MP (2007) Excitability constraints on voltage-gated sodium channels. PLoS Comput Biol 3(9):e177. doi:10.1371/journal.pcbi.0030177
233. Magee JC, Johnston D (1995) Characterization of single voltage-gated Na^+ and Ca^{++} channels in apical dendrites of rat CA1 pyramidal neurons. J Physiol 487:67–90
234. O'Dell TJ, Alger BE (1991) Single calcium channels in rat and guinea-pig hippocampal neurons. J Physiol 436:739–767
235. Gutman GA, Chandy KG, Adelman JP, Lazdunski M, Mckinnon D, Pardo LA, Robertson GA, Rudy B, Sanguinetti MC, Stuhmer W, Wang X (2005) International union of pharmacology. LIII. Nomenclature and molecular relationships of voltage-gated ion channels: potassium channels. Pharmacol Rev 57:473–508
236. Storm JF (1993) Functional diversity of K^+ currents in hippocampal pyramidal neurons. Semin Neurosci 5:79–92
237. Delmas P, Brown DA (2005) Pathways modulating neural KCNQ/M (Kv7) potassium channels. Nat Rev Neurosci 6:850–862
238. Metz AE, Spruston N, Martina M (2007) Dendritic D-type potassium currents inhibit the spike afterdepolarisation in rat hippocampal CA1 pyramidal neurons. J Physiol 581:175–187
239. Hoffman DA, Johnston D (1998) Downregulation of transient K^+ channels in dendrites of hippocampal CA1 pyramidal neurons by activation of PKA and PKC. J Neurosci 18:3521–3528
240. Wilke RA, Hsu S-F, Jackson MB (1998) Dopamine D4 receptor mediated inhibition of potassium current in neurohypophysial nerve terminals. J Pharmacol Exp Ther 284(2):542–548
241. Lambe EK, Aghajanian GK (2001) The role of Kv1.2-containing potassium channels in serotonin-induced glutamate release from thalamocortical terminals in rat frontal cortex. J Neurosci 21:9955–9963
242. Dong Y, White F (2003) Dopamine D1-class receptors selectively modulate a slowly inactivating potassium current in rat medial prefrontal cortex pyramidal neurons. J Neurosci 23:2686–2695
243. Wu R, Barish ME (1999) Modulation of a slowly inactivating potassium current, I_D, by metabotropic glutamate receptor activation in cultured hippocampal pyramidal neurons. J Neurosci 19:6825–6837
244. Halliwell JV, Paul R, Adams PR (1982) Voltage-clamp analysis of muscarinic excitation in hippocampal neurons. Brain Res 250:71–92
245. Colino A, Halliwell JV (1987) Differential modulation of three separate K-conductances in hippocampal CA1 neurons by serotonin. Nature 328:73–77
246. Charpak S, Gahwiler BH, Do KQ, Knopfel T (1990) Potassium conductances in hippocampal neurons blocked by excitatory amino-acid transmitters. Nature 347:765–767

247. Moore SD, Madamba SG, Joels M, Siggins GR (1988) Somatostatin augments the M-current in hippocampal neurons. Science 239:278–280
248. Vacher H, Mohapatra DP, Trimmer JS (2008) Localization and targeting of voltage-dependent ion channels in mammalian central neurons. Physiol Rev 88:1407–1447
249. Bekkers JM (2000) Properties of voltage-gated potassium currents in nucleated patches from large layer 5 cortical pyramidal neurons of the rat. J Physiol 525:593–609
250. Solomon JS, Nerbonne JM (1993) Hyperpolarisation-activated currents in isolated superior colliculus-projecting neurons from rat visual cortex. J Physiol 462:393–420
251. Williams SR, Stuart GJ (2003) Role of dendritic synapse location in the control of action potential output. Trends Neurosci 26:147–154
252. Berger T, Senn W, Luscher HR (2003) Hyperpolarisation activated current I_h disconnects somatic and dendritic spike initiation zones in layer V pyramidal neurons. J Neurophysiol 90:2428–2437
253. Sík A, Smith RL, Freund TF (2000) Distribution of chloride channel-2-immunoreactive neuronal and astrocytic processes in the hippocampus. Neuroscience 101:51–65
254. Jentsch TJ, Gunther W, Pusch M, Schwappach B (1995) Properties of voltage gated chloride channels of the ClC gene family. J Physiol 482:19S–25S
255. Lai HC, Jan LY (2006) The distribution and targeting of neuronal voltage-gated ion channels. Nat Rev Neurosci 7:548–562
256. Tzingounis AV, Nicoll RA (2008) Contribution of KCNQ2 and KCNQ3 to the medium and slow afterhyperpolarisation currents. Proc N Y Acad Sci 105:19974–19979
257. See Ref. [125]
258. Maingret F, Coste B, Hao J, Giamarchi A, Allen D, Crest M, Litchfield DW, Adelman JP, Delmas P (2008) Neurotransmitter modulation of small-conductance Ca^{2+}-activated K^+ channels by regulation of Ca^{2+} gating. Neuron 59(3):439–449
259. Stackman RW, Hammond RS, Linardatos E, Gerlach A, Maylie J, Adelman JP, Tzounopoulos T (2002) Small conductance Ca^{2+}-activated K^+ channels modulate synaptic plasticity and memory encoding. J Neurosci 22:10163–10171
260. Raley-Susman KM, Cragoe EJ Jr, Sapolsky RM, Kopito RR (1991) Regulation of intracellular pH in cultured hippocampal neurons by an amiloride-insensitive Na^+/H^+ exchanger. J Biol Chem 266:2739–2745
261. Kupfermann I (1980) Role of cyclic nucleotides in excitable cells. Annu Rev Physiol 42:629–641
262. Hong M, Ross WN (2007) Priming of intracellular calcium stores in rat CA1 pyramidal neurons. J Physiol 584:75–87
263. Muller W, Connor JA (1991) Dendritic spines as individual neuronal compartments for synaptic $Ca2^+$ responses. Nature 354:73–76
264. Higley MJ, Sabatini BL (2008) Calcium signaling in dendrites and spines: practical and functional considerations. Neuron 59:902–913
265. Brorson JR, Bleakman D, Gibbons SJ, Miller RJ (1991) The properties of intracellular calcium stores in cultured rat cerebellar neurons. J Neurosci 17:4024–4043
266. Bayer KU, Schulman H (2001) Regulation of signal transduction by protein targeting: the case for CaMKII. Biochem Biophys Res Commun 289:917–923
267. Chin D, Means AR (2000) Calmodulin: a prototypical calcium sensor. Trends Cell Biol 10:322–328
268. Wayman GA, Wei J, Wong S, Storm DR (1996) Regulation of type I adenylyl cyclase by calmodulin kinase IV in vivo. Mol Cell Biol 16:6075–6082
269. Wang J, Chen S, Siegelbaum SA (2001) Regulation of hyperpolarisation-activated HCN channel gating and cAMP modulation due to interactions of COOH terminus and core transmembrane regions. J Gen Physiol 118:237–250
270. Lucas KA, Pitari GM, Kazerounian S, Ruiz-Stewart I, Park J, Schulz S, Chepenik KP, Waldman SA (2000) Guanylyl cyclases and signaling by cyclic GMP. Pharmacol Rev 52:375–413

271. Conti M (2000) Phosphodiesterases and cyclic nucleotide signaling in endocrine cells. Mol Endocrinol 14:1317–1327
272. Menniti FS, Faraci WS, Schmidt CJ (2006) Phosphodiesterases in the CNS: targets for drug development. Nat Rev Drug Discov 5:660–670
273. Conti M, Beavo J (2007) Biochemistry and physiology of cyclic nucleotide phosphodiesterases: essential components in cyclic nucleotide signaling. Annu Rev Biochem 76:481–511
274. McDermott M, Wakelam MJO, Morris AJ (2004) Phospholipase D. Biochem Cell Biol 82:225–253
275. Neves SR, Ram PT, Iyengar R (2002) G protein pathways. Science 296:1636–1639
276. Alexander SPH, Mathie A, Peters JA (2009) Guide to receptors and channels (GRAC), 4th edn. Br J Pharmacol 158(suppl 1):S5–S101
277. Zhang S, Coso OA, Lee C, Gutkind JS, Simonds WF (1996) Selective activation of effector pathways by brain specific G protein b5. J Biol Chem 271:33575–33579
278. Pierce KL, Premont RT, Lefkowitz RJ (2002) Seven-transmembrane receptors. Nat Rev Neurosci 3:639–650
279. Fimia GM, Sassone-Corsi P (2001) Cyclic AMP signalling. J Cell Sci 114:1971–1972
280. Ferguson SSG (2001) Evolving concepts in G protein-coupled receptor endocytosis: the role in receptor desensitization and signaling. Pharmacol Rev 53:1–24
281. Tang W-J, Gilman AG (1992) Adenylyl cyclases. Cell 70:869–872
282. Kaupp UB, Reinhard Seifert R (2002) Cyclic nucleotide-gated ion channels. Physiol Rev 82:769–824
283. Cooper DMF (2003) Regulation and organization of adenylyl cyclases and cAMP. Biochem J 375:517–529
284. Montminy M (1997) Transcriptional regulation by cyclic AMP. Annu Rev Biochem 66:807–822
285. Thelen M, Didichenko SA (1997) G-protein coupled receptor-mediated activation of PI 3-kinase in neutrophils. Ann N Y Acad Sci 832:368–382
286. Herlitze S, Garcia DE, Mackie K, Hille B, Scheuer T, Catterall WA (1996) Modulation of Ca^{2+} channels by G-protein beta gamma subunits. Nature 380:258–262
287. Delmas P, Crest M, Brown DA (2004) Functional organization of PLC signaling microdomains in neurons. Trends Neurosci 27:41–47
288. Simpson CS, Morris BJ (1995) Induction of c-fos and zif/268 gene expression in rat striatal neurons, following stimulation of D1-like dopamine receptors, involves protein kinase A and protein kinase C. Neuroscience 68:97–106
289. Tanaka C, Nishizuka Y (1994) The protein kinase C family for neuronal signaling. Annu Rev Neurosci 17:551–567
290. Hall A (1998) Rho GTPases and the actin cytoskeleton. Science 279:509–514
291. Kvachnina E, Liu G, Dityatev A, Renner U, Dumuis A, Richter DW, Dityateva G, Schachner M, Voyno-Yasenetskaya TA, Ponimaskin EG (2005) 5-HT7 receptor is coupled to G subunits of heterotrimeric G12-protein to regulate gene transcription and neuronal morphology. J Neurosci 25:7821–7830
292. Ponimaskin EG, Profirovic J, Vaiskunaite R, Richter DW, Voyno-Yasenetskaya TA (2002) 5-Hydroxytryptamine 4(a) receptor is coupled to the G alpha subunit of heterotrimeric G13 protein. J Biol Chem 277:20812–20819
293. Thomas DR, Melotto S, Massagrande M, Gribble AD, Jeffrey P, Stevens AJ, Deeks NJ, Eddershaw PJ, Fenwick SH, Riley G, Stean T, Scott CM, Hill MJ, Middlemiss DN, Hagan JJ, Price GW, Forbes IT (2003) SB-656104-A, a novel selective 5-HT7 receptor antagonist, modulates REM sleep in rats. Br J Pharmacol 139:705–714
294. Mansour SJ, Matten WT, Hermann AS, Candia JM, Rong S, Fukasawa K, Vande Woude GF, Ahn NG (1994) Transformation of mammalian cells by constitutively active MAP kinase kinase. Science 265:966–970
295. Ferrell JE, Bhatt RR (1997) Mechanistic studies of the dual phosphorylation of mitogen-activated protein kinase. J Biol Chem 272:19008–19016
296. Adams JA (2001) Kinetic and catalytic mechanisms of protein kinases. Chem Rev 101:2271–2290

297. Barford D, Das AK, Egloff M-P (1998) The structure and mechanism of protein phosphatases: insights into catalysis and regulation. Annu Rev Biophys Biomol Struct 27:133–164
298. Maston GA, Evans SK, Green MR (2006) Transcriptional regulatory elements in the human genome. Annu Rev Genomics Hum Genet 7:29–59
299. Wang Z, Burge CR (2008) Splicing regulation: from a parts list of regulatory elements to an integrated splicing code. RNA 14:802–813
300. Bramham CR, Wells DG (2007) Dendritic mRNA: transport, translation and function. Nat Rev Neurosci 8:776–789
301. Steward O, Schuman EM (2001) Protein synthesis at synaptic sites on dendrites. Annu Rev Neurosci 24:299–325
302. Starokadomskyy PL (2007) Protein splicing. Mol Biol 41:278–293
303. Guhaniyogi J, Brewer G (2001) Regulation of mRNA stability in mammalian cells. Gene 265:11–23
304. Xi H, Shulha HP, Lin JM, Vales TR, Fu Y, Bodine DM, McKay RDG, Chenoweth JG, Tesar PJ, Furey TS, Ren B, Weng Z, Crawford GE (2007) Identification and characterization of cell type-specific and ubiquitous chromatin regulatory structures in the human genome. PLoS Genet 3:e136
305. Shepherd JD, Huganir RL (2007) The cell biology of synaptic plasticity: AMPA receptor trafficking. Annu Rev Cell Dev Biol 23:613–643
306. Leslie JH, Nedivi E (2011) Activity-regulated genes as mediators of neural circuit plasticity. Prog Neurobiol 94:233–237
307. Loebrich S, Nedivi E (2009) The function of activity-regulated genes in the nervous system. Physiol Rev 89:1079–1103
308. Nestler EJ, Human SE (2002) Regulation of gene expression. In: Davis KL, Charney D, Coyle JT, Nemeroff C (eds) Neuropharmacology: the fifth generation of progress. Lippincott, Williams, & Wilkins, Philadelphia
309. Lin Y-S, Carey M, Ptashne M, Green MR (1990) How different eukaryotic transcriptional activators can cooperate promiscuously. Nature 345:359–361
310. Carey M, Lin Y-S, Green MR, Ptashne M (1990) A mechanism for synergistic activation of a mammalian gene by GAL4 derivatives. Nature 345:361–364
311. Perez-Cadahia B, Drobic B, Davie JR (2011) Activation and function of immediate-early genes in the nervous system. Biochem Cell Biol 89:61–73
312. Spiegelman BM, Heinrich R (2004) Biological control through regulated transcriptional coactivators. Cell 119:157–167
313. Mayr B, Montminy M (2001) Transcriptional regulation by the phosphorylation-dependent factor CREB. Nat Rev Mol Cell Biol 2:599–609
314. Lyons MR, West AE (2011) Mechanisms of specificity in neuronal activity-regulated gene transcription. Prog Neurobiol 94:259–295
315. Lonze BE, Ginty DD (2002) Function and regulation of CREB family transcription factors in the nervous system. Neuron 35:605–623
316. Ch'ng TH, Uzgil B, Lin P, Avilyakulov NK, O'Dell TJ, Martin KC (2012) Activity-dependent transport of the transcriptional coactivator CRTC1 from synapse to nucleus. Cell 150:207–221
317. Kornhauser JM, Cowan CW, Shaywitz AJ, Dolmetsch RE, Griffith EC, Hu LS, Haddad C, Xia Z, Greenberg ME (2002) CREB transcriptional activity in neurons is regulated by multiple, calcium-specific phosphorylation events. Neuron 34:221–233
318. Knoll B, Nordheim A (2009) Functional versatility of transcription factors in the nervous system: the SRF paradigm. Trends Neurosci 32:432–442
319. Besnard A, Galan-Rodriguez B, Vanhoutte P, Caboche J (2011) Elk-1 a transcription factor with multiple facets in the brain. Front Neurosci 5:35
320. Posern G, Treisman R (2006) Actin' together: serum response factor, its cofactors and the link to signal transduction. Trends Cell Biol 16:588–596
321. Flavell SW, Greenberg ME (2008) Signaling mechanisms linking neuronal activity to gene expression and plasticity of the nervous system. Annu Rev Neurosci 31:563–590

322. Tao X, West AE, Chen WG, Corfas G, Greenberg ME (2002) A calcium-responsive transcription factor, CaRF, that regulates neuronal activity-dependent expression of BDNF. Neuron 33:383–395
323. West AE (2011) Biological functions and transcriptional targets of CaRF in neurons. Cell Calcium 49:290–295
324. Pfenning AR, Kim T-K, Spotts JM, Hemberg M, Su D, West AE (2010) Genome-wide identification of calcium-response factor (CaRF) binding sites predicts a role in regulation of neuronal signaling pathways. PLoS One 5(5):e10870
325. Blau J, Xiao H, McCracken S, O'Hare P, Greenblatt J, Bentley D (1996) Three functional classes of transcriptional activation domain. Mol Cell Biol 16:2044–2055
326. Kindler S, Wang H, Richter D, Tiedge H (2005) RNA transport and local control of translation. Annu Rev Cell Dev Biol 21:223–245
327. Beaumont V, Zhong N, Fletcher R, Froemke RC, Zucker RS (2001) Phosphorylation and local presynaptic protein synthesis in calcium- and calcineurin-dependent induction of crayfish long-term facilitation. Neuron 32:489–501
328. Sutton MA, Schuman EM (2005) Local translational control in dendrites and its role in long-term synaptic plasticity. J Neurobiol 64:116–131
329. Nayak AS, Zastrow DJ, Lickteig R, Zahniser NR, Browning MD (1998) Maintenance of late-phase LTP is accompanied by PKA dependent increase in AMPA receptor synthesis. Nature 394:680–683
330. Steward O, Wallace CS, Lyford GL, Worley PF (1998) Synaptic activation causes the mRNA for IEG Arc to localize selectively near activated postsynaptic sites on dendrites. Neuron 21:741–751
331. Sonenberg N, Dever TE (2003) Eukaryotic translation initiation factors and regulators. Curr Opin Struct Biol 13:56–63
332. Browne GJ, Proud CG (2002) Regulation of peptide-chain elongation in mammalian cells. Eur J Biochem 269:5360–5368
333. Mugnier P, Tuite MF (1999) Translation termination and its regulation in eukaryotes: recent insights provided by studies in yeast. Biochemistry (Mosc) 64:1360–1366
334. Messina V, Di Sauro A, Pedrotti S, Adesso L, Latina A, Geremia R, Rossi P, Sette C (2010) Differential contribution of the MTOR and MNK pathways to the regulation of mRNA translation in meiotic and postmeiotic mouse male germ cells. Biol Reprod 83:607–615
335. Bianchini A, Loiarro M, Bielli P, Busa R, Paronetto MP, Lorenzio F, Geremia R, Sette C (2008) Phosphorylation of eIF4E by MNKs supports protein synthesis, cell cycle progression and proliferation in prostate cancer cells. Carcinogenesis 29:2279–2288
336. Turjanski AG, Vaque JP, Gutkind JS (2007) MAP kinases and the control of nuclear events. Oncogene 26:3240–3253
337. Johnson GL, Razvan Lapadat R (2002) Mitogen-activated protein kinase pathways mediated by ERK, JNK, and p38 protein kinases. Science 298:1911–1912
338. Sweatt JD (2001) The neuronal MAP kinase cascade: a biochemical signal integration system subserving synaptic plasticity and memory. J Neurochem 76:1–10
339. Thomas GM, Rumbaugh GR, Harrar DB, Huganir RL (2005) Ribosomal S6 kinase 2 interacts with and phosphorylates PDZ domain-containing proteins and regulates AMPA receptor transmission. Proc Natl Acad Sci USA 102:15006–15011
340. Davis S, Vanhoutte P, Pages C, Caboche J, Laroche S (2000) The MAPK/ERK cascade targets both Elk-1 and cAMP response element-binding protein to control long-term potentiation-dependent gene expression in the dentate gyrus in vivo. J Neurosci 20:4563–4572
341. Lo RS, Wotton D, Massague J (2001) Epidermal growth factor signaling via Ras controls the Smad transcriptional co-repressor TGIF. EMBO J 20:128–136
342. Schaeffer HJ, Catling AD, Eblen ST, Collier LS, Krauss A, Michael J, Weber MJ (1998) MP1: a MEK binding partner that enhances enzymatic activation of the MAP kinase cascade. Science 281:1668–1671
343. Owens DM, Keyse SM (2007) Differential regulation of MAP kinase signalling by dual-specificity protein phosphatases. Oncogene 26:3203–3213

344. Muda M, Theodosiou A, Rodrigues N, Boschert U, Camps M, Gillieron C, Davies K, Ashworthi A, Steve Arkinstall S (1996) The dual specificity phosphatases M3/6 and MKP-3 are highly selective for inactivation of distinct mitogen-activated protein kinases. J Biol Chem 271:27205–27208
345. Balan V, Leicht DT, Zhu J, Balan K, Kaplun A, Singh-Gupta V, Qin J, Ruan H, Comb MJ, Guri Tzivion G (2006) Identification of novel in vivo Raf-1 phosphorylation sites mediating positive feedback Raf-1 regulation by extracellular signal-regulated kinase. Mol Biol Cell 17:1141–1153
346. Correa SAL, Eales KL (2012) The role of p38 MAPK and its substrates in neuronal plasticity and neurodegenerative disease. J Signal Transduct 2012(2012): Article ID 649079, 12pp. doi:10.1155/2012/649079
347. Xia Z, Storm DR (2005) The role of calmodulin as a signal integrator for synaptic plasticity. Nat Rev Neurosci 6:267–276
348. Griffith LC (2004) Regulation of calcium/calmodulin-dependent protein kinase II activation by intramolecular and intermolecular interactions. J Neurosci 24:8394–8398
349. Braun AP, Schulman H (1995) The multifunctional calcium/calmodulin-dependent protein kinase: from form to function. Annu Rev Physiol 57:417–445
350. Gaudilliere B, Konishi Y, de la Iglesia N, Yao G-l, Bonni A (2004) A CaMKII-NeuroD signaling pathway specifies dendritic morphogenesis. Neuron 41:229–241
351. Barria A, Malinow R (2005) NMDA receptor subunit composition controls synaptic plasticity by regulating binding to CaMKII. Neuron 48:289–301
352. Soderling TR (1999) The Ca^{2+}-calmodulin-dependent protein kinase cascade. Trends Biol Sci 24:232–236
353. Gradin HM, Marklund U, Larsson N, Chatila TA, Gullberg M (1997) Ca^{2+}/calmodulin-dependent kinase IV/Gr-dependent phosphorylation of oncoprotein 18. Mol Cell Biol 17:3459–3467
354. Ageta-Ishihara N, Takemoto-Kimura S, Nonaka M, Adachi-Morishima A, Kanzo Suzuki K, Kamijo S, Fujii H, Mano T, Blaeser F, Chatila TA, Mizuno H, Hirano T, Tagawa Y, Okuno H, Bito H (2009) Control of cortical axon elongation by a GABA-driven Ca^{2+}/calmodulin-dependent protein kinase cascade. J Neurosci 29:13720–13729
355. Tokumitsu H, Enslen H, Soderling TR (1995) Characterization of a Ca^{2+}/calmodulin-dependent protein kinase cascade. J Biol Chem 270:19320–19324
356. Berridge MJ, Bootman MD, Roderick HL (2003) Calcium signalling: dynamics, homeostasis and remodelling. Nat Rev Mol Cell Biol 4:517–529
357. Ishida A, Kameshita I, Fujisawa H (1998) A novel protein phosphatase that dephosphorylates and regulates Ca^{2+}/calmodulin-dependent protein kinase II. J Biol Chem 273:1904–1910
358. Abrams TW (2009) Cyclic AMP (cAMP) role in learning and memory. In: Squire LR (ed) Encyclopedia of neuroscience. Elsevier, Amsterdam
359. Mellor J, Nicoll RA, Schmitz D (2002) Mediation of hippocampal mossy fiber long-term potentiation by presynaptic I_h channels. Science 295:143–147
360. Levitan IB (1994) Modulation of ion channels by protein phosphorylation and dephosphorylation. Annu Rev Physiol 56:193–212
361. Seino S, Shibasaki T (2005) PKA-dependent and PKA-independent pathways for cAMP-regulated exocytosis. Physiol Rev 85:1303–1342
362. Wilson G, Garthwaite J (2009) Nitric oxide. In: Squire LR (ed) Encyclopedia of neuroscience. Elsevier, Amsterdam
363. Domek-Lopacinska K, Strosznajdner JB (2005) Cyclic GMP metabolism and its role in brain physiology. J Physiol Pharmacol 56(Supplement 2):15–34
364. Mochly-Rosen D (1995) Localization of protein kinases by anchoring proteins: a theme in signal transduction. Science 268:247–251
365. Coffer PJ, Jin J, Woodgett JR (1998) Protein kinase B (c-Akt): a multifunctional mediator of phosphatidylinositol 3-kinase activation. Biochem J 335:1–13
366. Nave BT, Ouwens DM, Withers DJ, Alessi DR, Shepherd PR (1999) Mammalian target of rapamycin is a direct target for protein kinase B: identification of a convergence point for opposing effects of insulin and amino-acid deficiency on protein translation. Biochem J 344:427–431
367. Hay N, Sonenberg N (2004) Upstream and downstream of mTOR. Genes Dev 18:1926–1945

368. Proud CG, Denton RM (1997) Molecular mechanisms for the control of translation by insulin. Biochem J 328:329–341
369. Jacinto E, Facchinetti V, Liu D, Soto N, Wei S, Jung SY, Huang Q, Qin J, Su B (2006) SIN1/MIP1 maintains rictor-mTOR complex integrity and regulates Akt phosphorylation and substrate specificity. Cell 127:125–137
370. Zinzalla V, Stracka D, Oppliger W, Hall MN (2011) Activation of mTORC2 by association with the ribosome. Cell 144:757–768
371. Hou L, Klann E (2004) Activation of the phosphoinositide 3-kinase-akt-mammalian target of rapamycin signaling pathway is required for metabotropic glutamate receptor-dependent long-term depression. J Neurosci 24:6352–6361
372. Crabtree GR (2001) Calcium, calcineurin, and the control of transcription. J Biol Chem 276:2313–2316
373. Rusnak F, Mertz P (2000) Calcineurin: form and function. Physiol Rev 80:1483–1521
374. Wu G-Y, Deisseroth K, Tsien RW (2001) Activity-dependent CREB phosphorylation: convergence of a fast, sensitive calmodulin kinase pathway and a slow, less sensitive mitogen-activated protein kinase pathway. Proc Natl Acad Sci USA 98:2808–2813
375. Scheper GC, Proud CG (2002) Does phosphorylation of the cap-binding protein eIF4E play a role in translation initiation? Eur J Biochem 269:5350–5359
376. Sweatt JD (2004) Mitogen-activated protein kinases in synaptic plasticity and memory. Curr Opin Neurobiol 14:311–317
377. Filippa N, Sable CL, Filloux C, Hemmings B, Van Obberghen E (1999) Mechanism of protein kinase B activation by cyclic AMP-dependent protein kinase. Mol Cell Biol 19:4989–5000
378. Schmitt JM, Wayman GA, Nozaki N, Soderling TR (2004) Calcium activation of ERK mediated by calmodulin kinase I. J Biol Chem 279:24064–24072
379. Vossler MR, Yao H, York RD, Pan M-G, Rim CS, Stork PJS (1997) cAMP activates MAP kinase and Elk-1 through a B-raf- and Rap1-dependent pathway. Cell 89:73–82
380. Mischak H, Seitz T, Janosch P, Eulitz M, Steen H, Schellerer M, Philipp A, Kolch W (1996) Negative regulation of Raf-1 by phosphorylation of serine 621. Mol Cell Biol 16:5409–5418
381. Ota KT, Monsey MS, Wu MS, Young GJ, Schafe GE (2010) Synaptic plasticity and NO-cGMP-PKG signaling coordinately regulate ERK-driven gene expression in the lateral amygdala and in the auditory thalamus following Pavlovian fear conditioning. Learn Mem 17:221–235
382. Manning G, Whyte DB, Martinez R, Hunter T, Sudarsanam S (2002) The protein kinase complement of the human genome. Science 298:1912–1934
383. Weng G, Bhalla US, Iyengar R (1999) Complexity in biological signaling systems. Science 284:92–96
384. Shabb JB (2001) Physiological substrates of cAMP-dependent protein kinase. Chem Rev 101:2381–2411
385. Goetz T, Wulff P, Wisden W (2009). $GABA_A$ receptors: molecular biology, cell biology, and pharmacology. In: Squire LR (ed) Encyclopedia of neuroscience. Elsevier, Amsterdam
386. Reynolds DS, Rosahl TW, Cirone J, O'Meara GF, Haythornthwaite A, Newman RJ, Myers J, Sur C, Howell O, Rutter AR, Atack J, Macaulay AJ, Hadingham KL, Hutson PH, Belelli D, Lambert JJ, Dawson GR, McKernan R, Whiting PJ, Keith A, Wafford KA (2003) Sedation and anesthesia mediated by distinct GABAA receptor isoforms. J Neurosci 23:8608–8617
387. Cadd G, McKnight GS (1989) Distinct patterns of cAMP-dependent protein kinase gene expression in mouse brain. Neuron 3:71–79
388. Skalhegg BS, Tasken K (1997) Specificity in the cAMP/PKA signaling pathway. Differential expression, regulation, and subcellular localization of subunits of PKA. Front Biosci 2:D331–D342
389. Pidoux G, Tasken K (2010) Specificity and spatial dynamics of protein kinase A signaling organized by A-kinase-anchoring proteins. J Mol Endocrinol 44:271–284

390. Feliciello A, Gottesman ME, Enrico V, Avvedimento EV (2001) The biological functions of A-kinase anchor proteins. J Mol Biol 308:99–114
391. Morrison DK, Davis RJ (2003) Regulation of MAP kinase signaling modules by scaffold proteins in mammals. Annu Rev Cell Dev Biol 19:91–118
392. Sato M, Blumer JB, Simon V, Lanier SM (2006) Accessory proteins for G proteins: partners in signaling. Annu Rev Pharmacol Toxicol 46:151–187
393. Feng W, Tu J, Yang T, Vernon PS, Allen PD, Worley PF, Pessah IN (2002) Homer regulates gain of RyR1. J Biol Chem 277:44722–44730
394. Korade Z, Kenworthy A (2008) Lipid rafts, cholesterol, and the brain. Neuropharmacology 55:1265–1273
395. Hering H, Lin C-C, Sheng M (2003) Lipid rafts in the maintenance of synapses, dendritic spines, and surface AMPA receptor stability. J Neurosci 23:3262–3271
396. Suzuki T, Du F, Tian Q-B, Zhang J, Endo S (2008) Ca^{++}/calmodulin-dependent protein kinase IIa clusters are associated with stable lipid rafts and their formation traps PSD-95. J Neurochem 104:596–619
397. Simons K, Toomre D (2003) Lipid rafts and signal transduction. Nat Rev Mol Cell Biol 1:31–41
398. Cooper DMF (2005) Compartmentalization of adenylate cyclase and cAMP signalling. Biochem Soc Trans 33:1319–1322
399. Helmchen F, Svoboda K, Denk W, Tank DW (1999) In vivo dendritic calcium dynamics in deep-layer cortical pyramidal neurons. Nat Neurosci 2:989–996
400. Polsky A, Mel BW, Schiller J (2005) Computational subunits in thin dendrites of pyramidal cells. Nat Neurosci 7:621–627
401. Williams SR, Stuart GJ (2000) Site independence of EPSP time course is mediated by dendritic Ih in neocortical pyramidal neurons. J Neurophysiol 83:3177–3182
402. Magee JC (1999) Dendritic I_h normalizes temporal summation in hippocampal CA1 neurons. Nat Neurosci 2:508–514
403. Kollo M, Holderith NB, Nusser Z (2006) Novel subcellular distribution pattern of A-type K^+ channels on neuronal surface. J Neurosci 26:2684–2691
404. Kampa BM, Stuart GJ (2006) Calcium spikes in basal dendrites of layer 5 pyramidal neurons during action potential bursts. J Neurosci 26:7424–7432
405. Trimmer JS, Rhodes KJ (2004) Localization of voltage-gated ion channels in mammalian brain. Annu Rev Physiol 66:477–519
406. Mckeown L, Swanton L, Robinson P, Jones OT (2008) Surface expression and distribution of voltage-gated potassium channels in neurons. Mol Membr Biol 25:332–343
407. Migliore M, Shepherd GM (2002) Emerging rules for the distributions of active dendritic conductances. Nat Rev Neurosci 3:362–370
408. Markram H, Toleo-Rodriguez M, Wang Y, Gupta A, Silberberg G, Wu C (2004) Interneurons of the neocortical inhibitory system. Nat Rev Neurosci 5:793–807
409. Gulyas AI, Miles R, Hajos N, Freund TF (1993) Precision and variability in postsynaptic target selection of inhibitory cells in the hippocampal CA3 region. Eur J Neurosci 5:1729–1751; Markram H, Toledo-Rodriguez M, Wang Y, Gupta A, Silberberg G, Wu C (2004) Interneurons of the neocortical inhibitory system. Nat Rev Neurosci 5:793–807
410. Howard A, Tamas G, Soltesz I (2005) Lighting the chandelier: new vistas for axo-axonic cells. Trends Neurosci 28:310–316
411. Somogyi P, Tamas G, Lujan R, Buhl EH (1998) Salient features of synaptic organisation in the cerebral cortex. Brain Res Rev 26:113–135
412. Wang Y, Toledo-Rodriguez M, Gupta A, Wu C, Silberberg G, Luo J, Markram H (2004) Anatomical, physiological and molecular properties of Martinotti cells in the somatosensory cortex of the juvenile rat. J Physiol 561:65–90
413. Chen X, Johnston D (2005) Constitutively active G-protein-gated inwardly rectifying K^+ channels in dendrites of hippocampal CA1 pyramidal neurons. J Neurosci 25:3787–3792
414. Scholz KP, Miller RJ (1991) $GABA_B$ receptor-mediated inhibition of Ca^{2+} currents and synaptic transmission in cultured rat hippocampal neurones. J Physiol 444:669–686

415. Perez-Garci E, Gassmann M, Bettler B, Larkum ME (2006) The $GABA_B$1b isoform mediates long-lasting inhibition of dendritic Ca^{2+} spikes in layer 5 somatosensory pyramidal neurons. Neuron 50:603–616
416. Hestrin S, Galarreta M (2005) Electrical synapses define networks of neocortical GABAergic neurons. Trends Neurosci 28:304–309
417. Villalobos C, Beique J-C, Gingrich JA, Andrade R (2005) Serotonergic regulation of calcium-activated potassium currents in rodent prefrontal cortex. Eur J Neurosci 22:1120–1126
418. Zhang Z-w, Arsenault D (2005) Gain modulation by serotonin in pyramidal neurones of the rat prefrontal cortex. J Physiol 566:379–394
419. Freund TF, Buzsaki G (1996) Interneurons of the hippocampus. Hippocampus 6:347–470
420. Giocomo LM, Hasselmo ME (2007) Neuromodulation by glutamate and acetylcholine can change circuit dynamics by regulating the relative influence of afferent input and excitatory feedback. Mol Neurobiol 36:184–200
421. Umbriaco D, Garcia S, Beaulieu C, Descarries L (1995) Relational features of acetylcholine, noradrenaline, serotonin and GABA axon terminals in the stratum radiatum of adult rat hippocampus (CA1). Hippocampus 5:605–620
422. Alkondon M, Pereira EFR, Edson X, Albuquerque EX (1998) α-Bungarotoxin- and methyllycaconitine-sensitive nicotinic receptors mediate fast synaptic transmission in interneurons of rat hippocampal slices. Brain Res 810:257–263
423. Khiroug L, Giniatullin R, Klein RC, Fayuk D, Yakel JL (2003) Functional mapping and Ca^{2+} regulation of nicotinic acetylcholine receptor channels in rat hippocampal CA1 neurons. J Neurosci 23:9024–9031
424. Gulledge AT, Stuart GJ (2005) Cholinergic inhibition of neocortical pyramidal neurons. J Neurosci 25:10308–10320
425. Smiley JF, Morrell F, Mesulam MM (1997) Cholinergic synapses in human cerebral cortex: an ultrastructural study in serial sections. Exp Neurol 144:361–368
426. Frotscher M, Leranth C (1985) Cholinergic innervation of the rat hippocampus as revealed by choline acetyltransferase immunocytochemistry: a combined light and electron microscopic study. J Comp Neurol 239:237–246
427. Mrzljak L, Levey AI, Goldman-Rakic PS (1993) Association of ml and m2 muscarinic receptor proteins with asymmetric synapses in the primate cerebral cortex: morphological evidence for cholinergic modulation of excitatory neurotransmission. Proc Natl Acad Sci USA 90:5194–5198
428. Gulyás AI, Hájos N, Freund TF (1996) Interneurons containing calretinin are specialized to control other interneurons in the rat hippocampus. J Neurosci 16:3397–3411; Barinka F, Druga R (2010) Calretinin expression in the mammalian neocortex: a review. Physiol Rev 59:665–677
429. Ludkiewicz B, Wójcik S, Spodnik E, Domaradzka-Pytel B, Mory J (2000) Cholinergic endings on various neurones containing calcium binding proteins and glutamic acid decarboxylase in the hippocampus of the rat. Folia Morphol 59:263–270
430. Lawrence JJ (2008) Cholinergic control of GABA release: emerging parallels between neocortex and hippocampus. Trends Neurosci 31:317–327
431. Medin T, Owe SG, Rinholm JE, Larsson M, Sagvolden T, Storm-Mathisen J, Bergersen LH (2011) Dopamine D5 receptors are localized at asymmetric synapses in the rat hippocampus. Neuroscience 192:164–171
432. Smith WB, Starck SR, Roberts RW, Schuman EM (2005) Dopaminergic stimulation of local protein synthesis enhances surface expression of GluR1 and synaptic transmission in hippocampal neurons. Neuron 45:765–779
433. Lavin A, Grace AA (2001) Stimulation of D1-type dopamine receptors enhances excitability in prefrontal cortical pyramidal neurons in a state-dependent manner. Neuroscience 104:335–346
434. Tseng KY, O'Donnell P (2004) Dopamine–glutamate interactions controlling prefrontal cortical pyramidal cell excitability involve multiple signaling mechanisms. J Neurosci 24:5131–5139

435. Bragin A, Jando G, Nadasdy Z, Hetke J, Wise K, Buzsaki G (1995) Gamma (40–100 Hz) oscillation in the hippocampus of the behaving rat. J Neurosci 15:47–60
436. Soltesz I, Deschenes M (1993) Low- and high-frequency membrane potential oscillations during theta activity in CA1 and CA3 pyramidal neurons of the rat hippocampus under ketamine-xylazine anesthesia. J Neurophysiol 70:97–116
437. Somogyi P, Klausberger T (2005) Defined types of cortical interneurone structure space and spike timing in the hippocampus. J Physiol 562:9–26
438. Klausberger T, Magill PJ, Marton LF, Roberts JDB, Cobden PM, Buzsaki G, Somogyi P (2003) Brain-state- and cell-type-specific firing of hippocampal interneurons in vivo. Nature 421:844–848
439. Rodriguez-Molina VM, Aertsen A, Heck DH (2007) Spike timing and reliability in cortical pyramidal neurons: effects of EPSC kinetics, input synchronization and background noise on spike timing. PLoS One 2(3):e319. doi:10.1371/journal.pone.0000319
440. Derkach V, Barria A, Soderling TR (1999) Ca^{2+}/calmodulin-kinase II enhances channel conductance of alpha-amino-3-hydroxy-5-methyl-4-isoxazolepropionate type glutamate receptors. Proc Natl Acad Sci USA 96:3269–3274
441. Archibald K, Perry MJ, Molnar E, Henley JM (1998) Surface expression and metabolic half-life of AMPA receptors in cultured rat cerebellar granule cells. Neuropharmacology 37:1345–1353
442. Man H-Y, Lin JW, Ju WH, Ahmadian G, Liu L, Becker LE, Sheng M, Wang YT (2000) Regulation of AMPA receptor-mediated synaptic transmission by clathrin-dependent receptor internalization. Neuron 25:649–662
443. Turrigiano GG (2000) AMPA receptors unbound: membrane cycling and synaptic plasticity. Neuron 26:5–8
444. Miyashiro K, Dichter M, Eberwine J (1994) On the nature and differential distribution of mRNAs in hippocampal neurites: implications for neuronal functioning. Proc Natl Acad Sci 91:10800–10804
445. Grooms SY, Noh K-M, Regis R, Bassell GJ, Bryan MK, Carroll RC, Zukin RS (2006) Activity bidirectionally regulates AMPA receptor mRNA abundance in dendrites of hippocampal neurons. J Neurosci 26:8339–8351
446. Correia SS, Bassani S, Brown TC, Lise M-F, Backos DS, El-Husseini A, Passafaro M, Esteban JA (2008) Motor protein–dependent transport of AMPA receptors into spines during long-term potentiation. Nat Neurosci 11:457–466
447. Sudhof TC, Rothman JE (2009) Membrane fusion: grappling with SNARE and SM proteins. Science 323:474–477
448. Ahmad M, Polepalli JS, Goswami D, Yang X, Kaeser-Woo YJ, Sidhof TC, Malenka RC (2012) Postsynaptic synaptin controls AMPA receptor exocytosis during LTP. Neuron 73:260–267
449. Yudowski GA, Puthenveedu MA, Leonoudakis D, Panicker S, Thorn KS, Beattie EC, von Zastrow M (2007) Real-time imaging of discrete exocytic events mediating surface delivery of AMPA receptors. J Neurosci 27:11112–11121
450. Wang Z, Edwards JG, Riley ND, Provance W Jr, Karcher R, Xiang-dong Li, Davison IG, Ikebe M, Mercer JA, Kauer JA, Ehlers MD (2008) Myosin Vb mobilizes recycling endosomes and AMPA receptors for postsynaptic plasticity. Cell 135:535–548
451. Toa-Cheng J-H, Crocker VT, Winters CA, Azzam R, Chludzinski J, Reese TS (2011) Trafficking of AMPA receptors at plasma membranes of hippocampal neurons. J Neurosci 31:4834–4843
452. Gerges NZ, Backos DS, Rupasinghe CM, Spaller MR, Esteban JA (2006) Dual role of the exocyst in AMPA receptor targeting and insertion into the postsynaptic membrane. EMBO J 25:1623–1634
453. Jaskolski F, Mayo-Martin B, Jane D, Henley JM (2009) Dynamin-dependent membrane drift recruits AMPA receptors to dendritic spines. J Biol Chem 284:12491–12503
454. Braithwaite SP, Xia H, Malenka RC (2002) Differential roles for NSF and GRIP/ABP in AMPA receptor cycling. Proc Natl Acad Sci USA 99:7096–7101

455. Chowdhury S, Shepherd JD, Okuno H, Lyford G, Petralia RS, Plath N, Kuhl D, Huganir RL, Worley PF (2006) Arc/Arg3.1 interacts with the endocytic machinery to regulate AMPA receptor trafficking. Neuron 52:445–459
456. Bosch M, Hayashi Y (2012) Structural plasticity of dendritic spines. Curr Opin Neurobiol 22:383–388
457. Honkura N, Matsuzaki M, Noguchi J, Ellis-Davies GC, Kasai H (2008) The subspine organization of actin fibers regulates the structure and plasticity of dendritic spines. Neuron 57:719–729
458. Okamato KI, Nagai T, Miyawaki A, Hayashi Y (2004) Rapid and persistent modulation of actin dynamics regulates post-synaptic reorganization underlying bi-directional plasticity. Nat Neurosci 7:1104–1112
459. Kardos R, Pozsonyi K, Nevalainen E, Lappalainen P, Nyitrai M, Hild G (2009) The effects of ADF/cofilin and profilin on the conformation of the ATP-binding cleft of monomeric actin. Biophys J 96:2335–2343
460. Hall A (1994) Small GTP-binding proteins and the regulation of the actin cytoskeleton. Annu Rev Cell Biol 10:31–54
461. Ponimaskin E, Voyno-Yasenetskaya T, Richter DW, Schachner M, Dityatev A (2007) Morphogenic signaling in neurons via neurotransmitter receptors and small GTPases. Mol Neurobiol 35:278–287
462. Watanabe N, Madaule P, Reid T, Ishizaki T, Watanabe G, Kakizuka A, Saito Y, Nakao K, Jockusch BM, Narumiya S (1997) p140mDia, a mammalian homolog of Drosophila diaphanous, is a target protein for Rho small GTPase and is a ligand for profilin. EMBO J 16:3044–3056
463. Lin Y-C, Koleske AJ (2010) Mechanisms of synapse and dendrite maintenance and their disruption in psychiatric and neurodegenerative disorders. Annu Rev Neurosci 33:349–378
464. Morgan SL, Teyler TJ (2001) Electrical stimuli patterned after the theta-rhythm induced multiple forms of LTP. J Neurophysiol 86:1289–1296
465. Raymond CR, Redman SJ (2006) Spatial segregation of neuronal calcium signals encodes different forms of LTP in rat hippocampus. J Physiol 570:97–111
466. Kauderer BS, Kandel ER (2000) Capture of a protein synthesis-dependent component of long-term depression. Proc Natl Acad Sci USA 97:13342–13347
467. Huber KM, Kayser MS, Bear MF (2000) Role for rapid dendritic protein synthesis in hippocampal mGluR-dependent LTD. Science 288:1254–1257
468. Lindecke A, Korte M, Zagrebelsky M, Horejschi V, Elvers M, Darius Widera D, Prullage M, Pfeiffer J, Kaltschmidt B, Kaltschmidt C (2006) Long-term depression activates transcription of immediate early transcription factor genes: involvement of serum response factor/Elk-1. Eur J Neurosci 24:555–563
469. Colbran RJ (2004) Protein phosphatases and calcium/calmodulin-dependent protein kinase II-dependent synaptic plasticity. J Neurosci 24:8404–8409
470. Chen X, Winters C, Azzam R, Li X, Galbraith JA, Richard D, Leapman RD, Reese TS (2008) Organization of the core structure of the postsynaptic density. Proc Natl Acad Sci USA 105:4453–4458
471. Serge A, Lawrence L, Hemar A, Choquet D (2002) Receptor activation and homer differentially control the lateral mobility of metabotropic glutamate receptor 5 in the neuronal membrane. J Neurosci 22:3910–3920
472. Codazzi F, Di Cesare A, Chiulli N, Albanese A, Meyer T, Zacchetti D, Grohovaz F (2006) Synergistic control of protein kinase Cγ activity by ionotropic and metabotropic glutamate receptor inputs in hippocampal neurons. J Neurosci 26:3404–3411
473. Krementsov DN, Krementsova EB, Trybus KM (2004) Myosin V: regulation by calcium, calmodulin and the tail domain. J Cell Biol 164:877–886
474. Man H-Y, Wang Q, Lu W-Y, Ju W, Ahmadian G, Liu L, D'Souza S, Wong TP, Taghibiglou C, Lu J, Becker LE, Pei L, Liu F, Wymann MP, MacDonald JF, Wang YT (2003) Activation of PI3-kinase is required for AMPA receptor insertion during LTP of mEPSCs in cultured hippocampal neurons. Neuron 38:611–624

475. Tao-Cheng J-H, Crocker VT, Winters CA, Azzam R, Chludzinski J, Reese TS (2011) Trafficking of AMPA receptors at plasma membranes of hippocampal neurons. J Neurosci 31:4834–4843
476. Tanaka J-i, Horiike Y, Matsuzaki M, Miyazaki T, Ellis-Davies GCR, Kasai H (2008) Protein synthesis and neurotrophin-dependent structural plasticity of single dendritic spines. Science 319:1683–1687
477. Ostroff LE, Fiala JC, Allwardt B, Harris KM (2002) Polyribosomes redistribute from dendritic shafts into spines with enlarged synapses during LTP in developing rat hippocampal slices. Neuron 35:535–545
478. Fortin DA, Davare MA, Srivastava T, Brady JD, Nygaard S, Derkach VA, Soderling TR (2010) Long-term potentiation-dependent spine enlargement requires synaptic Ca^{2+}-permeable AMPA receptors recruited by CaM-kinase I. J Neurosci 30:11565–11575
479. Ackermann M, Matus A (2003) Activity-induced targeting of profilin and stabilization of dendritic spine morphology. Nat Neurosci 6:1194–1200
480. Leslie JH, Nedivi E (2011) Activity-regulated genes as mediators of neural circuit plasticity. Prog Neurobiol 94:233–237; Adams JP, Dudek SM (2005) Late-phase long-term potentiation: getting to the nucleus. Nat Rev Neurosci 6:737–743
481. Silva AJ, Kogan JH, Frankland PW, Kida S (1998) CREB and memory. Annu Rev Neurosci 21:127–148
482. Sanchez-Huertas C, Rico B (2011) CREB-dependent regulation of GAD65 transcription by BDNF/TrkB in cortical interneurons. Cereb Cortex 21:777–788
483. Hong SJ, Li H, Becker KG, Dawson VL, Dawson TM (2004) Identification and analysis of plasticity-induced late-response genes. Proc Natl Acad Sci USA 101:2145–2150
484. Patterson SL, Grover LM, Schwartzkroin PA, Bothwell M (1992) Neurotrophin expression in rat hippocampal slices: a stimulus paradigm inducing LTP in CA1 evokes increases in BDNF and NT-3 mRNAs. Neuron 9:1081–1088
485. Xia Z, Dudek H, Miranti CK, Greenberg ME (1996) Calcium influx via the NMDA receptor induces immediate early gene transcription by a MAP kinase/ERK-dependent mechanism. J Neurosci 16:5425–5436
486. Dolmetsch RE, Pajvani U, Fife K, Spotts JM, Greenberg ME (2001) Signaling to the nucleus by an L-type calcium channel-calmodulin complex through the MAP kinase pathway. Science 294:333–339
487. Fernandez de Sevilla D, Buno W (2010) The muscarinic long-term enhancement of NMDA and AMPA receptor-mediated transmission at Schaffer collateral synapses develop through different intracellular mechanisms. J Neurosci 30:11032–11042
488. Albrecht C, von der Kammer H, Mayhaus M, Klaudiny J, Schweizer M, Nitsch RM (2000) Muscarinic acetylcholine receptors induce the expression of the immediate early growth regulatory gene CYR61. J Biol Chem 275:28929–28936
489. Leirch B, Szostek A, Lin R, Shevtsova O (2009) Subcellular distribution of L-type calcium channel subtypes in rat hippocampal neurons. Neuroscience 164:641–657
490. Hardingham GE, Arnold FJL, Bading H (2001) A calcium microdomain near NMDA receptors: on switch for ERK-dependent synapse-to-nucleus communication. Nat Neurosci 4:565–566
491. Cardinaux J-R, Notis JC, Zhang Q, Vo N, Craig JC, Fass DM, Brennan RG, Goodman RH (2000) Recruitment of CREB binding protein is sufficient for CREB-mediated gene activation. Mol Cell Biol 20:1546–1552
492. Anagnostaras SG, Murphy GG, Hamilton SE, Mitchell SL, Rahnama NP, Nathanson NM, Silva AJ (2003) Selective cognitive dysfunction in acetylcholine M1 muscarinic receptor mutant mice. Nat Neurosci 6:51–58
493. Knoll B (2011) Serum response factor mediated gene activity in physiological and pathological processes of neuronal motility. Front Mol Neurosci 4:49
494. Young JZ, Nguyen PV (2005) Homosynaptic and heterosynaptic inhibition of synaptic tagging and capture of long-term potentiation by previous synaptic activity. J Neurosci 25:7221–7231

495. Vickers CA, Dickson KS, Wyllie DJA (2005) Induction and maintenance of late-phase long-term potentiation in isolated dendrites of rat hippocampal CA1 pyramidal neurones. J Physiol 568:803–813
496. Malenka RC (1994) Synaptic plasticity in the hippocampus: LTP and LTD. Cell 78:535–538
497. Feldman DE (2012) The spike-timing dependence of plasticity. Neuron 75:556–571
498. McNaughton BL, Douglas RM, Goddard GV (1978) Synaptic enhancement in fascia dentata: cooperativity among coactive afferents. Brain Res 157:277–293
499. Sjöström PJ, Turrigiano GG, Nelson SB (2001) Rate, timing, and cooperativity jointly determine cortical synaptic plasticity. Neuron 32:1149–1164
500. Hirano T (1990) Depression and potentiation of the synaptic transmission between a granule cell and a Purkinje cell in rat cerebellar culture. Neurosci Lett 119:141–144
501. Hoffman DA, Magee JC, Colbert CM, Johnson D (1997) K^+ channel regulation of signal propagation in dendrites of hippocampal pyramidal neurons. Nature 387:869–875
502. Stuart GJ, Hausser M (2001) Dendritic coincidence detection of EPSPs and action potentials. Nat Neurosci 4:63–71
503. Delgado JY, Coba M, Anderson CNG, Thompson KR, Gray EE, Heusner CL, Martin KC, Grant SGN, O'Dell TJ (2007) NMDA receptor activation dephosphorylates AMPA receptor glutamate receptor 1 subunits at threonine 840. J Neurosci 27:13210–13221
504. Staubli U, Otaky N (1994) Serotonin controls the magnitude of LTP induced by theta bursts via an action on NMDA-receptor-mediated responses. Brain Res 643:1–16; Kulla A, Manahan-Vaughan D (2002) Modulation by serotonin 5-HT_4 receptors of long-term potentiation and depotentiation in the dentate gyrus of freely moving rats. Cereb Cortex 12:150–162
505. Stanton PK, Sarvey JM (1985) Depletion of norepinephrine, but not serotonin, reduces long-term potentiation in the dentate gyrus of rat hippocampal slices. J Neurosci 5:2169–2176; Stanton PK, Sarvey JM (1987) Norepinephrine regulates long-term potentiation of both the population spike and dendritic EPSP in hippocampal dentate gyrus. Brain Res Bull 18:115–119
506. Myhrer T (2003) Neurotransmitter systems involved in learning and memory in the rat: a meta-analysis based on studies of four behavioral tasks. Brain Res Rev 41:268–287
507. Szabo A, Somogyi J, Cauli B, Lambolez B, Somogyi P, Lamsa KP (2012) Calcium-permeable AMPA receptors provide a common mechanism for LTP in glutamatergic synapses of distinct hippocampal interneuron types. J Neurosci 32:6511–6516
508. Bannai H, Levi S, Schweizer C, Inoue T, Launey T, Racine V, Sibarita J-B, Mikoshiba K, Triller A (2009) Activity-dependent tuning of inhibitory neurotransmission based on $GABA_AR$ diffusion dynamics. Neuron 62:670–682
509. Holmgren CD, Zilberter Y (2001) Coincident spiking activity induced long-term changes in inhibition of neocortical pyramidal cells. J Neurosci 21:8270–8277
510. Gaiarsa J-L, Caillard O, Ben-Ari Y (2002) Long-term plasticity at GABAergic and glycinergic synapses: mechanisms and functional significance. Trends Neurosci 25:564–570
511. Kullman DM, Lamsa KP (2007) Long-term synaptic plasticity in hippocampal interneurons. Nat Rev Neurosci 8:687–699
512. Caporale N, Dan Y (2008) Spike timing-dependent plasticity: a Hebbian learning rule. Annu Rev Neurosci 31:25–46
513. Sjostrom PJ, Turrigiano GG, Nelson SB (2003) Neocortical LTD via coincident activation of presynaptic NMDA and cannabinoid receptors. Neuron 39:641–654
514. Chen X, Johnson D (2006) Voltage-gated ion channels in dendrites of hippocampal pyramidal neurons. Eur J Physiol 453:397–401
515. Zhang W, Linden DJ (2003) The other side of the engram: experience-driven changes in neuronal intrinsic excitability. Nat Rev Neurosci 4:885–900
516. Rosenkranz JA, Frick A, Johnston D (2009) Kinase-dependent modification of dendritic excitability after long-term potentiation. J Physiol 587:115–125
517. Sjostrom PJ, Rancz EA, Roth A, Hausser M (2008) Dendritic excitability and synaptic plasticity. Physiol Rev 88:769–840

518. Frick A, Johnston D (2005) Plasticity of dendritic excitability. J Neurobiol 64:100–115
519. Zhang J, Shapiro MS (2012) Activity-dependent transcriptional regulation of M-Type (Kv7) K^+ channels by AKAP79/150-mediated NFAT actions. Neuron 76:1133–1146
520. Shah MM, Hammond RS, Hoffman DA (2010) Dendritic ion channel trafficking and plasticity. Trends Neurosci 33:307–316
521. Barta P, Dazzan P (2003) Hemispheric surface area: sex, laterality and age effects. Cereb Cortex 13(4):364–370
522. Peters A, Morrison JH, Rosene DL, Hyman BT (1998) Are neurons lost from the primate cerebral cortex during normal aging? Cereb Cortex 8:295–300
523. Gross CG, Rocha-Miranda CE, Bender DB (1972) Visual properties of neurons in inferotemporal cortex of the macaque. J Neurophysiol 35(96–111):1972
524. Ito M, Tamura H, Fujita I, Tanaka K (1995) Size and position invariance of neuronal responses in monkey inferotemporal cortex. J Neurophysiol 73:218–226
525. Hubel DH (1995) Eye, brain and vision. Scientific American Library/Scientific American Books, New York
526. Tanaka K (2003) Columns for complex visual object features in the inferotemporal cortex: clustering of cells with similar but slightly different stimulus selectivities. Cereb Cortex 13:90–99
527. Zhou F-M, Hablitz JJ (1996) Morphological properties of intracellularly labeled layer I neurons in rat neocortex. J Comp Neurol 376:198–213
528. Glickfeld LL, Roberts JD, Somogyi P, Scanziani M (2009) Interneurons hyperpolarise pyramidal cells along their entire somatodendritic axis. Nat Neurosci 12:21–23
529. Woodruff A, Xu Q, Anderson SA, Yuste R (2009) Depolarising effect of neocortical chandelier neurons. Front Neural Circuits 3:15
530. Khirug S, Yamada J, Afzalov R, Voipio J, Khiroug L, Kaila K (2008) GABAergic depolarisation of the axon initial segment in cortical principal neurons is caused by the Na–K–2Cl cotransporter NKCC1. J Neurosci 28:4635–4639
531. Tamas G, Buhl EH, Lorinz A, Somogyi P (2000) Proximally targeted GABAergic synapses and gap junctions synchronize cortical interneurons. Nat Neurosci 3:366–371
532. Komatsu Y (1996) $GABA_B$ receptors, monoamine receptors, and postsynaptic inositol trisphosphate-induced Ca^{2+} release are involved in the induction of long-term potentiation at visual cortical inhibitory synapses. J Neurosci 16:6342–6352
533. Hilgetag C-C, Burns GAPC, O'Neill MA, Scannell JW, Malcolm P, Young MP (2000) Anatomical connectivity defines the organization of clusters of cortical areas in the macaque and the cat. Philos Trans R Soc Lond B 355:91–110
534. Risold PY, Thompson RH, Swanson LW (1997) The structural organization of connections between hypothalamus and cerebral cortex. Brain Res Rev 24:197–254
535. Sah P, Faber ESL, De Armentia ML, Power J (2003) The amygdaloid complex: anatomy and physiology. Physiol Rev 83:803–834
536. Bacon SJ, Headlam AJN, Gabbott PLA, Smith AD (1996) Amygdala input to medial prefrontal cortex (mPFC) in the rat: a light and electron microscope study. Brain Res 720:211–219
537. Willins DL, Deutch AY, Roth BL (1997) Serotonin 5-HT2A receptors are expressed on pyramidal cells and interneurons in the rat cortex. Synapse 27:79–82
538. Wilson MA, Molliver ME (1991) The organization of serotonergic projections to cerebral cortex in primates: regional distribution of axon terminals. Neuroscience 44:537–553
539. De Almeida J, Palacios JM, Mengod G (2008) Distribution of 5-HT and DA receptors in primate prefrontal cortex: implications for pathophysiology and treatment. Prog Brain Res 172:101–115
540. Houser CR, Crawford GD, Salvaterra PM, Vaugn JE (1985) Immunocytochemical localization of choline acetyltransferase in rat cerebral cortex: a study of cholinergic neurons and synapses. J Comp Neurol 234:17–34
541. Thomson AM, Bannister AP (2003) Interlaminar connections in the neocortex. Cereb Cortex 13:5–14

542. Huang ZJ, Di Cristo G, Ango F (2007) Development of GABA innervation in the cerebral and cerebellar cortices. Nat Rev Neurosci 8:673–686
543. Dantzker JL, Callaway EM (2000) Laminar sources of synaptic input to cortical inhibitory interneurons and pyramidal neurons. Nat Neurosci 3:701–707
544. Beaulieu C, Colonnier M (1989) Number of neurons in individual laminae of areas 3B, 4γ, and 6aα of the cat cerebral cortex: a comparison with major visual areas. J Comp Neurol 279:228–234
545. Lubke J, Feldmeyer D (2007) Excitatory signal flow and connectivity in a cortical column: focus on barrel cortex. Brain Struct Funct 212:3–17
546. Guillery RW, Feig SL, Lozsadi DA (1998) Paying attention to the thalamic reticuar nucleus. Trends Neurosci 21:28–32
547. Amunts K, Schleicher A, Ditterich A, Zilles K (2003) Broca's region: cytoarchitectonic asymmetry and developmental changes. J Comp Neurol 465:72–89
548. Knecht S, Drager B, Deppe M, Bobe L, Lohmann H, Floel A, Ringelstein E-B, Henningsen H (2000) Handedness and hemispheric language dominance in healthy humans. Brain 123:2512–2518
549. Joseph R (1988) The right cerebral hemisphere: emotion, music, visual-spatial skills, body-image, dreams, and awareness. J Clin Psychol 44:630–673
550. Scheperjans F, Palomero-Gallagher N, Grefkes C, Schleicher A, Zilles K (2005) Transmitter receptors reveal segregation of cortical areas in the human superior parietal cortex: relations to visual and somatosensory regions. Neuroimage 28:362–379
551. Haxby JV, Grady CL, Horwitz B, Ungerleider LG, Mishkin M, Carsons RE, Herscovitch P, Schapiro MB, Rapoport SI (1991) Dissociation of object and spatial visual processing pathways in human extrastriate cortex. Proc Natl Acad Sci USA 88:1621–1625
552. Milner AD, Perrett DI, Johnston RS, Benson PJ, Jordan TR, Heeley DW, Bettucci D, Mortara F, Mutani R, Terazzi E, Davidson DLW (1991) Perception and action in "Visual Form Agnosia". Brain 114:405–428; James TW, Culham JG, Keith Humphrey GK, Milner AD, Goodale MA (2003) Ventral occipital lesions impair object recognition but not object-directed grasping: an fMRI study. Brain 126:2463–2475
553. Jakobson LS, Archibald YM, Carey DP, Goodale MA (1991) A kinematic analysis of reaching and grasping movements in a patient recovering from optic ataxia. Neuropsychologia 29(8):803–809
554. Mikami A, Newsome WT, Wurtz RH (1986) Motion selectivity in Macaque visual cortex. II. Spatiotemporal range of directional interactions in MT and V1. J Neurophysiol 55:1328–1339
555. Richer F, Martinez M, Robert M, Bouvier G, Saint-Hilaire J-M (1993) Stimulation of human somatosensory cortex: tactile and body displacement perceptions in medial region. J Exp Brain Res 93(1):173–176
556. Nudo RJ (1999) Recovery after damage to motor cortical areas. Curr Opin Neurobiol 9:740–747
557. Hari R, Forss N, Avikainen S, Kirveskari E, Salenius S, Rizzolatti G (1998) Activation of human primary motor cortex during action observation: a neuromagnetic study. Proc Natl Acad Sci USA 95:15061–15065
558. Dijkerman HC, de Haan EHF (2007) Somatosensory processes subserving perception and action. Behav Brain Sci 30:189–239
559. Berndt RS, Caramazza A (1980) A redefinition of the syndrome of Broca's aphasia: implications for a neuropsychological model of language. Appl Psycholinguist 1:225–278
560. Dronkers NF, Wilkins DP, Van Valin RD Jr, Redfern BB, Jaeger JJ (2004) Lesion analysis of the brain areas involved in language comprehension. Cognition 92:145–177
561. Binder JR, Frost JA, Hammeke TA, Cox RW, Rao SM, Prieto T (1997) Human brain language areas identified by functional magnetic resonance imaging. J Neurosci 17:353–362
562. Thompson-Schill SL (2003) Neuroimaging studies of semantic memory: inferring "how" from "where". Neuropsychologia 41:280–292
563. Grossman M, Koenig P, Kounios J, McMillan C, Work M, Peachie Moore P (2006) Category-specific effects in semantic memory: category–task interactions suggested by fMRI. Neuroimage 30:1003–1009

564. Vandenberghe R, Price C, Wise R, Josephs O, Frackowiak RS (1996) Functional anatomy of a common semantic system for words and pictures. Nature 383:254–256
565. Lebreton K, Desgranges B, Landeau B, Baron J-C, Eustache F (2001) Visual priming within and across symbolic format using a tachistoscopic picture identification task: a PET study. J Cogn Neurosci 13:670–686
566. Miller EK, Cohen JD (2001) An integrative theory of prefrontal cortex function. Annu Rev Neurosci 24:167–202
567. Badre D, D'Esposito M (2009) Is the rostro-caudal axis of the frontal lobe hierarchical? Nat Rev Neurosci 10:659–669
568. Duncan J, Owen AM (2000) Common regions of the human frontal lobe recruited by diverse cognitive demands. Trends Neurosci 23:475–483
569. Kanwisher N, Stanley D, Harris A (1999) The fusiform face area is selective for faces not animals. Neuroreport 10:183–187
570. De Renzi E (1997) Prosopagnosia. In: Feinberg TE, Farah MJ (eds) Behavioral neurology and neuropsychology. McGraw-Hill, New York, pp 245–255
571. Whiteley AM, Warrington EK (1977) Prosopagnia: a clinical, psychological, and anatomical study of three patients. J Neurol Neurosurg Psychiatry 40:395–403
572. Gloning I, Gloning K, Jellinger K, Quatember R (1970) A case of "prosopagnosia" with necropsy findings. Neuropsychologia 8:199–204
573. Schultz TS, Grelotti DJ, Klin A, Kleinman J, Van der Gaag C, Marois R, Skudlarski P (2003) The role of the fusiform face area in social cognition: implications for the pathobiology of autism. Philos Trans R Soc Lond B 358:415–427
574. Critchley HD, Daly EM, Bullmore ET, Williams SCR, Van Amelsvoort T, Robertson DM, Rowe A, Phillips M, McAlonan G, Howlin P, Murphy DGM (2000) The functional neuroanatomy of social behaviour: changes in cerebral blood flow when people with autistic disorder process facial expressions. Brain 123:2203–2212
575. Cabeza R, Nyberg L (2000) Imaging cognition II: an empirical review of 275 PET and fMRI studies. J Cogn Neurosci 12:1–47
576. Mistry RB, Isaac JTR, Crabtree JW (2008) Two differential frequency-dependent mechanisms regulating tonic firing of thalamic reticular neurons. Eur J Neurosci 27:2643–2656
577. MacDonald KD, Fifkova E, Jones MS, Barth DS (1998) Focal stimulation of the thalamic reticular nucleus induces focal gamma waves in cortex. J Physiol 79:474–477
578. Vertes RP, Albo Z, Viana Di Prisco G (2001) Theta-rhythmically firing neurons in the anterior thalamus: implications for mnemonic functions of Papez's circuit. Neuroscience 104:619–625
579. Canolty RT, Edwards E, Dalal SS, Soltani M, Nagarajan SS, Kirsch HE, Berger MS, Barbaro NM, Knight RT (2006) High gamma power is phase-locked to theta oscillations in human neocortex. Science 313:1626–1628
580. Buzsaki G, Bickford RG, Ponomareff G, Thal LJ, Mandel R, Gage FH (1988) Nucleus basalis and thalamic control of neocortical activity in the freely moving rat. J Neurosci 8:4007–4026
581. Timofeev I, Bazhenov M (2005) Mechanisms and biological role of thalamocortical oscillations. In: Columbus F (ed) Trends in chronobiology research. Nova Science Publishers, New York
582. Arcelli P, Frassoni C, Regondi MC, De Biasi S, Spreafico R (1997) GABAergic neurons in mammalian thalamus: a marker of thalamic complexity? Brain Res Bull 42:27–37
583. Barroso-Chinea P, Castle M, Aymerich MS, Lanciego JL (2008) Expression of vesicular glutamate transporters 1 and 2 in the cells of origin of the rat thalamostriatal pathway. J Chem Neuroanat 35:101–107
584. Castro-Alamancos MA, Calcagnotto ME (1999) Presynaptic long-term potentiation in corticothalamic synapses. J Neurosci 19:9090–9097
585. Vogt BA, Rosene DL, Peters A (1981) Synaptic termination of thalamic and callosal afferents in cingulate cortex of the rat. J Comp Neurol 201:265–283
586. Viaene AN, Petrof I, Sherman SM (2011) Synaptic properties of thalamic input to layers 2/3 and 4 of primary somatosensory and auditory cortices. J Neurophysiol 105:279–292

587. Sherman SM (2012) Thalamocortical interactions. Curr Opin Neurobiol 22:575–579
588. Petrof I, Sherman SM (2009) Synaptic properties of the mammillary and cortical afferents to the anterodorsal thalamic nucleus in the mouse. J Neurosci 29:7815–7819
589. Sherman SM (2005) Thalamic relays and cortical functioning. Prog Brain Res 149:107–126
590. McCormick DA (1992) Neurotransmitter actions in the thalamus and cerebral cortex and their role in neuromodulation of thalamocortical activity. Prog Neurobiol 39:337–388
591. Middleton FA, Strick PL (2001) Cerebellar projections to the prefrontal cortex of the primate. J Neurosci 21(2):700–712
592. Sherman SM, Guillery RW (1996) Functional organization of thalamocortical relays. J Neurophysiol 76:1367–1395
593. Sanchez-Gonzalez MA, Garcia-Cabezas MA, Rico B, Cavada C (2005) The primate thalamus is a key target for brain dopamine. J Neurosci 25:6076–6083
594. Van der Werf YD, Witter MP, Groenewegen HJ (2002) The intralaminar and midline nuclei of the thalamus. Anatomical and functional evidence for participation in processes of arousal and awareness. Brain Res Rev 39:107–140
595. Parton A, Husain M (2004) Spatial neglect. Adv Clin Neurosci Rehabil 4:17–18
596. Karnath H-O, Himmelbach M, Rorden C (2002) The subcortical anatomy of human spatial neglect: putamen, caudate nucleus and pulvinar. Brain 125:350–360
597. Van Der Werf YD, Jolles J, Witter MP, Uylings HBM (2003) Contributions of thalamic nuclei to declarative memory functioning. Cortex 39:1047–1062
598. Rees G (2009) Visual attention: the thalamus at the centre? Curr Biol 19:R213–R214
599. O'Connor DH, Fukui MM, Pinsk MA, Kastner S (2002) Attention modulates responses in the human lateral geniculate nucleus. Nat Neurosci 5:1203–1209
600. McAlonan K, Cavanaugh J, Wurtz RH (2006) Attentional modulation of thalamic reticular neurons. J Neurosci 26:4444–4450
601. Mayo JP (2009) Intrathalamic mechanisms of visual attention. J Neurophysiol 101:1123–1125
602. Jensen O, Kaiser J, Lachaux J-P (2007) Human gamma-frequency oscillations associated with attention and memory. Trends Neurosci 30:317–324
603. Jones EG (2001) The thalamic matrix and thalamocortical synchrony. Trends Neurosci 24:595–601
604. Llinas RR, Leznik E, Urbano FJ (2002) Temporal binding via cortical coincidence detection of specific and nonspecific thalamocortical inputs: a voltage-dependent dye-imaging study in mouse brain slices. Proc Natl Acad Sci USA 99:449–454
605. Graybiel AM (1990) Neurotransmitters and neuromodulators in the basal ganglia. Trends Neurosci 13:244–254
606. Nambu A (2007) Globus pallidus internal segment. Prog Brain Res 160:135–150
607. Battaglini PP, Squatrito S, Galletti C, Maioli MG, Riva Sanseverino E (1982) Bilateral projections from the visual cortex to the striatum in the cat. Exp Brain Res 47:28–32
608. Fearnley JM, Lees AJ (1991) Ageing and Parkinson's disease: substantia nigra regional selectivity. Brain 114:2283–2301
609. Walker FO (2007) Huntingdon's disease. Lancet 369:218–228
610. Temel Y, Blokland A, Steinbusch HWM (2005) The functional role of the subthalamic nucleus in cognitive and limbic circuits. Prog Neurobiol 76(6):393–413
611. Singer HS, Reiss AL, Brown JE et al (1993) Volumetric MRI changes in basal ganglia of children with Tourette's syndrome. Neurology 43:950–956
612. Singer HS, Minzer K (2003) Neurobiology of Tourette's syndrome: concepts of neuroanatomic localization and neurochemical abnormalities. Brain Dev 25(S1):S70–S84
613. Brown LL, Schneider JS, Lidsky TI (1997) Sensory and cognitive functions of the basal ganglia. Curr Opin Neurobiol 7:157–163
614. Packard MG, Knowlton BJ (2002) Learning and memory functions of the basal ganglia. Annu Rev Neurosci 25:563–593
615. Dubois D, Pillon B (1997) Cognitive deficits in Parkinson's disease. J Neurol 244:2–8
616. Matison R, Mayeux R, Rosen J, Fahn S (1982) 'Tip-of-the-tongue' phenomenon in Parkinson's disease. Neurology 32:567–570

617. Ikemoto S, Glazier BS, Murphy JM, McBride WJ (1997) Role of dopamine D1 and D2 receptors in the nucleus accumbens in mediating reward. J Neurosci 17(21):8580–8587; Smith KS, Tindell AJ, Aldridge JW, Berridge KC (2009) Ventral palladium roles in reward and motivation. Behav Brain Res 196:155–167; Geisler S, Wise RA (2008) Functional implications of glutamatergic projections to the ventral tegmental area. Rev Neurosci 19:227–244
618. Boraud T, Brown P, Goldberg JA, Graybiel AM, Magill PJ, Bolam JP, Ingham CA (2005) Oscillations in the basal ganglia: the good, the bad and the unexpected. In: Magill PJ (ed) The basal ganglia VIII. Springer, New York, pp 3–24
619. Bergman H, Feingold A, Nini A, Raz A, Slovin H, Abeles M, Vaadia E (1998) Physiological aspects of information processing in the basal ganglia of normal and parkinsonian primates. Trends Neurosci 21:32–38
620. Sharott A, Moll CKE, Engler G, Denker M, Grün S, Engel AK (2009) Different subtypes of striatal neurons are selectively modulated by cortical oscillations. J Neurosci 29:4571–4585
621. Kubota Y, Kawaguchi Y (2000) Dependence of GABAergic synaptic areas on the interneuron type and target size. J Neurosci 20:375–386
622. Ibanez-Sandoval O, Tecuapetla F, Unal B, Shah F, Koos T, Tepper JM (2010) Electrophysiological and morphological characteristics and synaptic connectivity of tyrosine hydroxylase-expressing neurons in adult mouse striatum. J Neurosci 30:6999–7016
623. English DF, Ibanez-Sandoval O, Stark E, Tecuapetla F, Buzsaki G, Diesseroth K, Tepper JM, Koos T (2011) GABAergic circuits mediate the reinforcement-related signals of striatal cholinergic interneurons. Nat Neurosci 15:123–130
624. Yan Z, Flores-Hernandez J, Surmeier DJ (2001) Coordinated expression of muscarinic receptor messenger RNAs in striatal medium spiny neurons. Neuroscience 103:1017–1024
625. Kincaid AE, Zheng T, Wilson CJ (1998) Connectivity and convergence of single corticostriatal axons. J Neurosci 18:4722–4731
626. Smith Y, Bennett BD, Bolam JP, Parent A, Sadikot AF (1994) Synaptic relationships between dopaminergic afferents and cortical or thalamic input in the sensorimotor territory of the striatum in monkey. J Comp Neurol 344:1–19
627. Freund TF, Powell JF, Smith AD (1984) Tyrosine hydroxylase-immunoreactive boutons in synaptic contact with identified striatonigral neurons, with particular reference to dendritic spines. Neuroscience 13:1189–1215
628. Moss J, Bolam JP (2008) A dopaminergic axon lattice in the striatum and its relationship with cortical and thalamic terminals. J Neurosci 28(44):11221–11230
629. Graybiel AM, Ragsdale CW Jr (1978) Histochemically distinct compartments in the striatum of human, monkey, and cat demonstrated by acetylthiocholinesterase staining. Proc Natl Acad Sci USA 75:5723–5726
630. Levey AI, Hersch SM, Rye DB, Sunahara RK, Niznik HB, Kitt CA, Price DL, Maggio R, Brann MR, Ciliax BJ (1993) Localization of D1 and D2 dopamine receptors in brain with subtype-specific antibodies. Proc Natl Acad Sci USA 90:8861–8865
631. Langer LF, Graybiel AM (1989) Distinct nigrostriatal projection systems innervate striosomes and matrix in the primate striatum. Brain Res 498:344–350
632. Alexander GE, DeLong MR (1985) Microstimulation of the primate neostriatum. I. Physiological properties of striatal microexcitable zones. J Neurophysiol 53:1401–1416
633. West MO, Carelli RM, Cohen SM, Gardner JP, Pomerantz M, Chapin JK, Woodward DJ (1990) A region in the dorsolateral striatum of the rat exhibiting single unit correlations with specific locomotor limb movements. J Neurophysiol 64:1233–1246
634. Tippett LJ, Waldvogel HJ, Thomas SJ, Hogg VM, van Roon-Mom W, Synek BJ, Graybiel AM, Faull RLM (2007) Striosomes and mood dysfunction in Huntington's disease. Brain 130:206–221
635. Reiner A, Jiao Y, Del Mar N, Laverghetta AV, Lei WL (2003) Differential morphology of pyramidal tract-type and intratelencephalically projecting-type corticostriatal neurons and their intrastriatal terminals in rats. J Comp Neurol 457:420–440
636. Reiner A, Hart NM, Lei W, Deng Y (2010) Corticostriatal projection neurons – dichotomous types and dichotomous functions. Front Neuroanat 4:142

637. Gerfen CR (1989) The neostriatal mosaic: striatal patch-matrix organization is related to cortical lamination. Science 246:385–388
638. Ragsdale CW, Graybiel AM (1990) A simple ordering of neocortical areas established by the compartmental organization of their striatal projections. Proc Natl Acad Sci USA 87:6196–6199
639. Parthasarathy HB, Schall JD, Graybiell AM (1992) Distributed but convergent ordering of corticostriatal projections: analysis of the frontal eye field and the supplementary eye field in the macaque monkey. J Neurosci 12(11):4468–4488
640. Ballion B, Mallet N, Bezard E, Lanciego JL, Gonon F (2008) Intratelencephalic corticostriatal neurons equally excite striatonigral and striatopallidal neurons and their discharge activity is selectively reduced in experimental parkinsonism. Eur J Neurosci 27:2313–2321
641. Zheng T, Wilson CJ (2002) Corticostriatal combinatorics: the implications of corticostriatal axonal arborizations. J Neurophysiol 87:1007–1017
642. Wilson CJ (1992) Dendritic morphology, inward rectification and the functional properties of neostriatal neurons. In: McKenna T, Davis J, Zornetzer SF (eds) Single neuron computation. Academic, San Diego
643. Groenewegen HJ, Vermeulen-Van der Zee E, te Kortschot A, Witter MP (1987) Organization of the projections from the subiculum to the ventral striatum in the rat. A study using anterograde transport of phaseolus vulgaris leucoagglutinin. Neuroscience 23:103–120
644. Ragsdale CW, Graybiel AM (1988) Fibers from the basolateral nucleus of the amygdala selectively innervate striosomes in the caudate nucleus of the cat. J Comp Neurol 269:506–522
645. Kita H, Kwai ST (1990) Amygdaloid projections to the frontal cortex and the striatum in the rat. J Comp Neurol 298:40–49
646. Sershen H, Hashim A, Lajtha A (2000) Serotonin-mediated striatal dopamine release involves the dopamine uptake site and the serotonin receptor. Brain Res Bull 53(3):353–357
647. Stoof JC, Drukarch B, De Boer P, Westerink BHC, Groenewegen HJ (1992) Regulation of the activity of striatal cholinergic neurons by dopamine. Neuroscience 47:755–770; Aosaki T, Kiuchi K, Kawaguchi Y (1998) Dopamine D1-like receptor activation excites rat striatal large aspiny neurons in vitro. J Neurosci 18(14):5180–5190
648. Blomeley C, Bracci E (2005) Excitatory effects of serotonin on rat striatal cholinergic interneurones. J Physiol 569:715–721
649. Pisani A, Bonsi P, Centonze D, Martorana A, Fusco F, Sancesario G, De Persis C, Bernardi G, Calabresi P (2003) Activation of β1-adrenoceptors excites striatal cholinergic interneurons through a cAMP-dependent, protein kinase-independent pathway. J Neurosci 15:5272–5282
650. Mavridis M, Degryse A-D, Lategan AJ, Marien MR, Colpaert FC (1991) Effects of locus coeruleus lesions on Parkinsonian signs, striatal dopamine and substantia nigra cell loss after 1-methyl-4-phenyl-1,2,3,6-tetrahydropyridine in monkeys: a possible role for the locus coeruleus in the progression of Parkinson's disease. Neuroscience 41:507–523
651. Tepper JM, Bolam JP (2004) Functional diversity and specificity of neostriatal interneurons. Curr Opin Neurobiol 14:685–692
652. Parent A, Hazrati L-N (1995) Functional anatomy of the basal ganglia. I. The cortico-basal ganglia-thalamo-cortical loop. Brain Res Rev 20:91–127
653. Tepper JM, Koós T, Wilson CJ (2004) GABAergic microcircuits in the neostriatum. Trends Neurosci 27:662–669
654. Chuhma N, Tanaka KF, Hen R, Rayport S (2011) Functional connectome of the striatal medium spiny neuron. J Neurosci 31:1183–1192
655. Parent A, Hazrati L-N (1995) Functional anatomy of the basal ganglia. II. The place of subthalamic nucleus and external pallidum in basal ganglia circuitry. Brain Res Rev 20:128–154
656. Sato F, Parent M, Levesque M, Parent A (2000) Axonal branching pattern of neurons of the subthalamic nucleus in primates. J Comp Neurol 424:142–152
657. Nakano K, Hasegawa Y, Tokushige A, Nakagawa S, Kayahara T, Mizuno N (1990) Topographical projections from the thalamus, subthalamic nucleus and pedunculopontine tegmental nucleus to the striatum in the Japanese monkey, *Macaca fuscata*. Brain Res 537:54–68

658. Levesque J-C, Parent A (2005) GABAergic interneurons in human subthalamic nucleus. Mov Disord 20(5):574–584
659. Hazrati L-N, Parent A (1993) Striatal and subthalamic afferents to the primate pallidum: interactions between two opposite chemospecific neuronal systems. Prog Brain Res 99:89–104
660. Kita H (2007) Globus pallidus external segment. Prog Brain Res 160:111–133
661. Cooper AJ, Stanford IM (2002) Calbindin D-28k positive projection neurones and calretinin positive interneurones of the rat globus pallidus. Brain Res 929:243–251
662. Cheramy A, Leviel V, Glowinski J (1981) Dendritic release of dopamine in the substantia nigra. Nature 289:537–542
663. Mitchell SJ, Richardson RT, Baker FH, DeLong MR (1987) The primate globus pallidus: neuronal activity related to direction of movement. Exp Brain Res 68:491–505
664. Gardiner TW, Kitai ST (1992) Single-unit activity in the globus pallidus and neostriatum of the rat during performance of a trained head movement. Exp Brain Res 88:517–530
665. Hajos M, Greenfield SA (1994) Synaptic connections between pars compacta and pars reticulata neurones: electrophysiological evidence for functional modules within the substantia nigra. Brain Res 660:216–224
666. Tepper JM, Abercrombie ED, Bolam JP (2007) Basal ganglia macrocircuits. Prog Brain Res 160:3–7
667. Tritsch NX, Ding JB, Sabatini BL (2012) Dopaminergic neurons inhibit striatal output through non-canonical release of GABA. Nature 490:262–266
668. Yung WH, Hausser MA, Jack JJB (1991) Electrophysiology of dopaminergic and non-dopaminergic neurones of the guinea-pig substantia nigra pars compacta in vitro. J Physiol 436:643–667
669. Hebb MO, Robertson HA (2000) Identification of a subpopulation of substantia nigra pars compacta γ-aminobutyric acid neurons that is regulated by basal ganglia activity. J Comp Neurol 416:30–44
670. Groenewegen HJ, Wright CI, Beijer AV, Voorn P (1999) Convergence and segregation of ventral striatal inputs and outputs. Ann N Y Acad Sci 877:49–63
671. Zahm DS (1999) Functional-anatomical implications of the nucleus accumbens core and shell subterritories. Ann N Y Acad Sci 877:113–128
672. Sadikot AF, Sasseville R (1997) Neurogenesis in the mammalian neostriatum and nucleus accumbens: parvalbumin-immunoreactive GABAergic interneurons. J Comp Neurol 389:193–211
673. Wright CI, Beijer AVJ, Groenewegen HJ (1996) Basal amygdaloid complex afferents to the rat nucleus accumbens are compartmentally organized. J Neurosci 15(5):1877–1893
674. Pennartz CMA, Groenewegen HJ, Lopes Da Silva FH (1994) The nucleus accumbens as a complex of functionally distinct neuronal ensembles: an integration of behavioural, electrophysiological and anatomical data. Prog Neurobiol 47:719–761
675. Haber SN, Groenewegen HJ, Grove EA, Nauta WJH (1985) Efferent connections of the ventral pallidum: evidence of a dual striato pallidofugal pathway. J Comp Neurol 235:322–335
676. Kretschmer BD (2000) NMDA receptor antagonist-induced dopamine release in the ventral pallidum does not correlate with motor activation. Brain Res 859:147–156; Geisler S, Derst C, Veh RW, Zahm DS (2007) Glutamatergic afferents of the ventral tegmental area in the rat. J Neurosci 27(21):5730–5743
677. Kretschmer BD (2000) NMDA receptor antagonist-induced dopamine release in the ventral pallidum does not correlate with motor activation. Brain Res 859:147–156
678. Smith KS, Tindell AJ, Aldridge JW, Berridge KC (2009) Ventral pallidum roles in reward and motivation. Behav Brain Res 196:155–167
679. Johnson SW, North RA (1992) Two types of neurone in the rat ventral tegmental area and their synaptic inputs. J Physiol 450:455–468
680. Geisler S, Wise RA (2008) Functional implications of glutamatergic projections to the ventral tegmental area. Rev Neurosci 19(4–5):227–244

681. Ullsperger M, von Cramon DY (2003) Error monitoring using external feedback: specific roles of the habenular complex, the reward system, and the cingulate motor area revealed by functional magnetic resonance imaging. J Neurosci 23:4308–4314
682. Ungless MA, Magill PJ, Bolam JP (2004) Uniform inhibition of dopamine neurons in the ventral tegmental area by aversive stimuli. Science 303:2040–2042
683. Beckstead RM, DoMesick VB, Nauta WJH (1979) Efferent connections of the substantia nigra and ventral tegmental areas in the rat. Brain Res 175:191–217
684. Carr DB, Sesack SR (2000) Projections from the rat prefrontal cortex to the ventral tegmental area: target specificity in the synaptic associations with mesoaccumbens and mesocortical neurons. J Neurosci 20:3864–3873
685. Lovinger DM (2010) Neurotransmitter roles in synaptic modulation, plasticity and learning in the dorsal striatum. Neuropharmacology 58:951–961
686. Martinez-Gonzalez C, Bolam JP, Mena-Segovia J (2011) Topographical organization of the pedunculopontine nucleus. Front Neuroanat 5:22
687. Good CH, Lupica CR (2010) Afferent-specific AMPA receptor subunit composition and regulation of synaptic plasticity in midbrain dopamine neurons by abused drugs. J Neurosci 30:7900–7909
688. Albin RL, Young AB, Penney JB (1989) The functional anatomy of basal ganglia disorders. Trends Neurosci 12:366–375
689. Bolam JP, Brown MTC, Moss J, Magill PJ (2009) Basal ganglia: internal organization. In: Squire LR (ed) Encyclopedia of neuroscience. Elsevier, Amsterdam
690. Gerfen CR, Surmeier DJ (2011) Modulation of striatal projection systems by dopamine. Annu Rev Neurosci 34:441–466
691. Floresco SB, Todd CL, Grace AA (2001) Glutamatergic afferents from hippocampus to the nucleus accumbens regulate activity of ventral tegmental area dopamine neurons. J Neurosci 21:4915–4922
692. Hyland BI, Reynolds JNJ, Hay J, Perk CG, Miller R (2002) Firing modes of midbrain dopamine cells in the freely moving rat. Neuroscience. Elsevier, Amsterdam
693. Overton PG, Clark D (1997) Burst firing in midbrain dopaminergic neurons. Brain Res Rev 25:312–334
694. Grace AA, Bunney BS (1979) Paradoxical GABA excitation of nigral dopaminergic cells: indirect mediation through reticulata inhibitory neurons. Eur J Pharmacol 59:211–218
695. Goto Y, Otani S, Grace AA (2007) The Yin and Yang of dopamine release: a new perspective. Neuropharmacology 53:583–587
696. Otake K, Nakamura Y (2000) Possible pathways through which neurons of the shell of the nucleus accumbens influence the outflow of the core of the nucleus accumbens. Brain Dev 22:S17–S26
697. Alexander GE, Mahlon R, DeLong MR, Strick PL (1986) Parallel organization of functionally segregated circuits linking basal ganglia and cortex. Annu Rev Neurosci 9:357–381
698. Tekin S, Jeffrey L, Cummings JL (2002) Frontal–subcortical neuronal circuits and clinical neuropsychiatry: an update. J Psychosom Res 53:647–654
699. Crutcher MD, DeLong MR (1984) Single cell studies of the primate putamen. II. Relations to direction of movement and pattern of muscular activity. Exp Brain Res 53:244–258
700. Alexander GE, Crutcher MD (1990) Functional architecture of basal ganglia circuits: neural substrates of parallel processing. Trends Neurosci 13(7):266–271
701. Jaeger D, Gilman S, Wayne Aldridge JW (1995) Neuronal activity in the striatum and pallidum of primates related to the execution of externally cued reaching movements. Brain Res 694:111–127
702. Haber SN, Fudge JL, McFarland NR (2000) Striatonigrostriatal pathways in primates form an ascending spiral from the shell to the dorsolateral striatum. J Neurosci 20:2369–2382
703. McDonald AJ (2003) Is there an amygdala and how far does it extend? An anatomical perspective. Ann N Y Acad Sci 985:1–21
704. Kemppainen S, Jolkkonen E, Pitkanen A (2002) Projections from the posterior cortical nucleus of the amygdala to the hippocampal formation and parahippocampal region in rat. Hippocampus 12:735–755

705. Swanson LW (2003) The amygdala and its place in the cerebral hemisphere. Ann N Y Acad Sci 985:174–184
706. Meunier M, Bachevalier J, Murray EA, Malkova L, Mishkin M (1999) Effects of aspiration vs. neurotoxic lesions of the amygdala on emotional responses in monkeys. Eur J Neurosci 11:4403–4418
707. Fanselow MS, Gale GD (1994) The amygdala, fear, and memory. Proc Natl Acad Sci USA 91:11771–11776
708. Ohman A (2005) The role of the amygdala in human fear: automatic detection of threat. Psychoneuroendocrinology 30(10):953–958
709. Bechara A, Tranel D, Damasio H, Adolphs R, Rockland C, Damasio AR (1995) Declarative knowledge relative to the amygdala and hippocampus in humans. Science 269:1115–1118
710. Gallagher M, Holland PC (1994) The amygdala complex: multiple roles in associative learning and attention. Proc Natl Acad Sci USA 91:11771–11776
711. Ekman P, Friesen WV (1971) Constants across cultures in the face and emotion. J Pers Soc Psychol 17(2):124–129
712. Lanteaume L, Khalfa S, Regis J, Marquis P, Chauvel P, Bartolomei F (2007) Emotion induction after direct intracerebral stimulations of human amygdala. Cereb Cortex 17:1307–1313
713. Fanselow MS, Joseph E, LeDoux JE (1999) Why we think plasticity underlying viewpoint Pavlovian fear conditioning occurs in the basolateral amygdala. Neuron 23:229–232
714. Baxter MG, Murray EA (2002) The amygdala and reward. Nat Rev Neurosci 3:563–573
715. Málková L, Gaffan D, Murray EA (1997) Excitotoxic lesions of the amygdala fail to produce impairments in visual learning for auditory secondary reinforcement but interfere with reinforcer devaluation effects in rhesus monkeys. J Neurosci 17:6011–6020
716. LeDoux JE, Cicchetti P, Xagoraris A, Romanski LM (1990) The lateral amygdaloid nucleus: sensory interface of the amygdala in fear conditioning. J Neurosci 10:1062–1069
717. Adolphs R, Tranel D, Buchanan AW (2005) Amygdala damage impairs emotional memory for gist but not details of complex stimuli. Nat Neurosci 8:512–518
718. Buchanan AW, Tranel D, Adolphs R (2003) A specific role for the human amygdala in olfactory memory. Learn Mem 10:319–325
719. Likhtik E, Pelletier JG, Popescu AT, Pare D (2006) Identification of basolateral amygdala projection cells and interneurons using extracellular recordings. J Neurophysiol 96:3257–3265
720. Nishijo H, Ono T, Nishino H (1988) Single neuron responses in amygdala of alert monkey during complex sensory stimulation with affective significance. J Neurosci 8:3570–3583
721. Duguid I, Sjostrom PJ (2006) Novel presynaptic mechanisms for coincidence detection in synaptic plasticity. Curr Opin Neurobiol 16:312–322
722. Humeau Y, Shaban H, Bissiere S, Luthi A (2003) Presynaptic induction of heterosynaptic associative plasticity in the mammalian brain. Nature 426:841–845
723. Humeau Y, Herry C, Kemp N, Shaban H, Fourcaudot E, Bissiere S, Luthi A (2005) Dendritic spine heterogeneity determines afferent-specific Hebbian plasticity in the amygdala. Neuron 45:119–131
724. Samson RD, Pare D (2005) Activity-dependent synaptic plasticity in the central nucleus of the amygdala. J Neurosci 25:1847–1855
725. Walker DL, Davis M (2002) The role of amygdala glutamate receptors in fear learning, fear-potentiated startle, and extinction. Pharmacol Biochem Behav 71:379–392
726. Petrovich GD, Canteras NS, Swanson LW (2001) Combinatorial amygdalar inputs to hippocampal domains and hypothalamic behavior systems. Brain Res Rev 38(2001):247–289
727. Finch DM (1996) Neurophysiology of converging synaptic inputs from the rat prefrontal cortex, amygdala, midline thalamus, and hippocampal formation onto single neurons of the caudate/putamen and nucleus accumbens. Hippocampus 6(5):495–512
728. McDonald AJ (1991) Topographical organization of amygdaloid projections to the caudatoputamen, nucleus accumbens and related striatal-like areas of the rat brain. Neuroscience 44:15–33
729. Pare D, Duvarci S (2012) Amygdala microcircuits mediating fear expression and extinction. Curr Opin Neurobiol 22:717–723

730. Viviani D, Charlet A, van den Burg E, Robinet C, Hurni N, Abatis M, Magara F, Stoop R (2011) Oxytocin selectively gates fear responses through distinct outputs from the central amygdala. Science 333:104–107
731. Ciocchi S, Herry C, Grenier F, Wolff SBE, Letzkus JJ, Vlachos I, Ehrlich I, Sprengel R, Deisseroth K, Stadler MB, Muller C, Luthi A (2010) Encoding of conditioned fear in central amygdala inhibitory circuits. Nature 468:277–282
732. Knobloch HS, Charlet A, Hoffmann LC, Eliava M, Khrulev S, Cetin AH, Osten P, Schwarz MK, Seeburg PH, Stoop R, Grinevich V (2012) Evoked axonal oxytocin release in the central amygdala attenuates fear response. Neuron 73:553–566
733. Gozzi A, Jain A, Giovanelli A, Bertollini C, Crestan V, Schwarz AJ, Tsetsenis T, Ragozzino D, Gross CT, Bifone A (2010) A neural switch for active and passive fear. Neuron 67:656–666
734. Saper CB (2004) Hypothalamus. In: Paxinos G, Mai JK (eds) The human nervous system. Elsevier, San Diego, pp 513–550
735. Simerly RB (2004) Anatomical substrates of hypothalamic integration. In: Paxinos G (ed) The rat nervous system. Elsevier, San Diego, pp 335–368
736. Berger B, Esclapez M, Alvarez C, Meyer G, Catala M (2001) Human and monkey fetal brain development of the supramammillary-hippocampal projections: a system involved in the regulation of theta activity. J Comp Neurol 429:515–529
737. Swaab DF (2006) The human hypothalamus in metabolic and episodic disorders. Prog Brain Res 153:3–45
738. Palmini A, Chandler C, Andermann F, Costa da Costa J, Paglioli-Neto E, Polkey C, Rosenblatt B, Montes J, Martínez JV, Farmer JP, Sinclair B, Aronyk K, Paglioli E, Coutinho L, Raupp S, Portuguez M (2002) Resection of the lesion in patients with hypothalamic hamartomas and catastrophic epilepsy. Neurology 58:1338–1347
739. Tanaka Y, Miyazawa Y, Akaoka F, Yamada T (1997) Amnesia following damage to the mammillary bodies. Neurology 48:160–165
740. Rolls ET, Burton MJ, Mora R (1976) Hypothalamic neuronal responses associated with the sight of food. Brain Res 111:53–66
741. Panatier A, Gentles SJ, Bourque CW, Oliet SHR (2006) Activity-dependent synaptic plasticity in the supraoptic nucleus of the rat hypothalamus. J Physiol 573:711–721
742. Rao Y, Liu Z-W, Borok E, Rabenstein RL, Shanabrough M, Lu M, Picciotto MR, Horvath TL, Gao X-B (2007) Prolonged wakefulness induces experience-dependent synaptic plasticity in mouse hypocretin/orexin neurons. J Clin Invest 117:4022–4033
743. Rempel-Clower NL, Barbas H (1998) Topographic organization of connections between the hypothalamus and prefrontal cortex in the rhesus monkey. J Comp Neurol 398:393–419
744. Mesulam M-M, Mufson EJ, Levey AI, Wainer BH (1983) Cholinergic innervation of cortex by the basal forebrain: cytochemistry and cortical connections of the septal area, diagonal band nuclei, nucleus basalis (substantia innominata), and hypothalamus in the rhesus monkey. J Comp Neurol 214:170–197
745. Carmichael ST, Price JL (1995) Limbic connections of the orbital and medial prefrontal cortex in macaque monkeys. J Comp Neurol 363:615–641
746. Smeltzer MD, Curtis JT, Aragona BJ, Wang Z (2006) Dopamine, oxytocin, and vasopressin receptor binding in the medial prefrontal cortex of monogamous and promiscuous voles. Neurosci Lett 394:146–151
747. Johnson AK, Gross PM (1993) Sensory circumventricular organs and brain homeostatic pathways. FASEB J 7:678–686
748. Qin C, Luo M (2009) Neurochemical phenotypes of the afferent and efferent projections of the mouse medial habenula. Neuroscience 161:827–837
749. Ren J, Qin C, Hu F, Tan J, Qiu L, Zhao S, Feng G, Luo M (2010) Habenula "Cholinergic" neurons corelease glutamate and acetylcholine and activate postsynaptic neurons via distinct transmission modes. Neuron 69:445–452
750. Kiss J, Csaki A, Bokor H, Kocsis K, Kocsis B (2002) Possible glutamatergic/aspartatergic projections to the supramammillary nucleus and their origins in the rat studied by selective [3H]D-aspartate labelling and immunocytochemistry. Neuroscience 111:671–691

751. Lecourtiera L, Kelly PH (2007) A conductor hidden in the orchestra? Role of the habenular complex in monoamine transmission and cognition. Neurosci Biobehav Rev 31:658–672
752. Sperlágh B, Maglóczky Z, Vizi ES, Freund TF (1996) The triangular septal nucleus as the major source of ATP release in the rat habenula: a combined neurochemical and morphological study. Neuroscience 86:1195–1207
753. Araki M, McGeer PL, McGeer EG (1984) Retrograde HRP tracing combined with a pharmacohistochemical method for GABA transaminase for the identification of presumptive GABAergic projections to the habenula. Brain Res 304:271–277
754. Kalen P, Pritzel M, Nieoullon A, Wiklund L (1986) Further evidence for excitatory amino acid transmission in the lateral habenular projection to the rostral raphe nuclei: lesion-induced decrease of high affinity glutamate uptake. Neurosci Lett 68:35–40
755. Christoph GR, Leonzio RJ, Wilcox KS (1986) Stimulation of the lateral habenula inhibits dopamine-containing neurons in the substantia nigra and ventral tegmental area of the rat. J Neurosci 6:613–619
756. Park MR (1987) Monosynaptic inhibitory postsynaptic potentials from lateral habenula recorded in dorsal raphe neurons. Brain Res Bull 19:581–586
757. Kalen P, Strecker RE, Rosengren E, Bjorklund A (1989) Regulation of striatal serotonin release by the lateral habenula-dorsal raphe pathway in the rat as demonstrated by in vivo microdialysis: role of excitatory amino acids and GABA. Brain Res 492:187–202
758. Hikosaka O (2010) The habenula: from stress evasion to value-based decision-making. Nat Rev Neurosci 11:503–513
759. Morris JS, Smith KA, Cowen PJ, Friston KJ, Dolan RJ (1999) Covariation of activity in habenula and dorsal raphe nuclei following tryptophan depletion. Neuroimage 10:163–172
760. Glickstein M, Doron K (2008) Cerebellum: connections and functions. Cerebellum 7:589–594
761. Andersen BB, Korbo L, Pakkenberg B (1992) A quantitative study of the human cerebellum with unbiased stereological techniques. J Comp Neurol 326:549–560
762. Voogd J (2004) Cerebellum and precerebellar nuclei. In: Paxinos G, Mai JK (eds) The human nervous system. Elsevier, San Diego
763. Soteropoulos DS, Baker SN (2006) Cortico-cerebellar coherence during a precision grip task in the monkey. J Neurophysiol 95:1194–1206
764. Holmes G (1939) The cerebellum of man. Brain 62:1–30
765. Ravizza SM, McCormick CA, Schlerf JE, Justus T, Richard B, Ivry RB, Julie A, Fiez JA (2006) Cerebellar damage produces selective deficits in verbal working memory. Brain 129:306–320
766. Chen SH, Desmond JE (2005) Cerebrocerebellar networks during articulatory rehearsal and verbal working memory tasks. Neuroimage 24:332–338
767. Ackermann H, Mathiak K, Reicker A (2007) The contribution of the cerebellum to speech production and speech perception: clinical and functional imaging data. Cerebellum 6:202–213; Mathiak K, Hertrich I, Grodd W, Ackermann H (2002) Cerebellum and speech perception: a functional magnetic resonance imaging study. J Cogn Neurosci 14:902–912
768. Bracke-Tolkmitt R, Linden A, Canavan AGM, Rockstroh B, Scholz E, Wessel K, Diener H-C (1989) The cerebellum contributes to mental skills. Behav Neurosci 103(2):442–446
769. Ekerot C-F, Jörntell H (2001) Parallel fibre receptive fields of Purkinje cells and interneurons are climbing fibre-specific. Eur J Neurosci 13:1303–1310
770. Palay SL, Chan-Palay V (1974) Cerebellar cortex. Springer, Berlin
771. Nunzi MG, Russo M, Mugnaini E (2003) Vesicular glutamate transporters VGLUT1 and VGLUT2 define two subsets of unipolar brush cells in organotypic cultures of mouse vestibulocerebellum. Neuroscience 122:359–371
772. Munoz DG (1990) Monodendritic neurons: a cell type in the human cerebellar cortex identified by chromogranin A-like immunoreactivity. Brain Res 528:335–338
773. Kinney GA, Overstreet LS, Slater NT (1997) Prolonged physiological entrapment of glutamate in the synaptic cleft of cerebellar unipolar brush cells. J Neurophysiol 78:1320–1333

774. Mugnaini E, Sekerkova G, Martina M (2011) The unipolar brush cell: a remarkable neuron finally receiving deserved attention. Brain Res Rev 66:220–245
775. Uusisaari M, De Schutter E (2011) The mysterious microcircuitry of the cerebellar nuclei. J Physiol 589:3441–3457
776. Pijpers A, Apps R, Pardoe J, Voogd J, Rugrok TJH (2006) Precise spatial relationships between mossy fibers and climbing fibers in rat cerebellar cortical zones. J Neurosci 24:12067–12080
777. D'Angelo E, De Philippi G, Rossi P, Taglietti V (1995) Synaptic excitation of individual rat cerebellar granule cells in situ: evidence for the role of NMDA receptors. J Physiol 484:397–413
778. Dino MR, Schuerger RJ, Liu Y-B, Slater NT, Mugnaini E (2000) Unipolar brush cell: a potential feed forward excitatory interneuron of the cerebellum. Neuroscience 98(4): 625–636; Uusisaari M, De Schutter E (2011) The mysterious microcircuitry of the cerebellar nuclei. J Physiol 589:3441–3457
779. Sugihara I, Wu HS, Shinoda Y (2001) The entire trajectories of single olivocerebellar axons in the cerebellar cortex and their contribution to cerebellar compartmentalization. J Neurosci 21:7715–7723
780. Teune TM, van der Burg J, de Zeeuw CI, Voogd J, Ruigrok TJH (1998) Single Purkinje cell can innervate multiple classes of projection neurons in the cerebellar nuclei of the rat: a light microscopic and ultrastructural triple-tracer study in the rat. J Comp Neurol 392:164–178
781. Zheng N, Raman IM (2010) Synaptic inhibition, excitation, and plasticity in neurons of the cerebellar nuclei. Cerebellum 9:56–66
782. Brodal P, Bjaalie JG (1992) Organization of the pontine nuclei. Neurosci Res 13:83–118
783. Albus K, Donate-Oliver F, Sanides D, Fries W (1981) The distribution of pontine projection cells in visual and association cortex of the cat: an experimental study with horseradish peroxidase. J Comp Neurol 201:175–189
784. Knowlton BJ, Thompson JK, Thompson RF (1993) Projections from the auditory cortex to the pontine nuclei in the rabbit. Behav Brain Res 56:23–30
785. Ramnani N, Behrens TEJ, Johansen-Berg H, Richter MC, Pinsk MA, Andersson JLR, Rudebeck P, Ciccarelli O, Richter W, Thompson AJ, Gross CG, Robson MD, Kastner S, Matthews PM (2006) The evolution of prefrontal inputs to the cortico-pontine system: diffusion imaging evidence from macaque monkeys and humans. Cereb Cortex 16:811–818
786. Mihailoff GA, Kosinski RJ, Azizi SA, Border BG (1989) Survey of noncortical afferent projections to the basilar pontine nuclei: a retrograde tracing study in the rat. J Comp Neurol 282:617–643
787. Mihailoff GA, McArdle CB, Adams CE (1981) The cytoarchitecture, cytology, and synaptic organization of the basilar pontine nuclei in the rat. I. Nissl and Golgi studies. J Comp Neurol 195:181–201
788. Boesten AJP, Voogd J (1975) Projections of the dorsal column nuclei and the spinal cord on the inferior olive in the cat. J Comp Neurol 161(2):215–237
789. Saint-Cyr JA (1983) The projection from the motor cortex to the inferior olive in the cat. An experimental study using axonal transport techniques. Neuroscience 10(3):667–684
790. Watson TC, Jones MW, Apps R (2009) Electrophysiological mapping of novel prefrontal – cerebellar pathways. Front Integr Neurosci 3:18
791. Sedgwick EM, Williams TD (1966) Afferent connexions to single units in the inferior olive of the cat. Nature 212:1370–1371
792. Hoge GJ, Davidson KGV, Yasumura T, Castillo PE, Rash JE, Pereda AE (2011) The extent and strength of electrical coupling between inferior olivary neurons is heterogeneous. J Neurophysiol 105:1089–1101
793. De Zeeuw CI, Simpson JI, Hoogenraad CC, Galjart N, Koekkoek SKE, Ruigrok TJH (1998) Microcircuitry and function of the inferior olive. Trends Neurosci 21:391–400
794. Llinás R, Muhlethaler M (1988) Electrophysiology of guinea-pig cerebellar nuclear cells in the in vitro brain stem-cerebellar preparation. J Physiol 404:241–258

795. Aizenman CD, Linden DJ (1999) Regulation of the rebound depolarisation and spontaneous firing patterns of deep nuclear neurons in slices of ret cerebellum. J Neurophysiol 82:1697–1709; De Schutter E, Steuber V (2009) Patterns and pauses in Purkinje cell simple spike trains: experiments, modeling and theory. Neuroscience 162:816–826
796. Hoover JE, Strick PL (1999) The organization of cerebellar and basal ganglia outputs to primary motor cortex as revealed by retrograde transneuronal transport of herpes simplex virus type 1. J Neurosci 19:1446–1463
797. Morishima M, Kawaguchi Y (2006) Recurrent connection patterns of corticostriatal pyramidal cells in frontal cortex. J Neurosci 26:4394–4405
798. See Ref. [797]
799. Nambu A, Yoshida S, Jinnai K (1988) Projection on the motor cortex of thalamic neurons with pallidal input in the monkey. Exp Brain Res 71:658–662
800. Bostan AC, Strick PL (2010) The cerebellum and basal ganglia are interconnected. Neuropsychol Rev 20:261–270
801. Oscarsson O (1979) Functional units of the cerebellum – sagittal zones and microzones. Trends Neurosci 2:143–145
802. Apps R (1990) Columnar organisation of the inferior olive projection to the posterior lobe of the rat cerebellum. J Comp Neurol 302:236–254
803. Ito M (2001) Cerebellar long-term depression: characterization, signal transduction, and functional roles. Physiol Rev 81:1143–1195
804. Apps R, Garwicz M (2005) Anatomical and physiological foundations of cerebellar information processing. Nat Rev Neurosci 6:297–311
805. Ekerot C-F, Jörntell H, Garwicz M (1995) Functional relation between corticonuclear input and movements evoked on microstimulation in cerebellar nucleus interpositus anterior in the cat. Exp Brain Res 106:365–376
806. Vogt KE, Canepari M (2010) On the induction of postsynaptic granule cell-Purkinje neuron LTP and LTD. Cerebellum 9:284–290
807. Hartell NA (1996) Strong activation of parallel fibers produces localized calcium transients and a form of LTD that spreads to distant synapses. Neuron 16:601–610
808. Hansel C, Linden DJ, D'Angelo E (2001) Beyond parallel fiber LTD: the diversity of synaptic and nonsynaptic plasticity in the cerebellum. Nat Neurosci 4:467–475
809. Morishita W, Sastry BR (1996) Postsynaptic mechanisms underlying long-term depression of GABAergic transmission in neurons of the deep cerebellar nuclei. J Neurophysiol 76:59–68
810. Ouardouz M, Sastry BR (2000) Mechanisms underlying LTP of inhibitory synaptic transmission in the deep cerebellar nuclei. J Neurophysiol 84:1414–1421
811. D'Errico A, Prestori F, D'Angelo E (2009) Differential induction of bidirectional long-term changes in neurotransmitter release by frequency-coded patterns at the cerebellar input. J Physiol 587:5843–5857
812. D'Angelo E, Rossi P, Gall D, Prestori F, Nieus T, Maffei A, Sola E (2005) Long-term potentiation of synaptic transmission at the mossy fiber-granule cell relay of cerebellum. Prog Brain Res 148:69–80
813. McCormick DA, Lavond DG, Clark GA, Kettner RE, Rising CE, Thompson RF (1981) The engram found? Role of the cerebellum in classical conditioning of nictitating membrane and eyelid responses. Bull Psychon Soc 18:103–105
814. Lincoln JS, McCormick DA, Thompson RF (1982) Ipsilateral cerebellar lesions prevent learning of the classically conditioned nictitating membrane/eyelid response. Brain Res 242:190–193
815. Medina JF, Nores WL, Ohyama T, Mauk MD (2000) Mechanisms of cerebellar learning suggested by eyelid conditioning. Curr Opin Neurobiol 10:717–724
816. Steinmetz JE (2000) Brain substrates of classical eyeblink conditioning: a highly localized but also distributed system. Behav Brain Res 110:13–24
817. Moyer JR, Deyo RA, Disterhoft JF (1990) Hippocampectomy disrupts trace eye-blink conditioning in rabbits. Behav Neurosci 104:243–252

818. Gerwig M, Haerter K, Hajjar K, Dimitrova A, Maschke M, Kolb FP, Thilmann AF, Gizewski ER, Timmann D (2006) Trace eyeblink conditioning in human subjects with cerebellar lesions. Exp Brain Res 170:7–21
819. Niewiadomska G, Baksalerska-Pazera M, Riedel G (2009) The septo-hippocampal system, learning and recovery of function. Prog Neuropsychopharmacol Biol Psychiatry 33(2009):791–805
820. Lyness SA, Zarow C, Chui HC (2003) Neuron loss in key cholinergic and aminergic nuclei in Alzheimer disease: a meta-analysis. Neurobiol Aging 24:1–23
821. Hepler DJ, Olton DS, Wenk GL, Coyle JT (1985) Lesions in nucleus basalis magnocellularis and medial septal area of rats produce qualitatively similar memory impairments. J Neurosci 5:866–873
822. Winkler J, Suhr ST, Gage FH, Thal LJ, Fisher LJ (1995) Essential role of neocortical acetylcholine in spatial memory. Nature 375:484–487
823. Bartus RT, Flicker C, Dean RL, Pontecorvo M, Figueiredo JC, Fisher SK (1985) Selective memory loss following nucleus basalis lesions: long term behavioral recovery despite persistent cholinergic deficiencies. Pharmacol Biochem Behav 23:125–135
824. Baxter MG, Bucci DJ, Gorman LK, Wiley RG, Gallagher M (1995) Selective immunotoxic lesions of basal forebrain cholinergic cells: effects on learning and memory in rats. Behav Neurosci 109:714–722
825. McGaughy J, Dalley JW, Morrison CH, Everitt BJ, Robbins TW (2002) Selective behavioral and neurochemical effects of cholinergic lesions produced by intrabasalis infusions of 192 IgG-saporin on attentional performance in a five-choice serial reaction time task. J Neurosci 22:1905–1913
826. Dalley JW, Theobald DE, Bouger P, Chudasama Y, Cardinal RN, Robbins TW (2004) Cortical cholinergic function and deficits in visual attentional performance in rats following 192 IgG-saporin-induced lesions of the medial prefrontal cortex. Cereb Cortex 14:922–932
827. Warburton EC, Koder T, Cho K, Massey PV, Duguid G, Barker GRI, Aggleton JP, Bashir ZA, Malcolm W, Brown MW (2003) Cholinergic neurotransmission is essential for perirhinal cortical plasticity and recognition memory. Neuron 38:987–996
828. Ramanathan D, Tuszynski MH, Conner JM (2009) The basal forebrain cholinergic system is required specifically for behaviorally mediated cortical map plasticity. J Neurosci 29:5992–6000
829. Goard M, Dan Y (2009) Basal forebrain activation enhances cortical coding of natural scenes. Nat Neurosci 12:1444–1449
830. Buchanan KA, Petrovic MM, Chamberlain SEL, Marrion NV, Mellor JR (2010) Facilitation of long-term potentiation by muscarinic M1 receptors is mediated by inhibition of SK channels. Neuron 68:948–963
831. Brocher S, Artola A, Wolf Singer W (1992) Agonists of cholinergic and noradrenergic receptors facilitate synergistically the induction of long-term potentiation in slices of rat visual cortex. Brain Res 573:27–36
832. Detari L, Rasmusson DD, Semba K (1999) The role of basal forebrain neurons in tonic and phasic activation of the cerebral cortex. Prog Neurobiol 58:249–277
833. Walker LC, Koliatsos VE, Kitt CA, Richardson RT, Rokaeus A, Price DL (1989) Peptidergic neurons in the basal forebrain magnocellular complex of the rhesus monkey. J Comp Neurol 280:272–282
834. Manns ID (2001) The role of basal forebrain neurons in the modulation of cortical activity: a physiological and anatomical examination. Ph.D. thesis, Department of Neurology and Neurosurgery, McGill University, Montreal. http://digitool.Library.McGill.CA:80/webclient/DeliveryManager?application=DIGITOOL-3&owner=resourcediscovery&custom_att_2=simple_viewer&forebear_coll=&user=GUEST&pds_handle=&pid=37653&con_lng=ENG&search_terms=&adjacency=N&rd_session=http://digitool.Library.McGill.CA:80/R/29SIHEV2V547EVBVCH9FNP2HDAPNXLQMVALHEQHLF3GT1YJJBS-03192
835. Colom LV (2006) Septal networks: relevance to theta rhythm, epilepsy and Alzheimer's disease. J Neurochem 96:609–623

836. Penley SC, Hinman JR, Sabolek HR, Escabi MA, Markus EJ, Chrobak JJ (2011) Theta and gamma coherence across the septotemporal axis during distinct behavioral states. Hippocampus 22:1164–1175
837. Manns ID, Alonso A, Jones BE (2000) Discharge properties of juxtacellularly labeled and immunohistochemically identified cholinergic basal forebrain neurons recorded in association with the electroencephalogram in anesthetized rats. J Neurosci 20:1505–1518
838. Manns ID, Alonso A, Jones BE (2000) Discharge profiles of juxtacellularly labeled and immunohistochemically identified GABAergic basal forebrain neurons recorded in association with the electroencephalogram in anesthetized rats. J Neurosci 20:9252–9263
839. Colom LV, Castaneda MT, Reyna T, Hernandez S, Garrido-Sanabria E (2005) Characterization of medial septal glutamatergic neurons and their projection to the hippocampus. Synapse 58:151–164
840. Bigl V, Woolf NJ, Butcher LL (1982) Cholinergic projections from the basal forebrain to frontal, parietal, temporal, occipital, and cingulate cortices: a combined fluorescent tracer and acetylcholinesterase analysis. Brain Res Bull 8:727–749
841. Bickford ME, Gunluk AE, Van Horn SC, Sherman SM (1994) GABAergic projection from the basal forebrain to the visual sector of the thalamic reticular nucleus in the cat. J Comp Neurol 348:481–510
842. Haring JH, Wang RY (1986) The identification of some sources of afferent input to the rat nucleus basalis magnocellularis by retrograde transport of horseradish peroxidase. Brain Res 366:152–158
843. Risold PY (2004) The septal region. In: Paxinos G (ed) The rat nervous system. Elsevier, San Diego
844. Lee S-H, Yang Dan Y (2012) Neuromodulation of brain states. Neuron 76:209–222
845. Carter ME, Yizhar O, Chikahisa S, Nguyen H, Adamantidis A, Nishino S, Deisseroth K, de Lecea L (2010) Tuning arousal with optogenetic modulation of locus coeruleus neurons. Nat Neurosci 13:1526–1533
846. Monti JM (1993) Involvement of histamine in the control of the waking state. Life Sci 53:1331–1338
847. Monti JM, Jantos H (2008) The roles of dopamine and serotonin, and of their receptors, in regulating sleep and waking. Prog Brain Res 172:625–646
848. McCormick DA, Wang Z, Huguenard J (1993) Neurotransmitter control of neocortical neuronal activity and excitability. Cereb Cortex 3:387–398
849. Tully K, Li Y, Tsvetkov E, Bolshakov VY (2007) Norepinephrine enables the induction of associative long-term potentiation at thalamo-amygdala synapses. Proc N Y Acad Sci 104:14146–14150
850. Tully K, Bolshakov VY (2010) Emotional enhancement of memory: how norepinephrine enables synaptic plasticity. Mol Brain 3:15
851. Staubli U, Otaky N (1994) Serotonin controls the magnitude of LTP induced by theta bursts via an action on NMDA-receptor-mediated responses. Brain Res 643:10–16
852. Zhong P, Liu W, Gu Z, Yan Z (2008) Serotonin facilitates long-term depression induction in prefrontal cortex via p38 MAPK/Rab5-mediated enhancement of AMPA receptor internalization. J Physiol 586:4465–4479
853. Kojic L, Gu Q, Douglas RM, Cynader MS (1997) Serotonin facilitates synaptic plasticity in kitten visual cortex: an in vitro study. Dev Brain Res 101:299–304
854. Brown RE, Fedorov NB, Haas HL, Reymann KG (1995) Histaminergic modulation of synaptic plasticity in area CA1 of rat hippocampal slices. Neuropharmacology 34:181–190
855. Kuo M-C, Dringenberg HC (2008) Histamine facilitates in vivo thalamocortical long-term potentiation in the mature visual cortex of anesthetized rats. Eur J Neurosci 27:1731–1738
856. Kolomiets B, Marzo A, Caboche J, Vanhoutte P, Otani S (2009) Background dopamine concentration dependently facilitates long-term potentiation in rat prefrontal cortex through postsynaptic activation of extracellular signal-regulated kinases. Cereb Cortex 19:2708–2718
857. Li S, Cullen WK, Anwyl R, Rowan MJ (2003) Dopamine-dependent facilitation of LTP induction in hippocampal CA1 by exposure to spatial novelty. Nat Neurosci 6:526–531

858. Pasquier DA, Kemper TL, Forbes WB, Morgane PJ (1977) Dorsal raphe, substantia nigra and locus coeruleus: interconnections with each other and the neostriatum. Brain Res Bull 2:323–339
859. Ericson H, Blomqvist A, Köhler C (1989) Brainstem afferents to the tuberomammillary nucleus in the rat brain with special reference to monoaminergic innervation. J Comp Neurol 281:169–192
860. Brown RE, Sergeeva OA, Eriksson KS, Haas HL (2002) Convergent excitation of dorsal raphe serotonin neurons by multiple arousal systems (orexin/hypocretin, histamine and noradrenaline). J Neurosci 22:8850–8859
861. Haas HL, Sergeeva OA, Selbach O (2008) Histamine in the nervous system. Physiol Rev 88:1183–1241
862. Moore RY, Bloom FE (1979) Central catecholamine neuron systems: anatomy and physiology of the norepinephrine and epinephrine systems. Annu Rev Neurosci 2:113–168
863. Sara SJ (2009) The locus coeruleus and noradrenergic modulation of cognition. Nat Rev Neurosci 10:211–223
864. Foote SL, Morrison JH (1987) Extrathalamic modulation of cortical function. Annu Rev Neurosci 10:67–95
865. Loughlin SE, Foote SL, Fallon JH (1982) Locus coeruleus projections to cortex: topography, morphology and collateralization. Brain Res Bull 9:287–294
866. Lewis DA, Campbell MJ, Foote SL, Goldstein M, Morrison JH (1987) The distribution of tyrosine hydroxylase-immunoreactive fibers in primate neocortex is widespread but regionally specific. J Neurosci 7(1):279–290
867. Simpson KL, Altman DW, Wang L, Kirifides ML, Lin RC, Waterhouse BD (1997) Lateralization and functional organization of the locus coeruleus projection to the trigeminal somatosensory pathway in rat. J Comp Neurol 385:135–147
868. Wang Z, David A, McCormick DA (1993) Control of firing mode of corticotectal and corticopontine layer V burst-generating neurons by norepinephrine, acetylcholine, and 1S,3R-ACPD. J Neurosci 13:2199–2216
869. Samuels ER, Szabadi E (2008) Functional neuroanatomy of the noradrenergic locus coeruleus: its roles in the regulation of arousal and autonomic function part II: physiological and pharmacological manipulations and pathological alterations of locus coeruleus activity in humans. Curr Neuropharmacol 6:254–285
870. Aston-Jones G, Rajkowski J, Jonathan Cohen J (1999) Role of locus coeruleus in attention and behavioral flexibility. Biol Psychiatry 46:1309–1320
871. Ordway GA (1997) Pathophysiology of the locus coeruleus in suicide. Ann N Y Acad Sci 836:233–252
872. German DC, Manaye KF, White CL, Woodward DJ, McIntire DD, Smith WK, Kalaria RN, Mann DMA (1992) Disease-specific patterns of locus coeruleus cell loss. Ann Neurol 32(5):667–676
873. Hornung J-P (2004) Raphe nuclei. In: Paxinos G, Mai JK (eds) The human nervous system. Elsevier, Amsterdam
874. Bobillier P, Seguin S, Petitjean F, Salvert D, Touret M, Jouvet M (1976) The raphe nuclei of the cat brain stem: a topographical atlas of their efferent projections as revealed by autoradiography. Brain Res 113:449–486
875. Andrade R, Nicoll RA (1987) Pharmacologically distinct actions of serotonin on single pyramidal neurones of the rat hippocampus recorded in vitro. J Physiol 394:99–124
876. Qing-Ping W, Nakai Y (1994) The dorsal raphe: an important nucleus in pain modulation. Brain Res Bull 34:575–585
877. Monti JM (2011) Serotonin control of sleep-wake behavior. Sleep Med Rev 15:269–281
878. Meltzer HY (1999) The role of serotonin in antipsychotic drug action. Neuropsychopharmacology 21:106S–115S
879. Köhler C, Swanson LW, Haglund L, J-Yen W (1985) The cytoarchitecture, histochemistry and projections of the tuberomammillary nucleus in the rat. Neuroscience 16:85–110
880. Ericson H, Blomqvist A, Köhler C (1991) Origin of neuronal inputs to the region of the tuberomammillary nucleus of the rat brain. J Comp Neurol 311:45–64

881. Haas H, Panula P (2003) The role of histamine and the tuberomamillary nucleus in the nervous system. Nat Rev Neurosci 4:121–130
882. Martinez-Mir MI, Pollard H, Morea J, Arrang JM, Ruat M, Traiffort E, Schwartz JC, Palacios JM (1990) Three histamine receptors (H1, H2 and H3) visualized in the brain of human and non-human primates. Brain Res 526:322–327
883. Brown RE, Stevens DR, Haas HL (2001) The physiology of brain histamine. Prog Neurobiol 63:637–672
884. Bjorklund A, Dunnett SB (2007) Dopamine neuron systems in the brain: an update. Trends Neurosci 30:194–202
885. Dobi A, Margolis EB, Wang H-L, Harvey BK, Morales M (2010) Glutamatergic and nonglutamatergic neurons of the ventral tegmental area establish local synaptic contacts with dopaminergic and nondopaminergic neurons. J Neurosci 30:218–229
886. Gasbarri A, Sulli A, Packard MG (1997) The dopaminergic mesencephalic projections to the hippocampal formation in the rat. Prog Neuropsychopharmacol Biol Psychiatry 21:1–22
887. Cho YT, Fudge JL (2010) Heterogeneous dopamine populations project to specific subregions of the primate amygdala. Neuroscience 165:1501–1518
888. Berger B, Trottier S, Verney C, Gaspar P, Alvarez C (1988) Regional and laminar distribution of the dopamine and serotonin innervation in the macaque cerebral cortex: a radioautographic study. J Comp Neurol 273:99–119
889. Kröner S, Rosenkranz JA, Grace AA, Barrionuevo G (2005) Dopamine modulates excitability of basolateral amygdala neurons in vitro. J Neurophysiol 93:1598–1610
890. Choi WS, Machida CA, Ronnekleiv OK (1995) Distribution of dopamine D1, D2, and D5 receptor mRNAs in the monkey brain: ribonuclease protection assay analysis. Mol Brain Res 31:86–94
891. Wang GJ, Volkow ND, Fowler JS, Ding YS, Logan J, Gatley SJ, MacGregor RR, Wolf AP (1995) Comparison of two PET radioligands for imaging extrastriatal dopamine transporters in human brain. Life Sci 57:PL187–PL191
892. Lodge DJ, Buffalari DM, Grace AA (2009) Dopamine: CNS pathways and neurophysiology. In: Squire LR (ed) Encyclopedia of neuroscience. Elsevier, Amsterdam
893. Abi-Dargham A, Mawlawi O, Lombardo I, Gil R, Martinez D, Huang Y, Hwang DR, Keilp J, Kochan L, Van Heertum R, Gorman JM, Laruelle M (2002) Prefrontal dopamine D1 receptors and working memory in schizophrenia. J Neurosci 22:3708–3719
894. Schott BH, Seidenbecher CI, Fenker DB, Lauer CJ, Bunzeck N, Bernstein H-J, Tischmeyer W, Gundelfinger ED, Heinze H-J, Duzel E (2006) The dopaminergic midbrain participates in human episodic memory formation: evidence from genetic imaging. J Neurosci 26:1407–1417
895. Sesack SR, Grace AA (2010) Cortico-basal ganglia reward network: microcircuitry. Neuropsychopharmacology 35:27–47
896. Tritsch NX, Sabatini BL (2012) Dopaminergic modulation of synaptic transmission in cortex and striatum. Neuron 76:33–50
897. Canto CB, Wouterlood FG, Witter MP (2008) What does the anatomical organization of the entorhinal cortex tell us? Neural Plast 2008:381243
898. Blaizot X, Martinez-Marcos A, Arroyo-Jimenez M, Marcos P, Artacho-Pérula E, Muñoz M, Chavoix C, Ricardo Insausti R (2004) Parahippocampal gyrus in the baboon: anatomical, cytoarchitectonic and magnetic resonance imaging (MRI) studies. Cereb Cortex 14:231–246
899. O'Mara SM, Commins S, Anderson M, Gigg J (2001) The subiculum: a review of form, physiology and function. Prog Neurobiol 64:129–155
900. Insausti R, Amaral DG (2004) Hippocampal formation. In: Paxinos G, Mai JK (eds) The human nervous system. Elsevier, Amsterdam
901. Amaral D, Lavenex P (2006) Hippocampal neuroanatomy. In: Andersen P, Morris R, Amaral D, Bliss T, O'Keefe J (eds) The hippocampus book. Oxford University Press, Oxford
902. Kullmann DM (2011) Interneuron networks in the hippocampus. Curr Opin Neurobiol 21:709–716
903. Fischer Y, Wittner L, Freund TF, Gähwiler BH (2002) Simultaneous activation of gamma and theta network oscillations in rat hippocampal slice cultures. J Physiol 539:857–868

904. Siapas AG, Lubenov EV, Wilson MA (2005) Prefrontal phase locking to hippocampal theta oscillations. Neuron 46:141–151
905. Bliss TVP, Lomo T (1973) Long-lasting potentiation of synaptic transmission in the dentate area of the anaesthetized rabbit following stimulation of the perforant path. J Physiol 232:331–356
906. Kaibara T, Leung LS (1993) Basal versus apical dendritic long-term potentiation of commissural afferents to hippocampal CA1: a current-source density study. J Neurosci 13:2391–2404
907. Maccaferri G, Tóth K, McBain CJ (1998) Target-specific expression of presynaptic mossy fiber plasticity. Science 279:1368–1371
908. Acsady L, Kamondi A, Sik A, Freund T, Buzsaki G (1998) GABAergic cells are the major postsynaptic targets of mossy fibers in the rat hippocampus. J Neurosci 18:3386–3403
909. Suzuki WA (1996) Neuroanatomy of the monkey entorhinal, perirhinal and parahippocampal cortices: organization of cortical inputs and interconnections with amygdala and striatum. Semin Neurosci 8:3–12
910. Amaral DG, Ishizuka N, Claiborne B (1990) Neurons, numbers and the hippocampal network. Prog Brain Res 83:1–11
911. Ribak CE, Seress L, Amaral DG (1985) The development, ultrastructure and synaptic connections of the mossy cells of the dentate gyrus. J Neurocytol 14:835–857
912. Buckmaster PS, Strowbridge BW, Kunkel DD, Schmiege DL, Schwartzkroin PA (1992) Mossy cell axonal projections to the dentate gyrus molecular layer in the rat hippocampal slice. Hippocampus 2:349–362
913. Mody I (2002) The GAD-given right of dentate gyrus granule cells to become GABAergic. Epilepsy Curr 2:143–145
914. Vertes RP, McKenna JT (2000) Collateral projections from the supramammillary nucleus to the medial septum and hippocampus. Synapse 38:281–293
915. Sloviter RS (1983) "Epileptic" brain damage in rats induced by sustained electrical stimulation of the perforant path. I. Acute electrophysiological and light microscopic studies. Brain Res Bull 10:675–697
916. Amaral DG, Witter MP (1989) The three-dimensional organization of the hippocampal formation: a review of anatomical data. Neuroscience 31:571–591
917. Wyss JM, Swanson LW, Cowan WM (1979) A study of subcortical afferents to the hippocampal formation in the rat. Neuroscience 4:463–476
918. Bayat M, Hasandeh GR, Barzroodipour M, Javadi M (2005) The effect of low protein diet on thalamic projections of hippocampus in rat. Neuroanatomy 4:43–48
919. Shibata H (1993) Direct projections from the anterior thalamic nuclei to the retrohippocampal region in the rat. J Comp Neurol 337:431–445
920. Allen GV, Hopkins DA (1989) Mamillary body in the rat: topography and synaptology of projections from the subicular complex, prefrontal cortex, and midbrain tegmentum. J Comp Neurol 286:311–336
921. Veazey RB, Amaral DG, Cowan WM (1982) The morphology and connections of the posterior hypothalamus in the cynomolgus monkey (*Macaca fascicularis*). 11. Efferent connections. J Comp Neurol 207:135–156
922. Siegel A, Edinger H, Ohgami S (1974) The topographical organization of the hippocampal projection to the septal area: a comparative neuroanatomical analysis in the gerbil, rat, rabbit, and cat. J Comp Neurol 157:359–378
923. Lindsey BW, Tropepe V (2006) A comparative framework for understanding the biological principles of adult neurogenesis. Prog Neurobiol 80:281–307
924. Schmidt-Hieber C, Jonas P, Bischofberger J (2004) Enhanced synaptic plasticity in newly generated granule cells of the adult hippocampus. Nature 429:184–187
925. Corkin S (2002) What's new with the amnesic patient H.M.? Nat Rev Neurosci 3:153–160
926. Wickelgren WA (1968) Sparing of short-term memory in an amnesic patient: implications for strength theory of memory. Neuropsychologia 6:235–244

927. Gabrieli JDE, Corkin S, Mickel SF, Growden JH (1993) Intact acquisition and long-term retention of mirror tracing skill in Alzheimer's disease and in global amnesia. Behav Neurosci 107:899–910
928. Phelps EA (2006) Emotion and cognition: insights from studies of the human amygdala. Annu Rev Psychol 57:27–53
929. Caulo M, Van Hecke J, Toma L, Ferretti A, Tartaro A, Colosimo C, Romani GL, Uncini A (2005) Functional MRI study of diencephalic amnesia in Wernicke-Korsakoff syndrome. Brain 128:1584–1594
930. Dolcos F, LaBar KS, Cabeza R (2004) Interaction between the amygdala and the medial temporal lobe memory system predicts better memory for emotional events. Neuron 42:855–863
931. Berti A, Arienta C, Papagno C (1990) A case of amnesia after excision of the septum pellucidum. J Neurol Neurosurg Psychiatry 53:922–924
932. Coward LA (2001) The recommendation architecture: lessons from the design of large scale electronic systems for cognitive science. J Cogn Syst Res 2(2):111–156
933. Carruthers P (2006) The architecture of the mind. Oxford University Press, Oxford
934. Hutchinson JB, Uncapher MR, Wagner AD (2009) Posterior parietal cortex and episodic retrieval: convergent and divergent effects of attention and memory. Learn Mem 16:343–356
935. Tehovnik EJ, Sommer MA, Chou IH, Slocum WM, Schiller PH (2000) Eye fields in the frontal lobes of primates. Brain Res Rev 32:413–448
936. Krubitzer LA, Kass JH (1990) The organization and connections of somatosensory cortex in marmosets. J Neurosci 10:952–974
937. Yeo BTT, Krienen FM, Sepulcre J, Sabuncu MR, Lashkari D, Hollinshead M, Roffman JL, Smoller JW, Zöllei L, Polimeni JR, Fischl B, Liu H, Buckner RL (2011) The organization of the human cerebral cortex estimated by intrinsic functional connectivity. J Neurophysiol 106:1125–1165
938. Sereno AB, Maunsell JHR (1998) Shape selectivity in primate lateral intraparietal cortex. Nature 395:500–503
939. Ipata AE, Gee AL, Bisley JW, Goldberg ME (2009) Neurons in the lateral intraparietal area create a priority map by the combination of disparate signals. Exp Brain Res 192:479–488
940. Stuphorn V, Taylor TL, Schall JD (2000) Performance monitoring by the supplementary eye field. Nature 408:857–860
941. Chen LL, Walton MMG (2005) Head movement evoked by electrical stimulation in the supplementary eye field of rhesus monkey. J Neurophysiol 94:4502–4519
942. Voogd J, Schraa-Tam CKL, van der Geest JN, De Zeeuw CI (2012) Visuomotor cerebellum in human and nonhuman primates. Cerebellum 11:392–410
943. Jonides J, Schumacher EH, Smith EE, Koeppe RA, Awh E, Reuter-Lorenz PA, Marshuetz C, Willis CR (1998) The role of parietal cortex in verbal working memory. J Neurosci 18:5026–5034
944. Smith EE, Jonides J, Marshuetz C, Koeppe RA (1998) Components of verbal working memory: evidence from neuroimaging. Proc Natl Acad Sci USA 95:876–882
945. Floresco SB, Braaksma DN, Phillips AG (1999) Thalamic–cortical–striatal circuitry subserves working memory during delayed responding on a radial arm maze. J Neurosci 19:11061–11071
946. Watanabe Y, Funahashi S (2012) Thalamic mediodorsal nucleus and working memory. Neurosci Biobehav Rev 36:134–142
947. Rowe JB, Stephan KE, Friston K, Frackowiak RSJ, Richard E, Passingham RE (2005) The prefrontal cortex shows context specific changes in effective connectivity to motor or visual cortex during the selection of action or colour. Cereb Cortex 15:85–95
948. Shu Y, Hasenstaub A, McCormick DA (2003) Turning on and off recurrent balanced cortical activity. Nature 423:288–293
949. Baker MC (2001) The atoms of language: the mind's hidden rules of grammar. Basic Books, New York

950. Rempel-Clower NL, Zola SM, Squire LS, Amaral DG (1996) Three cases of enduring memory impairment after bilateral damage limited to the hippocampal formation. J Neurosci 16:5233–5255
951. Zola-Morgan S, Squire LS, Amaral DG (1986) Human amnesia and the medial temporal region: enduring memory impairment following a bilateral lesion limited to field CA1 of the hippocampus. J Neurosci 6(10):2950–2967
952. Graff-Radford NR, Tranel D, Van Hoesen GW, Brandt JP (1990) Diencephalic amnesia. Brain 113:1–25
953. The Avian Brain Nomenclature Consortium (2005) Avian brains and a new understanding of vertebrate brain evolution. Nat Rev Neurosci 6:151–159
954. Coward LA, Smith LS (2004) In: Hussain A, Aleksander I (eds) Simulation of a proposed binding model. Brain inspired cognitive systems 2004. University of Stirling, Stirling

Index

L.A. Coward, *Towards a Theoretical Neuroscience: from Cell Chemistry to Cognition*, Springer Series in Cognitive and Neural Systems 8, DOI 10.1007/978-94-007-7107-9,

The manufacturer's authorised representative in the EU is Springer Nature Customer Service Centre GmbH, Europaplatz 3, 69115 Heidelberg, Germany. If you have any concerns regarding our products, please contact ProductSafety@springernature.com

Printed and bound by CPI Group (UK) Ltd, Croydon, CR0 4YY
30/09/2026
02225445-0013